Neuropsychology

HUMAN BRAIN FUNCTION
Assessment and Rehabilitation

SERIES EDITORS:

Antonio Puente, *University of North Carolina at Wilmington*
North Carolina

Gerald Goldstein, *VA Pittsburgh Healthcare System and University of Pittsburgh*
Pittsburgh, Pennsylvania

Erin D. Bigler, *Brigham Young University, Provo, Utah*

NEUROIMAGING I: Basic Science
Edited by Erin D. Bigler

NEUROIMAGING II: Clinical Applications
Edited by Erin D. Bigler

NEUROPSYCHOLOGY
Edited by Gerald Goldstein, Paul David Nussbaum,
and Sue R. Beers

REHABILITATION
Edited by Gerald Goldstein and Sue R. Beers

A Continuation Order Plan is available for this series. A continuation order will bring delivery of each new volume immediately upon publication. Volumes are billed only upon actual shipment. For further information please contact the publisher.

Neuropsychology

Edited by

Gerald Goldstein

*VA Pittsburgh Healthcare System
and University of Pittsburgh
Pittsburgh, Pennsylvania*

Paul David Nussbaum

*Lutheran Affiliated Services
Aging Research and Education Center
Mars, Pennsylvania*

and

Sue R. Beers

*School of Medicine
University of Pittsburgh
Pittsburgh, Pennsylvania*

PLENUM PRESS • NEW YORK AND LONDON

Library of Congress Cataloging-in-Publication Data

Neuropsychology / edited by Gerald Goldstein, Paul David Nussbaum, and
   Sue R. Beers.
         p.   cm. -- (Human brain function)
      Includes bibliographical references and index.
      ISBN 0-306-45646-X
      1. Clinical neuropsychology. 2. Neuropsychological tests.
   I. Goldstein, Gerald, 1931-   . II. Nussbaum, Paul David.
   III. Series.
      [DNLM: 1. Neuropsychology. 2. Neuropsychological Tests.
   3. Brain--physiopathology. 4. Brain Diseases--diagnosis.   WL 103.5
   N4941 1997]
   RC386.6.N48N492   1997
   616.8--dc21
   DNLM/DLC
   for Library of Congress   97-42120
                                                            CIP

ISBN 0-306-45646-X

© 1998 Plenum Press, New York
A Division of Plenum Publishing Corporation
233 Spring Street, New York, N.Y. 10013

http://www.plenum.com

10 9 8 7 6 5 4 3 2 1

Printed in the United States of America

Contributors

KENNETH M. ADAMS, *Psychology Service, VA Medical Center, and Department of Psychiatry–Division of Psychology, University of Michigan Medical Center, Ann Arbor, Michigan 48109-0840*

DANIEL N. ALLEN, *Psychology Service (116B), VA Pittsburgh Healthcare System, Highland Drive Division, Pittsburgh, Pennsylvania 15206-1297*

IDA SUE BARON, *Neuropsychology Consulting, Potomac, Maryland 20854*

JEFFREY T. BARTH, *Department of Psychiatric Medicine, University of Virginia Medical School, Charlottesville, Virginia 22908*

RICHARD A. BERG, *Coastal Rehabilitation Hospital and Wilmington Health Associates, Wilmington, North Carolina 28401*

ROBERT A. BORNSTEIN, *Neuropsychology Program, The Ohio State University, Columbus, Ohio 43210*

BRUCE M. CAPLAN, *Department of Rehabilitation Medicine, Thomas Jefferson University Hospital, Philadelphia, Pennsylvania 19107*

GORDON J. CHELUNE, *Department of Psychiatry and Psychology, Cleveland Clinic Foundation (P-57), Cleveland, Ohio 44195*

DEAN C. DELIS, *Psychology Service (116B), VA Medical Center, San Diego, California 92161*

C. DELLA MORA, *Department of Psychiatry/Neuropsychology, The Ohio State University, Columbus, Ohio 43210*

ROBERT DIAMOND, *Department of Psychiatric Medicine, University of Virginia Medical School, Charlottesville, Virginia 22908*

MICHELLE C. DOLSKE, *Florida Hospital, Medical Psychology Section, Orlando, Florida 32804*

GERARD A. GIOIA, *Pediatric Psychology/Neuropsychology, Mt. Washington Pediatric Hospital, Baltimore, Maryland 21209*

ANTHONY J. GIULIANO, *University of Hartford, West Hartford, Connecticut 06117; and Memory Disorders Clinic, Southwestern Vermont Medical Center, Bennington, Vermont 05201*

v

GERALD GOLDSTEIN, *VA Pittsburgh Healthcare System, Highland Drive Division (151R), Pittsburgh, Pennsylvania 15206-1297*

KATHLEEN Y. HAALAND, *Research and Psychology Services, Albuquerque VA Medical Center, and Departments of Psychiatry, Neurology, and Psychology, University of New Mexico, Albuquerque, New Mexico 87108*

DEBORAH L. HARRINGTON, *Research Service, Albuquerque VA Medical Center, and Departments of Psychology and Neurology, University of New Mexico, Albuquerque, New Mexico 87108*

NICOLE ENGLUND HEFFRON, *Psychology Service (116B), VA Pittsburgh Healthcare System, Highland Drive Division, Pittsburgh, Pennsylvania 15206-1297*

ROCK A. HEYMAN, *Department of Neurology, University of Pittsburgh School of Medicine, Pittsburgh, Pennsylvania 15261*

DOUG JOHNSON-GREENE, *Department of Physical Medicine and Rehabilitation, Johns Hopkins University School of Medicine, Baltimore, Maryland 21239*

JOEL H. KRAMER, *Department of Psychiatry, University of California Medical Center, San Francisco, California 94143*

LISA A. MORROW, *Department of Psychiatry, Western Psychiatric Institute and Clinic, and University of Pittsburgh School of Medicine, Pittsburgh, Pennsylvania 15213*

RICHARD I. NAUGLE *Department of Psychiatry and Psychology, Cleveland Clinic Foundation (P-57), Cleveland, Ohio 44195*

PAUL DAVID NUSSBAUM, *Aging Research and Education Center, Lutheran Affiliated Services/St. John Specialty Care Center, Mars, Pennsylvania 16046*

SARAH ROMANS, *Department of Rehabilitation Medicine, Thomas Jefferson University Hospital, Philadelphia, Pennsylvania 19107; and Drexel University, Philadelphia, Pennsylvania 19104*

CHRISTOPHER M. RYAN, *Department of Psychiatry, Western Psychiatric Institute and Clinic, Pittsburgh, Pennsylvania 15213*

CAROL J. SCHRAMKE, *Psychology Service (116B), VA Pittsburgh Healthcare System, Highland Drive Division, Pittsburgh, Pennsylvania 15206-1297*

DON J. SIEGEL, *Western Psychiatric Institute and Clinic, University of Pittsburgh Medical Center, Pittsburgh, Pennsylvania 15213*

RANDY J. SMITH, *Department of Psychology, Naval Medical Center, Portsmouth, Virginia 23708*

DAVID G. SPRENKEL, *Psychology Service (116B), VA Pittsburgh Healthcare System, Highland Drive Division, Pittsburgh, Pennsylvania 15206-1297*

ROBERT M. STOWE, *Departments of Neurology and Psychiatry, University of Pittsburgh School of Medicine, and VA Pittsburgh Healthcare System, Highland Drive Division, Pittsburgh, Pennsylvania 15206-1297*

H. GERRY TAYLOR, *Department of Pediatrics, Case Western Reserve University, and Rainbow Babies and Children's Hospital, Cleveland, Ohio 44106-6038*

NILS R. VARNEY, *Psychology Service, VA Medical Center, Iowa City, Iowa 52246*

KEITH OWEN YEATES, *Department of Pediatrics, The Ohio State University, and Children's Hospital, Columbus, Ohio 43205*

Preface

In this volume of the series *Human Brain Function: Assessment and Rehabilitation* we cover the area of how brain function is assessed with behavioral or neuropsychological instruments. These assessments are typically conducted by clinical neuropsychologists or behavioral neurologists, and so we made an effort to present the somewhat differing approaches to these two related disciplines. Clinical neuropsychologists are psychologists who typically utilize standardized tests, while behavioral neurologists are physicians who generally assess brain function as part of the clinical neurological evaluation. Both approaches have much to offer.

The basic assumption of neuropsychological assessment is that the brain is the organ of behavior, and therefore, the condition of the brain may be evaluated with behavioral measures. Neuropsychological tests are those measures found by research to be particularly sensitive to alterations in brain function. An adequate neuropsychological test is a procedure that can be related to some objective measure of alteration in brain function. Over the years, these objective measures have changed, but generally involve documentation through direct observation of brain tissue, or through histological, pathological, neuroimaging, or other laboratory procedures. The methods described in the first two volumes of this series describe the neuroimaging procedures that are often used in the validation of neuropsychological tests.

In recent years, the field of neuropsychological assessment has become increasingly sophisticated and specialized. Because of these rapid changes, this book has been organized in a way that attempts to reflect the breadth and complexity of the field. One important consideration involves neuropsychological assessment across the life span. Procedures used to assess infants and young children, older children, adults, and elderly adults cannot be the same, and must reflect changes in brain function in the individual over time. Neuropsychological assessment is now specialized on a developmental basis as reflected in Chapters 2, 3, 4, and 5, each of which covers a different period in the life span.

The clinical aspect of neuropsychological assessment has also become specialized. Brain–behavior relationships in different clinical disorders vary extensively from each other, interacting not only with the development epoch during which the disorder appears, but also with the nature of the neuropathology. Some disorders, such as autism or Alzheimer's disease, only appear during certain periods in life. Other disorders, such as brain trauma, may appear at any time in life but may have differing consequences depending on developmental level. Chapters 6 through 14 provide detailed descriptions of neuropsychological aspects of a number of disorders frequently evaluated with neuropsychological tests.

A neuropsychological assessment typically consists of an evaluation of a number of domains of function, notably abstract reasoning and problem-solving, memory, language, spatial abilities, and motor skills. The neuropsychological literature contains extensive information concerning relationships both between brain function and the various domains and between basic research conceptualizations and development of clinical assessment methodologies and specific procedures. It has been said that the neuropsychological experiment of today is the neuropsychological test of tomorrow. In clinical practice, it is sometimes more useful to obtain an overview of each of the domains, while at other times it is more pertinent to concentrate the evaluation on a particular domain. However, both general and specialized assessment benefit from psychometrically and clinically sophisticated evaluation of the domains, regardless of whether they are tested in a brief or intensive manner. Chapters 14 through 19 provide presentations concerning specialized assessment in the major domains. Finally, we have a chapter that provides an extensive overview of how assessments are accomplished by behavioral neurologists and neuropsychiatrists.

We made a substantial effort to select authors who had what we viewed as among the best scientific or professional reputations in the topic covered by their chapters. The contributions of these authors uniformly fully lived up to their reputations. We express our indebtedness for the time, thought, and effort that must have gone into this writing, and can assure our readers that they can look forward to reading works reflecting superior expertise and scholarship. The lengthy preparation of this volume was sustained by continued support from Eliot Werner and Mariclaire Cloutier of Plenum Press. We offer our personal thanks, and hope that they are pleased by the product. We would like to acknowledge the support of the Department of Veterans Affairs for the work involved in preparation of this book.

Gerald Goldstein
Sue R. Beers
Paul David Nussbaum

Contents

CHAPTER 8

Evaluation of Cerebrovascular Disease
C. Della Mora and Robert A. Bornstein

CHAPTER 9

Evaluation of Demyelinating and Degenerative Disorders
*Daniel N. Allen, David G. Sprenkel, Rock A. Heyman, Carol J. Schramke,
and Nicole Englund Heffron*

CHAPTER 10

Assessment following Neurotoxic Exposure
Lisa A. Morrow

CHAPTER 11

Assessing Medically Ill Patients: Diabetes Mellitus as a Model Disease
Christopher M. Ryan

CHAPTER 12

Evaluation of Neoplastic Processes
Richard A. Berg

CHAPTER 13

Evaluation of Patients with Epilepsy
Michelle C. Dolske, Gordon J. Chelune, and Richard I. Naugle

CHAPTER 14

Evaluation of Neuropsychiatric Disorders
Doug Johnson-Greene and Kenneth M. Adams

PART III   SPECIALIZED ASSESSMENT

CHAPTER 15

Neuropsychological Assessment of Abstract Reasoning
Gerald Goldstein

CHAPTER 20

Assessment Methods in Behavioral Neurology and Neuropsychiatry
Robert M. Stowe

Neuropsychology

Introduction to Neuropsychological Assessment

GERALD GOLDSTEIN

Neuropsychological assessment is the clinical practice of using tests and other behavioral evaluation instruments to determine the status of brain function. It is based on the assumption that the brain is the organ of behavior, and so the status of the brain can be evaluated through the use of behavioral measures. Over many years of research, particular procedures have been found to have particular sensitivities to alterations in brain function, and these procedures have come to be known as neuropsychological tests. A neuropsychological test may therefore be defined as a behavioral procedure that is particularly sensitive to the condition of the brain. While any purposeful behavior involves the brain, neuropsychological tests provide the clearest demonstrations of behaviors that indicate that the brain is functioning normally, or that something is wrong with it. There are many factors that can produce brain dysfunction including genetic endowment, developmental abnormalities, physical injury, exposure to toxic or infectious agents, systemic diseases (e.g., vascular disorders, cancer, metabolic disorders), and progressive disorders that specifically affect central nervous system tissue, such as multiple sclerosis.

The clinical practice of neuropsychology involves an integration of knowledge bases from the disciplines of psychology, psychometrics, neuroscience, clinical neurology, and psychiatry. In this volume of the *Human Brain Function: Assessment and Rehabilitation* series, we present an introduction to how neuropsychologists assess brain function. First, we will look at the area from a developmental perspective, illustrating how neuropsychologists evaluate young preschool children, older children, younger adults, and elderly adults. The brain is an evolving organ and functions very differently at different stages in the life span. Assessment issues are different during these stages, as are the prevalences and characteristics of the various brain disorders. Obviously, the behavioral evaluation of a 3-year-old child cannot use the same materials and methods as the evaluation of a 40-year-old adult.

GERALD GOLDSTEIN   VA Pittsburgh Healthcare System, Highland Drive Division (151R), Pittsburgh Pennsylvania 15206-1297.

Neuropsychology, edited by Goldstein *et al.* Plenum Press, New York, 1998.

It has therefore been necessary to develop different tests and methods for neuro-psychological evaluations across the life span, and consequently we now have tests and test batteries specifically designed for infants and young children, older children, adults, and the elderly. Part I of this volume is organized on the basis of these considerations.

Neuropsychologists and neurologists studying behavior have been greatly concerned with localization of function in the brain. The first great discovery of neuropsychology is thought to be Paul Broca's identification of the relationship between language and the left hemisphere of the brain. Neuropsychological tests are often used to assist in determining localization of brain damage, and that practice continues even after the development of the relatively new neuroimaging procedures. However, contemporary views of brain function tend to conceptualize localization in interaction with a number of developmental and pathological considerations. Localization of function in the brain of an infant is not the same as it is in the adult. Localization in women is not the same as it is in men. Furthermore, the neurobehavioral characteristics of disease or destruction of the very same brain regions may vary substantially with the particular pathological process. We have therefore chosen to introduce neuropsychological assessment on the basis of these different processes, with localization treated within the contexts of those processes.

Part II of this book provides material about the major disorders of brain function written by individuals with particular expertise in the neuropsychological aspects of these various disorders. Some of these disorders have major interactions with the developmental phenomena considered in Part I, and this interaction occurs in two ways. Some forms of brain dysfunction are present from before birth, some are typically acquired during adulthood, and some occur only in old age. In some cases, such as Huntington's disease, the pathological agent is present before birth, but the symptoms of the disorder do not appear until later in life. In the second type of interaction, a particular type of brain damage may have differing outcomes depending on when it was acquired. There is an extensive literature concerning differences in outcome between brain trauma which occurs early and that which occurs late in life. It has also been noted that elderly individuals may suffer more profound consequences from head trauma than younger individuals. What has come to be known as the "Kennard Principle" (Kennard & McCulloch, 1943) indicates that brain damage that occurs early in life is more benign than the same brain damage occurring later. However, Hebb (1949) asserted some time ago that the issue is more complex than that, in that early brain damage affects later performance of language-related skills, but not sensory and motor capacities.

Some disorders, notably head trauma, are not developmental in nature, but are associated with cultural considerations. Thus, at least in Western culture, head trauma is more common in young male adults than it is in other demographic groups. Similarly, age of exposure to carcinogenic and other environmental agents and use of substances that may be associated with acquisition of brain dysfunction determines the time of onset of clinical changes. The brain is a part of the body, and so general health status is a very important consideration in neuropsychology. The condition of the heart and of the vascular, pulmonary, endocrine, and autoimmune systems are all crucial determinants of the status of brain function. The brain can also be affected by cancer directly, and by cancer occurring elsewhere in the body. Part II of this volume therefore covers the major process areas, developmental disorders, trauma, vascular disorders, cancer, exposure to environmental toxins, and the diseases that specifically affect the nervous system. There are also chapters on epilepsy and on neuropsychiatric disorders. While epilepsy can be a product of numerous disorders, it has its own neuropsychological literature and requires separate consideration. In recent years, the study of neuropsychiatric disorders has been solidly incorporated into neuropsychology. This is largely because of recent discoveries indicating that several of these disorders previously thought to be the result of life experience actually have a neurobiological basis. That conclusion is particularly clear in the cases of autism and schizophrenia, but there is also impressive evidence for neurobiological bases for certain mood disorders, obsessive–compulsive disorder, alcoholism, and attention deficit–hyperactivity disorder. Several of these disorders are developmental in nature, but with widely varying ages of onset of clinical symptoms.

One can divide neuropsychological assessment into two areas: comprehensive and specialized assessment. Comprehensive assessment generally employs standard test batteries, notably the Halstead–Reitan or Luria–Nebraska batteries. A comprehensive assessment typically evaluates all of the areas evaluated by specialized assessments, but not in as much detail. In actual practice, some neuropsychologists only do comprehensive assessments, while others do not do them at all. Members of this latter group may administer a brief screening procedure aimed at identifying relevant areas for more targeted specialized assessments, or may do only specialized assessments in response to the referral question. This matter is controversial in the field, but the more productive approach is probably highly related to the setting in which one works, and the nature of its patient population. The comprehensive assessment methods are covered in Part I, particularly in Chapter 4.

Part III is devoted largely to specialized assessment. As a framework, the major brain functions are typically divided into modalities and domains. The major modalities are motor function and the senses of vision, hearing, touch, and rarely, smell. The major domains are the cognitive abilities and include abstract reasoning and intellectual function, memory, language, spatial abilities, and motor skills. Thus, Part III contains chapters on several of these domains and modalities, again written by experts in the various specialized areas. We have not included separate chapters for the modalities of vision, hearing, touch, and smell, since in clinical neuropsychology, they are typically studied in interaction with some domain, for example, visual perception of language. However, motor skills have received extensive study as an independent area, and so are discussed in a separate chapter. Finally, Part III contains a chapter on how assessments are conducted by behavioral neurologists. Behavioral neurologists typically do specialized evaluations based on their initial examination and review of the history, but they use a somewhat different conceptual framework and methodological approach from clinical neuropsychologists, providing an interesting and important contrast.

The aim of specialized assessment is often to identify a syndrome and specify its probable basis in abnormal brain function. The basic purpose of identifying a syndrome is to characterize the deficit and make a formulation concerning possible neurological correlates. For example, in the case of memory, the diagnostic question often involves whether the patient has amnesia, and if so, what type. Thus, there is an association between the domain and a class of abnormal syndromes, illustrated in Table 1. As will be seen in Part III, this table is a gross oversimplification, but is only meant to suggest the association of certain cognitive domains with different neurobehavioral syndromes.

TABLE 1. BRAIN FUNCTION DOMAINS AND ASSOCIATED ABNORMAL SYNDROMES

Domain	Syndrome
Abstraction/intellectual function	Dementia
Memory	Amnesia
Language	Aphasia, alexia, acalculia
Spatial abilities	Constructional apraxia, visuospatial defects
Motor skills	Apraxia

The neuropsychology of the various modalities and domains involves applications from the knowledge base concerning the domain or modality itself, its neurobiological substratum, and the functional changes that take place as a result of brain damage or injury. Thus, the neuropsychology of memory involves application of the experimental psychology of memory, our knowledge of how memory is represented in the brain, and the changes that take place in memory as a consequence of brain damage, these changes being characterized as the amnesic syndromes. Similarly, disorders of speech, language, reading, writing, and mathematical abilities are understood in terms of linguistics and the psychology of language, and of what is known about the relationship between the brain and language. Lacking definitive knowledge of how the brain really works, neuropsychologists have constructed elegant conceptual models of how the brain processes information within and across modalities and domains.

Probably the most useful model of neuropsychological interpretation is described in Reitan and Wolfson's (1993) four approaches to assessment: level of performance, pathognomonic signs, pattern of performance, and comparison of the left and right sides of the body. The last approach can also include comparisons between the anterior and posterior parts of the brain, or between cortical and subcortical structures, but our knowledge base remains strongest for right versus left comparisons. A comprehensive assessment ideally uses all four approaches. Some forms of specialized assessment rely heavily on pathognomonic signs, behaviors that are almost exclusively seen in brain-damaged patients and that have some specific, often localizing significance. Other forms of specialized assessment use both pathognomonic signs and performance profiles or patterns. Specialized evaluations rarely use the general level of performance approach, which is generally reflected in some kind of summary index of impairment based on tests of varying domains and modalities. While, in our view, all of these approaches are important, relative emphases on any of them may relate to the setting in which one practices and the clinical characteristics of the clientele in that practice. In the light of contemporary patterns of health care provision, the distinction between being in a primary-care, "first-line" setting and a specialized tertiary-care practice is very important.

To summarize, neuropsychological assessment provides information concerning the status of brain function across the life span. It does so primarily by testing those functions and abilities that seem most sensitive to brain dysfunction. These functions and abilities may be evaluated in a comprehensive, specialized, or combined manner. The areas typically assessed in a neuropsychological evaluation include the ability to reason and conceptualize; to remember; to speak and understand spoken and written language; to attend to and perceive the environment accurately through the senses of vision, hearing, touch, and smell; to construct objects in two- or three-dimensional space; and to perform skilled, purposive movements. Clinical neuropsychology in particular has the task of identifying in individual patients the level and pattern of disruption of these abilities as a result of brain dysfunction.

REFERENCES

Hebb, D. O. (1949). *Organization of behavior: A neuropsychological theory.* New York: John Wiley & Sons.

Kennard, M. A., & McCulloch, W. S. (1943). Motor response to stimulation of cerebral cortex in absence of areas 4 and 6 (*Macaca mulatta*). *Journal of Neurophysiology, 6,* 181–190.

Reitan, R. M., & Wolfson, D. (1993). *The Halstead-Reitan neuropsychological test battery: Theory and clinical interpretation* (2nd ed.). Tucson, AZ: Neuropsychology Press.

Developmental Considerations

I

2

Neuropsychology of Infants and Young Children

Ida Sue Baron and Gerard A. Gioia

This chapter is intended to provide a framework for the neurobehavioral investigation and assessment of infants and young children (i.e., those under 5 years of age), and to discuss the relevance of those assessments to clinical interventions with at-risk children. A selected survey of the increasing knowledge base regarding brain–behavior relationships that is especially pertinent in this age range is included. We also emphasize three functional domains of particular interest in this age range: attention, memory, and executive function. Our goal is to challenge clinicians to develop concepts about brain–behavior relationships in this age group, and to provide a basis for clinicians to generate hypotheses in their own clinical evaluations and effect appropriate interventions. Our objective is not to provide an exhaustive listing of tests and measures, but instead to consider the functional domains for which a judicious selection of tests can be made. We hope to convince the reader that neuropsychological services provided in early childhood do not depend merely on the administration of test measures but instead on the knowledge of unique brain–behavior relationships that are applicable within this age range, their direct application to these early periods of development, and their relevance beyond these early years.

"Neuropsychology" of Infants and Young Children?

Whether one should refer to a "neuropsychology" of infants and young children is perhaps controversial to some. How does one reliably assess and evaluate brain–behavior relationships in the newborn, neonate, or very young child? If this is even possible, how practical would such evaluation be? Is the methodology used by pediatric neuropsychologists applicable to the youngest ages and, if so, with

Ida Sue Baron   Neuropsychology Consulting, Potomac, Maryland 20854.   Gerard A. Gioia Pediatric Psychology/Neuropsychology, Mt. Washington Pediatric Hospital, Baltimore, Maryland 21209.

Neuropsychology, edited by Goldstein *et al.* Plenum Press, New York, 1998.

what degree of reliability or validity? Which variables traceable to the very youngest ages will result in long-lasting (i.e., adult) cognitive compromise? What interventions can be applied in these very early years to lessen the impact of early insults? Since these and other related questions remain largely unanswered, many practitioners are naturally reluctant to endorse terminology that may be misleading or inappropriate.

What is *not* controversial is the fact that pediatric medical and psychological specialists frequently encounter infants and young children whose developmental delays or cognitive deficiencies are attributable to underlying neurodevelopmental abnormality or to documented neurological disease or disorder that occurred in the earliest stages of growth and development. The increased recognition that these etiological factors exist and have an important influence on the child's later cognitive outcome is a result of a number of converging developments. These include (1) a better definition of the unique developmental concepts that are applicable to infants and young children, (2) finely detailed analyses of normal and abnormal brain development from experimental studies of nonhumans, (3) major advances in neurodiagnostic techniques, (4) an expanding clinical and research literature on human developmental studies, and (5) an increase in societal attention to the needs of infants and young children, in part emphasized by preschool screening and intervention programs.

DEVELOPMENTAL CONCEPTS UNIQUE TO INFANTS AND YOUNG CHILDREN

The cognitive and social–emotional development of infants and very young children has unique features. This age range is associated with less differentiation of some functional areas, the presence of early developmental constructs that are less dominant than in older children, and an increased variability of performance compared to older children. Understanding the critical precursor behaviors during these early years allows the neuropsychologist to generate early predictions about later patterns of strength and weakness, for example, in attention, memory, and executive function domains.

Two concepts that are especially critical to understanding brain development in the very young child are symbolic representation and imitation in learning. Knowing these concepts helps us better understand developmental progress and provides a stronger foundation for understanding cognitive and social development. The social environment also has a special role.

Symbolic representation is the representation of both the external and internal world through symbols and has been discussed in relation to important neuropsychological domains, including executive self-regulation (Goldman-Rakic, 1987) and memory (Schacter & Moscovitch, 1984). Goldman-Rakic (1987) articulated a neurobiological model of cognitive development, positing that the ability to represent the world symbolically serves as the cornerstone for working memory and internalized executive problem-solving. The child who can work with symbols is able to regulate his or her behavior more effectively by bridging the past knowledge and experience with a future goal or need. The development of the declarative memory system is dependent on the development of the symbolic function that occurs between 8 and 12 months of age (Moscovitch, 1984), when the external world is represented and manipulated through the memory process by the child. An example of a disorder of symbolic representation can be found in infants who have sustained severe brain injury with shearing. The developmental outcome in an infant that cannot represent the world in symbolic ways is mental retardation.

The concept of *imitation* in the learning process is also a key developmental concept for infants and young children. Although social learning theory (Bandura, 1977) has long posited modeling as an important element in learning, imitation behaviors are significantly more pronounced and overt in infants and young children. From a neuropsychological perspective, imitation can be an adaptive form of stimulus-bound behavior, which can be considered pathological at older ages. Functional imitative behavior in the infant may be viewed as environmentally focused, stimulus-bound behavior which is adaptive for learning in this early developmental period. Imitation arguably becomes more voluntary with development but appears to be a mandatory and compulsory developmental feature for the infant. Its function appears to be to direct the infant's attention to the external world to learn important "social" behaviors (Vygotsky, 1978). Goldman-Rakic (1987) posits that as the ability to symbolize develops together with the executive control functions (e.g., inhibitory, strategic planning), greater self-regulation and voluntary control over this imitative behavior emerges. Damage to brain systems subserving these developing functions (e.g., orbital medial and dorsolateral frontal) results in an increase in, or the abnormal persistence of, this stimulus-bound behavior. Examples of imitation disorder can be found in children with pervasive developmental disorder (autism), who have significant problems attending to extrapersonal behavior. As a result, they are not able to use the social arena as a critical learning vehicle.

Nonhuman Experimental Studies

How normal brain development proceeds is essential knowledge for professionals concerned with understanding infant and young child development. Laboratory studies of nonhumans have contributed substantially to our knowledge about normal and abnormal human brain function. Our understanding of the abnormalities that can occur and that may explain an individual's neurobehavioral dysfunctioning has broadened considerably as a result of these studies. Recovery of function following early brain injury has been well documented (Goldman & Lewis, 1978). These studies have clarified our understanding of the crucial stages involved in central nervous system growth, development, and organization. Laboratory investigations have also correlated different patterns of clinical symptoms with differences in the timing of these developmental stages and with aberrations in the normal sequence of development. There are four general stages of brain development: (1) cell division, (2) migration, (3) proliferation, and (4) pruning or segregation (Rakic, 1991). The reader is referred to more detailed summaries of current knowledge about these stages and normal brain development, along with discussions of anomalies and deviations from normal development (see Baron, Fennell, & Voeller, 1995; Spreen, Risser, & Edgell, 1995).

Advances in Neurodiagnostic Techniques

Major scientific and technological advances have made it easier to correlate behavioral symptomatology suspected as having a neurological basis with actual neuroanatomic (structural) and neurophysiological abnormalities. Many of these advances have been applied to the study of infants and young children, such as techniques to examine the fetus in utero and to monitor development in the peri-natal period. For example, real-time ultrasonography is useful for determining the presence, timing, and course of intraventricular hemorrhage in a preterm infant.

Some techniques visualize anatomy, providing objective confirmation of structural abnormality. These include neuroradiological imaging procedures such as computed tomography (CT) and magnetic resonance imaging (MRI). However, gross brain structure may appear normal despite functional behavioral abnormality that suggests to the clinician that there is an underlying neurological etiology. This may be particularly perplexing when a formal neuropsychological evaluation has documented dysfunction. In contrast, structural abnormality may be visualized but neurobehavioral dysfunction may not be documented with formal evaluation; the behavioral impact of the noted neurological insult may not be detected until the child matures, or methods may not currently exist for the reliable and valid assessment of the behavior. For example, agenesis of the corpus callosum may only be detected in a child incidentally, perhaps on a CT scan after a closed head injury (Fischer, Ryan, & Dobyns, 1992). It is not typically found as a result of a clinically obvious interhemispheric processing deficit (Bruyer, Dupuis, Ophoven, Rectens, & Reynaert, 1985; Ettlinger, Blakemore, Milner, & Wilson, 1974; Pirozzolo, Pirozzolo, & Ziman, 1979; Saul & Sperry, 1968; Temple & Vilarroya, 1990). However, in some cases mild abnormality may be observed clinically but not associated with an underlying brain anomaly; for example, nonspecific delayed development evident in infancy may be responsive to early intervention (Teeter & Hynd, 1981) or a lowered intelligence quotient (IQ) may be noted in childhood (Field, Ashton, & White, 1978).

Besides subtle deficits whose etiology is not discernible without neurodiagnostic techniques, gross anatomic abnormalities may also exist that remain undetected in a medically asymptomatic child or one whose only manifestation is developmental delay. For example, ventriculomegaly may be a marker of a slowly insidious progressive hydrocephalic condition. This condition may remain undetected during the critical early years since symptoms can be minimal and alternative behavioral or socioemotional explanations may suffice even when symptoms are present. Whether early intervention (i.e., ventricular shunting) may reverse a slowly downward course and stabilize or reverse a presumed progressive condition remains an area of ongoing investigation (Fletcher *et al.*, 1992; Wills, 1993). In hydrocephalus, as with other treatable conditions of childhood, early detection and treatment are critical. Consequently, it is necessary to recognize that neuroanatomic visualization and neurobehavioral assessment may be mutually exclusive procedures that together allow for more complete delineation of neurobehavioral integrity.

Other neurodiagnostic advances include positron emission tomography (PET) measurements of brain glucose metabolism or of blood flow, functional magnetic resonance imaging (fMRI), measurement of cerebral blood flow, single photon emission computed tomography (SPECT), as well as improved electroencephalography (EEG) techniques that allow the evaluation of a child's brain integrity and neurobehavioral state in a relatively noninvasive manner. For example, power spectral and discriminant analyses of EEG recordings at term and 3 months post-term differentiated risk from nonrisk infants, and infants with neonatal medical complications who had good versus poor developmental outcomes (Richards, Parmelee, & Beckwith, 1986). Event-related potentials (ERP) are useful for measurement of vision and audition in the first year of life (Salaptek & Nelson, 1985). This electrophysiological technique is a synchronized portion of the ongoing EEG record which is time-locked to the onset of an event in the infant's environment. It is used to provide information about when the child perceives or encodes information or decides to respond. More recently, the application of electrophysiological measures has been recommended to study developmental neurocognition (i.e., memory, attention, language, spatial abilities, and emotion) in normal and clinical populations (Molfese, 1992). Nelson and deRegnier (1992) investigated the neural correlates of recognition memory in the first year of life with ERP methodology. Using this tool, the authors were able to document the progression of encoding into recognition memory from 4 months to 12 months of age. At the earlier stage, the infant can encode a single stimulus and retain it in memory with frequent presentation. The infant progresses to the point of encoding and retaining more than one piece of information of both a familiar and unfamiliar nature. By 12 months, a relatively greater selective attention and stability of memory encoding was noted, with the infant preferring partially familiar events and ignoring highly familiar and highly novel events. Molfese and Wetzel (1992) extended these findings to the second year, showing differential short- and long-term recognition memory with the involvement of the frontal cortex in novelty recognition. Evoked response potentials have also been instrumental in establishing early lateralization of language in infants (Molfese & Betz, 1988). However, extension of some procedures to infant populations may not be recommended because of the associated high risk or the concomitant procedures, such as sedation. Other procedures require adaptation (e.g., electron beam computed tomography involves a fast scanning speed that lessens blurring in images of moving children, thereby reducing the need for sedation) or hold promise for future neurobehavioral correlations. For example, the variability of the pattern of EEG asymmetry between infants may be a marker of differences in temperament, and may prove helpful in research on the maturation of frontal lobe function for cognitive and emotional behaviors and the importance of this brain region in the generation of mature emotion experience (Fox, 1991).

Advances in molecular biology have made possible the investigation of the genetic contribution, although how gene defects affect development is as yet unknown. The "eventual understanding of the effects of these abnormalities on development depends on three different sources of information: the neuropsychological profiles of the various syndromes; the precise structural and functional anomalies of the brain that are characteristic of each syndrome; and the specific ways in which genes affect brain development and structure, and thus alter normal cognition" (Baron *et al.*, 1995, p. 71). Genetic analyses and genetic linkage studies to date have resulted in a wider recognition that some conditions originally assumed to be unique occurrences may in fact be part of a pattern of associated conditions within a family's gene pool. For example, epilepsy may be a clinical manifestation of periventricular heterotopias, and these developmental defects may be genetically determined (Huttenlocher, Taravath, & Mohtahedi, 1994); some forms of dyslexia are transmitted genetically (Pennington & Smith, 1983); autism is associated with specific genetic conditions (Folstein & Piven, 1991); and agenesis of the corpus callosum may have a genetic (autosomal recessive) basis (Toriello & Carey, 1988).

Before procedures and advances such as these were available, disturbances in neurological integrity were not readily recognized as an explanation for, or a contribution to, the behavioral symptomatology of clinical concern. Misattribution of a nonneurological etiology was far more likely in clinical settings, with the unfortunate result that inappropriate or inadequate treatments resulted. In those instances where a neurological etiology was recognized, it was not always possible to appreciate the extent of perinatal insult that would have impact on growth and development. With improved neurodiagnostic methods it has become possible to document more accurately an abnormality that is suspected and to have unsuspected neurological etiological factors distinguished.

Despite all these advances we often will not know, or have any way of judging, the status of the very young child's brain. Current empirical investigational methods are often restricted to experimental protocols and not available to clinical practitioners. For example, it is not routine to investigate or search for the existence of neuromigrational anomalies or hormonal abnormalities or fluctuations, nor is it routine to investigate genetic inheritance. Yet, it is becoming apparent that these influences have a direct impact on the developmental course and behavior of the maturing organism and should be considered when development is not proceeding normally.

CLINICAL STUDIES

The opportunity to investigate the conditions that influence stages of growth and development from gestation to infancy to early childhood and their impact on eventual neurobehavioral outcome is especially challenging. It is precisely these investigations that are being addressed now by researchers in neuroscience, psychology, and related fields concerned with neurodevelopment. Empirical data have only recently been more widely collected to verify or negate claims of a clear causal connection between certain medical conditions and later cognitive functioning. Misconceptions have been perpetuated in the absence of such data and with reliance on single unsubstantiated or incomplete case reports and through well-meaning but unfounded inferences derived from methodologically poorly designed studies.

The pediatric clinical literature has expanded greatly in its coverage of the wide variety of medical circumstances that can negatively affect the developing human brain. As a consequence, populations of children who are at risk for cognitive impairment have been identified, and formal investigations have provided insight into the influences of the many factors that influence the success or failure of cognitive development. That infants at risk are more likely to have learning difficulties than infants not at risk is well established (Astbury, Orgill, Bajuk, & Yu, 1990). The first year of life is a critical period for brain growth. One study found head growth and head circumference at 8 months of age to be measures with strong prediction value for later IQ at 3 years of age (Hack & Breslau, 1986). Factors that interfere in this critical period obviously have the potential to result in lasting cognitive impairments, whether subtle or overt. For example, that brain development can be negatively affected by neurotoxic substances is well established, and there is a greater negative influence on the immature brain than on the more mature brain. As a result, the developing brain may be particularly vulnerable to toxic substances, more so than the neonatal brain, and known prenatal exposure may therefore be predictive of a child at risk (Yeates & Mortensen, 1994).

The link between structural abnormality and unique behavioral aberration has become clearer as technological advances have allowed for even more finely tuned discrimination than that obtained from study of gross structural anatomy. This was made dramatically apparent by several early neuropathological studies of dyslexia. The classic diagnosis of dyslexia was based on psychoeducational features until advances in neurological diagnosis enabled identification of associated neuropathological mechanisms and anatomic abnormalities. Neuromigrational abnormalities were initially found in the brains of dyslexic young adults on autopsy (Galaburda, 1989; Galaburda & Kemper, 1979), and additional studies confirmed the importance of these results. Postmortem examinations of four male developmental dyslexics revealed cerebral cortex abnormalities including symmetry of the temporal plana, neuronal ectopias, and architectonic dysplasias in the perisylvian regions (mostly in the left hemisphere) (Galaburda, Sherman, Rosen, Aboitiz, & Geschwind, 1985), and the brains of three females with developmental dyslexia also were found to have symmetrical temporal plana, focal architectonic dysplasias, and molecular layer ectopias (Humphreys, Kaufmann, & Galaburda, 1990). Further, cerebrocortical microdysgenesis was rarely found in the brains of normal males, but when present there were fewer abnormalities and they were present in different locations than in developmental dyslexics (Kaufmann, & Galaburda, 1989).

Other neuropsychological disorders have also been correlated with a wide range of neurodevelopmental abnormalities. For example, a dysplastic microgyrus was found in the left insular cortex of a patient and was associated with developmental dysphasia and oromotor apraxia (Cohen, Campbell, & Yaghmai, 1989). Certain forms of epilepsy are thought to be a clinical manifestation of periventricular heterotopias (Huttenlocher *et al.*, 1994). Inhibition of stages of brain development (e.g., cell division and migration) was documented in a histopathological study following methylmercury exposure (Clarkson, 1991). A neuropathologic study of the brains of four neonates exposed to ethanol during gestation found errors in migration of neuronal and glial elements; two of the children had hydrocephalus (Clarren, Alvord, Sumi, Streissguth, & Smith, 1978). Cocaine-induced vascular compromise disrupts fetal growth and increases the likelihood of structural anomalies while cocaine's action at the postsynaptic junction places the infant at risk for neurobehavioral abnormalities (Chasnoff, 1991), although developmental index scores may not be sufficient to delineate the long-term outcome of drug and alcohol exposure on the developing brain (Chasnoff, Griffith, Freier, & Murray, 1992). Lead as a neurotoxicant has been shown to affect specific neurodevelopmental processes of synapse formation and dendritic arborization (Goldstein, 1992) as well as the sensitivity of presynaptic terminals to neurotransmitter release (Silbergeld, 1992). These neurodevelopmental factors are believed to be among the agents responsible for the neuropsychological impairments in the specific subdomains of attention, memory, and executive function (Bellinger, 1995).

Such findings may seem to suggest that a predestined behavioral outcome is to be expected, based on the actual neural and structural development that occurs in the neonate. That is, interference in any of the four stages of brain development must result in some neurobehavioral manifestation of the process gone awry. However, neurobehavioral outcomes are not so simply predicted. The complexity of maturation is mediated in part by the influences of environmental experience, which are not yet well understood. Outcome is further influenced by a multiplicity of factors, including the nature and degree of abnormality, temporal factors in development, plasticity, and the timing and application of appropriately targeted early therapeutic interventions.

NATURE AND DEGREE OF ABNORMALITY. The nature and degree of abnormality will affect normal neural growth and maturational outcome. For example, an insult to the developing brain may be the result of toxic exposure (e.g., alcohol abuse, lead exposure, maternal drug use), nutritional deprivation, trauma, or environmental circumstances. Early birth and very low birth weight have received much attention and are often accompanied by a number of neonatal complications that may contribute to eventual cognitive compromise, including intraventricular hemorrhage (IVH), hyaline membrane disease and associated respiratory distress syndrome, hyperbilirubinemia, asphyxia, bronchopulmonary dysplasia, and apnea. The effects of prenatal or perinatal oxygen deprivation are perhaps most commonly cited as a main pathogenetic factor in neurodevelopmental problems. The relative contributions of each of these, and of combinations of these, must therefore be considered. For example, early studies of low birth weight (LBW) children predicted an increased risk for specific learning disabilities and behavioral problems in these children compared to normal birth weight children (Caputo, Goldstein, & Taub, 1979; Nelson & Broman, 1977; Siegel, 1983). These impairments appeared to correlate negatively with birth weight; that is, lower birth weight was associated with greater cognitive impairment (Breslau *et al.*, Hack *et al.*, 1991; Klein, Hack, Gallagher, & Fanaroff, 1985; Vohr and Garcia Coll, 1985). However, early studies did not make the discrimination between premature children with IVH and those without such a complication. Yet, a high percentage of children have this additional complicating feature, and the presence and severity of IVH have now been associated with poorer developmental outcome and an increased incidence of cognitive impairment (Sostek, 1992). Thus, it is necessary to consider the multiple influences on the developing brain to weigh the contribution of the variety of factors that can have a deleterious impact. It is also important to consider other stressors and factors, sometimes subtle, that could interfere with the normal developmental course, including intrauterine effects, adverse effects during delivery, prenatal and perinatal trauma or exposure, and complicating illness (Harland & Coren, 1996).

Despite the numerous factors that can negatively affect brain development and that are associated with neurodevelopmental delay or late-effect cognitive disruption, it is impressive that the young brain can effectively withstand a number of critical insults. This is sometimes perceived as counterintuitive. For example, the immature brain is remarkably resistant to the potentially negative effects of hypoxia (Robinson, 1981); birth asphyxia cannot necessarily be assumed to be the decisive factor affecting outcome in the prenatal or perinatal period (Nelson, 1991).

TEMPORAL FACTORS. Given the ongoing schedule of postnatal neurodevelopment, the age of the child at the time of exposure and the chronicity can have a significant effect on the type of neuropsychological damage and dysfunction. Given the ongoing development of brain connectivity and the progressively greater reliance on the more complex cognitive and behavioral functions during school age, adolescence, and young adulthood, the actual damage or dysfunction resulting from brain injury or disorder may not be fully realized until years after the child's exposure. For example, a study of Turner's syndrome subjects supported the idea that spatial abilities and cognitive functioning may be influenced by hormones during the prenatal or perinatal period and that early developed cognitive patterns may be unaffected by hormonal fluctuations seen later in development (Shucard, Shucard, Clopper, & Schachter, 1992). Hormonal influences on brain function have received greater attention in recent years. Some studies examined relationships that exist between measures of brain functioning and patterns of cognitive abilities. It has been noted, for example, that an overproduction of naturally occurring hormones may affect the developing brain negatively. An autosomal recessive disorder, congenital adrenal hyperplasia, is such a case. This disorder results from an overproduction of androgen that affects the female fetus. Intrauterine exposure to high androgen levels was associated with learning difficulties (Nass & Baker, 1991) and a higher incidence of sinistrality in affected females (Nass *et al.*, 1987; Tirosh, Rod, Cohen, & Hochberg, 1993). Prenatal testosterone exposure in utero has also been considered a factor in placing a child at risk. Dyslexia, left-handedness, and autoimmune and allergic disorders were also hypothesized to be associated with such exposure (Geschwind & Behan, 1982). Criticism of Geschwind and Behan's theory of the correlation of developmental disorder and left-handedness, the preponderance of learning disorder (LD) in boys, and the potential role of prenatal immune factors in LD has been more recent (Berenbaum & Denburg, 1995). The timing of exposure to lead has also been found to be associated with differential neuropsychological dysfunction. Shaheen (1984) reports greater deficits in language-related functions with exposure prior to the age of 2 years whereas others report a "window of vulnerability" for general intellectual and language dysfunction with exposure between 18 and 35 months of age (Jordan, Shapiro, Kunin, & Zelinsky, 1995).

PLASTICITY. The notion of brain plasticity has been of interest to researchers and clinicians alike for decades. A broad review of the issues related to the effects of injury to the human cortex is presented by Kolb (1995). The outcome of injury is the result of the underlying plastic potential of the brain and varies with the neurodevelopmental stage at the time of insult, the type of lesion, the severity of the lesion, the behavior being measured, the age of the individual at assessment, and other factors. It has been proposed that the brain modifies itself through change at the synapse, that is, by alterations in the axon terminal, spine density, dendritic arborization, or structure of the existing synapse (Kolb, 1995). The neural processes occurring in the recovery from brain injury are thought to be similar to the processes involved in learning from experience, which result in the production of new synapses, the loss of old synapses (pruning), and the modification of existing synapses. Adaptation in response to insult is also an impressive finding. For example, support for the theory of transfer and crowding after early left-side brain damage (Milner, 1974), and therefore for language development in an undamaged right hemisphere associated with secondary impairment or extinction of visual processes, was recently obtained in a study of children with congenital spastic hemiplegia (Korkman & von Wendt, 1995). There are, however, limitations on plasticity. Recovery is reduced if neural cell mitosis is incomplete and neural migration has not proceeded when interference occurs. Recovery is optimal when interference occurs during the active phase of cortical connectivity. The time-limited process of plasticity is illustrated by a study of language recovery in children who had brain injury before or after the age of 5 years, with the former exhibiting a significantly better outcome (Rasmussen & Milner, 1977).

EARLY THERAPEUTIC INTERVENTION. Since outcome may be ameliorated by well-timed and appropriate treatment, the prevailing belief is that dysfunction should be identified early to allow for optimal outcome. There is great interest in examining which, if any, early intervention strategies will result in a significantly improved neurobehavioral outcome. The neurorehabilitation program described by Bagnato and Mayes (1986) for infants and preschool children with both acquired and congenital brain injuries serves to highlight the importance of developmentally appropriate early intervention directed specifically at the cognitive, language, perceptual, social-emotional, motor, and self-care areas of need. The intervention outcomes reportedly demonstrate highly significant improvements in developmental patterns for both groups of children, beyond the levels expected due to maturation alone. Each group exhibited its own pattern of development/recovery which further highlighted the need for population- and etiology-specific intervention approaches. Another example of an approach taken in the interest of examining the influences of early developmental factors and applying those techniques that alter outcome is found in the recent research on children with language learning disorder. Clinicians often find frequent ear infections and fluid blockage (otitis media with effusion [OME]) reported in the history of a child with language learning problems. In one study, early cognitive deficit at 3 years of age in children was correlated with duration of OME and may have resolved by 5 years of age with cognitive stimulation serving as the compensatory strategy. However, language skills continued to lag at 5 years (Baldwin, Owen, & Johnson, 1996). The notion that an early medical problem may directly affect sound discrimination perception or that there are important temporal associations made in infancy that may be a basis for later language learning disability has been postulated (Tallal, Curtiss, & Allard, 1991; Tallal, Miller, & Fitch, 1993). Further, it has been proposed that there is a possibility for remediation of language (reading) disabilities generally thought to be intractable (Merzenich *et al.*, 1996; Tallal, 1980; Tallal *et al.*, 1996). This controversial suggestion remains to be replicated and validated but is of considerable heuristic interest with respect to hypotheses about other neurobehavioral outcomes and their etiologies.

SOCIETAL ATTENTION AND ENVIRONMENTAL INFLUENCES

The importance of social environment and early experiences in neurodevelopment is increasingly recognized. Due to developmentally adaptive aspects of imitation (stimulus-boundedness), infants and young children rely on the environmental context for the appropriate acquisition of knowledge and cognitive stimulation. Children's cognitive function may be more strongly influenced by social and environmental factors than that of adults (Plomin, 1989). The brain may be thought of as a "dependent variable" that is shaped in part by the facilitative stimulation that is experienced (Bakker, 1984). Further, whether a young child will respond and demonstrate knowledge acquisition is often dependent on "optimal" environmental conditions. Thus, this context must be taken into account when considering a child's demonstrated strengths and weaknesses and their generalizability from the evaluation setting to the real world. Assessment of the sequelae of brain insult may be confounded by these important factors (Taylor, Schatschneider, & Rich, 1991), and consideration of these influences is needed to avoid simplistic hypotheses about brain–behavior functioning (Fletcher & Taylor, 1984).

Early intervention programs for the infant and early childhood populations have focused more attention on the cognitive, behavioral, and social–emotional needs of children. An enormously successful outcome of the greater attention that has been directed toward younger children within these naturalistic settings is that there has been greater comparative assessment of function between children at even earlier ages than previously considered useful. Comparisons of developmental rate and acquisition of skills among age peers have become more routine in these formal early developmental settings. As a consequence of these and associated factors, primary practitioners have been made more aware of risk factors, making it easier to identify children at risk earlier. As a result of this educative process, the child neuropsychologist is increasingly asked to consider neurobehavioral dysfunction secondary to neurological and nonneurological etiologies. Data about how congenital or early acquired conditions may affect the neurobehavioral course of newborns or young children and their later neuropsychological integrity are now rapidly being expanded. Pediatric neuropsychologists now have a knowledge base that enables exploration of these conditions, as well as tools to accomplish some of these goals. They also remain restricted by limitations due to the complex maturational issues and the imprecise way one must predict future outcome based on minimal indices of current limited behavioral repertoires. Early intervention programs have provided an important stimulus for observation and for heuristics.

Environmental factors, such as parenting skills and socioeconomic status, have been recognized as greatly influencing the development of the infant (Waber, Bauermeister, Cohen, Ferber, & Wolff, 1981; Waber, Carlson, Mann, Merola, & Moylan, 1984). In an early study, high socioeconomic status was associated with reducing effects of LBW and anoxia, while poor environmental factors were associated with increasing these effects (Sameroff & Chandler, 1975). Research has increasingly highlighted the importance of social and environmental factors for the very young child. Decline in cognitive status, as assessed by IQ scores, was associated with socioeconomic status in a study that concluded that environmental deficits and stresses impair early cognitive and psychosocial development for both full-term and premature infants, the latter being the most vulnerable (Escalona, 1982). A study comparing preterm and full-term infants during the first 5 years of life found that children who were developmentally delayed overcame their deficits in more stimulating environments while normal 1-year-olds in less stimulating environments were delayed at 3 years of age (Siegel, 1984). In a study of early predictors of neurodevelopmental outcome of very low birth weight infants at 3 years of age, researchers found that IQ is strongly influenced by environmental factors while neurological status tended to be more stable. Children classified as neurologically abnormal remained within an abnormal range despite improvements with age (Ross, Lipper, & Auld, 1986). Significant correlations were found for Beery Visual Motor Integration scores and maternal education and social class in 5-year-olds with a history of very low birth weight (Klein et al., 1985).

CLINICAL EVALUATION OF INFANTS AND YOUNG CHILDREN

Approaches to clinical practice advanced in recent years are mostly directed at older children and adolescents (Rourke, 1989). There has not been a similar focus on the infant and young child. To date, theoretical models developed for older children and adults have had little impact on clinical practice with young children. The absence of a model of neuropsychological development that spans the entire age range is limiting and in part responsible for the relative lack of attention directed to the very youngest children. A comprehensive model would (1) provide continuity in understanding developmental progress, or lack thereof, in those disorders that affect neurological development; (2) encourage a broader consideration of function and a more diverse selection of methodologies, that is, not restricting assessment to general cognitive development instruments alone; (3) allow for earlier definition and differentiation of functions during an active stage of

IDA SUE
BARON
AND
GERARD A.
GIOIA

developmental gain or delay; (4) encourage early and specific recommendations that can directly influence the developmental course and reduce the impact of an obstacle to development; (5) provide a basis for measurement of the effectiveness of treatment recommendations; (6) encourage the elaboration of existing knowledge about the natural history of normal and abnormal brain–behavior relationships in the early years; and (7) stimulate the development of innovations and techniques that can lead to better science and practice, for example, to evaluate etiological factors more precisely or to increase understanding of later concomitants of early injury or illness. The power that such a model would offer explains why the focus should be increasingly on the very youngest children and on a comprehensive understanding of the full life span.

While the approach to the neuropsychological examination of infants and young children is similar in its general form to that used in the evaluation of older individuals, there are differences. These revolve around two factors: (1) the existence of a body of knowledge regarding the cognitive and social–emotional development processes in this age group and brain–behavior relationships in normal and abnormal developmental conditions; and (2) relatively less reliance on standardized test measures to assess all desired functional areas. As noted above, the child should not be assessed without reference to the external environment and emotional context within which development is proceeding. Integration of ecological data has too often been neglected.

There are limitations in clinical practice that contribute to the idea that infants are beyond the scope of the clinical child neuropsychologist. A prerequisite in the clinical evaluation of infants and young children is an appropriate knowledge and experience with children who have had "normal" neuropsychological development. This is particularly true when one attempts to evaluate those functional areas where clinical instruments are not available or are less well developed. The pediatric neuropsychologist must rely on the creative application of the key developmental concepts through informed observation, directed interview of the caretakers, and the use of more naturalistic activities that highlight the critical behaviors. Limitations include the unavailability of tests for the functional area of concern, the limited generalizability to performance at a later age, the limited sampling of behavior that is possible since children still must "grow into their disability," and inappropriate or limited normative data. Therefore, nonstandardized evaluation and experimental techniques may be preferable to available standardized instrumentation. Clinical inferential and quantitative integration is needed (Wilson & Risucci, 1986). In addition, with the collection of multiple sources of information, including the "real world" contexts of the youngster, the hypotheses generated about the functional strengths and weaknesses and the resulting diagnostic formulation can translate directly into the intervention context of the home and preschool environments. While evaluations of a child at different time points are subject to inaccuracy due to behavioral variability, there is even a greater chance of such variability within this age span. The likelihood of misconstruing one behavioral sample as accurate is often high. Clinical influences that may hinder valid assessment include, for example, fatigue, hunger, stranger anxiety, and temperament.

The beliefs traditionally held about the capacities of the very young child must change, be reversed, or be modified as research results undo long-held misconceptions. Abilities previously thought to be beyond the young child's repertoire may in fact be well within his or her grasp and these can be demonstrated experimentally.

More recent studies have attempted to partition cognitive domains in an effort to understand relatively preserved areas of function and areas of particular vulnerability. It is now more common to obtain data not only about a child's level of intelligence (i.e., mental age compared to chronological age) but to examine more discrete functions such as attention/alertness, executive function/mental shifting, learning and memory, affective social interaction and understanding, communication/language readiness, visuomotor/visuoperceptive development, and numerical skills.

Visual motor integration has been a natural area of investigation. Many developmental tests include items about eye–hand coordination, and measures of drawing and fine motor skill have been well represented in the standardized tests available for very young children. This domain has been explored in studies of preterm and LBW infants. Early reports of visuomotor and motor delays have been supported by more recent studies which consistently find motor delays that extend across follow-up measurement periods. Motor delays were more common than mental delays (Sostek, 1992), and while improvements were noted between 1 and 2 years of age these deficits persisted and were found at 5 years of age. Comparing infants with minor hemorrhages with infants with major hemorrhages, Sostek (1992) found that the former had better motor performances and exhibited greater readiness for school. Visual motor deficit was present irrespective of degree of IVH, and preterm infants who had no hemorrhage performed poorly compared to full-term controls.

In the evaluation of infants and young children, all sources of reliable and valid data available to the pediatric neuropsychologist must be used, including the history, direct and indirect observations of behavior, and performance on selected tests (Rudel, Holmes, & Pardes, 1988). A clinical investigation begins with a careful history-taking. One must consider the many factors that may alone, or in combination, affect behavior and outcome. A thorough history-taking includes an exploration of events that precede birth, factors that surround the time of birth, and developmentally relevant issues that extend from delivery up to the present time. An outline of the types of questions that are generally asked in a clinical history interview is presented in Table 1. Further information is typically elicited from hospital and primary physicians medical records, interviews with close family members, and incidental historical records such as baby diaries kept by parents. The information one obtains may provide essential clues about neurological integrity, and therefore about neurobehavioral status.

Although the evaluation of an individual at any age should never rest solely upon test performance, work with infants and young children demands to an even greater degree the use of multiple sources of data. At the present time, the development of specific neuropsychological tests for infants and young children lags significantly behind the research knowledge base. This current state of affairs should not, however, inhibit the pediatric neuropsychologist from undertaking a competent neuropsychological evaluation. It has become more common to assess not only a child's general level of intelligence or developmental level but also more discrete functions. Child neuropsychologists, like their colleagues who study adults, are considering how general domains of function may be broken down even more discretely into subdomains. The further breakdown of behavioral domains into fractionated elements is a natural progression because of the growing knowledge about brain–behavior relationships.

An example of this progression exists in the literature about preterm and LBW

TABLE 1. SAMPLE OUTLINE OF CHILD HISTORY QUESTIONNAIRE

Basic identifying data:
Child's name
Date of birth
Date of evaluation
Person referring for evaluation
Person filling out the questionnaire
Child's pediatrician and address

Referral information:
Reason for referral
Circumstances/factors judged responsible for this problem
Child's strengths
Child's weaknesses
Do parents agree about the nature and causes of the problem?

Family information:
Address
Telephone
Parents (name, age, education, marital status)
Child's natural, adopted, or foster status
Siblings (name, age)
Others living in home
Approximate family income
Father's occupation
Mother's occupation
Significant family or marital conflict

Pregnancy, birth history, neonatal period:
Age of mother at delivery
Health problems of mother during pregnancy
Length of labor and any complications
Delivery type (vaginal, Cesarean) and any complications
Term length (full, premature, number of weeks gestation)
Birth weight and height
Condition of baby (e.g., baby breathed spontaneously, Apgar scores)
Type of nursery (e.g., normal newborn, pediatric intensive care)
Days until discharge from the hospital after birth
Medical problems after discharge (e.g., jaundice, fever)
Any problems in the first few months

Developmental history:
Motor
   Age at first accomplishment (e.g., sat alone, crawled, walked alone)
   Was child slow to develop motor skills or awkward compared to siblings/friends (e.g., running, skipping, climbing, biking, playing ball)?
   Handedness (right, left, both); family history of left-handedness
   Need for physical therapy or occupational therapy
Language
   Age at first accomplishment (e.g., first word, put two to three words together)
   Speech/language delays/problems (e.g., stutters, difficult to understand, poor comprehension)
   Oralmotor problems (e.g., late drooling, poor sucking, poor chewing)
   Was child slow to learn (e.g., the alphabet, name colors, count)
   Languages spoken in home
   Provision of speech/language therapy

TABLE 1. (*Continued*)

Toileting
    Age when toilet trained
    Associated problems (e.g., bedwetting, urine accidents, soiling)
Social behavior
Relationships with other children; with adults
Ability to begin and maintain friendships
Understanding of gestures, nonverbal stimuli, social cues
Appropriateness of sense of humor

Medical history
Results of vision check
Results of hearing check
Serious illnesses/injuries/hospitalizations/surgeries
Head injuries (e.g., date, type, loss of consciousness?, changes in behavior)
Current medications and reasons

Personal history
Febrile seizures
Epilepsy
Lead poisoning/toxic ingestion
Asthma or allergies
Loss of consciousness
Abdominal pains/vomiting, and when they occur
Headaches, and when they occur
Frequent ear infections
Sleep difficulties
Eating difficulties
Tics/twitching
Repetitive/stereotyped movements
Impulsivity
Temper tantrums
Nail biting
Clumsiness
Head banging
Self-injurious behavior

Family history
Learning difficulty
Neurological illness
Seizures
Psychiatric disorder
Instances of similar problem in any family member

Educational history
Current school and address
Grade and type of placement (e.g., regular, resource, special education, emotionally disturbed)
Grades skipped or repeated
Teachers reported problem areas (e.g., reading, spelling, arithmetic, writing, attention/concentration,
    social adjustment)
Problems with hyperactivity or inattention in the classroom

Prior psychological history
Previous contact with a social agency, psychologist, psychiatrist, clinic, or private agency

children. There has been increased survival of preterm infants of very early gestational age. Concomitantly, LBW survival rates have increased along with advances in perinatal care. However, along with these successes in viability of the neonate, medical and psychological morbidity rates have increased compared to full-term infants. Of particular interest to pediatric neuropsychologists is the increased risk for cognitive impairment among LBW children. Early studies relied on a measure of general intelligence but were supplanted by studies that attempted to evaluate specific cognitive domains. Studies that relied solely upon measures of general intellectual function (e.g., studies using standardized infant assessment measures that calculate an IQ equivalent score) did not detect subtle deficit or early indications of a deficit. Such deficit could become more overt as the child matures into the deficit (Baron *et al.*, 1995) or as sufficient time elapses to lessen the likelihood of underestimate or overestimate of developmental delay, that is, attainment of school age (Astbury *et al.*, 1990). These studies illustrate the point that the methodology chosen will influence conclusions about residual deficit. For example, while developmental quotients were not sufficient to discriminate deficit in a group of preterm infants with minor hemorrhage who were compared with preterm infants with no hemorrhage, a study of visual information processing in such children found evidence of deficit in response to novelty in the infants who had hemorrhage (Glass and Sostek, 1986).

Although it was once thought that the capacities of the very young child were quite limited, it is now well understood that the infant has cognitive abilities that can be demonstrated with appropriate methodological techniques. An examination of the literature on three domains, namely attention, memory, and executive function, provides data that support this idea.

ATTENTION

The construct of attention has been found to comprise several interrelated elements that the pediatric neuropsychologist can consider in the clinical evaluation (Cooley & Morris, 1990; Mirsky, 1989). For example, attention may be conceptualized as involving more specific components, such as the ability to initiate, sustain, inhibit, and shift as proposed by Denckla (1989), or the ability to focus/execute (scan the stimulus field and respond), sustain (be vigilant, attend for a time interval), encode (sequential registration, recall, and mental manipulation of information), and shift (be flexible) as proposed by Mirsky (1989). In particular, Mirsky's multicomponent model is a useful heuristic for the assessment of attention as it relates to other cognitive processes. The ability to focus and sustain attention is especially relevant to the study of attention in the infant and the preschool child. The shift dimension of attention will be discussed in the context of executive function.

The focus aspect of attention is neuropsychologically related to the perceptual processes as the child initially engages his or her cognitive system to make meaning out of a stimulus. Problems in the focus of attention are apparent when the child cannot engage with the material, gives it fleeting attention, and/or appears to have great difficulty beginning the task despite a variety of efforts by the clinician or caretaker to involve the child in the task. Problems with the focus of attention can be observed in either of the "content" domains, that is, the language-related domain (with an apparent lack of understanding of the language information) or the visual/nonverbal domain (engaging in only brief interaction with the objects or pictures). Although strict normative values are not available to guide the assessment of sustained attention, the pediatric neuropsychologist can nevertheless observe the extent to which the child is able to consistently complete an age-appropriate task (taking normal variability into account, of course). This can be observed in the context of natural play activities or in the completion of formal tasks (e.g., Animal Pegs subtest of the Wechsler Preschool or Primary Scales of Intelligence–Revised). The degree to which the examiner or caretaker is required to redirect the child back to the task after having initially been engaged can serve as a barometer for the sustain aspect of attention.

The data sources for attention in infants and young children are largely the examiner's direct observations, the indirect observations as reported by a reliable caretaker, and/or the rating of behaviors related to attention (e.g., Conners' scales; preschool adaptation of the DuPaul [1990] ADHD Rating Scale [Gioia, 1996]) (see Table 2). Interestingly, the Child Behavior Checklist (CBCL) for ages 2 to 3 years does not include a separable attention dimension, suggesting that the normal variability was too great to discriminate this as an independent behavioral cluster. The CBCL for 4 to 18 years does include a specific behavioral cluster which can be used for the late preschool age. Regarding specific performance measures, assessment of attentional processes with the continuous performance test (CPT) paradigm has not been particularly useful so far. The bounds of normal variability appear to be too great for 4- and 5-year-olds on the existing measures to make these useful for many children. Several measures reported in the research literature (Weissberg, Ruff, & Lawson, 1990) use more age-appropriate stimuli (e.g., animals as opposed to alphanumeric or geometric stimuli) and may result in useful clinical measures for attentional processes.

TABLE 2. ASSESSMENT TOOLS: ATTENTION

Test name	Age	Type of measure	Comments
Bayley Scales of Infant Development – Behavior Rating Scale	1–42 months	Examiner behavior observation	Attention/arousal, Orientation/engagement Factors (subscales) represent mix of behaviors including attention.
Conners' Scale	3+ years	Behavior rating scale; parent, teacher forms	Scale also includes disruptive behaviors such as hyperactivity and impulsivity.
ADHD Rating Scale	3+ years	Behavior rating scale; parent, teacher forms	Items are the nine *DSM-IV* criteria for inattention.
Child Behavior Check-list – Attention Problems Scale	4+ years	Behavior rating scale; parent	Mixture of behaviors including attention, behavioral control, and task performance
Gordon Diagnostic Test	4+ years	5-minute continuous performance test (visual)	Uses 1, 0 stimuli to assess vigilance (omission), impulsivity (commissions), and reaction time
Test of Variables of Attention (TOVA)	4+ years	11-minute continuous performance test (visual)	Uses box stimuli to assess vigilance (omission), impulsivity (commission), reaction time, and variability

IDA SUE
BARON
AND
GERARD A.
GIOIA

MEMORY

The acquisition, retention, and retrieval of new information is a domain of much importance in infants and young children at risk for neuropsychological dysfunction (e.g., those with acquired brain injuries, neurotoxic exposure, developmental language disorders, hydrocephalus, or a history of hypoxia).

Memory research with young children has provided some examples of functioning analogous to that described for older children and adults and has resulted in a new appreciation for the capacity of the very young child. For example, in their first year, children can retain information about object location (Ashmead & Perlmutter, 1980; Baillargeon, 1993). Event recall (e.g., of an ordered reproduction of events) has been shown to be possible in the months following the event. Accurate long-term recall over a 1-week delay was found in 13-month-olds (Bauer & Hertsgaard, 1993). Children ages 21 to 29 months were tested for information presented when they were 8 months younger, and novel events were recalled over the long delay, even by the youngest children (Bauer, Hertsgaard, & Dow, 1994). Children aged 3 years were found to have well-organized representations of familiar events (Hudson, 1986; Slackman, Hudson, & Fivush, 1986). Young children can recall events from when they were 2 years old (Fivush, Gray, & Fromhoff, 1987; Fivush & Hamond, 1989, 1990) although adults cannot remember events occurring before 3 or 4 years of age (Sheingold & Tenney, 1982). Further, the complexity of children's event representations increases with age (Kuebli & Fivush, 1995). It is suggested that what determines event recall is what the child is asked to remember, the number of exposures to the event, and the availability of cues and reminders of the event, much as in the older child and adult (Bauer, 1996).

Despite the attention the construct of memory has received in the neuropsychological literature (Squires & Butters, 1992) and the developmental literature (Bornstein & Sigman, 1986; Fivush & Hudson, 1990; Schneider & Pressley, 1988; Wellman, 1988), application to the clinical assessment of infants and young children is still in the early stages. Nevertheless, a number of useful concepts can be applied and several formal measures will be reviewed in the neuropsychological evaluation process with this age group. Specifically, tasks of infant habituation/recognition memory, immediate memory span, and verbal learning/memory (list, story passage, names) are available (see Table 3).

The revision of the Bayley Scales of Infant Development (Bayley, 1993) includes items, for 1- to 3-month-old infants, that assess recognition memory through the ability to habituate to auditory and visual stimuli using the novelty preference paradigm, that is, novel stimuli elicit more attention than familiar ones (Olson & Strauss, 1984). The infant is presented a stimulus to which he or she attends and eventually habituates. A novel stimulus is then presented with subsequent observation of whether a greater amount of time is spent attending to the novel versus the familiar stimulus. The paradigm postulates that memory for the stimulus has been established when the infant spends less time attending to the familiar versus the novel stimulus, because a memorial representation of the familiar stimulus has been formed and stored. The early foundations of memory foundation can be assessed with these items, albeit in a limited manner, in 1- to 3-month-old infants. Butterbaugh (1988) reviewed the various research tasks that have possible clinical use given their relatively higher predictive validity for later cognitive development. Although an abundance of research regarding infant visual recognition memo-

TABLE 3. ASSESSMENT TOOLS: MEMORY

Test name	Age	Type of measure
Habituation		
Bayley Scales of Infant Development–Recognition Memory	1–3 months	Novelty preference stimuli
Immediate memory span (selected tests)		
DAS Number Recall	2 years	Digit span (forward)
SB4 Sentence Memory	2 years	Sentence span
DAS Picture Recognition		Picture (object) recognition of increasing number of stimuli
MSCA Picture Memory	2½ years	Picture name recall; six-item card
K-ABC Spatial memory	5 years	Recall of picture location on matrix; simultaneous presentation of x-y items
SB4 Bead Memory	2 years	Memory for picture of beads, of increasing quantity
Active learning and memory tasks		
DAS Object Recall	4 years	Three-trial verbal recall of 20 picture names; immediate learning trials; delayed recall
MSCA Story Memory	2½ years	Immediate verbal recall of story passage

Note: DAS-Differential Abilities Scale; MSCA-McCarthy Scales of Children's Abilities; SB4-Stanford Binet Intelligence Scale, Fourth Edition.

ry/habituation exists (Fagan & Singer, 1983; Mayes & Kessen, 1989; Moscovitch, 1984), few measures are formally available in this area. Bornstein and Ludemann (1989) offer a naturalistic task of visual recognition memory for older infants which could be considered for application in the clinical setting.

For the preschool population, the assessment of *immediate memory span*, though somewhat limited in its scope, is the most developed area. Many of the major test batteries incorporate tasks of immediate recall, including tasks of digit span (e.g., Differential Abilities Scale [DAS] [Elliott, 1990], Kaufman Assessment Battery for Children [K-ABC] [Kaufman & Kaufman, 1983], McCarthy Scales of Children's Abilities [MSCA] [McCarthy, 1972]); sentence span (e.g., Stanford Binet Intelligence Scale, Fourth Edition [SB4] [Thorndike, Hayes, & Sattler, 1986], MSCA); and motor/hand movements (e.g., MSCA, K-ABC). Other tasks of immediate memory involving visual material include recognition memory for spatial location (e.g., K-ABC), faces (K-ABC), and pictures (DAS) and recall of picture names (MSCA) and bead configurations (SB4). Use of these various measures allows an assessment of the amount of information that can be processed at any one time, an important precursor to more active, strategic learning and memory functions.

To date, only a few *active learning/memory* tasks are available for the preschool population, including memory for story passages (MSCA) and pictorial list material (DAS Object Recall). Only the DAS Object Recall task includes a delayed recall condition to tap memory storage and retrieval. All other tasks include only an immediate recall condition which significantly limits the ability to understand the critical storage and retrieval aspects of memory.

EXECUTIVE FUNCTION

Although a discussion of the executive control functions of preschool children may appear to be premature to some, more recent literature of the development of these regulatory functions across the age span has recognized their potential import (Welsh & Pennington, 1988; Welsh, Pennington, & Grossier, 1991). The executive control functions are a collection of component abilities with the purpose of exercising regulatory control over thought and behavior in the service of goal-oriented problem-solving. A significantly greater degree of external environmental support is required by infants and young children to demonstrate age-appropriate goal-oriented problem-solving but we believe, nevertheless, that it is important to consider these component functions. The data sources for these functions are observation of the child particularly during active nonroutine problem-solving; guided interviews of caretakers regarding inhibitory abilities, flexibility with changes in routines or expectations, and ability to initiate play activities; and review of behavior ratings on specific behavior rating scales. Research investigations indicate differential developmental trajectories for the various subcomponents of the executive functions (Passler, Isaac, & Hynd, 1985; Welsh et al., 1991) which can assist the clinician in directing the focus of the assessment to the elements that are relevant for the particular developmental period. The key executive control functions for infants and young children include initiating, inhibiting, and shifting/flexibility. For example, toddlers demonstrate an increasing ability to inhibit a prepotent response in the presence of an attractive stimulus (Vaughan, Kopp, & Krakow, 1984) as illustrated with the use of developmentally relevant tasks (e.g., telephone tasks, food reward task, gift delivery tasks) that could be applied in more clinical settings. The child's inhibitory capacity is a critical function that becomes more important through the preschool years and, as such, requires explicit assessment. Clinically, the child who is poorly inhibited or who is not responsive to limits that facilitate inhibitory control is at risk for later problems with self-regulation (Mayes, 1991). Similarly, the infant's or young child's abilities to accept environmental change and transition from one activity to the next are behaviors that are relevant to the executive subdomain of shifting/flexibility. Observation of the child under changing stimulus conditions during the evaluation and directed interview of caretakers regarding the behavioral/emotional response under changing conditions in the home setting can reveal potential problems in flexibly shifting mental set. Specific populations where the problems of executive regulation of cognition and behavior are potentially relevant include children who have sustained brain injury (Ylvisaker, Szekeres, & Hartwick, 1991), lead poisoning (Bellinger, 1995), and in utero exposure to drugs and alcohol (Streissguth et al., 1984).

CONCLUSION

As a result of medical, psychological, and societal changes, the child neuropsychologist is increasingly asked to provide clinical services to younger children: to observe, to evaluate, to recommend treatment options, to educate other caregivers. For some, there are opportunities to participate in research protocols that influence medical and psychological care and that increase our knowledge of normal and abnormal brain development. There are expanding domains of functional behavior, increased cross-discipline research and communication, and a need for even greater training and education of child neuropsychologists who have an interest in infants and young children.

More recently, child neuropsychologists have directed their attention to the nature and degree of neurobehavioral deficit and to the variety or range of neurological factors and conditions that may be implicated as causal. Understanding etiology often has the added significance of leading to more efficacious treatments. Focus has therefore naturally turned as well to the assessment, management, and treatment of a wider variety of neurobehavioral deficits in the very youngest age ranges along with evaluation of late effects through longitudinal study.

REFERENCES

Ashmead, D., & Perlmutter, M. (1980). Infant memory in everyday life. In M. Perlmutter (Ed.), *New directions for child development: Children's memory* (pp. 1–16). San Francisco: Jossey-Bass.

Astbury, J., Orgill, A. A., Bajuk, B., & Yu, V. Y. (1990). Neurodevelopmental outcome, growth and health of extremely low-birthweight survivors: How soon can we tell? *Developmental Medicine and Child Neurology, 32,* 582–589.

Bagnato, S. J., & Mayes, S. D. (1986). Patterns of developmental and behavioral progress for young brain-injured children during interdisciplinary intervention. *Developmental Neuropsychology, 2,* 213–240.

Baillargeon, R. (1993). The object concept revisited: New directions in the investigation of infants' physical knowledge. In C. Granrud (Ed.), *Visual perception and cognition in infancy* (pp. 265–315). Hillsdale, NJ: Erlbaum.

Bakker, D. J. (1984). The brain as dependent variable. *Journal of Clinical Neuropsychology, 6,* 1–16.

Baldwin, C. D., Owen, M. J., & Johnson, D. L. (1996, May). Effects of early otitis media with effusion (OME) on cognitive development at 3 and 5 years. [Abstract 763]. Presented at the Pediatric Academic Societies' 1996 annual meeting, Washington, DC.

Bandura, A. (1977). *Social learning theory.* Cliffs, NJ: Prentice-Hall.

Baron, I. S., Fennell, E. B., & Voeller, K. (1995). *Pediatric neuropsychology in the medical setting.* New York: Oxford University Press.

Bauer, P. J. (1996). What do infants recall of their lives? Memory for specific events by one- to two-year-olds. *American Psychologist, 51,* 29–41.

Bauer, P. J., & Hertsgaard, L. A. (1993). Increasing steps in recall of events: Factors facilitating immediate and long-term memory in 13.5- and 16.5-month-old children. *Child Development, 64,* 1204–1223.

Bauer, P. J., Hertsgaard, L. A., & Dow, G. A. (1994). After 8 months have passed: Long-term recall of events by 1- to 2-year-old children. *Memory, 2,* 353–382.

Bayley, N. (1993). *Bayley Scales of Infant Development,* Manual (2nd ed.). New York: Psychological Corporation.

Bellinger, D. (1995). Lead and neuropsychologic function in children: Progress and problems in establishing brain–behavior relationships. In M. Tramontana & S. R. Hooper (Eds.), *Advances in child neuropsychology* (Vol. 3). New York: Springer Verlag.

Berenbaum, S. A., & Denburg, S. D. (1995). Evaluating the empirical support for the role of testosterone in the Geschwind–Behan–Galaburda model of cerebral lateralization: Commentary on Bryden, McManus, and Bulman–Fleming. *Brain & Cognition, 27,* 79–83.

Bornstein, M. H., & Ludemann, P. M. (1989). Habituation at home. *Infant Behavior and Development, 12,* 525–529.

Bornstein, M. H., & Sigman, M. D. (1986). Continuity in mental development from infancy. *Child Development, 57,* 251–274.

Breslau, N., Del Dotto, J. E., Brown, G. G., Kumar, S., Ezhuthachan, S., Hufnagle, K. G., & Peterson, E. L. (1994). A gradient relationship between low birth weight and IQ at age 6 years. *Archives of Pediatric & Adolescent Medicine, 148,* 377–383.

Bruyer, R., Dupuis, M., Ophoven, E., Rectem, D., & Reynaert, C. (1985). Anatomical and behavioral study of a case of asymptomatic callosal agenesis. *Cortex, 21,* 417–430.

Butterbaugh, G. J. (1988). Selected psychometric and clinical review of neurodevelopmental infant tests. *Clinical Neuropsychologist, 2,* 350–364.

Caputo, D., Goldstein, K., & Taub, H. (1979). The development of prematurely born children through middle childhood. In T. M. Field, A. Sostek, & H. H. Shuman (Eds.), *Infants born at risk: Behavior and development*. Jamaica, NY: Spectrum.

Chasnoff, I. J. (1991). Cocaine and pregnancy: Clinical and methodologic issues. *Clinics in Perinatology, 18*(1), 113–123.

Chasnoff, I. J., Griffith, D. R., Frier, C., & Murray, J. (1992). Cocaine/polydrug use in pregnancy: Two-year follow-up. *Pediatrics, 89*(2), 284–289.

Clarkson, T. (1991). Methylmercury. *Fundamental and Applied Toxicology, 16,* 20–21.

Clarren, S. K., Alvord, E. C., Sumi, S. M., Streissguth, A., & Smith, D. (1978). Brain malformations related to prenatal exposure to ethanol. *Journal of Pediatrics, 92,* 64–67.

Cohen, M., Campbell, R., & Yaghmai, F. (1989). Neuropathological abnormalities in developmental dysphasia. *Annals of Neurology, 25,* 567–570.

Cooley, E. L., & Morris, R. D. (1990). Attention in children: A neuropsychological based model for assessment. *Developmental Neuropsychology, 6,* 239–274.

Denckla, M. B. (1989). Executive function and the overlap zone between attention deficit hyperactivity disorder and learning disabilities. *International Pediatrics, 4,* 155–160.

Elliott, C. (1990). *Differential Abilities Scale*. San Antonio, TX: Psychological Corporation.

Escalona, S. K. (1982). Babies at double hazard: Early development of infants at biologic and social risk. *Pediatrics, 70,* 670–676.

Ettlinger, G., Blakemore, C. B., Milner, A. D., & Wilson, J. (1974). Agenesis of the corpus callosum: A further behavioural investigation. *Brain, 97,* 225–234.

Fagan, J. F., & Singer, L. T. (1983). Infant recognition memory as a measure of intelligence. In L. P. Lipsitt (Ed.), *Advances in infancy research* (Vol. 2, pp. 31–78). Norwood, NJ: Ablex.

Field, M., Ashton, R., & White, K. (1978). Agenesis of the corpus callosum: Report of two preschool children and review of the literature. *Developmental Medicine and Child Neurology, 20,* 47–61.

Fischer, M., Ryan, S. B., & Dobyns, W. B. (1992). Mechanism of interhemispheric transfer and patterns of cognitive function in acallosal subjects of normal intelligence. *Archives of Neurology, 42,* 271–277.

Fivush, R., Gray, J. T., & Fromhoff, F. A. (1987). Two-year-olds talk about the past. *Cognitive Development, 2,* 393–410.

Fivush, R., & Hamond, N. R. (1989). Time and again: Effects of repetition and retention interval on 2 year olds' event recall. *Journal of Experimental Child Psychology, 47,* 259–273.

Fivush, R., & Hamond, N. R. (1990). Autobiographical memory across the preschool years: Toward reconceptualizing childhood amnesia. In R. Fivush & J. A. Hudson (Eds.), *Knowing and remembering in young children* (pp. 223–248). New York: Cambridge University Press.

Fivush, R., & Hudson, J. A. (1990). *Knowing and remembering in young children*. New York: Cambridge University Press.

Fletcher, J. M., Bohan, T. P., Brandt, M. E., Brookshire, B. L., Beaver, S. R., Francis, D. J., Davidson, K. C., Thompson, N. M., & Miner, M. E. (1992). Cerebral white matter and cognition in hydrocephalic children. *Archives of Neurology, 49,* 818–824.

Fletcher, J. M., & Taylor, H. G. (1984). Neuropsychological approaches to children: Towards a developmental neuropsychology. *Journal of Clinical Neuropsychology, 6,* 39–56.

Folstein, S. E., & Piven, J. (1991). Etiology of autism: Genetic influences. *Pediatrics, 87* (Suppl.), 767–773.

Fox, N. (1991). If it's not left, it's right: Electroencephalograph asymmetry and the development of emotion. *American Psychologist, 46,* 863–872.

Galaburda, A. M. (1989). Ordinary and extraordinary brain development: Anatomical variation in developmental dyslexia. *Annals of Dyslexia, 39,* 67.

Galaburda, A. M., & Kemper, T. (1979). Cytoarchitectonic abnormalities in developmental dyslexia: A case study. *Annals of Neurology, 6,* 94–100.

Galaburda, A. M., Sherman, G. F., Rosen, G. D., Aboitiz, F., & Geschwind, N. (1985). Developmental dyslexia: Four consecutive cases with cortical anomalies. *Annals of Neurology, 18,* 222–233.

Geschwind, N., & Behan, P. (1982). Left-handedness: Association with immune disease, migraine, and developmental learning disorder. *Proceedings of the National Academy of Science, 79,* 5097–5100.

Gioia, G. A. (1996). *Preschool adaptation of ADHD rating scale*. Unpublished scale.

Glass, P., & Sostek, A. (1986). Information processing and interventricular hemorrhage. *Infant Behavior and Development, 9,* 142.

Goldman, P. S., & Lewis, M. (1978). Developmental biology of brain damage and experience. In C. W. Cotman (Ed.), *Neuronal plasticity* pp. 291–310). New York: Raven Press.

Goldman-Rakic, P. S. (1987). Circuitry of primate prefrontal cortex and regulation of behavior by representational memory. In V. B. Mountcastle (Ed.), *Handbook of physiology, Section 1: The nervous system: Vol. 5, Part 1, Higher functions of the brain.* Bethesda, MD: American Physiology Association.

Goldstein, G. W. (1992). Developmental neurobiology of lead toxicity. In H. L. Needleman (Ed.), *Human lead exposure.* Ann Arbor, MI: CRC Press.

Hack, M., & Breslau, N. (1986). Very low birth weight infants: Effects of brain growth during infancy on intelligence quotient at three years of age. *Pediatrics, 22,* 196–202.

Hack, M., Breslau, N., Weissman, B., Aran, D., Klein, N., & Borawski, E. (1991). Effect of very low birthweight and subnormal head size on cognitive abilities at school age. *New England Journal of Medicine, 325,* 231–237.

Harland, R. E., & Coren, S. (1996). Adult sensory capacities as a function of birth risk factors. *Journal of Clinical and Experimental Neuropsychology, 18,* 394–405.

Hudson, J. A. (1986). Memories are made of this. General event knowledge and development of autobiographic memory. In K. Nelson (Ed.), *Event knowledge, structure and function in development* (pp. 97–118). Hillsdale, NJ: Erlbaum.

Humphreys, P., Kaufmann, W. E., & Galaburda, A. M. (1990). Developmental dyslexia in women: Neuropathological findings in three cases. *Annals of Neurology, 28,* 727–738.

Huttenlocher, P. R., Taravath, S., & Mohtahedi, S. (1994). Periventricular heterotopia and epilepsy. *Neurology, 44,* 51–55.

Jordan, K., Shapiro, E., Kunin, A., & Zelinsky, D. (1995). The effects of lead overburden on neuropsychological performance: The mediating role of age at first burden and duration of burden. *Journal of International Neuropsychological Society, 1,* 124.

Kaufman, A. S., & Kaufman, N. L. (1983). *K-ABC: Kaufman Assessment Battery for Children.* Circle Pines, MN: American Guidance Service.

Kaufmann, W. E., & Galaburda, A. M. (1989). Cerebrocortical microdysgenesis in neurologically normal subjects: A histopathological study. *Neurology, 39,* 238–244.

Klein, N., Hack, M., Gallagher, J., & Fanaroff, A. V. (1985). Preschool performance of children with normal intelligence who were very low-birth-weight infants. *Pediatrics, 75,* 531–537.

Kolb, B. (1995). *Brain plasticity and behavior.* Mahwah, NJ: Lawrence Erlbaum Associates.

Korkman, M., & von Wendt, L. (1995). Evidence of altered dominance in children with congenital spastic hemiplegia. *Journal of the International Neuropsychological Society, 1,* 261–270.

Kuebli, J., & Fivush, R. (1995). Children's representation and recall of event alternatives. *Journal of Experimental and Child Psychology, 58,* 25–45.

Mayes, L. C., & Kessen, W. (1989). Maturational changes in measures of habituation. *Infant Behavior and Development, 12,* 437–450.

Mayes, S. D. (1991). Play assessment of preschool hyperactivity. In C. E. Schaefer, K. Gitlin, & A. Sandgrund (Eds.), *Play diagnosis and assessment.* New York: Wiley and Sons.

McCarthy, D. A. (1972). *Manual for the McCarthy Scales of Children's Abilities.* New York: Psychological Corporation.

Merzenich, M. M., Jenkins, W. M., Johnston, P., Schreiner, C., Miller, S., & Tallal, P. (1996). Temporal processing deficits of language-learning impaired children ameliorated by training. *Science, 271,* 77–81.

Milner, B. (1974). Hemispheric specialization: Scope and limits. In F. O. Schmitt & F. G. Worden (Eds.), *The neurosciences: Third study program* (pp. 75–89). Cambridge, MA: MIT Press.

Mirsky, A. F. (1989). The neuropsychology of attention: Elements of a complex behavior. In E. Perecman (Ed.), *Integrating theory and practice in clinical neuropsychology* (pp. 75–91). Hillsdale, NJ: Erlbaum.

Molfese, D. L. (1992). Electrophysiological correlates of developmental neurocognition: An overview. *Developmental Neuropsychology, 8,* 115–118.

Molfese, D. L., & Betz, J. C. (1988). Electrophysiological indices of the early development of lateralization for language and cognition and their implications for predicting later development. In D. L. Molfese & S. J. Segalowitz (Eds.), *Brain lateralization in children* (pp. 171–190). New York: Guilford.

Molfese, D. L., & Wetzel, W. F. (1992). Short- and long-term auditory recognition memory in 14-month old human infants: Electrophysiological correlates. *Developmental Neuropsychology, 8,* 135–160.

Moscovitch, M. (1984). *Infant memory: Its relation to normal and pathological memory in humans and other animals.* New York: Plenum.

Nass, R., & Baker, S. (1991). Learning disabilities in children with congenital adrenal hyperplasia. *Journal of Child Neurology, 6,* 306–312.

Nass, R., Baker, S., Speiser, P., Virdis, R., Balsamo, A., Cacciari, E., Loche, A., Dumic, M., & New, M.

(1987). Hormones and handedness: Left-hand bias in female congenital adrenal hyperplasia patients. *Neurology, 37,* 711–715.

Nelson, C. A., & deRegnier, R. (1992). Neural correlates of attention and memory in the first year of life. *Developmental Neuropsychology, 8,* 119–134.

Nelson, K. B. (1991). Prenatal and perinatal factors in the etiology of autism. *Pediatrics, 87* (Suppl.), 761–766.

Nelson, K. B., & Broman, S. H. (1977). Perinatal risk factors in children with serious motor and mental handicaps. *Annals of Neurology, 2,* 371–377.

Olson, G. M., & Strauss, M. S. (1984). A theory of infant memory. In M. Moscovitch (Ed.), *Infant memory.* New York: Plenum Press.

Passler, M. A., Isaac, W., & Hynd, G. W. (1985). Neuropsychological development of behavior attributed to frontal lobe functioning. *Developmental Neuropsychology, 1,* 349–370.

Pennington, B. F., & Smith, S. D. (1983). Genetic influences on learning disabilities and speech and language disorders. *Child Development, 54,* 369–387.

Pirozzolo, F. J., Pirozzolo, P. H., & Ziman, R. B. (1979). Neuropsychological assessment of callosal agenesis: Report of a case with normal intelligence and absence of the disconnection syndrome. *Clinical Neuropsychology, 1,* 13–16.

Plomin, R. (1989). Environment and genes: Determinants of behavior. *American Psychologist, 44,* 105–111.

Rakic, P. (1991). Development of the primate cerebral cortex. In M. Lewis (Ed.), *Child and adolescent psychiatry* (pp. 11–28). Baltimore, MD: Williams and Wilkins.

Rasmussen, T., & Milner, B. (1977). The role of early left-brain injury in determining lateralization of cerebral speech functions. *Annals of the New York Academy of Sciences, 299,* 355–359.

Richards, J. E., Parmelee, A. H., & Beckwith, L. (1986). Spectral analysis of infant EEG and behavioral outcome at age five. *Electroencephalography and Clinical Neurophysiology, 64,* 1–11.

Robinson, R. (1981). Equal recovery in child and adult brain? *Developmental Medicine and Child Neurology, 23,* 379–383.

Ross, G., Lipper, E. G., & Auld, P. A. M. (1986). Early predictors of neurodevelopmental outcome of very low birthweight infants at three years. *Developmental Medicine and Child Neurology, 28,* 171–179.

Rourke, B. P. (1989). Nonverbal learning disabilities: The syndrome and the model. New York: Guilford.

Rudel, R., Holmes, J. B., & Pardes, J. (1988). *Assessment of developmental learning disorders.* New York: Basic Books.

Salaptek, P., & Nelson, C. A. (1985). Event-related potentials and visual development. In G. Gottlieb & N. A. Krasnegor (Eds.), *Measurement of audition and vision in the first year of postnatal life: A methodological overview* (pp. 419–453). Norwood, NJ: Ablex.

Sameroff, A. J., & Chandler, M. J. (1975). Reproductive risk and the continuum of caretaking causality. In F. D. Horowitz, E. M. Hetherington, Jr., J. Scarr-Salaptek, & G. M. Siegel (Eds.), *Child development research* (Vol. 4, pp. 187–244). Chicago: Chicago University Press.

Saul, R. E., & Sperry, R. W. (1968). Absence of commissurotomy symptoms with agenesis of the corpus callosum. *Neurology, 18,* 307.

Schacter, D. L., & Moscovitch, M. (1984). Infants, amnesia, and dissociable memory systems. In M. Moscovitch (Ed.), *Infant memory: Its relation to normal and pathological memory in humans and other animals* (pp. 173–216). New York: Plenum.

Schneider, W., & Pressley, M. (1988). *Memory development between two and twenty.* New York: Springer-Verlag.

Shaheen, S. J. (1984). Neuromaturation and behavioral development: The case of childhood lead poisoning. *Developmental Neuropsychology, 20,* 542–550.

Sheingold, K., & Tenney, Y. J. (1982). Memory for a salient childhood event. In U. Neisser (Ed.), *Memory observed* (pp. 201–212). San Francisco: Freeman.

Shucard, D. W., Shucard, J. L., Clopper, R. R., & Schachter, M. (1992). Electrophysiological and neuropsychological indices of cognitive processing deficits in Turner syndrome. *Developmental Neuropsychology, 8,* 299–323.

Siegel, L. S. (1983). The prediction of possible learning disabilities in preterm and full-term children. In T. M. Field & A. Sostek (Eds.), *Infants born at risk: Physiological, perceptual, and cognitive processes* (pp. 295–315). New York: Grune and Stratton.

Siegel, L. S. (1984). Home environment influences on cognitive development in preterm and full term children during the first five years. In A. W. Gottsfried (Ed.), *Home environment and early cognitive development* (pp. 197–233). New York: Academic Press.

Silbergeld, E. (1992). Mechanisms of lead neurotoxicity, or looking beyond the lamppost. *FASEB Journal, 6,* 3201–3206.

Slackman, E. A., Hudson, J. A., & Fivush, R. (1986). Actions, actors, links, and goals: The structure of children's event representations. In K. Nelson (Ed.), *Event knowledge, structure and function in development* (pp. 47–69). Hillsdale, NJ: Erlbaum.

Sostek, A. M. (1992). Prematurity as well as intraventricular hemorrhage influence developmental outcome at 5 years. In S. L. Friedman & M. D. Sigman (Eds.), *The psychological development of low birthweight children* (pp. 259–274). Norwood, NJ: Ablex.

Spreen, O., Risser, A., & Edgell, D. (1995). *Developmental neuropsychology.* New York: Oxford University Press.

Squires, L., & Butters, N. (Eds.). (1992). *Neuropsychology of memory* (2nd ed.). New York: Guilford Press.

Streissguth, A. P., Martin, D. C., Barr, H. M., Sandman, B. M., Kirchner, G. L., & Darby, B. L. (1984). Intrauterine alcohol and nicotine exposure: Attention and reaction time in 4 year old children. *Developmental Psychology, 20,* 533–541.

Tallal, P. (1980). Auditory temporal perception, phonics, and reading disabilities in children. *Brain and Language, 9,* 182–198.

Tallal, P., Curtiss, S., & Allard, L. (1991). Otitis media in language-impaired and normal children. *Journal of Speech-Language Pathology and Audiology, 15,* 33–41.

Tallal, P., Miller, S. L., Bedi, G., Byma, G., Wang, X., Nagarajan, S., Shreiner, C., Jenkins, W. M., Merzenich, M. M. (1996). Language comprehension in language-learning impaired children improved with acoustically modified speech. *Science, 271,* 81–84.

Tallal, P., Miller, S., & Fitch, R. H. (1993). Neurobiological basis of speech: A case for the preeminence of temporal processing. *Annals of the New York Academy of Science, 682,* 27–47.

Taylor, H. G., Schatschneider, C., & Rich, D. (1991). Sequelae of Haemophilus influenzae meningitis: Implications for the study of brain disease and development. In M. Tramontana & S. Hooper (Eds.), *Advances in child neuropsychology* (Vol. 1, pp. 50–108). New York: Springer-Verlag.

Teeter, A., & Hynd, G. (1981). Agenesis of the corpus callosum: A developmental study during infancy. *Clinical Neuropsychology, 3,* 29–32.

Temple, C. M., & Vilarroya, O. (1990). Perceptual and cognitive perspective taking in two siblings with callosal agenesis. *British Journal of Developmental Psychology, 8,* 3.

Thorndike, R. L., Hayes, E. P., & Sattler, J. M. (1986). *Stanford-Binet Intelligence Scale, fourth edition.* Chicago: Riverside Publishing Company.

Tirosh, E., Rod, R., Cohen, A., & Hochberg, Z. (1993). Congenital adrenal hyperplasia and cerebral lateralizations. *Pediatric Neurology, 9,* 198–201.

Toriello, H. V., & Carey, J. C. (1988). Corpus callosum agenesis, facial anomalies, Robin sequence, and other anomalies: A new autosomal recessive syndrome? *American Journal of Medical Genetics, 31,* 17–23.

Vaughan, B. E., Kopp, C. B., & Krakow, J. B. (1984). The emergence and consolidation of self-control from eighteen to thirty months of age: Normative trends and individual differences. *Child Development, 55,* 990–1004.

Vohr, B. R., & Garcia Coll, C. T. (1985). Neurodevelopmental and school performance of very low-birth weight infants: A seven year longitudinal study. *Pediatrics, 76,* 345–350.

Vygotsky, L. S. (1978). *Mind in society.* Cambridge, MA: Harvard University Press.

Waber, D. P., Bauermeister, M., Cohen, C., Ferber, R., & Wolff, P. (1981). Behavioral correlates of physical and neuromotor maturity in adolescents from different environments. *Developmental Psychobiology, 14,* 513–522.

Waber, D. P., Carlson, D., Mann, M., Merola, J., & Moylan, P. (1984). SES-related aspects of neuropsychological performance. *Child Development, 55,* 1878–1886.

Weissberg, R., Ruff, H., & Lawson, K. R. (1990). The usefulness of reaction time tasks in studying attention and organization of behavior in young children. *Journal of Developmental and Behavioral Pediatrics, 11,* 59–64.

Wellman, H. M. (1988). The early development of memory strategies. In F. E. Weinert & M. Perlmutter (Eds.), *Memory development: Universal changes and individual differences.* Hillsdale, NJ: Lawrence Erlbaum Associates.

Welsh, M. C., & Pennington, B. F. (1988). Assessing frontal lobe functioning in children: Views from developmental psychology. *Developmental Neuropsychology, 4,* 119–230.

Welsh, M. C., Pennington, B. F., & Grossier, D. B. (1991). A normative-developmental study of executive function: A window on prefrontal function in children. *Developmental Neuropsychology, 7,* 131–149.

IDA SUE
BARON
AND
GERARD A.
GIOIA

Wills, K. E. (1993). Neuropsychological functioning in children with spina bifida and/or hydrocephalus. *Journal of Child Clinical Psychology, 22,* 247–265.

Wilson, B. C., & Risucci, D. A. (1986). A model for clinical-quantitative classification. Generation 1: Application to language-disordered preschool children. *Brain and Language, 27,* 281–309.

Yeates, K. O., & Mortensen, M. E. (1994). Acute and chronic neuropsychological consequences of mercury vapor poisoning in two early adolescents. *Journal of Clinical and Experimental Neuropsychology, 16,* 209–222.

Ylvisaker, M., Szekeres, S. F., & Hartwick, P. (1991). Cognitive rehabilitation following traumatic brain injury in children. In M. Tramontana & S. Hooper (Eds.), *Advances in child neuropsychology* (Vol. 1, pp. 168–218). New York: Springer-Verlag.

3

Neuropsychological Assessment of Older Children

KEITH OWEN YEATES AND H. GERRY TAYLOR

INTRODUCTION

In recent years, the field of child neuropsychology has undergone tremendous growth. There has been a burgeoning interest in the neuropsychological assessment of children with disorders of the central nervous system, systemic medical illnesses, neurodevelopmental and related learning disabilities, and psychiatric disorders (Taylor & Fletcher, 1990). Recent surveys indicate that a significant proportion of neuropsychologists now devote most of their clinical services to children and adolescents (Putnam & DeLuca, 1990; Slay & Valdivia, 1988; Sweet & Moberg, 1990). The special concerns of these neuropsychologists are reflected in the substantial and growing literature devoted to the scientific and applied aspects of child neuropsychology (e.g., Baron, Fennell, & Voeller, 1995; Obrzut & Hynd, 1986a, 1986b; Pennington, 1991; Reynolds & Fletcher-Janzen, 1989; Rourke, Bakker, Fisk, & Strang, 1983; Rourke, Fisk, & Strang, 1986; Spreen, Risser, & Edgell, 1995; Yeates, Ris, & Taylor, in press.

The purpose of this chapter is to provide a general introduction to the neuropsychological assessment of school-age children and adolescents. We begin with a brief overview of the history of child neuropsychological assessment. Our next step is to describe a set of general principles that help to conceptualize neuropsychological assessment. The principles acknowledge that child development is driven by a complex interplay of multiple forces, including, but not limited to, brain function (Bernstein & Waber, 1990; Fletcher & Taylor, 1984; Taylor & Fletcher, 1990). We then describe more concretely the methods and procedures of neuropsychological assessment and examine the reliability and validity of those methods. Next, we briefly examine some of the recent clinical and scientific applications of child

KEITH OWEN YEATES   Department of Pediatrics, The Ohio State University, and Children's Hospital, Columbus, Ohio 43205.   H. GERRY TAYLOR   Department of Pediatrics, Case Western University, and Rainbow Babies and Children's Hospital, Cleveland, Ohio 44106-6038.

Neuropsychology, edited by Goldstein *et al.* Plenum Press, New York, 1998.

neuropsychological assessment. Finally, we conclude with a discussion of the limitations and future prospects of child neuropsychology.

HISTORICAL OVERVIEW

The field of child neuropsychology has been influenced by a variety of related disciplines. The evolution of child neuropsychological assessment has involved not only the downward extension of the methods of adult neuropsychological assessment, but also the influences of developmental psychology, child clinical psychology, and pediatric neurology, among others. This rich but diverse pool of contributions makes it difficult to trace the history of child neuropsychological assessment. Much like the children who are its subject matter, the field has not always followed a strictly linear course of development.

Roughly speaking, however, the history of child neuropsychological assessment can be divided into three eras. The first, dating to about 1940, has been called the medical era (Kessler, 1980), because it generally involved observations of children with various medical conditions by physicians who then theorized about possible relationships between the status of the central nervous system and behavior. Morgan (1896), for example, provided one of the earliest descriptions of dyslexia, or what he termed "congenital word-blindness," and its potential association with abnormalities involving the angular gyrus of the left hemisphere. Similarly, the concept of brain-based hyperactivity, or "organic drivenness," arose from studies by neurologists and psychiatrists of children who had been victims of the encephalitis outbreak that occurred in Europe and North America around the time of World War I (Kahn & Cohen, 1934). Other conditions that were studied included birth injury, head trauma, Down's syndrome, and cerebral palsy (cf. Reitan & Wolfson, 1992).

The advances of the medical era were limited by the anecdotal nature of clinical reports. The second era of child neuropsychological assessment began around the time of World War II with the collaborative efforts by A. A. Strauss, Heinz Werner, and their colleagues. They introduced the use of experimental techniques for studying the behavior of children with purported brain injuries and coined the term "minimal brain injury" to describe a cluster of behaviors that they argued were characteristic of brain-injured children more generally (Strauss & Lehtinen, 1947; Strauss & Werner, 1941). Their work was very influential and continued to hold sway for several decades, as reflected in the popularity of the notion of "minimal brain dysfunction" in the conceptualization of learning disabilities (Rutter, 1982; Taylor, 1983). Together with the growth of professional psychology following World War II and the continuing refinement of standardized techniques for assessing intellectual functioning in school-age children (Sattler, 1992), the work of Strauss and Werner ushered in psychometric and psychoeducational approaches to the study of child neuropsychology.

One of the major shortcomings of this second era was the relatively scant knowledge about and limited methods available for assessing brain function in children. Scientists and practitioners often made the unwarranted assumption that the presence of cognitive or behavioral deficits believed to be characteristic of children with documented brain injury were necessarily indicative of brain abnormalities in children without any known injury (Fletcher & Taylor, 1984). Ironically, around the time these shortcomings were becoming widely apparent, modern neuroimaging technology and other advances in cognitive neuroscience began to make feasible the third and current era of child neuropsychology, which has involved a more sophisticated approach to the study of brain–behavior relationships in children. Indeed, modern neuroscience methods have enabled studies of brain–behavior relationships not only in children with neurological disease, but also in those with other systemic medical illnesses; developmental disorders such as learning disabilities and attention deficit disorder; psychiatric disorders; and normal developmental status (Dennis & Barnes, 1994; Fletcher, 1994).

The transition to the most recent era of child neuropsychology also has been driven by attempts to ground child neuropsychological assessment in broader conceptual models (Bernstein & Waber, 1990; Taylor & Fletcher, 1990). These models acknowledge the contributions of developmental cognitive neuroscience (Diamond, 1991; Goldman-Rakic, 1987) and encourage a rapprochement with developmental psychology (Fischer & Rose, 1994; Welsh & Pennington, 1988). Thus, current approaches to child neuropsychological assessment rest on systemic models that conceive of development as a dynamic, adaptive process that results from reciprocal interactions within and across multiple levels of analysis (Ford & Lerner, 1992).

The goal of neuropsychological assessment, therefore, is not simply to document the presence of cognitive deficits and their possible association with known or suspected brain damage. Instead, the goal is to enhance children's adaptation, both now and in the future, by describing their cognitive and behavioral functioning and relating it both to biological, brain-based constraints and to the environmental forces operating in their lives to produce ecologically relevant outcomes. In the following section, we summarize a set of four principles that characterize this perspective on child neuropsychological assessment.

GENERAL PRINCIPLES

The neuropsychological assessment of school-age children and adolescents is not primarily a technological enterprise. That is, it is not defined on the basis of specific interview procedures or test instruments, because the latter methods are subject to substantial change and refinement over time. Instead, child neuropsychological assessment is based on a conceptual foundation and knowledge base, the application of which is grounded in an interest in understanding brain–behavior relationships for the purpose of enhancing children's adaptation. Thus, recent models of child neuropsychological assessment, whether characterized as biobehavioral (Taylor & Fletcher, 1990) or systemic–neurodevelopmental (Bernstein & Waber, 1990), have shared several general principles.

ADAPTATION

The first principle is that the central goal of assessment is to promote the adaptation of the child, rather than simply to document the presence or location of brain damage or dysfunction. Adaptation can be understood as resulting from the interactions between children and the contexts within which they develop, or as reflecting the functional relationship between children and their environments. Failures in adaptation, such as poor school performance or unsatisfactory peer relationships, are usually the presenting problems that bring children to the attention of clinical neuropsychologists. Neuropsychological assessment is useful largely to the extent that it helps explain those failures and facilitates more successful future outcomes. Indeed, the broader goal of assessment extends beyond the facilitation of learning and behavior in the immediate context of school and home to include the promotion of long-term adaptation to the demands of adult life. Surprisingly, relatively little effort has been devoted to establishing the ecological validity of neuropsychological assessment; as we will indicate later, this remains a very important concern for future research (Taylor & Schatschneider, 1992).

BRAIN AND BEHAVIOR

A second principle is that insight into children's adaptation can be gained through an analysis of brain–behavior relationships. Advances in the neurosciences over the past two decades have begun to yield a clearer appreciation of the relationship between the brain and behavior. Old notions of localization have been replaced by more dynamic models involving the interaction of multiple brain regions (Cummings, 1993; Damasio, 1989; Derryberry & Tucker, 1992; Goldman-Rakic, 1988; Mesulam, 1990). Although most of these models concern adults, recent advances in functional neuroimaging and related techniques have provided the opportunity to determine if similar models might apply to children. Future research should therefore eventually yield major advances in our understanding of the relationship between brain function and behavior in children, permitting us to move beyond the uncritical application of adult models of brain function (Fletcher & Taylor, 1984). Such advances are certain to hold significant implications for neuropsychological assessment. In the meantime, the assessment of brain–behavior relationships in children is already quite complex, and clearly depends on factors such as the age of the child, the specific cognitive skills and behaviors assessed, the type of disorder under consideration, and the nature of the documented or hypothesized brain impairment (Dennis, 1988).

CONTEXT

A third principle guiding the neuropsychological assessment of children is that environmental contexts help to constrain and determine behavior. Thus, the ability of neuropsychological assessment to determine whether brain impairment contributes to failures of adaptation rests on a careful examination of the influences of environmental or contextual variables that also influence behavior. The reasons for examining these influences are to rule out alternative explanations for a child's adaptive difficulties and to assess the nature of the child's environment and of the situational demands being placed on the child. In this regard, neuropsychological assessment is designed not so much to measure a child's specific cognitive skills, but to determine how the child applies those skills in particular environments. A close examination of how children's cognitive and behavioral profiles match the contextual demands of their environments permits the neuropsychologist to characterize the developmental risks facing children, and thereby to make more informed recommendations for intervention (Bernstein & Waber, 1990).

DEVELOPMENT

The final guiding principle is that assessment involves the measure of change, or development, across multiple levels of analysis. Developmental neuroscience has highlighted the multiple processes that characterize brain development (e.g., cell differentiation and migration, synaptogenesis, dendritic arborization and pruning, myelinization), as well as the timing of those processes (Nowakowski, 1987). Although less research has been conducted concerning developmental changes in children's environments, there is nevertheless a natural history of environments that is characteristic of most children in our culture (Holmes, 1987). Behavioral development, in turn, can be conceptualized as the result of the joint interplay of these biological and environmental timetables (Greenough & Black, 1992), and is characterized by the emergence, stabilization, and maintenance of new skills, as well as the loss of earlier ones (Dennis, 1988).

The neuropsychological assessment of children, therefore, requires an appreciation for the developmental changes that occur in brain, behavior, and context, because the interplay between these levels of analysis determines adaptational outcomes. Failures in adaptation often reflect a clash between different developmental timetables—biological and environmental—that results in a mismatch or lack of fit between children and the contexts within which they are expected to function. The developmental timing of a brain lesion, for example, has significant implications for children's neuropsychological outcomes, because an early lesion is more likely to affect the emergence of new skills and a later lesion is more likely to disrupt the performance of established skills (Dennis, 1988). Similarly, the expression of a brain lesion may vary according to the age at which a child is assessed, because developmental changes in brain structure can result in changes in the behaviors mediated by specific brain systems (Goldman, 1971). Finally, environmental demands and resources may moderate the expression of an underlying brain lesion, reflecting the importance of context as a determinant of adaptation (Bendersky & Lewis, 1994; Taylor, Drotar, *et al.*, 1995).

METHODS OF ASSESSMENT

The four general principles outlined above—adaptation, brain and behavior, context, and development—serve as the foundation for the specific methods of assessment used by child neuropsychologists. Although neuropsychological assessment is usually equated with the administration of a battery of tests designed to assess various cognitive skills, in actual practice most neuropsychologists draw on multiple sources of information and do not rely solely on the results of psychological tests. The most typical combination of methods involves the collection of historical information, behavioral observations, and psychological testing, which together permit a broader and more detailed characterization of neuropsychological functioning.

HISTORY

The careful collection of historical information, typically accomplished by a combination of questionnaires and parent interviews, is essential in neuropsychological assessment. A thorough history not only clarifies the nature of a child's presenting problems, but also assists in the determination of its source. That is, a careful history can help to determine if a child's presenting problems have a neuropsychological basis or if they are related primarily to psychosocial or environmental factors.

The collection of a detailed history begins with some very simple facts. Knowledge of a child's age, for instance, helps to place the entire assessment process in a developmental context. The child's address, together with other demographic information, sheds light on the environment within which a child has developed and currently is expected to function. Another basic item of information that is important is the referral source. A complete history, of course, involves much more than these specific facts. Additional information is required about (1) birth and early development, (2) past or current medical involvement, (3) family and social circumstances, and (4) school history.

BIRTH AND DEVELOPMENTAL HISTORY. Collection of information regarding a child's early development usually begins with the mother's pregnancy, labor, and delivery, and extends to the acquisition of developmental milestones. Information about such topics is useful in identifying early risk factors, as well as early indicators of anomalous development. The presence of early risk factors or developmental anomalies makes a stronger case for a constitutional or neuropsychological basis for a child's failures in adaptation.

Perinatal risk factors are often of particular importance. For example, maternal illnesses during pregnancy, such as rubella, are often associated with later neuropsychological deficits (Desmond et al., 1985). Complications associated with a child's birth, such as premature delivery or low birth weight, also can increase the risk of neuropsychological deficits during later childhood (Taylor, Hack, Klein, & Schatschneider, 1995). Finally, various early environmental deficiencies, such as malnutrition, also can affect later neuropsychological functioning (Zeskind & Ramey, 1981).

The early development of the child also warrants study, including interactions with parents, socialization with peers, gross and fine motor skills, receptive and expressive language skills, constructional skills (i.e., block/puzzle/picture play), attentional abilities, feeding and sleeping patterns, and development of hand preference. Delays or anomalies in these domains are often early precursors of later learning problems (Satz, Taylor, Friel, & Fletcher, 1978).

MEDICAL HISTORY. A child's medical history often contains predictors of neuropsychological functioning. Perhaps the most obvious predictor is the presence of some documented brain abnormality or insult. For instance, closed head injuries during childhood can clearly compromise cognitive and behavioral function (Levin, Benton, & Grossman, 1982). Similarly, seizure disorders are frequently associated with neuropsychological deficits (Seidenberg, 1989).

Other aspects of a child's medical history can be linked in a more indirect fashion to neuropsychological deficits. For example, frequent ear infections are occasionally linked with cognitive delays (Kavanagh, 1986). Indeed, even allergies and other autoimmune disorders may be predictors of later academic difficulties (Geschwind & Behan, 1982).

Another critical piece of medical information is whether the child is taking any medications. Almost any medication has the potential to affect a child's neuropsychological functioning, including his or her performance on tests of cognitive skills. Among those that are especially likely to do so are anticonvulsants (Trimble, 1987), stimulants (Brown & Borden, 1989), and other psychotropic medications (Cepeda, 1989).

FAMILY AND SOCIAL HISTORY. Recent studies suggest that genetic variation plays an important role in the etiology of learning problems (Pennington, 1991).

Hence, the collection of information regarding a family's history of academic difficulties is often relevant in establishing a possible familial basis for learning problems. Family history should also be collected regarding psychiatric disturbances, language disorders, and neurological illnesses, each of which can also signal a biological foundation for later neuropsychological deficits (Lewis, 1992).

On the environmental side of the ledger, a review of family history must also attend to socioeconomic factors. Parental education and occupation are reasonably good indicators of the amount of developmental stimulation and learning opportunities that a child has received. The latter opportunities, in turn, predict later childhood intellectual competence (Yeates, MacPhee, Campbell, & Ramey, 1983). Indeed, socioeconomic disadvantage is one of the prime risk factors for mild mental retardation (Ramey, MacPhee, & Yeates, 1982).

Information regarding a child's social history should also be collected. Routine questions regarding peer relationships and a child's capacity for sustained friendships are extremely important in neuropsychological assessment. Poor peer relationships in some cases are an outcome of nonverbal learning problems, and may be linked with a particular neuropsychological profile (Rourke, 1989). In other instances, difficulties with peer relationships may be a sign of the psychological stress and poor self-esteem secondary to other adaptational difficulties, such as academic failure (Taylor, 1988).

EDUCATIONAL HISTORY. A complete school history includes information regarding a child's current grade placement, any grade repetitions or special education programs, and changes in school placement. Information about school history is usually available from parents. School personnel also can be contacted to obtain additional descriptions of a child's academic and behavioral difficulties at school. The value of school reports is that teachers and other school personnel are aware of the child's ability to meet educational demands and of how the child compares to peers. School reports often corroborate parental information, but can frequently add new or even contradictory impressions.

If a child has been evaluated for or received special educational services in the past, the timing of those services can serve as a valuable clue to the nature of a child's underlying learning problems (Holmes, 1987). In addition, the results of prior testing can be compared to the child's current test performance, providing evidence of either change or stability in neuropsychological performance and adaptive functioning across age.

Descriptions of the educational interventions provided by schools are also valuable. They help to characterize the academic demands that a child is being expected to meet, as well as the amount and type of support given to the child. If prior services are judged inadequate, then information about other resources is also needed to make practical recommendations.

BEHAVIORAL OBSERVATIONS

Behavioral observations of the child are the second critical source of information available to the neuropsychologist. Qualitative observations are extremely important, not only in interpreting the results of neuropsychological testing (Kaplan, 1988), but also in judging the adequacy of social, communicative, problem-solving, and sensorimotor skills that may not be amenable to standardized testing.

Behavioral observations are often noteworthy to the extent that they involve alterations in the examiner's usual responses to a child (Holmes, 1988). That is, changes in the examiner's usual style of interaction may signal anomalies in a child's functioning. For instance, the need for the clinician to modify his or her utterances—through simplification, repetition, or reduced rate of speech—may signal a language disorder. Similarly, if the clinician must use verbal prompts more frequently than usual to keep a child on task, then the child may have attention problems. Rigorous observation, though, must be referenced to certain basic domains of functioning to which neuropsychologists routinely attend. A typical list of domains would include the following:

- Mood and affect
- Motivation and cooperation
- Social interaction
- Attention and activity level
- Response style
- Speech, language, and communication
- Sensory and motor skills
- Physical appearance

The first two domains listed above often color the entire process of assessment. Clearly, if a child is depressed and sad, or angry and oppositional, then he or she is likely to be less motivated or cooperative than other children. A lack of full cooperation does not necessarily invalidate psychological testing, but must be taken into account in interpreting findings. Test-taking behavior also serves a valuable diagnostic function. A failure to engage in testing, for example, may indicate that an emotional disorder is contributing to the child's learning or behavior problems.

Even if affect or mood is appropriate, subtle disturbances of motivation or cooperation may be noteworthy. The child who is compliant but unenthusiastic, who follows directions but does not initiate many actions spontaneously, may be demonstrating the sort of indifference sometimes associated with frontal lobe pathology (Stuss & Benson, 1984). Alternatively, a lack of enthusiasm about the testing itself, or even outright resistance to testing, may signal the anxiety and avoidance that so often characterize children with learning disabilities when they are presented with formal school-like tasks.

Observations of social interactions with both parents and the examiner are also useful. A child who is oppositional with parents but responds to the examiner's methods of setting limits may not be managed very effectively at home. On the other hand, if the child is consistently oppositional, then a more generalized behavioral disturbance may be present. Similarly, a child who is appropriate with his or her parents but socially awkward or even inappropriate with the examiner may have a subtle nonverbal learning disability that only becomes evident in unfamiliar settings (Rourke, 1989).

Within the context of testing, the child's capacity to regulate attention and activity level deserves especially close appraisal. Attention disorders may not be apparent during structured, one-on-one testing; hence, behavior ratings from multiple sources are needed to help establish the existence of broad-based attentional problems (Barkley, 1990). Observations of inattention and motor disinhibition during an evaluation, however, provide important information regarding children's capacity to modulate their behavior under the stress of performance demands, as well as their responsiveness to various contingencies designed to increase on-task behavior.

A child's response style also provides a rich source of hypotheses regarding neuropsychological deficits, over and above that afforded by reference to the level of test performance per se. Qualitatively different errors, for instance, have been associated with different types of brain dysfunction (Kaplan, 1988). More generally, the rate and organization of responses to test demands offer valuable clues regarding the nature of the child's cognitive deficits. Observations of a child's response style may also offer clues as to strategies that the child is able to use to compensate for specific neuropsychological deficits.

Speech, language, and communication skills are also deserving of close observation. Disturbances in language skills are presumed to cause many failures in the acquisition of basic academic skills, such as reading and writing. Thus, neuropsychologists usually monitor several different facets of children's speech and language competencies, including the ease with which they engage in spontaneous conversation, their level of comprehension, the quality of their language expression in terms of both form and content, and the occurrence of more pathognomonic errors, such as naming problems and paraphasias.

Disorders of communication may be present in children whose speech and language appears grossly intact. For instance, even when speech is fluent and language expression is appropriate, children may display deficits in pragmatic language (Dennis & Barnes, 1993). Neuropsychological assessment should include opportunities for informal conversation, which permit observation of children's discourse skills, their ability to maintain a topic of conversation and engage in reciprocal turn-taking, and their appreciation of paralinguistic features such as intonation, gesture, and facial expression.

Gross disturbances in sensory and motor functioning are also noteworthy, not only because they may interfere with the standardized administration of psychological tests, but also because of the purported association between neurological "soft signs" and cognitive functioning (Shaffer, O'Connor, Shafer, & Prupis, 1983). In this regard, asymmetries in sensory and motor function may be especially important, because they can sometimes assist with the localization of brain dysfunction and thereby provide support for the notion that a child's adaptive difficulties have a neurological basis.

The final category of observation is physical appearance. Major or minor physical anomalies are frequently a sign of more specific neuropathological syndromes. The more mundane aspects of a child's appearance also deserve note. These include size, dress, and hygiene. A child who is dressed in a disheveled manner, for instance, may also be disorganized in other spheres. On the other hand, being taller than one's peers is typically an advantage, especially for boys, and may prevent the potential negative consequences of mild neuropsychological deficits.

PSYCHOLOGICAL TESTING

Psychological testing is the third source of information about the child, and the source most often equated with neuropsychological assessment. The findings obtained from formal testing allow for normative comparisons. Formal testing also provides a context for making qualitative observations of response styles and problem-solving strategies under standardized conditions. Debates continue regarding the relative merits of standardized test batteries, such as the Halstead–Reitan Neuropsychological Test Battery, as opposed to more flexible approaches to assessment (Taylor & Fletcher, 1990; Reitan & Wolfson, 1992). In general, however, most child neuropsychologists administer a variety of tests that sample from a broad range of behavioral domains. The administration of a comprehensive battery provides converging evidence for specific deficits or problems and ensures an accurate portrayal of a child's overall profile of functioning. Test batteries typically assess the following domains:

- General cognitive ability
- Language abilities
- Visuoperceptual and constructional abilities
- Attention
- Learning and memory
- Executive functions
- Corticosensory and motor capacities
- Academic skills
- Emotional status, behavioral adjustment, and adaptive behavior

As stressed earlier, these domains are assessed through a review of the child's developmental, school, and psychosocial history and through behavioral observations, as well as by formal testing. Nevertheless, tests can assess a broad range of skills in each of the domains outlined above (Franzen & Berg, 1989; Lezak, 1995; Spreen & Strauss, 1991). In the following sections, we give examples of measures and briefly examine the rationale and limitations of measurement in each domain.

GENERAL COGNITIVE ABILITY. General cognitive ability is usually assessed using standardized intelligence tests, such as the Wechsler Intelligence Scale for Children–Third Edition (WISC-III) (Wechsler, 1991), the Stanford–Binet Intelligence Scale–Fourth Edition (Thorndike, Hagen, & Sattler, 1986), and the Kaufman Assessment Battery for Children (Kaufman & Kaufman, 1983). Intelligence tests are well-standardized measures that assess a variety of cognitive skills. They have been shown to have excellent psychometric properties, in terms of both reliability and validity (Sattler, 1992). They provide an estimate of a child's overall functioning that can be useful in describing research samples and in justifying placement recommendations in clinical practice.

On the other hand, intelligence tests are not, as often supposed, measures of learning potential. Performance reflects both biological and environmental influences (Liverant, 1960). Moreover, the tests were designed primarily to predict academic achievement. For the most part, the subtests that comprise intelligence tests were not developed to assess specific mental abilities. Intelligence tests also fail to measure many important skills, as reflected in the robust relationships between academic achievement and specific neuropsychological test performance, even after IQ is partialed from the correlation (Taylor, Fletcher, & Satz, 1982). Thus, intelligence tests are helpful in neuropsychological assessment, but by no means sufficient for evaluating children's cognitive abilities.

LANGUAGE ABILITIES. The study of aphasia and other acquired language disorders was one of the driving forces in the growth of neuropsychology this century. Thus, when testing language abilities, child neuropsychologists often draw on aphasia batteries, such as the Neurosensory Center Comprehensive Examination of Aphasia (Gaddes & Crockett, 1975), or associated tests, such as the Boston Naming Test (Kirk, 1992; Yeates, 1994). They also make use of tests used by speech pathologists and other psychologists, such as the Peabody Picture Vocabulary Test–Revised (Dunn & Dunn, 1981) and the Clinical Evaluation of Language Fundamentals–Revised (Semel, Wiig, & Secord, 1987).

Tests of language abilities are relevant for neuropsychological assessment because language skills are a critical determinant of academic success and social competence, and children with learning disorders and brain injuries frequently display language deficits (Ewing-Cobbs, Levin, Eisenberg, & Fletcher, 1987). As with intelligence tests, though, performance on language measures often reflects skills in addition to those for which the procedures were designed. For example, attention problems rather than comprehension deficits may interfere with a child's ability to follow oral directions, and a lack of familiarity with the to-be-named objects rather than retrieval problems may hamper performance on a picture naming test. As is often true in neuropsychological assessment, the interpretation of language test performance must take into account performance in other domains.

VISUOPERCEPTUAL AND CONSTRUCTIONAL ABILITIES. Tests of nonverbal skills typically fall into two categories, those that draw on visuoperceptual abilities without requiring any motor output, and those that also demand constructional skills and hence involve motor control and planning. Examples of the former include the Hooper Visual Organization Test (Kirk, 1992b) and the Judgment of Line Orientation Test (Benton, Sivan, Hamsher, Varney, & Spreen, 1994). Tests of the latter skills include the Developmental Test of Visual–Motor Integration (Beery, 1989) and the Rey–Osterrieth Complex Figure (Waber & Holmes, 1985, 1986), both of which require drawing skills, and the Three-Dimensional Block Construction Test (Spreen & Strauss, 1991).

Tests of nonverbal skills are important because nonverbal deficits have been found to be associated with poor performance in certain academic skills, particularly arithmetic, as well as with a heightened risk for psychosocial maladjustment, including poor peer relationships (Rourke, 1989). In addition, nonverbal deficits are especially common in children with brain injuries and other acquired neurological insults, suggesting that nonverbal skills are especially vulnerable to brain damage in children (Taylor, Barry, & Schatschneider, 1993). Once again, however, most tests of nonverbal abilities also draw on other skills, such as attention and organization. Many of the tests also demand substantial motor dexterity. Test interpretation thus requires an appreciation for a child's overall neuropsychological test profile.

ATTENTION. From a neuropsychological perspective, attention is a multidimensional construct that overlaps with the domain of "executive" functions discussed below (Mesulam, 1981; Mirsky, Anthony, Duncan, Ahearn, & Kellam, 1991; Taylor, Schatschneider, Petrill, Barry, & Owens, 1996). Neuropsychological assessment therefore usually involves tests that assess various aspects of attention, such as vigilance, selective and divided attention, the ability to shift set, and cognitive efficiency (Cooley & Morris, 1990). Exemplary procedures include the Gordon Diagnostic System, which is one of several continuous performance tests (Grant, Ilai, Nussbaum, & Bigler, 1990); the Contingency Naming Test (Taylor, Albo, Phebus, Sachs, & Bierl, 1987); the Trail Making Test (Reitan & Wolfson, 1992); and the Arithmetic, Digit Span, Coding, and Symbol Search subtests from the WISC-III (Wechsler, 1991).

Attention problems are one of the most common reasons for referral to a child neuropsychologist, and are central to the diagnosis of attention deficit–hyperactivity disorder (Barkley, 1990). Unfortunately, the relationship between formal tests of attention and the attentional behaviors about which parents and teachers complain is often modest at best (Barkley, 1991). The weak relationship may reflect the complex nature of attention as a cognitive construct and the extent to which compensatory abilities or situational factors influence the manifestation of attentional problems outside of formal testing. Despite conceptual advances, considerable uncertainty also remains as to the nature of attention, and the differentiation of its various dimensions is often more feasible conceptually than technologically (Cooley & Morris, 1990). More pragmatically, it is often difficult to distinguish inattention from a lack of motivation. Nevertheless, because attentional functioning is likely to moderate performance on many psychological tests, tests of attention remain an important component of neuropsychological assessment.

LEARNING AND MEMORY. Difficulties with learning and memory are another frequent cause of referral to child neuropsychologists. Although children with brain impairment do not usually demonstrate the dense amnesia characteristic of adults with neurological disorders such as Alzheimer's disease, they do show distinct patterns of deficits on tests of learning and memory (Yeates, Blumenstein, Patterson, & Delis, 1995; Yeates, Enrile, Loss, Blumenstein, & Delis, 1995). Despite the obvious importance of learning and memory for children's adaptation, and especially their school performance, there were until recently few instruments available for assessing these skills. Fortunately, a variety of measures are now available in this domain, including the California Verbal Learning Test–Children's Version (Delis, Kramer, Kaplan, & Ober, 1994), the Test of Memory and Learning (Reynolds & Bigler, 1994), and the Wide Range Assessment of Memory and Learning (Sheslow & Adams, 1990).

Performance on tests of memory and learning, however, is multiply determined. Performance on tests of verbal memory, for example, is affected by children's language competencies and attentional skills. A further limitation of measurement in this domain is that current tests of children's learning and memory do not necessarily reflect recent advances in the neuroscience of memory (Butters & Delis, 1995; Squire, 1987). Future research also is needed to explore the discriminant and construct validity of measures of learning and memory (Loring & Papanicolau, 1987).

EXECUTIVE FUNCTIONS. Executive functions are those involved in the planning, organization, regulation, and monitoring of goal-directed behavior (Denckla, 1994). In the past, the assessment of these skills and related abilities, such as problem-solving and abstract reasoning, has often been conducted informally, often by examining the quality of performance on tests falling within other measurement domains. Recently, however, a wider variety of tests has been employed to assess executive functions, including measures such as the Wisconsin Card Sorting Test (Heaton, Chelune, Talley, Kay, & Curtiss, 1993), the Tower of London (Anderson, Anderson, & Lajoie, 1996; Krikorian, Bartok, & Gay, 1994), and the Children's Category Test (Boll, 1992).

Executive functions would appear to play a critical role in determining a child's adaptive functioning. Deficits in executive functions are ubiquitous in children who demonstrate neuropsychological impairment, whether as a result of focal brain dysfunction or in association with developmental learning disorders. Considerable research has been devoted in recent years to clarification of the construct of executive function in children and to assessment of its various dimensions (Levin *et al.,*

1991; Taylor *et al.*, 1996; Welsh, Pennington, & Groisser, 1991). Unfortunately, the nature of executive function remains uncertain. A major shortcoming of the existing literature is the uncritical assumption that the broad array of skills subsumed under the executive function rubric are mediated by frontal brain systems. Another is the paucity of studies examining the ecological validity of purported measures of executive function.

CORTICOSENSORY AND MOTOR CAPACITIES. Tests of corticosensory and motor capacities usually involve standardized versions of various components of the traditional neurological examination. Relevant corticosensory skills include finger localization, stereognosis, graphesthesia, sensory extinction, and left–right orientation, for which a variety of standardized assessment procedures are available (Benton *et al.*, 1994; Reitan & Wolfson, 1992). In the motor domain, tests such as the Grooved Pegboard (Matthews & Kløve, 1964), the Purdue Pegboard (Rapin, Tourk, & Costa, 1966), and the Finger Tapping Test (Reitan & Wolfson, 1992) are typically used to assess motor speed and dexterity, although batteries such as the Bruininks–Oseretsky Test of Motor Proficiency (Bruininks, 1978) are available for assessment of a wider range of skills. In addition, tests of oculomotor control, motor overflow, alternating and repetitive movements, and other related skills are often used to assess "soft" neurological status (Denckla, 1985).

Tests of corticosensory and motor capacities are useful because they are sensitive to neurological disorders and can provide useful confirmatory evidence of localized brain dysfunction. Tests of corticosensory and motor capacities also may help to predict learning problems in younger children and to differentiate older children with different types of learning disorders (Casey & Rourke, 1992). Unfortunately, sensory and motor tasks have low ceilings, and many children perform well on them despite clear evidence of learning disorders or neurological impairment. Moreover, especially for young children, the assessment of sensory and motor abilities is often compromised by lapses in attention or motivation. Finally, the results of sensory and motor testing do not always carry obvious implications for treatment.

ACADEMIC SKILLS. Academic underachievement is probably the most common reason for a child to be referred for neuropsychological assessment. Hence, testing of academic skills is often critical for understanding a child's adaptive functioning. A wide variety of tests are available, including the Wide Range Achievement Test–Third Edition (Wilkinson, 1993), the Woodcock–Johnson Tests of Achievement–Revised (Woodcock & Mather, 1989), the Kaufman Test of Educational Achievement (Kaufman & Kaufman, 1985), and the Peabody Individual Achievement Test–Revised (Markwardt, 1992). Each of these test batteries assesses reading, writing, and math skills.

The assessment of academic skills provides information on the nature and severity of underachievement. Selective problems in achievement—for instance, in reading but not math, or in decoding but not sight word recognition—provide evidence of specific learning disabilities and can also offer insights into the types of deficits contributing to a child's learning problems. Qualitative analyses also can be useful. For instance, an analysis of errors in word recognition may show that a child has a deficit in phonic awareness, consistent with a specific reading disorder (Taylor, 1988). Unfortunately, few efforts have been made to standardize the results from qualitative analyses. Standardization is also lacking in the assessment of other pertinent academic abilities, such as study habits and the use of strategies for remembering and problem-solving. Thus, achievement tests are often used to determine whether a child is eligible for special education services, and their results may also carry more specific treatment implications. At the same time, the tests sometimes afford limited insight into the cognitive processes underlying underachievement.

EMOTIONAL STATUS, BEHAVIORAL ADJUSTMENT, AND ADAPTIVE BEHAVIOR. Adaptive failures frequently occur in domains other than academic performance. Children referred for neuropsychological assessment often demonstrate psychological distress, inappropriate or otherwise undesirable behavior, or other deficits in adaptive functioning, including poor daily living skills or social skills. A wide variety of formal checklists are available to assess emotional status and behavioral adjustment, including the Child Behavior Checklist (Achenbach, 1991), the Personality Inventory for Children (Wirt, Lachar, Klinedinst, & Seat, 1984), and the Conners' Rating Scales (Conners, 1990). Checklists are also available to assess other aspects of adaptive behavior, such as the Normative Adaptive Behavior Scale (Adams, 1986), as are more detailed semistructured interview procedures, such as the Vineland Adaptive Behavior Scales (Sparrow, Balla, & Ciccetti, 1984).

If the central goal of neuropsychological assessment is to promote adaptation, then the assessment of emotional and behavioral adjustment and of adaptive functioning in various other domains is obviously crucial. A careful analysis of deficits in adjustment and adaptive behavior can help to define the mismatch between a child's neuropsychological profile and the environmental demands with which he or she is expected to cope. However, relationships between neuropsychological skills and adjustment problems or deficits in adaptive behavior are not always as straightforward as they may seem. Premorbid behavioral difficulties and adaptive deficits may actually increase the risk of brain injury and associated neuropsychological deficits. Adjustment problems and adaptive difficulties may also be an indirect result of the frustration associated with consistent failures to cope with environmental demands. In other cases, behavioral difficulties or adaptive deficits may be a more direct manifestation of neuropsychological deficits (Rourke & Fuerst, 1991).

INTERPRETATION AND MANAGEMENT

The final step in any assessment, of course, is the integration of historical information, behavioral observations, and test results. The integration of the findings entails an interpretation of a child's presenting problems from a neuropsychological perspective. Treatment recommendations follow from this formulation, based on impressions regarding the reasons for the child's current adaptive problems and predictions regarding the adaptive risks anticipated in the future. Recommendations typically focus on remediating the child's weaknesses, promoting compensatory strategies, or modifying the environmental demands affecting the child. Thus, the clinician must weigh the findings carefully, considering from a developmental perspective both the biological and environmental constraints under which a child operates, with an eye to establishing whether a child's adaptive failures have a neuropsychological basis. Only then can the neuropsychologist make appropriate recommendations for management.

The clinical skills needed to integrate diverse pieces of information from a neuropsychological perspective are difficult to describe in explicit terms. Indeed, the ability of neuropsychologists to utilize the myriad pieces of information generated in neuropsychological assessment has been called into question (Dawes, Faust, & Meehl, 1989). Nevertheless, systematic research indicates that clinicians can use the information they gather to generate diagnostic formulations in a reliable fashion (Brown, Del Dotto, Fisk, Taylor, & Breslau, 1993; Waber, Bernstein, Kammerer, Tarbell, & Sallan, 1992).

For the neuropsychologist, the integration of findings typically proceeds in two steps. In the first step, behavioral observations and test results are typically evaluated with reference to the level of performance on each activity, the pattern of performance across activities, and the strategies or processes used to reach solutions on each activity. In the second step, the clinician searches through the array of findings for *diagnostic behavioral clusters* (Bernstein & Waber, 1990). A diagnostic behavioral cluster is a group of findings, not necessarily content- or domain-specific, that is consistent with current knowledge regarding both brain function in children and the nature of cognitive–behavioral relationships across development.

For clinical purposes, childhood brain function is defined in general terms, with reference to the three primary neuroanatomic axes of the brain: left-hemisphere/right-hemisphere; anterior/posterior; and cortical/subcortical (Bernstein & Waber, 1990). The three-axis model is used as an organizational heuristic that helps to integrate the data gleaned during neuropsychological assessment. Diagnostic behavioral clusters are thus defined in terms of their presumed neural substrates, but are not presumed to reflect underlying brain damage. For instance, a constellation of delays in the acquisition of language milestones, poor performance on language and reading tests, configurational approaches on constructional tasks, and right-sided sensorimotor deficits, within the context of otherwise normal functioning, might be conceptualized as implicating the left hemisphere, without assuming that there is any focal brain lesion or dysfunction.

The conceptualization of cognitive–behavioral relationships often proceeds along analogous lines. That is, neuropsychologists employ the knowledge base about cognitive and behavioral development to characterize what Taylor and Fletcher (1990) refer to as behavior–behavior relationships. For example, they might draw distinctions between verbal and nonverbal skills, perceptual input and behavioral output, or automatic and effortful processing. Similarly, they could correlate reading disabilities with deficits in phonological awareness, or arithmetic disabilities with poor visuospatial skills.

Following the isolation of a diagnostic behavioral cluster, the neuropsychologist can then assess the match between a child's neuropsychological profile and the context within which the child functions, relating the degree of fit to the child's current adaptive failures and forecasting the potential risks faced by the child in the future. The assessment of the risks faced by a specific child must take into account not only the child's neuropsychological profile, including both strengths and weaknesses, but also the particular characteristics of the child's environment, including his or her family, school, and community (Rourke, Fisk, & Strang, 1986). Only then can the clinician make informed treatment recommendations. In other words, recommendations for management do not arise directly from test results, but instead from the integration of historical information, behavioral observations, and test results in light of the four general principles outlined earlier.

RELIABILITY AND VALIDITY

The adequacy of psychological tests rests on essential psychometric properties, including reliability and validity. Reliability and validity are multidimensional concepts, but in essence refer to the extent to which test results are stable and meaningful (Franzen, 1989). A review of the reliability and validity of the tests used in neuropsychological assessment is beyond the scope of this chapter (for comprehensive reviews, see Lezak, 1995; Spreen & Strauss, 1991). Suffice it to say, however, that the psychometric properties of neuropsychological tests are often poorly documented, even for some of the most widely used instruments, such as the Halstead–Reitan Neuropsychological Test Battery (Reynolds, 1989; but see Brown, Rourke, & Cicchetti, 1989; Leckliter, Forster, Klonoff, & Knights, 1992).

An even more fundamental issue, however, is the reliability and validity of the neuropsychological assessment process as a whole (Matarazzo, 1990). This broader issue concerns both the consistency with which neuropsychologists interpret test results and the extent to which the results of neuropsychological assessments are related both to underlying neurological impairment and to children's adaptive functioning. In other words, can clinicians agree on the nature of a child's neuropsychological profile, and does that profile have both neurological and psychological validity (Taylor & Schatschneider, 1992)?

Several recent studies have addressed these questions directly. Brown and colleagues (1993) investigated the reliability of clinical ratings of performance on pediatric neuropsychological tests. Three neuropsychologists were asked to rate the functioning of normal and low birth weight children in seven neuropsychological domains (i.e., intelligence; auditory-perceptual/language functioning; visual perceptual/visuomotor functioning; haptic perceptual functioning; memory; attention; and global functioning). The ratings were based on the results of a standard battery of neuropsychological tests, and were made using a 7-point scale. Raters were blind to each other's ratings and to birth weight status. The results revealed good to excellent agreement between raters; intraclass correlations of ratings within domains ranged from .53 for attention to .85 for intelligence. Interestingly, linear modeling of the judgments of global functioning suggested that two raters emphasized auditory-perceptual/language functioning, visual-perceptual/visuomotor functioning, and haptic perceptual functioning in deriving global ratings, while the third emphasized intelligence and attention. These results suggest that neuropsychologists can produce reliable summary ratings of functioning in specific neuropsychological domains, although they may weigh the ratings differently when characterizing global functioning.

Another investigation studied the ability of clinicians to make reliable neuropsychological diagnoses (Waber *et al.*, 1992). Two judges were asked to rate severity of overall impairment and to make neuropsychological diagnoses based on test data and behavior ratings generated in assessments of children treated for either acute lymphoblastic leukemia or Wilms's tumor. Severity of impairment was rated using a 6-point scale, and neuropsychological diagnoses were structured in terms of the three axes of brain function described earlier. More specifically, for each axis, raters were asked if a localizing discrimination could be made and, if so, in what direction (e.g., left or right hemisphere, anterior or posterior). Raters were blind to each other's ratings and to medical diagnosis.

Ratings of severity of impairment showed excellent agreement ($r = .895$). The reliability of neuropsychological diagnoses was also satisfactory. The judges generally agreed about the possibility of localization; moreover, when they both indicated that localization was possible, they displayed excellent agreement as to direction. Disagreements arose primarily when one judge did not believe that localization was possible. In other words, when one judge believed localization was possible, the other typically agreed; however, when one judge believed localization was not possible, the other was likely to believe that it was. After disagreements were resolved through consensus, the final neuropsychological diagnostic profiles were found to be systematically related to age at diagnosis: Children treated before 36 months of age exhibited profiles implicating the right hemisphere, while those treated at older ages showed profiles thought to implicate the left hemisphere.

The results from these two studies, although preliminary, suggest that clinicians can make reliable summary ratings of neuropsychological test performance and of its presumed neural substrates. The findings from the second study also suggest that clinicians' ratings are systematically related to important developmental variations in outcome.

A more direct test of the validity of neuropsychological assessment was conducted by Taylor and Schatschneider (1992). Their purpose was to determine (1) if neuropsychological test results vary with the integrity of the central nervous system and (2) if the results also predict ecologically relevant outcomes such as academic skills or behavioral adjustment. The sample consisted of 113 school-age children who had been hospitalized early in life with *Haemophilus influenza* type b meningitis. Variables that provided an estimate of the degree of brain insult (e.g., presence/absence of neurological complications) were derived from the children's medical records pertaining to their acute-phase illness. Social variables also were assessed (e.g., family socioeconomic status). The children were administered a comprehensive neuropsychological test battery, which included measures of academic skills, and their parents provided ratings of behavioral adjustment and adaptive behavior. Even after controlling for social factors, acute-phase medical variables accounted for a significant proportion of variance in performance on a variety of neuropsychological tests. According to hierarchical regression analyses, the amount of additional variance accounted for by the medical variables ranged from 0% to 13%. Similarly, although social factors accounted for a significant proportion of variance in functional outcomes, significant additional variance was explained by neuropsychological test results. The amount of additional variance accounted for by neuropsychological test performance ranged from 3% to 28%.

Thus, the results indicated that child neuropsychological assessment is sensitive to neurological insult, even when it is relatively mild and occurs at a time far removed from the actual assessment. The results also indicated that neuropsychological test performance can predict meaningful variations in children's academic skills, behavioral adjustment, and adaptive functioning. Interestingly, however, the tests that were most sensitive to neurological insult were not always the same as those that were most predictive of functional outcomes. The authors suggested that the utility of neuropsychological tests should not be judged solely on the basis of their neurological validity, as has often been the case, but also in terms of their psychological validity, which is important because it is likely to have more immediate implications for clinical management.

CLINICAL AND SCIENTIFIC APPLICATIONS

Child neuropsychological assessment is a growing clinical enterprise. As noted earlier, recent surveys indicate that a significant proportion of neuropsychologists devote most of their clinical service to children and adolescents (Putnam & DeLuca, 1990; Slay & Valdivia, 1988; Sweet & Moberg, 1990). The concerns of these neuropsychologists are reflected in a rapidly expanding literature pertaining to neuropsychological assessment of children (e.g., Baron, Fennell, & Voeller, 1995; Rourke, Fisk, & Strang, 1986; Taylor & Fletcher, 1990). Child neuropsychologists are referred a wide variety of patients with disorders of the central nervous system, systemic medical illnesses, attention problems and learning disabilities, and psychiatric disorders. Clinical referrals originate from many different sources, including physicians, schools, attorneys, and parents. Child neuropsychologists also provide a broad array of clinical services. Although assessment often occupies the bulk of their time, they also engage in other activities, including school consultation, cognitive rehabilitation, behavior management, parent training, and child psychotherapy.

A recent survey of referral patients in hospital settings confirms the broad scope of patient populations and referral sources of relevance to child neuropsychologists (Yeates, Ris, & Taylor, 1995). The survey included 472 cases of children referred for neuropsychological assessment. Common diagnoses included learning disability, traumatic brain injury, attention deficit disorder, seizure disorder, primary psychiatric disorder, idiopathic mental retardation, brain tumor, leukemia, stroke, and encephalitis. Primary referral sources included neurologists, pediatricians, parents, physiatrists, neurosurgeons, oncologists, psychologists, rheumatologists, and psychiatrists. Because the findings were based on a small sample of programs in three large teaching hospitals, they almost certainly underestimate the range of referrals received by child neuropsychologists, whose practice settings also include non-teaching hospitals, schools, clinics, and a variety of private practice arrangements. Nevertheless, the results of the survey serve both to show that the practice of child neuropsychology is distinct from that of adult neuropsychology and to highlight the growing diversity of neuropsychological practice more generally (Puente, 1992; Rourke, 1991).

The diversity of the clinical activities in child neuropsychology is also characteristic of research in the field. Dennis and Barnes (1994) have described five different paradigms that are currently guiding research efforts in developmental neuropsychology. The cognitive dysfunction paradigm focuses on the neuropsychological performance of children selected because of anomalous behavior or cognition, such as those with reading problems, attention disorders, or language deficits. The systemic disease paradigm focuses on the effects on neuropsychological outcomes of various medical conditions, such as congenital hypothyroidism or acute lymphoblastic leukemia. The brain disorder paradigm emphasizes the neuropsychological consequences of specific brain disorders, such as congenital malformations or acquired pathology resulting from head injuries or brain tumor. The brain development paradigm focuses on the timetables that characterize brain–behavior relationships across time, in both young nonhuman primates and young children. Finally, the cognitive development paradigm examines the behavioral correlates of anomalies in brain development to test theories regarding the development of normal cognitive function.

Within these paradigms, research in child neuropsychology can also be characterized in terms of its relevance to the four general principles outlined earlier. That is, research is often concerned with questions pertaining to adaptation, the relationship between brain and behavior, the relevance of contextual variables, and the role of development. For instance, within the cognitive dysfunction paradigm, a large literature has evolved that is devoted to the psychosocial outcomes associated with learning disability subtypes (Rourke & Fuerst, 1991). Of particular interest is the proposed link between nonverbal learning disabilities and "internalizing" behavior problems such as anxiety and depression. This literature helps to describe how the presence of different types of learning disabilities affects children's broader adaptation. Research related to adaptation also has been conducted in other paradigms. For example, in the brain disorder paradigm, recent studies have examined the effects of pediatric closed head injuries on behavioral adjustment and family functioning (Fletcher, Ewing-Cobbs, Miner, Levin, & Eisenberg, 1990; Taylor, Drotar, *et al.*, in press).

Research in child neuropsychology has also been devoted to examining brain–behavior relationships, and recently has begun to capitalize on advances in neuroimaging. Within the systemic disease paradigm, for example, studies of children with Williams syndrome and Down's syndrome have shown that they display unique profiles of neuropsychological strengths and weaknesses (Bellugi, Wang, & Jernigan, 1994). Volumetric MRI analyses indicate that there are corresponding differences in the sizes of certain brain structures (Jernigan & Bellugi, 1994). Similarly, in the brain disorder paradigm, a recent study of children with closed head injuries showed that performance on the Tower of London, a test of executive function, was related to the volume of the frontal lobes but not to the volume of extrafrontal brain tissue (Levin *et al.*, 1994). This study is noteworthy because it shows that brain–behavior relationships are now amenable to study even in groups of children commonly regarded as suffering from diffuse brain impairment.

The relevance of contextual variables in the prediction of neuropsychological outcomes has received less attention than the preceding topics. Nonetheless, recent research demonstrates the importance of environmental variation. For instance, within the systemic disease paradigm, the effects of prematurity on cognitive and motor outcome during early childhood have been shown to vary as a function of both environmental and biological risk with environmental risk accounting for less variation in outcomes for children at highest biological risk (Bendersky & Lewis, 1994). Within the brain disorder paradigm, Taylor, Drotar, and co-workers (1995) are currently examining the role of family and environmental context in relation to outcomes following pediatric closed head injuries. Preliminary analyses indicate that the families of children with closed head injuries show more prolonged indications of distress than do those whose children suffered orthopedic injuries. In turn, the neuropsychological outcomes following closed head injuries are related not only to the severity of injury, but also to premorbid family functioning. The findings suggest that neuropsychological outcomes are jointly influenced by environmental contexts and brain status.

Finally, the role of development is a central concern in research in child neuropsychology (Fletcher & Taylor, 1984). In the brain disorder paradigm, for instance, there is a longstanding controversy regarding the effects of early versus late closed head injuries. Recent research suggests that early injuries are associated with more generalized and more lasting neuropsychological dysfunction than are later injuries (Ewing-Cobbs *et al.*, 1995; Moore & Anderson, 1995). Further study of this concern is likely to capitalize on the methodological advances entailed by growth curve analysis, which permits the modeling of individual recovery following brain injury (Fletcher *et al.*, 1991; Thompson *et al.*, 1994). Developmental issues have also been the focus of research within both the brain development and cognitive development paradigms. For example, studies of electroencephalography (EEG) coherence patterns in children have suggested that the brain's cortical regions undergo cyclical reorganization along both rostral-to-caudal and lateral-to-medial dimensions, and that these "growth spurts" show a timetable consistent with major milestones of cognitive and motor development (Thatcher, 1992). Similarly, studies of brain development in nonhumans have suggested that the growth of the brain entails both the addition of new pathways and the elimination of others. Some of these changes depend on specific environmental experiences, while others are likely to occur in most species-typical environments (Greenough & Black, 1992).

LIMITATIONS AND PROSPECTS FOR THE FUTURE

Child neuropsychology is characterized by a rich diversity in terms of both clinical practice and research activity. Like the children who are its subjects, the field is relatively young and is experiencing rapid growth. At the same time, the field also is faced with a number of difficult challenges. To begin with, the relevance of neuropsychological assessment to children's functional adaptation remains unclear. Although the ecological validity of neuropsychological assessment is receiving increased attention (e.g., Sbordone & Long, 1995), there is little empirical support for the contention that neuropsychological test findings contribute to more effective educational planning (Reschly & Graham, 1989). Experimental research designs may prove especially powerful in addressing this shortcoming. One way to fill this void would be to design studies to test the hypothesis that children with distinct neuropsychological profiles profit from different instructional techniques (Lyon & Flynn, 1991). Of course, the ecological validity of neuropsychological assessment must also be examined in terms of outcomes in addition to academic achievement. Relevant areas for future research will include social competence, daily living skills, and vocational attainment.

The nature of brain–behavior relationships in children is another important area for future research. The application of modern neuroimaging techniques to children is a relatively new endeavor. The correlation of volumetric neuroimaging analyses with neuropsychological functioning has already provided some interesting insights into brain–behavior relationships during childhood and adolescence. Studies of the changing nature of these relationships with development should prove especially revealing. Despite advances in neuroimaging, however, brain–behavior relationships in children are likely to remain relatively obscure unless the brain can be studied "on line," in terms of its functional rather than structural integrity. Until recently, available techniques for measuring brain function either lacked sufficient temporal or spatial resolution to be of much assistance in understanding brain–behavior relationships, or involved the administration of radioactive isotopes that in most cases precluded the participation of younger children (Chugani, 1994). With the advent of functional MRI, however, more fine-grained analyses of relationships between brain function and behavior should now become feasible in a much broader range of childhood samples (Binder & Rao, 1994).

At the same time that scientific advances are permitting us to delve more deeply into the nature of brain–behavior relationships, future research must also consider the broader historical and environmental context in which assessment occurs. Child neuropsychologists have often paid scant attention to the role of the environment as a determinant of children's neuropsychological functioning. Detailed analyses of the tests that comprise neuropsychological assessment batteries can help us understand how different test characteristics, including the manner in which they are administered, affect performance at different ages. At a more macroscopic level of analysis, examination of children's environments will permit a determination of how those environments constrain and moderate neuropsychological functioning. In this regard, procedures for evaluating children's environments and the role of the environment in determining developmental outcomes deserve utmost consideration (Wachs, 1992; Wachs & Plomin, 1991). Future studies of the relationship between neuropsychological functioning and environmental factors need to acknowledge not only the reciprocal relationship between children and their environments, but also the changing nature of that relationship across development (Scarr & McCartney, 1983).

The dynamic nature of the interaction between brain, behavior, and context during childhood and adolescence highlights the importance of a developmental approach to child neuropsychology (Dennis & Barnes, 1993). Unfortunately, clinical and research activity in child neuropsychology frequently does not reflect a developmental perspective. In clinical practice, for instance, tests developed for use with adults are often applied to children unthinkingly, as are adult models of brain–behavior relationships. In the research arena, age-related differences in neuropsychological functioning during childhood and adolescence are often ignored, and only recently have methods for modeling individual patterns of growth become widely available (Francis, Fletcher, Stuebing, Davidson, & Thompson, 1991). In the future, clinicians and researchers in child neuropsychology will need to seek a rapprochement with both developmental neuroscience (Diamond, 1991; Goldman-Rakic, 1987) and developmental psychology (Fischer & Rose, 1994; Welsh & Pennington, 1988). Contributions from those disciplines will provide the foundation for a more precise characterization of the timetables of brain and behavioral development, the environments in which development occurs, and the complex interplay among those levels of analysis.

REFERENCES

Achenbach, T. M. (1991). *Integrative guide for the 1991 CBCL/4-18, YSR, and TRF profiles*. Burlington, VT: University of Vermont Department of Psychiatry.

Adams, G. L. (1986). *Normative Adaptive Behavior Checklist examiner's manual*. San Antonio, TX: Psychological Corporation.

Anderson, P., Anderson, V., & Lajoie, G. (1996). The Tower of London test: Validation and standardization for pediatric populations. *The Clinical Neuropsychologist, 10*, 54–65.

Barkley, R. A. (1990). *Attention deficit hyperactivity disorder: A handbook for diagnosis and treatment*. New York: Guilford Press.

Barkley, R. A. (1991). The ecological validity of laboratory and analogue assessment methods of ADHD symptoms. *Journal of Abnormal Child Psychology, 19*, 149–178.

Baron, I. S., Fennell, E., & Voeller, K. (1995). *Pediatric neuropsychology in the medical setting*. New York: Oxford University Press.

Beery, K. E. (1989). *The Developmental Test of Visual-Motor Integration administration, scoring, and teaching manual (3rd rev.)*. Cleveland, OH: Modern Curriculum Press.

Bellugi, U., Wang, P. P., & Jernigan, T. L. (1994). Williams syndrome: An unusual neuropsychological profile. In S. H. Broman & J. Grafman (Eds.), *Atypical cognitive deficits in developmental disorders: Implications for brain function* (pp. 23–56). Hillsdale, NJ: Lawrence Erlbaum Associates.

Bendersky, M., & Lewis, M. (1994). Environmental risk, biological risk, and developmental outcome. *Developmental Psychology, 30,* 484–494.

Benton, A. L., Sivan, A. B., Hamsher, K. deS., Varney, N. R., & Spreen, O. (1994). *Contributions to neuropsychological assessment: A clinical manual (2nd ed.).* New York: Oxford University Press.

Bernstein, J. H., & Waber, D. P. (1990). Developmental neuropsychological assessment: The systemic approach. In A. A. Boulton, G. B. Baker, & M. Hiscock (Eds.), *Neuromethods: Vol. 17. Neuropsychology* (pp. 311–371). New York: Humana Press.

Binder, J. R., & Rao, S. M. (1994). Human brain mapping with functional magnetic resonance imaging. In A. Kertesz (Ed.), *Localization and neuroimaging in neuropsychology* (pp. 185–212). Orlando, FL: Academic Press.

Boll, T. (1992). *Children's Category Test manual.* New York: Psychological Corporation.

Brown, G., Del Dotto, J. E., Fisk, J. L., Taylor, H. G., & Breslau, N. (1993). Analyzing clinical ratings of performance on pediatric neuropsychology tests. *The Clinical Neuropsychologist, 7,* 179–189.

Brown, R. T., & Borden, K. A. (1989). Neuropsychological effects of stimulant medication on children's learning and behavior. In C. R. Reynolds & E. Fletcher-Janzen (Eds.), *Handbook of clinical child neuropsychology* (pp. 443–474). New York: Plenum.

Brown, S. J., Rourke, B. P., & Cicchetti, D. V. (1989). Reliability of tests and measures used in the neuropsychological assessment of children. *The Clinical Neuropsychologist, 3,* 353–368.

Bruininks, R. H. (1978). *Bruininks-Oseretsky Test of Motor Proficiency.* Circle Pines, MN: American Guidance Service.

Butters, N., & Delis, D. C. (1995). Clinical assessment of memory disorders in amnesia and dementia. *Annual Review of Psychology, 46,* 493–523.

Casey, J. E., & Rourke, B. P. (1992). Disorders of somatosensory perception in children. In I. Rapin & S. J. Segalowitz (Eds.), *Handbook of neuropsychology: Vol. 6, Child neuropsychology* (pp. 477–494). Amsterdam: Elsevier Science Publishers.

Cepeda, M. L. (1989). Nonstimulant psychotropic medication: Side effects on children's cognition and behavior. In C. R. Reynolds & E. Fletcher-Janzen (Eds.), *Handbook of clinical child neuropsychology* (pp. 475–488). New York: Plenum.

Chugani, H. T. (1994). Development of regional brain glucose metabolism in relation to behavior and plasticity. In G. Dawson & K. W. Fischer (Eds.), *Human behavior and the developing brain* (pp. 153–175). New York: Guilford Press.

Conners, C. K. (1990). *Conners' Rating Scales manual.* Toronto: Multi-Health Systems.

Cooley, E. L., & Morris, R. D. (1990). Attention in children: A neuropsychologically based model for assessment. *Developmental Neuropsychology, 6,* 239–274.

Cummings, J. L. (1993). Frontal-subcortical circuits and human behavior. *Archives of Neurology, 50,* 873–880.

Damasio, A. R. (1989). Time-locked multiregional retroactivation: A systems-level proposal for the neural substrates of recall and recognition. *Cognition, 33,* 25–62.

Dawes, R. M., Faust, D., & Meehl, P. E. (1989). Clinical versus actuarial judgment. *Science, 243,* 1668–1674.

Delis, D. C., Kramer, J. H., Kaplan, E., & Ober, B. A. (1994). *Manual for the California Verbal Learning Test for Children.* New York: Psychological Corporation.

Denckla, M. B. (1985). Revised neurological examination for soft signs. *Psychopharmacology Bulletin, 21,* 733–800, 1111–1124.

Denckla, M. B. (1994). Measurement of executive function. In G. R. Lyon (Ed.), *Frames of reference for the assessment of learning disabilities* (pp. 117–142). Baltimore, MD: Paul H. Brookes Publishing.

Dennis, M. (1988). Language and the young damaged brain. In T. Boll & B. K. Bryant (Eds.), *Clinical neuropsychology and brain function: Research, measurement, and practice* (pp. 85–123). Washington, DC: American Psychological Association.

Dennis, M., & Barnes, M. (1993). Oral discourse skills in children and adolescents after early-onset hydrocephalus: Linguistic ambiguity, figurative language, speech acts, and script-based inferences. *Journal of Pediatric Psychology, 18,* 639–652.

Dennis, M., & Barnes, M. (1994). Developmental aspects of neuropsychology: Childhood. In D. Zaidel (Ed.), *Handbook of perception and cognition* (pp. 219–246). New York: Academic Press.

Derryberry, D., & Tucker, D. M. (1992). Neural mechanisms of emotion. *Journal of Clinical and Consulting Psychology, 60,* 329–338.

Desmond, M. M., Wilson, G. S., Vorderman, A. L., Murphy, M. A., Thurber, S., Fisher, E. S., & Kroulik,

E. M. (1985). The health and educational status of adolescents with congenital rubella syndrome. *Developmental Medicine and Child Neurology, 27*, 721–727.

Diamond, A. (1991). Some guidelines for the study of brain–behavior relationships during development. In H. Levin, H. Eisenberg, & A. Benton (Eds.), *Frontal lobe function and dysfunction* (pp. 189–211). New York: Oxford University Press.

Dunn, L. M., & Dunn, L. M. (1981). *Peabody Picture Vocabulary Test–Revised manual for forms L and M.* Circle Pines, MN: American Guidance Service.

Ewing-Cobbs, L., Fletcher, J. M., Levin, H. S., Copeland, K., Francis, D., & Miner, M. (1995). Closed head injury in infants and preschoolers: A three year longitudinal neuropsychological follow-up. *Journal of the International Neuropsychological Society, 1*, 352.

Ewing-Cobbs, L., Levin, H. S., Eisenberg, H. M., & Fletcher, J. M. (1987). Language functions following closed-head injury in children and adolescents. *Journal of Clinical and Experimental Neuropsychology, 9*, 575–592.

Fischer, K. W., & Rose, S. P. (1994). Dynamic development of coordination of components in brain and behavior: A framework for theory and research. In G. Dawson & K. W. Fischer (Eds.), *Human behavior and the developing brain* (pp. 3–66). New York: Guilford Press.

Fletcher, J. M. (1994). Afterword: Behavior–brain relationships in children. In S. H. Broman & J. Grafman (Eds.), *Atypical cognitive deficits in developmental disorders: Implications for brain function* (pp. 297–326). Hillsdale, NJ: Lawrence Erlbaum Associates.

Fletcher, J. M., Ewing-Cobbs, L., Miner, M. E., Levin, H. S., & Eisenberg, H. M. (1990). Behavioral changes after closed-head injury in children. *Journal of Consulting and Clinical Psychology, 58*, 93–98.

Fletcher, J. M., Francis, D. J., Pequegnat, W., Raudenbush, S. W., Bornstein, M. H., Schmitt, F., Brouers, P., & Stover, E. (1991). Neurobehavioral outcomes in diseases of childhood: Individual change models for pediatric human immunodeficiency viruses. *American Psychologist, 46*, 1267–1277.

Fletcher, J. M., & Taylor, H. G. (1984). Neuropsychological approaches to children: Towards a developmental neuropsychology. *Journal of Clinical Neuropsychology, 6*, 39–56.

Ford, D. H., & Lerner, R. M. (1992). *Developmental systems theory: An integrative approach and the living systems framework.* Newbury Park, CA: Sage Publications.

Francis, D. J., Fletcher, J. M., Stuebing, K. K., Davidson, K. C., & Thompson, N. M. (1991). Analysis of change: Modeling individual growth. *Journal of Consulting and Clinical Psychology, 59*, 27–37.

Franzen, M. D. (1989). *Reliability and validity in neuropsychological assessment.* New York: Plenum Press.

Franzen, M., & Berg, R. (1989). *Screening children for brain impairment.* New York: Springer Publishing.

Gaddes, W. H., & Crockett, D. J. (1975). The Spreen-Benton aphasia tests: Normative data as a measure of normal language development. *Brain and Language, 2*, 257–280.

Geschwind, N., & Behan, P. (1982). Left-handedness: Association with immune disease, migraine, and developmental disorder. *Proceedings of the National Academy of Sciences, 79*, 5097–5100.

Goldman, P. S. (1971). Functional development of the prefrontal cortex in early life and the problem of neuronal plasticity. *Experimental Neurology, 32*, 366–387.

Goldman-Rakic, P. S. (1987). Development of cortical circuitry and cortical function. *Child Development, 58*, 601–622.

Goldman-Rakic, P. S. (1988). Topography of cognition: Parallel distributed networks in primate association cortex. *Annual Review of Neuroscience, 11*, 147–156.

Grant, M. L., Ilai, D., Nussbaum, N. L., & Bigler, E. D. (1990). The relationship between continuous performance tasks and neuropsychological tests in children with attention deficit hyperactivity disorder. *Perceptual and Motor Skills, 70*, 435–445.

Greenough, W. T., & Black, J. E. (1992). Induction of brain structure by experience: Substrates for cognitive development. In M. R. Gunnar & C. A. Nelson (Eds.), *The Minnesota symposia on child psychology: Vol. 24. Developmental behavioral neuroscience* (pp. 155–200). Hillsdale, NJ: Lawrence Erlbaum Associates.

Heaton, R. K., Chelune, C. J., Talley, J. L., Kay, G. G., & Curtiss, G. (1993). *Wisconsin Card Sorting Test manual, revised and expanded.* Odessa, FL: Psychological Assessment Resources.

Holmes, J. M. (1987). Natural histories in learning disabilities: Neuropsychological difference/environmental demand. In S. J. Ceci (Ed.), *Handbook of cognitive, social, and neuropsychological aspects of learning disabilities* (pp. 303–319). Hillsdale, NJ: Lawrence Erlbaum Associates.

Holmes, J. M. (1988). History and observations. In R. Rudel (Ed.), *Assessment of developmental learning disorders: A neuropsychological approach* (pp. 144–165). New York: Basic Books.

Jernigan, T. L., & Bellugi, U. (1994). Neuroanatomical distinctions between Williams and Down syndrome. In S. H. Broman & J. Grafman (Eds.), *Atypical cognitive deficits in developmental disorders: Implications for brain function* (pp. 57–66). Hillsdale, NJ: Lawrence Erlbaum Associates.

Kahn, E., & Cohen, L. H. (1934). Organic drivenness. *New England Journal of Medicine, 210,* 748–756.

Kaplan, E. (1988). A process approach to neuropsychological assessment. In T. Boll & B. K. Bryant (Eds.), *Clinical neuropsychology and brain function: Research, measurement, and practice* (pp. 125–168). Washington, DC: American Psychological Association.

Kaufman, A. S., & Kaufman, N. (1983). *Kaufman Assessment Battery for Children.* Circle Pines, MN: American Guidance Service.

Kaufman, A. S., & Kaufman, N. (1985). *Kaufman Test of Educational Achievement.* Circle Pines, MN: American Guidance Service.

Kavanagh, J. F. (Ed.). (1986). *Otitis media and child development.* Parkton, MD: York Press.

Kessler, J. W. (1980). History of minimal brain dysfunctions. In H. E. Rie & E. D. Rie (Eds.), *Handbook of minimal brain dysfunction: A critical view.* New York: John Wiley and Sons.

Kirk, U. (1992a). Confrontation naming in normally developing children: Word-retrieval or word knowledge? *The Clinical Neuropsychologist, 6,* 156–170.

Kirk, U. (1992b). Evidence for early acquisition of visual organization ability: A developmental study. *The Clinical Neuropsychologist, 6,* 171–177.

Krikorian, R., Bartok, J., & Gay, N. (1994). Tower of London procedure: A standard method and developmental data. *Journal of Clinical and Experimental Neuropsychology, 16,* 840–850.

Leckliter, I. N., Forster, A. A., Klonoff, H., & Knights, R. M. (1992). A review of reference group data from normal children for the Halstead–Reitan Neuropsychological Test Battery for Older Children. *The Clinical Neuropsychologist, 6,* 201–229.

Levin, H. S., Benton, A. L., & Grossman, R. G. (1982). *Neurobehavioral consequences of closed head injury.* New York: Oxford University Press.

Levin, H. S., Culhane, K. A., Hartmann, J., Evankovich, K., Mattson, A. J., Harward, H., Ringholz, G., Ewing-Cobbs, L., & Fletcher, J. M. (1991). Developmental changes in performance on tests of purported frontal lobe functioning. *Developmental Neuropsychology, 1991,* 377–395.

Levin, H. S., Mendelsohn, D., Lilly, M. A., Fletcher, J. M., Culhane, K. A., Chapman, S. B., Harward, K., Kusnerik, L., Bruce, D., & Eisenberg, H. M. (1994). Tower of London performance in relation to magnetic resonance imaging following closed head injury in children. *Neuropsychology, 8,* 171–179.

Lewis, B. A. (1992). Pedigree analysis of children with phonology disorders. *Journal of Learning Disabilities, 25,* 586–597.

Lezak, M. D. (1995). *Neuropsychological assessment* (3rd ed.). New York: Oxford University Press.

Liverant, S. (1960). Intelligence: A concept in need of re-examination. *Journal of Consulting Psychology, 24,* 101–110.

Loring, D. W., & Papanicolau, A. C. (1987). Memory assessment in neuropsychology: Theoretical considerations and practical utility. *Journal of Clinical and Experimental Neuropsychology, 9,* 340–358.

Lyon, G. R., & Flynn, J. M. (1991). Educational validation studies of subtypes of learning-disabled readers. In B. P. Rourke (Ed.), *Neuropsychological validation of learning disability subtypes* (pp. 233–242). New York: Guilford Press.

Markwardt, F. C. (1992). *Peabody Individual Achievement Test–Revised manual.* Circle Pines, MN: American Guidance Service.

Matarazzo, J. D. (1990). Psychological assessment versus psychological testing: Validation from Binet to the school, clinic, and courtroom. *American Psychologist, 45,* 999–1017.

Matthews, C. G., & Kløve, H. (1964). *Instruction manual for the Adult Neuropsychology Test Battery.* Madison, WI: University of Wisconsin Medical School.

Mesulam, M.-M. (1981). A cortical network for directed attention and unilateral neglect. *Annals of Neurology, 10,* 309–325.

Mesulam, M.-M. (1990). Large scale neurocognitive networks and distributed processing for attention, language, and memory. *Annals of Neurology, 28,* 597–613.

Mirsky, A. F., Anthony, B. J., Duncan, C. C., Ahearn, M. B., & Kellam, S. G. (1991). Analysis of the elements of attention: A neuropsychological approach. *Neuropsychology Review, 2,* 109–145.

Moore, C., & Anderson, V. (1995). Pediatric head injury: The relationship between age at injury and neuropsychological sequelae. *Journal of the International Neuropsychological Society, 1,* 334.

Morgan, W. P. (1896). A case of congenital word-blindness. *British Medical Journal, 2,* 1378.

Nowakowski, R. S. (1987). Basic concepts of CNS development. *Child Development, 58,* 568–595.

Obrzut, J. E., & Hynd, G. W. (Eds.). (1986a). *Child neuropsychology: Vol. 2. Clinical practice.* New York: Academic Press.

Obrzut, J. E., & Hynd, G. W. (Eds.). (1986b). *Child neuropsychology: Vol. 1. Theory and research.* New York: Academic Press.

Pennington, B. F. (1991). *Diagnosing learning disorders: A neuropsychological framework.* New York: Guilford Press.

Puente, A. E. (1992). The status of clinical neuropsychology. *Archives of Clinical Neuropsychology, 7,* 297–312.

Putnam, S. H., & DeLuca, J. W. (1990). The TCN professional practice survey: Part I. General practices of neuropsychologists in primary employment and private practice settings. *The Clinical Neuropsychologist, 4,* 199–244.

Ramey, C. T., MacPhee, D., & Yeates, K. O. (1982). Preventing developmental retardation: A general systems model. In L. A. Bond & J. M. Joffe (Eds.), *Facilitating infant and early childhood development* (pp. 343–401). Hanover, NH: University Press of New England.

Rapin, I., Tourk, L. M., & Costa, L. D. (1966). Evaluation of the Purdue Pegboard as a screening test for brain damage. *Developmental Medicine and Child Neurology, 8,* 45–54.

Reitan, R. M., & Wolfson, D. (1992). *Neuropsychological evaluation of older children.* South Tucson, AZ: Neuropsychology Press.

Reschly, D. J., & Graham, F. M. (1989). Current neuropsychological diagnosis of learning problems: A leap of faith. In C. R. Reynolds & C. Fletcher-Janzen (Eds.), *Handbook of clinical child neuropsychology* (pp. 503–520). New York: Plenum Press.

Reynolds, C. R. (1989). Measurement and statistical problems in neuropsychological assessment of children. In C. R. Reynolds & C. Fletcher-Janzen (Eds.), *Handbook of clinical child neuropsychology* (pp. 147–166). New York: Plenum Press.

Reynolds, C. R., & Bigler, E. D. (1994). *Test of Memory and Learning examiner's manual.* Austin, TX: Pro-Ed.

Reynolds, C. R., & Fletcher-Janzen, E. (Eds.). (1989). *Handbook of clinical child neuropsychology.* New York: Plenum.

Rourke, B. P. (1989). *Nonverbal learning disabilities: The syndrome and the model.* New York: Guilford Press.

Rourke, B. P. (1991). Human neuropsychology in the 1990's. *Archives of Clinical Neuropsychology, 6,* 1–14.

Rourke, B. P., Bakker, D. J., Fisk, J. L., & Strang, J. D. (1983). *Child neuropsychology: An introduction to theory, research, and clinical practice.* New York: Guilford Press.

Rourke, B. P., Fisk, J. L., & Strang, J. D. (1986). *Neuropsychological assessment of children: A treatment-oriented approach.* New York: Guilford Press.

Rourke, B. P., & Fuerst, D. R. (1991). *Learning disabilities and psychosocial functioning.* New York: Guilford Press.

Rutter, M. (1982). Syndromes attributed to "minimal brain dysfunction" in childhood. *American Journal of Psychiatry, 139,* 21–33.

Sattler, J. M. (1992). *Assessment of children* (3rd ed., rev.). San Diego, CA: Jerome M. Sattler Publisher.

Satz, P., Taylor, H. G., Friel, J., & Fletcher, J. M. (1978). Some developmental and predictive precursors of reading disabilities: A six year follow-up. In A. L. Benson & D. Pearl (Eds.), *Dyslexia: An appraisal of current knowledge* (pp. 313–347). New York: Oxford University Press.

Sbordone, R. J., & Long, C. J. (Eds.). (1995). *Ecological validity of neuropsychological testing.* Delray Beach, FL: St. Lucie Press.

Scarr, S., & McCartney, K. (1983). How people make their own environments: A theory of genotype environment effects. *Child Development, 54,* 424–435.

Seidenberg, M. (1989). Academic achievement and school performance of children with epilepsy. In B. P. Hermann & M. Seidenberg (Eds.), *Childhood epilepsies: Neuropsychological, psychosocial, and intervention aspects* (pp. 105–118). New York: Wiley.

Semel, E., Wiig, E. H., & Secord, W. (1987). *Clinical Evaluation of Language Fundamentals—Revised examiner's manual.* New York: Psychological Corporation.

Shaffer, D., O'Connor, P. A., Shafer, S. Q., & Prupis, S. (1983). Neurological "soft signs": Their origins and significance. In M. Rutter (Ed.), *Developmental neuropsychiatry* (pp. 144–163). New York: Guilford Press.

Sheslow, D., & Adams, W. (1990). *Wide Range Assessment of Memory and Learning administration manual.* Wilmington, DE: Jastak.

Slay, D. K., & Valdivia, L. (1988). Neuropsychology as a specialized health service listed in the National Register of Health Service Providers in Psychology. *Professional Psychology: Research and Practice, 19,* 323–329.

Sparrow, S. S., Balla, D. A., & Ciccetti, D. V. (1984). *Vineland Adaptive Behavior Scales.* Circle Pines, MN: American Guidance Service.

Spreen, O., Risser, A. H., & Edgell, D. (1995). *Developmental neuropsychology.* New York: Oxford University Press.

Spreen, O., & Strauss, E. (1991). *A compendium of neuropsychological tests: Administration, norms, and commentary.* New York: Oxford University Press.

Squire, L. R. (1987). *Memory and brain.* New York: Oxford University Press.

Strauss, A. A., & Lehtinen, L. (1947). *Psychopathology and education of the brain-injured child.* New York: Grune and Stratton.

Strauss, A. A., & Werner, H. (1941). The mental organization of the brain-injured mentally defective child. *American Journal of Psychiatry, 97,* 1194–1203.

Stuss, D. T., & Benson, D. F. (1984). Neuropsychological studies of the frontal lobes. *Psychological Bulletin, 95,* 3–28.

Sweet, J. J., & Moberg, P. J. (1990). A survey of practices and beliefs among ABPP and non-ABPP clinical neuropsychologists. *The Clinical Neuropsychologist, 4,* 101–120.

Taylor, H. G. (1983). MBD: Meaning and misconceptions. *Journal of Clinical Neuropsychology, 5,* 271–287.

Taylor, H. G. (1988). Learning disabilities. In E. J. Mash & L. G. Terdal (Eds.), *Behavioral assessment of childhood disorders* (2nd ed., pp. 402–450). New York: Guilford Press.

Taylor, H. G., Albo, V. C., Phebus, C. K., Sachs, B. R., & Bierl, P. G. (1987). Postirradiation treatment outcomes for children with acute lymphocytic leukemia. *Journal of Pediatric Psychology, 12,* 395–411.

Taylor, H. G., Barry, C., & Schatschneider, C. (1993). School-age consequences of *Haemophilus influenzae* type B meningitis. *Journal of Clinical Child Psychology, 22,* 196–206.

Taylor, H. G., Drotar, D., Wade, S., Yeates, K. O., Stancin, T., & Klein, S. (1995). Recovery from traumatic brain injury in children: The importance of the family. In S. Broman & M. Michel (Eds.), *Traumatic brain injury in children* (pp. 188–218). New York: Oxford University Press.

Taylor, H. G., & Fletcher, J. M. (1990). Neuropsychological assessment of children. In M. Hersen & G. Goldstein (Eds.), *Handbook of psychological assessment* (2nd ed., pp. 228–255). New York: Pergamon.

Taylor, H. G., Fletcher, J. M., & Satz, P. (1982). Component processes in reading disabilities: Neuropsychological investigation of distinct reading subskill deficits. In R. N. Malatesha & P. G. Aaron (Eds.), *Reading disorders: Varieties and treatments.* New York: Academic Press.

Taylor, H. G., Hack, M., Klein, N., & Schatschneider, C. (1995). Achievement in <750 gm birthweight children with normal cognitive abilities: Evidence for specific learning disabilities. *Journal of Pediatric Psychology, 20,* 703–719.

Taylor, H. G., & Schatschneider, C. (1992). Child neuropsychological assessment: A test of basic assumptions. *The Clinical Neuropsychologist, 6,* 259–275.

Taylor, H. G., Schatschneider, C., Petrill, S., Barry, C. T., & Owens, C. (1996). Executive dysfunction in children with early brain disease: Outcomes post *Haemophilus influenzae* meningitis. *Developmental Neuropsychology, 12,* 35–52.

Thatcher, R. W. (1992). Cyclic cortical reorganization during childhood. *Brain and Cognition, 20,* 24–50.

Thompson, N. M., Francis, D. J., Stuebing, K. K., Fletcher, J. M., Ewing-Cobbs, L., Miner, M. E., Levin, H. S., & Eisenberg, H. M. (1994). Motor, visual-spatial, and somatosensory skills after closed-head injury in children and adolescents: A study of change. *Neuropsychology, 8,* 333–342.

Thorndike, R. L., Hagen, E. P., & Sattler, J. M. (1986). *Stanford-Binet Intelligence Scale* (4th ed.). Chicago: Riverside Publishing.

Trimble, M. R. (1987). Anticonvulsant drugs and cognitive function: A review of the literature. *Epilepsia, 28* (Suppl. 3), 37–45.

Waber, D. P., Bernstein, J. H., Kammerer, B. L., Tarbell, N. J., & Sallan, S. E. (1992). Neuropsychological diagnostic profiles of children who received CNS treatment for acute lymphoblastic leukemia: The systemic approach to assessment. *Developmental Neuropsychology, 8,* 1–28.

Waber, D. P., & Holmes, J. M. (1985). Assessing children's copy productions of the Rey–Osterrieth Complex Figure. *Journal of Clinical and Experimental Neuropsychology, 7,* 264–280.

Waber, D. P., & Holmes, J. M. (1986). Assessing children's memory productions of the Rey–Osterrieth Complex Figure. *Journal of Clinical and Experimental Neuropsychology, 8,* 563–580.

Wachs, T. D. (1992). *The nature of nurture.* Newbury Park, CA: Sage Publications.

Wachs, T. D., & Plomin, R. (Eds.). (1991). *Conceptualization and measurement of organism-environment interaction.* Washington, DC: American Psychological Association.

Wechsler, D. (1991). *Manual for the Wechsler Intelligence Scale for Children Third Edition.* New York: Psychological Corporation.

Welsh, M. C., & Pennington, B. F. (1988). Assessing frontal lobe functioning in children: Views from developmental psychology. *Developmental Neuropsychology, 4,* 199–230.

Welsh, M. C., Pennington, B. F., & Groisser, D. B. (1991). A normative-developmental study of executive function: A window on prefrontal function in children. *Developmental Neuropsychology, 7,* 131–149.

Wilkinson, G. S. (1993). *Wide Range Achievement Test-Revision 3 manual.* Wilmington, DE: Jastak Associates.

Wirt, R. D., Lachar, D., Klinedinst, J. K., & Seat, P. D. (1984). *Multidimensional description of child personality: A manual for the Personality Inventory for Children* (Rev. ed.). Los Angeles: Western Psychological Services.

Woodcock, R., & Mather, N. (1989). *Woodcock–Johnson Tests of Achievement–Revised standard and supplemental batteries.* New York: Psychological Corporation.

Yeates, K. O. (1994). Comparison of developmental norms for the Boston Naming Test. *The Clinical Neuropsychologist, 8,* 91–98.

Yeates, K. O., Blumenstein, E., Patterson, C. M., & Delis, D. C. (1995). Verbal learning and memory following pediatric closed-head injury. *Journal of the International Neuropsychological Society, 1,* 78–87.

Yeates, K. O., Enrile, B. E., Loss, N., Blumenstein, E., & Delis, D. C. (1995). Verbal learning and memory in children with myelomeningocele. *Journal of Pediatric Psychology, 20,* 801–815.

Yeates, K. O., MacPhee, D., Campbell, F. A., & Ramey, C. T. (1983). Maternal IQ and home environment as determinants of early childhood intellectual competence: A developmental analysis. *Developmental Psychology, 19,* 731–739.

Yeates, K. O., Ris, M. D., & Taylor, H. G. (1995). Hospital referral patterns in pediatric neuropsychology. *Child Neuropsychology, 1,* 56–62.

Yeates, K. O., Ris, M. D., & Taylor, H. G. (in press). *Pediatric neuropsychology: Theory, research, and practice.* New York: Guilford Press.

Zeskind, P. S., & Ramey, C. T. (1981). Preventing intellectual and interactional sequelae of fetal malnutrition: A longitudinal, transactional, and synergistic approach to development. *Child Development, 52,* 213–218.

4

Neuropsychological Assessment of Adults

GERALD GOLDSTEIN

INTRODUCTION AND HISTORICAL BACKGROUND

This chapter provides a general introduction to the field of neuropsychological assessment and deals specifically with the extensive standard test batteries and individual tests used with adults. The focus of neuropsychological assessment has traditionally been on the brain-damaged patient, but there have been major extensions of the field to psychiatric disorders (Goldstein, 1986; Yozawitz, 1986), functioning of normal individuals (e.g., Gold, Berman, Randolph, Goldberg, & Weinberger, 1996), and normal aging (Goldstein & Nussbaum, 1996).

Perhaps the best definition of a neuropsychological test has been offered by Ralph Reitan, who describes it as a test that is sensitive to the condition of the brain. If performance on a test changes with a change in brain function, then the test is a neuropsychological test. However, neuropsychological assessment is not restricted to the use of only neuropsychological tests. It should also contain some tests that are generally insensitive to brain dysfunction, primarily because such tests are often useful for providing a baseline against which the extent of impairment associated with acquired brain damage can be measured.

The practice of neuropsychological assessment is roughly divided into two approaches. Some practitioners use standard comprehensive neuropsychological test batteries, while others use individual tests that do not constitute a formal battery. Sometimes the tests used vary from patient to patient depending on referral and diagnostic considerations, and sometimes essentially the same tests are always used by the practitioner, but the collection of tests does not constitute a standard battery. This chapter will emphasize standard batteries, since Part IV of this volume contains a great deal of information about the individual tests, most of which are specialized in nature. A standard comprehensive battery is a procedure

GERALD GOLDSTEIN   VA Pittsburgh Healthcare System, Highland Drive Division (151R), Pittsburgh, Pennsylvania 15206-1297.

Neuropsychology, edited by Goldstein *et al.* Plenum Press, New York, 1998.

that assesses all of the major functional areas generally affected by structural brain damage. We use the term in an ideal sense because none of the standard, commonly available procedures entirely achieves full comprehensiveness. Since brain damage most radically affects cognitive processes, most neuropsychological tests assess various areas of cognition, but perception and motor skills are also frequently evaluated. Thus, neuropsychological tests are generally thought of as assessment instruments for a variety of cognitive, perceptual, and motor skills. While the emphasis is on cognitive function, neuropsychologists in general practice typically add brief personality assessments using the Minnesota Multiphasic Personality Inventory (MMPI) or similar procedure, and some measure of academic achievement, such as the Revised Wide Range Achievement Test (Jastak & Wilkinson, 1984).

Neuropsychological assessment typically involves the functional areas of general intellectual capacity; memory; speed and accuracy of psychomotor activity; visual–spatial skills; visual, auditory, and tactile perception; language; and attention. Thus, a comprehensive neuropsychological assessment may be defined as a procedure that at least surveys all of these areas.

Neuropsychological tests have the same standardization requirements as all psychological tests. That is, there is the need for appropriate quantification, norms, and related test construction considerations, as well as the need to deal with issues related to validity and reliability. However, there are some special considerations regarding neuropsychological tests. Neuropsychological test batteries must be administered to brain-damaged patients who may have substantial cognitive impairment and severe physical disability. Thus, stimulus and response characteristics of the tests themselves, as well as of the test instructions, become exceedingly important considerations. In general, the test material should consist of salient stimuli that the patient can readily see or hear and understand. Instructions should not be unduly complex, and if the patient has a sensory deficit, it should be possible to give the instructions in an intact modality, without jeopardizing the use of established test norms. The opportunity should be available to repeat and paraphrase instructions until it is clear that they are understood. Similarly, the manner of responding to the test material (e.g., pressing a lever or writing on a multiple-choice form) should be within the patient's capabilities.

Neuropsychological assessment aims at specifying in as much detail as possible the functional deficits that exist in a manner that allows for mapping of these deficits onto known systems in the brain. There are several methods of achieving this goal, and not all neuropsychologists agree as to the most productive route. In general, some prefer to examine patients in what may be described as a linear manner, with a series of interlocking component abilities, while others prefer using more complex tasks in the form of standard, extensive batteries and interpretation through examination of performance configurations. The linear approach is best exemplified in the work of A. R. Luria and various collaborators (Luria, 1973), while the configurational approach is seen in the work of Ward Halstead (Halstead, 1947), Ralph Reitan (Reitan & Wolfson, 1993), and their many collaborators. In either case, however, the aim of the assessment is to determine the pattern of the patient's preserved and impaired functions and infer from this pattern the nature of the disturbed brain function. Individual tests and test batteries are really only of neuropsychological value if they can be analyzed by one of these two methods.

VALIDITY AND RELIABILITY

With regard to concurrent validity, the criterion used in most cases is the objective identification of some central nervous system lesion arrived at independently of the neuropsychological test results. Therefore, validation is generally provided by neurologists or neurosurgeons. Historically, identification of lesions of the brain has been problematic because, unlike many organs of the body, the brain cannot usually be visualized directly in the living individual. The major exceptions occur when the patient undergoes brain surgery or a brain biopsy. In the absence of these procedures, validation has been dependent on autopsy data or the various brain imaging techniques. Autopsy data are not always entirely usable for validation purposes, in that numerous changes may have taken place in the patient's brain between the time of testing and the time of examination of the brain. Currently, the new neuroimaging procedures, and the very extensive research associated with them (described in the first two volumes of this series) have made substantial progress toward resolution of this problem. Of the various imaging techniques, magnetic resonance imaging (MRI) is currently the most widely used. Cooperation among neuroradiologists, neurologists, and neuropsychologists has already led to the accomplishment of several important studies correlating MRI data with neuropsychological test results.

Within neuropsychological assessment, there appears to have been a progression regarding the relationship between level of inference and criterion. Early studies in the field as well as the development of new assessment batteries generally addressed themselves to the matter of simple presence or absence of structural brain damage. Thus, the first question raised had to do with the accuracy with which an assessment procedure could discriminate between brain-damaged and non-brain-damaged patients, as independently classified by the criterion procedure. In the early studies, the criterion used was generally clinical diagnosis, perhaps supported in some cases by neurosurgical data or some laboratory procedure such as a skull x-ray or an electroencephalogram (EEG). It soon became apparent, however, that many neuropsychological tests were performed at abnormal levels, not only by brain-damaged patients, but by patients with several of the functional psychiatric disorders. Since many neuropsychologists worked in neuropsychiatric rather than general medical settings, this matter became particularly problematic. Great efforts were then made to find tests that could discriminate between brain-damaged and psychiatric patients or, as sometimes put, between "functional" and "organic" conditions. There have been several reviews of this research (Goldstein, 1978; Heaton, Baade, & Johnson, 1978; Heaton & Crowley, 1981; Malec, 1978), all of which were critical of the early work in this field in light of current knowledge about several of the functional psychiatric disorders. The chronic schizophrenic patient was particularly problematic, since such patients often performed on neuropsychological tests in a manner indistinguishable from that of patients with generalized structural brain damage. This issue has been largely reformulated in terms of looking at the neuropsychological aspects of many of the functional psychiatric disorders (Goldstein, 1985), largely under the influence of the newer biological approaches to psychopathology.

The search for validity criteria has become increasingly precise with recent advances in the neurosciences as well as increasing opportunities to collect test data from various patient groups. One major conceptualization largely attributable to Reitan and co-workers is that localization does not always operate independently with regard to determination of behavioral change, but interacts with type of lesion or the specific process that produced the brain damage. It became apparent, through an extremely large number of studies (cf. Filskov & Boll, 1981), that there are many forms of type–locus interactions, and that level and pattern of performance on neuropsychological tests may vary greatly with the particular nature of the brain disorder. This development paralleled such advances in the neurosciences as the discovery of neurotransmitters and the relationship between neurochemical abnormalities and a number of the neurological disorders whose etiology had been unknown. We therefore have the beginnings of the development of certain neurochemical validating criteria (Davis, 1983). There has also been increasing evidence for a genetic basis for several mental and neurological disorders. The gene for Huntington's disease has been discovered, and there is growing evidence for a significant genetic factor contributing to the acquisition of certain forms of alcoholism (Steinhauer, Hill, & Zubin, 1987).

In general, the concurrent validity studies have been quite satisfactory, and many neuropsychological test procedures have been shown to be accurate indicators of many parameters of brain dysfunction. However, there is no clear agreement concerning what is the most accurate and reliable external criterion. We have taken the position (Goldstein & Shelly, 1982; Russell, Neuringer, & Goldstein, 1970) that no one method can be superior in all cases, and that the best criterion is generally the final medical opinion based on a comprehensive but pertinent evaluation, excluding, of course, behavioral data. In some cases, for example, the MRI scan may be relatively noncontributory, but there may be definitive laboratory findings based on examination of blood or cerebrospinal fluid. In some cases (e.g., Huntington's disease) the family history may be the most crucial part of the evaluation. No one method can stand out as superior in all cases when dealing with a variety of disorders. The diagnosis is often best established through the integration of data coming from a number of sources by an informed individual. A final problem to be mentioned here is that objective criteria do not yet exist for a number of neurological disorders, but even this problem appears to be undergoing a rapid stage of resolution. Most notable in this regard is the in vivo differential diagnosis of the degenerative diseases of old age, such as Alzheimer's disease. There is also no objective laboratory marker for multiple sclerosis, and diagnosis of that disorder continues to be made on a clinical basis. Only advances in the neurosciences will lead to ultimate solutions to problems of this type.

In clinical neuropsychology, predictive validity has mainly to do with course of illness. Will the patient get better, stay the same, or deteriorate? Generally, the best way to answer questions of this type is through longitudinal studies, but very few such studies have actually been done. There is, however, some early literature on recovery from stroke, much of which is attributable to the work of Meier and collaborators (Meier, 1974). Levin, Benton, and Grossman (1982) provide a discussion of recovery from closed head injury. Perhaps one of the most extensive efforts to establish the predictive validity of neuropsychological tests was accomplished by Paul Satz and various collaborators, involving the prediction of reading achievement in grade school based on neuropsychological assessments accomplished during kindergarten (Fletcher & Satz, 1980). At the other end of the age spectrum, there are currently several ongoing longitudinal studies contrasting normal elderly individuals with dementia patients (Danziger, 1983; Wilson & Kaszniak, 1983). We are beginning to understand more about the neurobiological precursors of dementia, but we do not yet have adequate clinical prognostic instruments (Pettegrew, Moosy, Withers, McKeag, & Panchalingam, 1988).

Data available suggest that neuropsychological tests can predict to some extent the degree of recovery or deterioration and the treatment outcome. Since many neurological disorders change over time, getting better or worse, and the treatment of neurological disorders is becoming an increasingly active field (Reisberg, Ferris, & Gershon, 1980), it is often important to have some knowledge of what will happen to the patient in the future in a specific rather than general way and to determine whether or not the patient is a good candidate for some form of treatment. Efforts have also been made to predict functional abilities involved in personal self-care and independent living on the basis of neuropsychological test performance, particularly in the case of elderly individuals (McCue, Rogers, Goldstein, & Shelly, 1987). The extent to which neuropsychological assessment can provide this prognostic information will surely be associated with the degree of its acceptance in clinical settings.

Studies of the construct validity of neuropsychological tests represent the bulk of basic clinical neuropsychological research. Neuropsychology abounds with constructs: short-term memory, attention, visual–spatial skills, psychomotor speed, motor engrams, and cell assemblies. Tests are commonly characterized by the construct they purport to measure: Test A is a test of long-term memory; Test B is a test of attention; Test C is a test of abstraction ability; Test D is a measure of biological intelligence, and so on. Sometimes we fail to recognize constructs as such because they are so well established, but concepts like memory, intelligence, and attention are in fact theoretical entities used to describe certain classes of observable behaviors.

An important way of establishing the construct validity of neuropsychological test batteries involves determining the capacity to classify cases into meaningful subtypes. In recent years, several groups of investigators have used classification statistics, notably R-type factor analysis and cluster analysis, in order to determine whether combinations of test scores from particular batteries classify cases in accordance with established diagnostic categories or into types that are meaningful from the standpoint of neuropsychological considerations. A great deal of effort has gone into establishing meaningful, empirically derived subtypes of learning disability (Rourke, 1985), and there has also been work done in the neuropsychologically based empirical classification of neuropsychiatric patients (Goldstein & Shelly, 1987; Goldstein, Shelly, McCue, & Kane, 1987; Schear, 1987).

While neuropsychological tests should ideally have reliability levels commensurate with other areas of psychometrics, there are some relatively unique problems. These problems are particularly acute when the test–retest method is used to determine the reliability coefficients. The basic problem is that this method really assumes the stability of the subject over testing occasions. When reliability coefficients are established through the retesting of adults over a relatively brief time period that assumption is a reasonable one, but it is not as reasonable in samples of brain-damaged patients who may be rapidly deteriorating or recovering. Indeed, it is generally thought to be an asset when a test reflects the appropriate changes. Another difficulty with the test–retest method is that many neuropsychological tests are not really repeatable because of substantial practice effects. The split-half method is seldom applicable, since most neuropsychological tests do not consist of lengthy lists of items, readily allowing for odd–even or other split-half comparisons. In consideration of these difficulties, the admittedly small number of reliability studies done with the standard neuropsychological test batteries have yielded perhaps surprisingly good results. Boll (1981) has reviewed reliability studies done with the Halstead–Reitan battery, while the test manual (Golden, Hammeke, & Purisch, 1980; Golden, Purisch, & Hammeke, 1985) reports reliability data for the Luria–Nebraska battery. In any event, it seems safe to say that most neuropsychological test developers have not been greatly preoccupied with the reliabilities of their procedures, but those who have studied the matter appear to have provided sufficient data to permit the conclusion that the standard, commonly used procedures are at least not so unreliable as to impair the validities of those procedures.

An Introduction to the Comprehensive Batteries

The Halstead–Reitan Battery

History. The history of this procedure and its founders has recently been reviewed by Reed and Reed (1997). They trace the beginnings of the battery to the special laboratory established by Halstead in 1935 for the study of neurosurgical patients. The first major report on the findings of this laboratory appeared in a book called *Brain and Intelligence: A Quantitative Study of the Frontal Lobes* (Halstead, 1947), suggesting that the original intent of Halstead's tests was describing frontal lobe function. In this book, Halstead proposed his theory of "biological intelligence" and presented what was probably the first factor analysis done with neuropsychological test data. Perhaps more significantly, however, the book contains descriptions of many of the tests now contained in the Halstead–Reitan battery. In historical perspective, Halstead's major contributions to neuropsychological assessment, in addition to his very useful tests, include the concept of the neuropsychological laboratory in which objective tests are administered in standard fashion and quantitatively scored, and the concept of the impairment index, a global rating of severity of impairment and probability of the presence of structural brain damage.

Reitan adopted Halstead's methods and various test procedures and with them established a laboratory at the University of Indiana. He supplemented these tests with a number of additional procedures in order to obtain greater comprehensiveness and initiated a clinical research program that is ongoing. The program began with a cross-validation of the battery and expanded into numerous areas, including validation of new tests added to the battery (e.g., the Trail Making Test), lateralization and localization of function, aging, and neuropsychological aspects of a wide variety of disorders such as alcoholism, hypertension, disorders of children, and mental retardation. Theoretical matters were also considered. Some of the major contributions included the concept of type–locus interaction (Reitan, 1966), the analysis of quantitative as opposed to qualitative deficits associated with brain dysfunction (Reitan, 1958, 1959), the concept of the brain–age quotient (Reitan, 1973), and the scheme for levels and types of inference in interpretation of neuropsychological test data (Reitan & Wolfson, 1993). In addition to the published research, Reitan and his collaborators developed a highly sophisticated method of blind clinical interpretation of the Halstead–Reitan battery that continues to be taught at workshops conducted by Dr. Reitan and associates.

The Halstead–Reitan battery, as the procedure came to be known over the years, also has a history. It has been described as a "fixed battery," but that is not actually the case. Lezak (1976) says in reference to this development: "This set of tests has grown by accretion and revision and continues to be revised" (p. 440). The tests that survived a long research history include the Category Test, the Tactual Performance Test, the Speech Perception Test, the Seashore Rhythm Test, and Finger Tapping. There have been numerous additions, including the various Wechsler Intelligence Scales, the Trail Making Test, a sub-battery of perceptual tests, the Reitan Aphasia Screening Test, the Klove Grooved Pegboard, and other tests that are used in some laboratories but not in others. Most recently, a procedure described as an "expanded Halstead–Reitan battery" has appeared that includes the original tests plus several additional ones, listed below (Heaton, Grant, & Matthews, 1991). Three major new methods have also been developed for scoring the battery and computing the impairment index (Reitan, 1991; Russell *et al.*, 1970; Russell & Starkey, 1993).

The Halstead–Reitan battery continues to be widely used as a clinical and research procedure. Numerous investigators use it in their research, and there have been several successful cross-validations done in settings other than Reitan's laboratory (Goldstein & Shelly, 1972; Vega & Parsons, 1967). In addition to the continuation of factor analytic work with the battery, several investigators have applied other forms of multivariate analysis to it in various research applications. Some of this research has been conducted relative to objectifying and even computerizing interpretation of the battery; the most well-known efforts are the Selz–Reitan rules for classification of brain function in older children (Selz & Reitan, 1979) and the Russell, Neuringer, and Goldstein "neuropsychological keys" (Russell *et al.*, 1970). The issue of reliability of the battery has been addressed, with reasonably successful results (Goldstein & Watson, 1989). Clinical interpretation of the battery continues to be taught at workshops and in numerous programs engaged in the training of professional psychologists. A detailed description of the battery is available elsewhere (Reitan & Wolfson, 1993).

STRUCTURE AND CONTENT. Although there are several versions of the Halstead–Reitan battery, the differences tend to be minor, and there appears to be a core set of procedures that essentially all versions of the battery contain. The battery must be administered in a laboratory containing specific equipment. It is probably best to plan on about 6 to 8 hours of patient time. Each test of the battery is independent and may be administered separately from the other tests. However, it is generally assumed that a certain number of the tests must be administered in order to compute an impairment index.

Scoring for the Halstead–Reitan varies with the particular test, such that individual scores may be expressed in time to completion, errors, number correct, or some form of derived score. These scores are often converted to standard scores or ratings so that they may be profiled. Russell and co-workers (1970) rate all of the tests contributing to the impairment index on a 6-point scale, the data being displayed as a profile of the ratings. They have also provided quantitative scoring systems for the Reitan Aphasia Screening Test and for the drawing of a Greek cross that is part of that test. However, some clinicians do not quantify those procedures, except in the form of counting the number of aphasic symptoms elicited. Heaton, Grant, and Matthews (1991) have developed an extensive demographically corrected score system and have published normative data corrected for sex, age, and education.

THEORETICAL FOUNDATIONS. There are really two theoretical bases for the Halstead–Reitan battery, one contained in *Brain and Intelligence* and related writings of Halstead; the other in numerous papers and chapters written by Reitan and various collaborators (e.g., Reitan, 1966; Reitan & Wolfson, 1993). Halstead was really the first to establish a human neuropsychology laboratory in which patients were administered objective tests, some of which were semiautomated, utilizing standard procedures and sets of instructions. His Chicago laboratory may have been the initial stimulus for the now common practice of administration of neuropsychological tests by trained technicians. Halstead was also the first to use sophisticated, multivariate statistics in the analysis of neuropsychological test data.

Reitan's program can be conceptualized as an effort to demonstrate the usefulness and accuracy of Halstead's tests and related procedures in clinical assessment of brain-damaged patients. Halstead's concept of a standard neuropsychological battery administered under laboratory conditions and consisting of objective, quantifiable procedures was maintained and expanded by Reitan. Both Halstead and Reitan shared what might be described as a Darwinian approach to neuropsychology. Halstead's discriminating tests are viewed as measures of adaptive abilities, of skills that ensured man's survival on the planet. Many neuropsychologists are now greatly concerned with the relevance of their test procedures to adaptation—the capacity to carry on functional activities of daily living and to live independently (Heaton & Pendleton, 1981). This general philosophy is somewhat different from the more traditional models emanating from behavioral neurology, in which there is a much greater emphasis on the more medical-pathological implications of behavioral test findings.

One could say that Reitan's great concern has always been with the empirical validity of test procedures. Such validity can only be established through the collection of large amounts of data obtained from patients with reasonably complete documentation of their medical/neurological conditions. Both presence and absence of brain damage had to be well documented, and if present, findings related to site and type of lesion had to be established. He has described his work informally as one large experiment, necessitating maximal consistency in the procedures used, and to some extent, in the methods of analyzing the data. Reitan and his various collaborators represent the group that was primarily responsible for the introduction of the standard battery approach to clinical neuropsychology. It is clear from reviewing the Reitan group's work that there is substantial emphasis on performing controlled studies with samples sufficiently large to allow for the application of conventional statistical procedures.

It would probably be fair to say that the major thrust of Reitan's research and writings has not been espousal of some particular theory of brain function, but rather an extended examination of the inferences that can be made from behavioral indices relative to the condition of the brain. There is a great emphasis on methods of drawing such inferences in the case of the individual patient. Thus, this group's work has always involved empirical research and clinical interpretation, with one feeding into the other. In this regard, there has been a formulation of inferential methods used in neuropsychology (Reitan & Wolfson, 1993) that provides a framework for clinical interpretation. Four methods are outlined: level of performance, pattern of performance, specific behavioral deficits (pathognomonic signs), and right–left comparisons. In other words, one examines whether or not the patient's general level of adaptive function is comparable to that of normal individuals, whether there is some characteristic performance profile that suggests impairment even though the average score may be within normal limits, whether there are unequivocal individual signs of deficits, and whether there is a marked discrepancy in functioning between the two sides of the body.

COMPONENT TESTS. Some form of lateral dominance examination is always administered, generally including tests for hand, foot, and eye dominance.

Halstead's Biological Intelligence Tests. There are five subtests in this section of the Halstead–Reitan battery developed by Halstead.

1. The Halstead Category Test: This test is a concept identification procedure in which the subject must discover the concept or principle that governs various series of geometric forms and verbal and numerical material. The apparatus for the test includes a display screen with four horizontally arranged numbered switches placed beneath it. The stimuli are on slides, and the examiner uses a control console to administer the procedure. The subject is asked to press the switch that the picture reminds him or her of, and is provided with additional instructions. The point of the test is to see how well the subject can learn the concept, idea, or principle that connects the pictures. If the correct switch is pressed, the subject will hear a pleasant chime, while wrong answers are associated with a rasping buzzer. The conventionally used score is the total number of errors for the seven groups of stimuli that form the test. Booklet forms (Adams & Trenton, 1981; DeFillippis, McCampbell, & Rogers, 1979) and abbreviated forms (Calsyn, O'Leary, & Chaney, 1980; Russell & Levy, 1987; Sherrill, 1987) of this test have been developed.

2. The Halstead Tactual Performance Test: This procedure uses a version of the Seguin–Goddard Formboard, but it is done blindfolded. The subject's task is to place all of the 10 blocks into the board, using only the sense of touch. The task is repeated three times, once with the preferred hand, once with the nonpreferred hand, and once with both hands, after which the board is removed. After removing the blindfold, the subject is asked to draw a picture of the board, filling in all of the blocks he or she remembers in their proper locations on the board. Scores from this test include time to complete the task for each of the three trials, total time, number of blocks correctly drawn, and number of blocks correctly drawn in their proper locations on the board.

3. The Speech Perception Test: The subject is asked to listen to a series of 60 sounds, each of which consists of a double *e* digraph with varying prefixes and suffixes (e.g., geend). The test is given in a four-alternative multiple-choice format, the task being to underline on an answer sheet the sound heard. The score is the number of errors.

4. The Seashore Rhythm Test: This test consists of 30 pairs of rhythmic patterns. The task is to judge whether the two members of each pair are the same or different and to record the response by writing an *S* or a *D* on an answer sheet. The score is either the number correct or the number of errors.

5. Finger Tapping: The subject is asked to tap his or her extended index finger on a typewriter key attached to a mechanical counter. Several series

of 10-second trials are run, with both the right and left hand. The scores are the average number of taps, generally over five trials, for the right and left hand.

Tests Added to the Battery by Reitan. Reitan added four components to the battery.

1. The Wechsler Intelligence Scales: This test is given according to manual instructions and is not modified in any way. Most clinicians use the most current revision of these scales, although much of the early research was done with the Wechsler–Bellevue (Wechsler, 1944) and the Wechsler Adult Intelligence Scale (WAIS) (Wechsler, 1955).
2. The Trail Making Test: In Part A of this procedure the subject must connect in order a series of circled numbers randomly scattered over a sheet of $8\frac{1}{2} \times 11$ paper. In part B, there are circled numbers and letters and the subject's task involves alternating between numbers and letters in serial order (e.g., 1 to A to 2 to B, etc.). The score is time to completion expressed in seconds for each part.
3. The Reitan Aphasia Screening Test: This test serves two purposes in that it contains both copying and language-related tasks. As an aphasia screening procedure, it provides a brief survey of the major language functions: naming, repetition, spelling, reading, writing, calculation, narrative speech, and right–left orientation. The copying tasks involve having the subject copy a square, Greek cross, triangle, and key. The first three items must each be drawn in one continuous line. The language section may be scored by listing the number of aphasic symptoms or by using the quantitative system developed by Russell and co-workers (1970). The drawings are either not formally scored or are rated through a matching to model system also provided by Russell and colleagues.
4. Perceptual Disorders: These procedures actually constitute a sub-battery and include tests of the subject's ability to recognize shapes by touch and identify numbers written on the fingertips, as well as tests of finger discrimination and visual, auditory, and tactile neglect. The number of errors is the score for all of these procedures.

Other Tests. The Halstead–Reitan battery was expanded further by other researchers to include more tests.

1. The Klove Grooved Pegboard Test: The subject must place pegs shaped like keys into a board containing recesses that are oriented in randomly varying directions. The test is administered twice, once with the right and once with the left hand. Scores are time to completion in seconds for each hand and errors for each hand, defined as the number of pegs dropped during performance of the task.
2. The Klove Roughness Discrimination Test: The subject must order four blocks covered with varying grades of sandpaper presented behind a blind with regard to degree of roughness. Time and error scores are recorded for each hand.
3. Visual Field Examination: Russell *et al.* (1970) include a formal visual field examination using a perimeter as part of their assessment procedure.
4. Tests in the expanded version (Heaton, Grant, & Matthews, 1991) include the Wisconsin Card Sorting, Thurstone Word Fluency, Story Memory, Fig-

ural Memory, Seashore Tonal Memory, Digit Vigilance, Peabody Individual Achievement, and Boston Naming tests, plus a part of the Boston Diagnostic Aphasia Examination.

STANDARDIZATION RESEARCH. The Halstead–Reitan battery, as a whole, meets rigorous validity requirements. Following Halstead's initial validation (1947) it was cross-validated by Reitan (1955) and in several other laboratories (Russell *et al.,* 1970; Vega & Parsons, 1967). Validity, in this sense, means that all component tests of the battery that contribute to the impairment index discriminate at levels satisfactory for producing usable cutoff scores for distinguishing between brain-damaged and non-brain-damaged patients. The major exceptions, the Time Sense and Flicker Fusion tests, have been dropped from the battery by most of its users. In general, the validation criteria for these studies consisted of neurosurgical and other definitive neurological data. It may be mentioned, however, that most of these studies were accomplished before the advent of the computed tomography (CT) scan, and it would probably now be possible to do more sophisticated validity studies, perhaps through correlating the extent of impairment with quantitative measures of brain damage (e.g., CT scan or MRI measures). Validity studies were also accomplished with tests added to the battery such as the Wechsler scales, the Trail Making Test, and the Reitan Aphasia Screening Test, with generally satisfactory results (Reitan, 1966).

By virtue of the level of inferences made by clinicians from Halstead–Reitan battery data, validity studies must obviously go beyond the question of presence or absence of brain damage. The first issue raised related to discriminative validity between patients with left hemisphere and right hemisphere brain damage. Such measures as Finger Tapping, the Tactual Performance Test, the perceptual disorders sub-battery, and the Reitan Aphasia Screening Test all were reported to have adequate discriminative validity in this regard. There have been very few studies, however, that go further and provide validity data related to more specific criteria such as localization and type of lesion. It would appear from one impressive study (Reitan, 1964) that valid inferences concerning prediction at this level must be made clinically, and one cannot call upon the standard univariate statistical procedures to make the necessary discriminations. This study provided the major impetus for Russell and co-workers' (1970) neuropsychological key approach, which was in essence an attempt to objectify higher-order inferences. The discriminative validity of the Halstead–Reitan battery in the field of psychopathology, mainly regarding schizophrenia, has been widely studied, and has been substantially reconceptualized since the discovery of numerous neurobiological abnormalities in schizophrenia. What we have now is really a neuropsychology of schizophrenia (Goldstein, 1986), to which the Halstead–Reitan battery has contributed.

Although there have been several studies of the predictive validity of neuropsychological tests with children (Fletcher & Satz, 1980; Rourke, 1985) and other studies with adults that did not use the full Halstead–Reitan battery (Meier, 1974), no major formal assessment of the predictive validity of the Halstead–Reitan battery has been accomplished with adults. Within neuropsychology, predictive validity has two aspects: (1) predicting everyday academic, vocational, and social functioning and (2) predicting course of illness. With regard to the first aspect, Heaton and Pendleton (1981) document the lack of predictive validity studies using extensive batteries of the Halstead–Reitan type. However, they do report one study (Newman, Heaton, & Lehman, 1978) in which the Halstead–Reitan successfully predicted employment status on 6-month follow-up. With regard to prediction of course of illness, there appears to be a good deal of clinical expertise, but no major formal studies in which the battery's capacity to predict whether the patient will get better, worse, or stay the same is evaluated. This matter is of particular significance in such conditions as head injury and stroke, since outcome tends to be quite variable in these conditions. The changes that occur during the early stages of these disorders are often the most significant ones related to prognosis (e.g., length of time unconscious).

In general, there has not been a great deal of emphasis on studies involving the reliability of the Halstead–Reitan battery, probably because of the nature of the tests themselves, particularly with regard to the practice effect problem, and because of the changing nature of those patients for whom the battery was developed. Goldstein and Watson (1989) provided a review of Halstead–Reitan battery reliability studies, as well as a test–retest study of their own, concluding that reliability levels were satisfactory in a number of different clinical groups. The Category Test can have its reliability assessed through the split-half method. In a study by Shaw (1966), a .98 reliability coefficient was obtained.

The Luria–Nebraska Neuropsychological Battery

HISTORY. This procedure was first reported on in 1978 in the form of two initial validity studies (Golden, Hammeke, & Purisch, 1978; Purisch, Golden, & Hammeke, 1978). Historically, Christensen, a student of the prominent Russian neurologist and neuropsychologist A. R. Luria, published a book called *Luria's Neuropsychological Investigation* (Christensen, 1975a). The book was accompanied by a manual and a kit containing test materials used by Luria and his co-workers (Christensen, 1975b, 1975c). Although some of Luria's procedures had previously appeared in English (Luria, 1966, 1970), they had never been presented in a manner that encouraged direct administration of the test items to patients. The materials published by Christensen did not contain information relevant to standardization of these items. There was no scoring system, norms, data regarding validity and reliability, or review of research accomplished with the procedure as a standard battery. This work was taken on by a group of investigators under the leadership of Charles J. Golden (Golden *et al.*, 1978; Purisch *et al.*, 1978). Thus, in historical sequence, Luria adopted or developed these items over the course of many years, Christensen published them in English but without standardization data, and finally Golden and collaborators provided quantification and standardization. Since that time, Golden's group as well as other investigators have produced a massive amount of studies with what is now known as the Luria–Nebraska Neuropsychological Battery. The battery was published in 1980 by Western Psychological Services (Golden, Hammeke, & Purisch, 1980a) and is now extensively used in clinical and research applications. An alternate form of the battery is now available (Golden *et al.*, 1985), as is a children's version (Golden, 1981).

STRUCTURE AND CONTENT. The battery contains 269 items, each of which may be scored on a 2- or 3-point scale. A score of 0 indicates normal performance. Some items may receive a score of 1, indicating borderline performance. A score of 2 indicates clearly abnormal performance. The items are organized into the categories provided in the Christensen kit (Christensen, 1975c), but while Christensen organized the items primarily to suggest how they were used by Luria, the Luria–

Nebraska version is presented as a set of quantitative scales. The raw score for each scale is the sum of the 0, 1, and 2 item scores. Thus, the higher the score, the poorer the performance. The scores for the individual items may be based on speed, accuracy, or quality of response. In some cases, two scores may be assigned to the same task, one for speed and the other for accuracy. These two scores are counted as individual items. For example, one of the items is a block counting task, with separate scores assigned for number of errors and time to completion of the task. In the case of time scores, blocks of seconds are associated with the 0, 1, and 2 scores. When quality of response is scored, the manual provides both rules for scoring and, in the case of copying tasks, illustrations of figures representing 0, 1, and 2 scores.

The 269 items are divided into 11 content scales, each of which may be administered individually. Since these scales contain varying numbers of items, raw scale scores are converted to T scores with a mean of 50 and a standard deviation of 10. These T scores are displayed as a profile on a form prepared for that purpose. In the alternate form of the battery, the names of the content scales have been replaced by abbreviations. Thus, we have Motor, Rhythm, Tactile, Visual, Receptive Speech, Expressive Speech, Writing, Reading, Arithmetic, Memory, and Intellectual Processes scales, which are referred to as the C1 through C11 scales in the alternate form.

In addition to these 11 content scales, there are three derived scales that appear on the standard profile form: the Pathognomonic, Left Hemisphere, and Right Hemisphere scales. The Pathognomonic scale contains items from throughout the battery found to be particularly sensitive to the presence or absence of brain damage. The Left and Right Hemisphere scales are derived from the Motor and Tactile scale items that involve comparisons between the left and right sides of the body. They therefore reflect sensorimotor asymmetries in the two sides of the body.

Several other scales have been developed by Golden and various collaborators, all of which are based on different ways of scoring the same 269 items. These special scales include new (empirically derived) right and left hemisphere scales (McKay & Golden, 1979a), a series of localization scales (McKay & Golden, 1979b), a series of factor scales (McKay & Golden, 1981), and double-discrimination scales (Golden, 1979). The new right and left hemisphere scales contain items from throughout the battery and are based on actual comparisons among patients with right hemisphere, left hemisphere, and diffuse brain damage. The localization scales are also empirically derived (McKay & Golden, 1979b), being based on studies of patients with localized brain lesions. There are frontal, sensorimotor, temporal, and parieto-occipital scales for each hemisphere. The factor scales are based on extensive factor analytic studies of the major content scales (e.g., Golden & Berg, 1983). The new right and left hemisphere, localization, and factor scales may all be expressed in T scores with a mean of 50. There are also two scales that provide global indices of dysfunction, and are meant as equivalents to the Halstead impairment index. They are called the Profile Elevation and Impairment scales.

The Luria–Nebraska procedure involves an age and education correction. It is accomplished through computation of a cutoff score for abnormal performance based on an equation that takes into consideration both age and education. The computed score is called the critical level and is equal to .214 (Age) + 1.47 (Education) + 68.8 (Constant). Typically, a horizontal line is drawn across the profile at the computed critical level point. The test user has the option of considering scores above the critical level, which may be higher or lower than 60, as abnormal.

As indicated above, extensive factor analytic studies have been accomplished, and the factor structure of each of the major scales has been identified. These analyses were based on item intercorrelations, rather than on correlations among the scales. It is important to note that most items on any particular scale correlate better with other items on that scale than they do with items on other scales (Golden, 1981). This finding lends credence to the view that the scales are at least somewhat homogeneous, and thus that the organization of the 269 items into those scales can be justified.

THEORETICAL FOUNDATIONS. As in the case of the Halstead–Reitan battery, one could present two theoretical bases for the Luria–Nebraska, one revolving around the use of Luria's name and the other around the Nebraska group, namely, Golden and his collaborators. This view is elaborated upon in Goldstein (1986). Luria himself had nothing to do with the development of the Luria–Nebraska battery, nor did any of his co-workers. The use of his name in the title of the battery is, in fact, somewhat controversial, and seems to have been essentially honorific in intent, recognizing his development of the items and the underlying theory for their application. Indeed, Luria died some time before publication of the battery but was involved in the preparation of the Christensen materials, which he endorsed. Furthermore, the method of testing employed by the Luria–Nebraska was not Luria's method, and the research done to establish the validity, reliability, and clinical relevance of the Luria–Nebraska was not done by Luria and his collaborators. Therefore, our discussion of the theory underlying the Luria–Nebraska battery will be based on the assumption that the only connecting link between Luria and that procedure is the set of Christensen items. In doing so, it becomes clear that the basic theory underlying the development of Luria–Nebraska is based on a philosophy of science that stresses empirical validity, quantification, and application of established psychometric procedures. Indeed, as pointed out elsewhere (Goldstein, 1982, 1986), it is essentially the same epistemology that characterizes the work of the Reitan group.

Thus, research done with the Luria–Nebraska battery determined (1) whether it discriminates between brain-damaged patients in general and normal controls; (2) whether it discriminates between patients with structural brain damage and those with schizophrenia; (3) whether the procedure has the capacity to lateralize and regionally localize brain damage; and (4) whether there are performance patterns specific to particular neurological disorders, such as alcoholic dementia or multiple sclerosis. Since this research was accomplished in recent years, it was able to benefit from the new brain imaging technology, notably the CT scan, and the application of high-speed computer technologies, allowing for extensive use of powerful multivariate statistical methods. With regard to methods of clinical inference, the same methods suggested by Reitan—level of performance, pattern of performance, pathognomonic signs, and right–left comparisons—are used with the Luria–Nebraska.

Adhering to our assumption that the Luria–Nebraska bears little resemblance to Luria's methods and theories, there seems little point in examining the theoretical basis for the substance of the Luria–Nebraska battery. For example, there is little point in examining the theory of language that underlies the Receptive Speech and Expressive Speech scales or the theory of memory that provides the basis for the Memory scale. We believe that the Luria–Nebraska battery is not a means of using Luria's theory and methods in English-speaking countries, but rather a standardized psychometric instrument with established validity for certain purposes and reliability. The choice of using items selected by Christensen (1975b) to illustrate Luria's testing methods was, in retrospect, probably less crucial than the research methods chosen to investigate the capabilities of this item set. Indeed, it is somewhat misleading to characterize these items as "Luria's tests," since many of them are standard items used by neuropsychologists and neurologists throughout the world. Surely, one cannot describe asking a patient to interpret proverbs or determine 2-point thresholds as being exclusively "Luria's tests." They are, in fact, venerable, widely used procedures.

STANDARDIZATION RESEARCH. There are published manuals for the Luria–Nebraska (Golden, Hammeke, & Purisch, 1980; Golden et al., 1985) that describe the battery in detail and provide information pertinent to validity, reliability, and norms. There are also several review articles (e.g., Golden, 1981; Purisch & Sbordone, 1986) that comprehensively describe the research done with the battery. Very briefly reviewing this material, satisfactory discriminative validity has been reported in studies directed toward differentiating miscellaneous brain-damaged patients from normal controls and from chronic schizophrenics. Cross-validations were generally successful, but Shelly and Goldstein (1983) could not fully replicate the studies involved with discriminating between brain-damaged and schizophrenic patients. Discriminative validity studies involving lateralization and localization achieved satisfactory results, but the localization studies were based on small samples. Quantitative indices from the Luria–Nebraska were found to correlate significantly with CT scan quantitative indices in alcoholic (Golden et al., 1981) and schizophrenic (Golden, Moses, et al., 1980) samples. There have been several studies of specific neurological disorders including multiple sclerosis (Golden, 1979), alcoholism (Chmielewski & Golden, 1980), Huntington's disease (Moses, Golden, Berger, & Wisniewski, 1981), and learning disability (McCue, Shelly, Goldstein, & Katz-Garris, 1984), all with satisfactory results in terms of discrimination.

The test manual reports reliability data. Test–retest reliabilities for the 13 major scales range from .78 to .96. The problem of interjudge reliability is generally not a major one for neuropsychological assessment, since most of the tests used are quite objective and have quantitative scoring systems. However, there could be a problem with the Luria–Nebraska, since the assignment of 0, 1, and 2 scores sometimes requires a judgment by the examiner. During the preliminary screening stage in the development of the battery, items in the original pool that did not attain satisfactory interjudge reliability were dropped. A 95% inter-rater agreement level was reported by the test constructors for the 282 items used in an early version of the battery developed after dropping those items. The manual contains means and standard deviations for each item based on samples of control, neurologically impaired, and schizophrenic subjects. An alternate form of the battery is available. To the best of our knowledge, there have been no predictive validity studies. It is unclear whether or not there have been studies that address the issue of construct validity. Stambrook (1983) suggested that studies involved with item–scale consistency, factor analysis, and correlation with other instruments are construct validity studies, but it does not appear to us that they apply to the validation of Luria's constructs. The attempt to apply Luria's constructs has not involved the empirical testing of specific hypotheses derived from Luria's theory. Thus, we appear to have diagnostic or discriminative validity established by a large number of studies. There also seems to be content validity, since the items correlate best with the scale to which they are assigned, but the degree of construct validity remains unclear. For example, there have been no studies of Luria's important construct of the functional system or of his hypotheses concerning the role of frontal lobe function in the programming, regulation, and verification of activity (Luria, 1973).

SUMMARY AND CONCLUSIONS

In the first part of this chapter, general problems in the area of standardization of comprehensive neuropsychological test batteries were discussed, while the second part contained brief reviews of the two most widely used procedures, the Halstead–Reitan and the Luria–Nebraska. These batteries have their advantages and disadvantages. The Halstead–Reitan is well established and detailed but is lengthy, cumbersome, and neglects certain areas, notably memory. The Luria–Nebraska is also fairly comprehensive and briefer than the Halstead–Reitan but is currently quite controversial and is thought to have major deficiencies in standardization and rationale, at least by some observers. We have taken the view that all of these standard batteries are screening instruments, but not in the sense of screening for presence or absence of brain damage. Rather, they may be productively used to screen a number of functional areas, such as memory, language, or visual–spatial skills, that may be affected by brain damage. With the development of the new imaging techniques in particular, it is important that the neuropsychologist not simply tell the referring agent what he or she already knows. The unique contribution of standard neuropsychological assessment is the ability to describe functioning in many crucial areas on a quantitative basis. The extent to which one procedure can perform this type of task more accurately and efficiently than other procedures will no doubt greatly influence the relative acceptability of these batteries by the professional community.

REFERENCES

Adams, R. L., & Trenton, S. L. (1981). Development of a paper-and-pen form of the Halstead Category test. *Journal of Consulting and Clinical Psychology, 49,* 298–299.

Boll, T. J. (1981). The Halstead–Reitan neuropsychology battery. In S. B. Filskov & T. J. Boll (Eds.), *Handbook of clinical neuropsychology* (pp. 577–607). New York: Wiley-Interscience.

Calsyn, D. A., O'Leary, M. R., & Chaney, E. F. (1980). Shortening the Category test. *Journal of Consulting and Clinical Psychology, 48,* 788–789.

Chmielewski, C., & Golden, C. J. (1980). Alcoholism and brain damage: An investigation using the Luria–Nebraska Neuropsychological Battery. *International Journal of Neuroscience, 10,* 99–105.

Christensen, A. L. (1975a). *Luria's neuropsychological investigation.* New York: Spectrum.

Christensen, A. L. (1975b). *Luria's neuropsychological investigation: Manual.* New York: Spectrum.

Christensen, A. L. (1975c). *Luria's neuropsychological investigation: Test cards.* New York: Spectrum.

Danziger, W. (1983, October). *Longitudinal study of cognitive performance in healthy and mildly demented (SDAT) older adults.* Paper presented at conference on Clinical Memory Assessment of Older Adults, Wakefield, MA.

Davis, K. (1983, October). *Potential neurochemical and neuroendocrine validators of assessment instruments.* Paper presented at conference on Clinical Memory Assessment of Older Adults, Wakefield, MA.

DeFillippis, N. A., McCampbell, E., & Rogers, P. (1979). Development of a booklet form of the Category Test: normative and validity data. *Journal of Clinical Neuropsychology, 1,* 339–342.

Filskov, S. B., & Boll, T. J. (1981). *Handbook of clinical neuropsychology.* New York: Wiley-Interscience.

Fletcher, J. M., & Satz, P. (1980). Developmental changes in the neuropsychological correlates of reading achievement: A six-year longitudinal follow-up. *Journal of Clinical Neuropsychology, 2,* 23–37.

Gold, J. M., Berman, K. F., Randolph, C., Goldberg, T. E., & Weinberger, D. R. (1996). PET validation of a novel prefrontal task: Delayed response alternation. *Neuropsychology, 10,* 3–10.

Golden, C. J. (1979). Identification of specific neurological disorders using double discrimination scales derived from the standardized Luria neuropsychological battery. *International Journal of Neuroscience, 10,* 51–56.

Golden, C. J. (1981). A standardized version of Luria's neuropsychological tests: A quantitative and qualitative approach to neuropsychological evaluation. In S. B. Filskov & T. J. Boll (Eds.), *Handbook of clinical neuropsychology* (pp. 608–642). New York: Wiley-Interscience.

Golden, C. J., & Berg, R. A. (1983). Interpretation of the Luria–Nebraska Neuropsychological Battery by item intercorrelation: The memory scale. *Clinical Neuropsychology, 5,* 55–59.

Golden, C. J., Graber, B., Blose, I., Berg, R., Coffman, J., & Block, S. (1981). Difference in brain densities between chronic alcoholic and normal control patients. *Science, 211,* 508–510.

Golden, C. J., Hammeke, T., & Purisch, A. (1978). Diagnostic validity of the Luria neuropsychological battery. *Journal of Consulting and Clinical Psychology, 46,* 1258–1265.

Golden, C. J., Hammeke, T., & Purisch, A. (1980). *The Luria–Nebraska battery manual.* Los Angeles: Western Psychological Services.

Golden, C. J., Moses, J. A., Zelazowski, R., Graber, B., Zatz, L. M., Horvath, T. B., & Berger, P. A. (1980). Cerebral ventricular size and neuropsychological impairment in young chronic schizophrenics. *Archives of General Psychiatry, 37,* 619–623.

Golden, C. J., Purisch, A., & Hammeke, T. (1985). *Luria–Nebraska Neuropsychological Battery Manual—Forms I and II.* Los Angeles: Western Psychological Services.

Goldstein, G. (1978). Cognitive and perceptual differences between schizophrenics and organics. *Schizophrenia Bulletin, 4,* 160–185.

Goldstein, G. (1982). *Overview: Clinical application of the Halstead–Reitan and Luria–Nebraska batteries.* Paper presented at the NE-RMEC Conference, Northport, NY.

Goldstein, G. (1985). The neuropsychology of schizophrenia. In I. Grant & K. M. Adams (Eds.), *Neuropsychological assessment of psychiatric disorders: Clinical methods and empirical findings* (pp. 147–171). New York: Oxford University Press.

Goldstein, G. (1986). The neuropsychology of schizophrenia. In I. Grant and K. M. Adams (Eds.), *Neuropsychological assessment of neuropsychiatric disorders* (pp. 147–171). New York: Oxford University Press.

Goldstein, G., & Nussbaum, P. D. (1996). The neuropsychology of aging. In J. G. Beaumont & J. Segent (Eds.), *The Blackwell dictionary of neuropsychology.* London: Wiley.

Goldstein, G., & Shelly, C. (1972). Statistical and normative studies of the Halstead Neuropsychological Test Battery relevant to a neuropsychiatric hospital setting. *Perceptual and Motor Skills, 34,* 603–620.

Goldstein, G., & Shelly, C. (1982). A further attempt to cross-validate the Russell, Neuringer and Goldstein neuropsychological keys. *Journal of Consulting and Clinical Psychology, 50,* 721–726.

Goldstein, G., and Shelly, C. (1987). The classification of neuropsychological deficit. *Journal of Psychopathology and Behavioral Assessment, 9,* 183–202.

Goldstein, G., Shelly, C., McCue, M., & Kane, R. L. (1987). Classification with the Luria–Nebraska Neuropsychological Battery: An application of cluster and ipsative profile analysis. *Archives of Clinical Neuropsychology, 2,* 215–235.

Goldstein, G., & Watson, J. R. (1989). Test–retest reliability of the Halstead–Reitan battery and the WAIS in a neuropsychiatric population. *The Clinical Neuropsychologist, 3,* 265–273.

Halstead, W. C. (1947). *Brain and intelligence: A quantitative study of the frontal lobes.* Chicago: The University of Chicago Press.

Heaton, R. K., Baade, L. E., & Johnson, K. L. (1978). Neuropsychological test results associated with psychiatric disorders in adults. *Psychological Bulletin, 85,* 141–162.

Heaton, R., & Crowley, T. (1981). Effects of psychiatric disorders and their somatic treatment on neuropsychological test results. In S. B. Filskov & T. J. Boll (Eds.), *Handbook of clinical neuropsychology* (pp. 481–525). New York: Wiley-Interscience.

Heaton, R. K., Grant, I., & Matthews, C. G. (1991). *Comprehensive norms for an expanded Halstead–Reitan battery.* Odessa, FL: Psychological Assessment Resources.

Heaton, R. K., & Pendleton, M. G. (1981). Use of neuropsychological tests to predict adult patients' everyday functioning. *Journal of Consulting and Clinical Psychology, 49,* 807–821.

Jastak, S., & Wilkinson, G. S. (1984). *The Wide Range Achievement Test–Revised.* Wilmington, DE: Jastak Associates, Inc.

Levin, H. S., Benton, A. L., & Grossman, R. G. (1982). *Neurobehavioral consequences of closed head injury.* New York: Oxford University Press.

Lezak, M. (1976). *Neuropsychological assessment* (1st ed.). New York: Oxford University Press.

Luria, A. R. (1966). *Higher cortical functions in man.* New York: Basic Books.

Luria, A. R. (1970). *Traumatic aphasia.* The Hague: Mouton and Co. Printers.

Luria, A. R. (1973). *The working brain.* New York: Basic Books.

Malec, J. (1978). Neuropsychological assessment of schizophrenia vs. brain damage: A review. *Journal of Nervous and Mental Disease, 166,* 507–516.

McCue, M., Rogers, J. C., Goldstein, G. and Shelly, C. (1987, August). *The relationship of neuropsychological skills and functional outcome in the elderly.* Paper presented at the annual meeting of the American Psychological Association, New York, NY.

McCue, M., Shelly, C., Goldstein, G., & Katz-Garris, L. (1984). Neuropsychological aspects of learning disability in young adults. *Clinical Neuropsychology, 6,* 229–233.

McKay, S., & Golden, C. J. (1979a). Empirical derivation of experimental scales for the lateralization of brain damage using the Luria–Nebraska Neuropsychological Battery. *Clinical Neuropsychology, 1,* 1–5.

McKay, S., & Golden, C. J. (1979b). Empirical derivation of experimental scales for localizing brain lesions using the Luria–Nebraska Neuropsychological Battery. *Clinical Neuropsychology, 1,* 19–23.

McKay, S. E., & Golden, C. J. (1981). The assessment of specific neuropsychological skills using scales derived from factor analysis of the Luria–Nebraska Neuropsychological Battery. *International Journal of Neuroscience, 14,* 189–204.

Meier, M. J. (1974). Some challenges for clinical neuropsychology. In R. M. Reitan & L. A. Davison (Eds.), *Clinical neuropsychology: Current status and applications* (pp. 289–323). Washington, DC: V. H. Winston and Sons.

Moses, J. A., Golden, C. J., Berger, P. A., & Wisniewski, A. M. (1981). Neuropsychological deficits in early, middle, and late stage Huntington's disease as measured by the Luria–Nebraska Neuropsychological Battery. *International Journal of Neuroscience, 14,* 95–100.

Newman, O. S., Heaton, R. K., & Lehman, R. A. W. (1978). Neuropsychological and MMPI correlates of patients' future employment characteristics. *Perceptual and Motor Skills, 46,* 635–640.

Pettegrew, J. W., Moosy, J., Withers, G., McKeag, D., & Panchalingam, K. (1988). ^{31}P Nuclear magnetic resonance study of the brain in Alzheimer's disease. *Journal of Neuropathology and Experimental Neurology, 47,* 235–248.

Purisch, A. D., Golden, C. J., & Hammeke, T. A. (1978). Discrimination of schizophrenic and brain-injured patients by a standardized version of Luria's neuropsychological tests. *Journal of Consulting and Clinical Psychology, 46,* 1266–1273.

Purisch, A. D., & Sbordone, R. J. (1986). The Luria–Nebraska Neuropsychological Battery. In G. Goldstein & R. E. Tarter (Eds.), *Advances in clinical neuropsychology* (Vol. 3, pp. 291–316). New York: Plenum Press.

Reed, J. C., & Reed, H. B. C. (1997). The Halstead-Reitan neuropsychological battery. In G. Goldstein & T. M. Incagnoli (Eds.), *Contemporary approaches to neuropsychological assessment* (pp. 93–129). New York: Plenum Press.

Reisberg, B., Ferris, S. H., & Gershon, S. (1980). Pharmacotherapy of senile dementia. In J. O. Cole & J. E. Barrett (Eds.), *Psychopathology in the aged* (pp. 233–261). New York: Raven Press.

Reitan, R. M. (1958). Qualitative versus quantitative mental changes following brain damage. *The Journal of Psychology, 46,* 339–346.

Reitan, R. M. (1959). Correlations between the trail making test and the Wechsler-Bellevue scale. *Perceptual and Motor Skills, 9,* 127–130.

Reitan, R. M. (1964). Psychological deficits resulting from cerebral lesions in man. In J. M. Warren & K. Akert (Eds.), *The frontal granular cortex and behavior* (pp. 295–312). New York: McGraw-Hill.

Reitan, R. M. (1966). A research program on the psychological effects of brain lesions in human beings. In N. R. Ellis (Ed.), *International review of research in mental retardation* (pp. 153–218). New York: Academic Press.

Reitan, R. M. (1973, August). Behavioral manifestations of impaired brain functions in aging. In J. L. Fozard (Chair), *Similarities and differences of brain-behavior relationships in aging and cerebral pathology.* Symposium presented at the American Psychological Association, Montreal, Canada.

Reitan, R. M. (1991). *The Neuropsychological Deficit Scale for Adults: Computer program, users manual.* Tucson, AZ: Neuropsychology Press.

Reitan, R. M., & Wolfson, D. (1993). *The Halstead–Reitan neuropsychological test battery: Theory and clinical interpretation* (2nd ed.). Tucson, AZ: Neuropsychology Press.

Rourke, B. P. (Ed.). (1985). *Neuropsychology of learning disabilities: Essentials of subtype analysis.* New York: Guilford Press.

Russell, E. W., & Levy, M. (1987). Revision of the Halstead Category Test. *Journal of Consulting and Clinical Psychology, 55,* 898–901.

Russell, E. W., Neuringer, C., & Goldstein, G. (1970). *Assessment of brain damage: A neuropsychological key approach.* New York: Wiley-Interscience.

Russell, E. W., & Starkey, R. I. (1993). *Halstead–Russell neuropsychological evaluation system: Manual and computer program.* Los Angeles: Western Psychological Services.

Schear, J. M. (1987). Utility of cluster analysis in classification of mixed neuropsychiatric patients. *Archives of Clinical Neuropsychology, 2,* 329–341.

Selz, M., & Reitan, R. M. (1979). Rules for neuropsychological diagnosis: Classification of brain function in older children. *Journal of Consulting and Clinical Psychology, 47,* 258–264.

Shaw, D. (1966). The reliability and validity of the Halstead Category test. *Journal of Clinical Psychology, 22,* 176–180.

Shelly, C., & Goldstein, G. (1983). Discrimination of chronic schizophrenia and brain damage with the Luria–Nebraska battery: A partially successful replication. *Clinical Neuropsychology, 5,* 82–85.

Sherrill, R. E., Jr. (1987). Options for shortening Halstead's Category test for adults. *Archives of Clinical Neuropsychology, 2,* 343–352.

Stambrook, M. (1983). The Luria–Nebraska Neuropsychological Battery: A promise that may be partly fulfilled. *Journal of Clinical Neuropsychology, 5,* 247–269.

Steinhauer, S. R., Hill, S. Y., & Zubin, J. (1987). Event-related potentials in alcoholics and their first-degree relatives. *Alcohol, 4,* 307–314.

Vega, A., & Parsons, O. (1967). Cross-validation of the Halstead–Reitan tests for brain damage. *Journal of Consulting Psychology, 31,* 619–625.

Wechsler, D. (1944). *The measurement of adult intelligence* (3rd ed.). Baltimore, MD: Williams & Wilkins.

Wechsler, D. (1955). *Wechsler Adult Intelligence Scale manual.* San Antonio, TX: The Psychological Corporation.

Wilson, R., & Kaszniak, A. (1983, October). *Progressive memory decline in progressive idiopathic dementia.* Paper presented at Conference on Clinical Memory Assessment of Older Adults, Wakefield, MS.

Yozawitz, A. (1986). Applied neuropsychology in a psychiatric center. In I. Grant and K. M. Adams (Eds.), *Neuropsychological assessment of neuropsychiatric disorders* (pp. 121–146). New York: Oxford University Press.

5

Neuropsychological Assessment of the Elderly

PAUL DAVID NUSSBAUM

INTRODUCTION

A rapid growth in the interest in brain–behavior relations has occurred during the past decade. This is particularly true with regard to the effects of aging on the central nervous system (CNS). Indeed, much attention has been dedicated to this area (Albert & Moss, 1988; Goldstein & Nussbaum, 1997; Katzman & Rowe, 1992; La Rue, 1992; Lovell & Nussbaum, 1994; Nussbaum, 1996), resulting in the establishment of geriatric neuropsychology as a specialized discipline of study. Several factors contribute to the continued growth of geriatric neuropsychology. First, neuropsychology continues to represent a primary area of professional growth within the larger area of clinical psychology. Second, the population of the United States is aging, with those over the age of 65 representing one of the fastest-growing segments of society (La Rue, 1992). Third, there is a critical need for health care providers of the older adult. Fourth, the emergence of Alzheimer's disease as one of the leading causes of death in the older adult population has generated much interest in the study of aging. Geriatric neuropsychology has the opportunity to contribute significantly to an understanding of the effects of aging on the CNS and to the development of new models of health care delivery for the older adult.

The goal of this chapter, therefore, is not merely to reflect the current status of neuropsychology and aging, but to generate ideas regarding the future direction of the field, in terms of both clinical and scientific endeavors. To meet this goal, this chapter first reviews the literature on the effects of aging on the CNS. Then, an overview of the current status of neuropsychological assessment of the older adult patient is presented. The chapter concludes with recommendations for the future direction of geriatric neuropsychology, addressing both scientific and clinical areas.

PAUL DAVID NUSSBAUM   Aging Research and Education Center, Lutheran Affiliated Services/St. John Specialty Care Center, Mars, Pennsylvania 16046.

Neuropsychology, edited by Goldstein *et al*. Plenum Press, New York, 1998.

For purposes of the present chapter, "elderly" and "older adults" will be defined as those age 65 and older.

AGING AND THE CENTRAL NERVOUS SYSTEM

A comprehensive review of the effects of aging on the CNS is certainly not possible within the current section. However, a brief review of the major effects of aging on the CNS is presented to facilitate understanding of the existence of particular changes in brain structure and function with advanced age and to provide a context for differentiating normal aging from disease-based changes in the CNS.

The medical community, of which neuropsychology is considered a member, has consistently described normal aging by differentiating it from a neuro-pathological condition such as Alzheimer's disease (AD). Indeed, the definition and understanding of normal aging have been obscured because of this disease-based approach. In relative terms, little research has been dedicated to investigation of normal changes in neuroanatomy and cognition when compared to similar efforts with diseases of late life. As a consequence, normal changes in structural and functional properties of the brain with age are not well understood. Future research will need to define normal age-related changes in the CNS using longitudinal methodology.

STRUCTURAL CHANGES IN CNS

Structural changes of the brain generally begin around the fifth decade of life, although there is variability to this process (Powers, 1994). Indeed, factors such as genetics, physical and emotional health, and environment certainly contribute to the onset and extent of structural change. The human brain weighs approximately 400 grams at birth and 1,450 grams at maturity; by the 75th year, it has lost approximately 100 grams (Kolb & Wishaw, 1990). Brain size is positively related to adult body size; however, no conclusive evidence has been found to relate brain size to human ability (Kolb & Wishaw, 1990). Brain weight has been shown to decrease across the life span although the onset of the decrease in brain weight varies from the third decade to the sixth or seventh decade (Dekaban & Sadowsky, 1978). Miller and Corsellis (1977) calculated the increases in adult brain weight and body length for male and female subjects over an 80-year period (1860 to 1940). They found an increase in brain weight of 52 grams for men and 23 grams for women. Using this information, Miller, Alston, and Corsellis (1980) measured the volume of cerebral hemispheres in males (n = 47) and females (n = 44) ages 20 to 98. There was no reported change in hemispheric volume until age 50, with a 2% decrease per decade for both genders thereafter. Interestingly, Miller *et al.* (1980) reported that the decrease in hemispheric volume was related to an increase in the ratio between gray and white matter, reflecting the white matter to be the primary area of volume loss. This is further supported by recent research using MRI technology that documents changes in the white matter of between 18% and 61% of healthy, non-neurologically impaired older adults (Coffey & Figiel, 1991).

At the neuronal level, senile plaques, amyloid deposits, and cholinergic deficits had traditionally been regarded as markers of neuropathology. However, recent evidence demonstrates that similar changes in neuronal integrity occur with normal aging, obscuring the differential between normal aging and disease (Powers, 1994). Powers asserts that although neuronal atrophy and lack of neuronal replication occur with normal aging, reorganization of synapses and dendritic aborizations is possible. Other neuronal components such as number and affinity of transmitter receptors, molecular facilitators of plasticity and reinnervation, and trophic factors are likely affected by the aging process. At the present time, there is no methodology to discriminate the effects of aging on neuronal change from disease states.

Cortical areas particularly vulnerable to the aging process include the hippocampus (Golomb *et al.*, 1993; Lim, Zipursky, Murphy, & Pfefferbaum; Powers, 1994; Squire, 1987) and prefrontal regions (Haug, *et al.*, 1983; Squire, 1987). Subcortical white matter and gray matter changes also have been documented in normal aging (Coffey & Figiel, 1991; Nussbaum, 1994) as have changes in corpus callosum (Doraiswamy *et al.*, 1991). These cortical and subcortical regions of neuronal vulnerability with normal aging relate to neuropsychological changes manifested by older adults, an issue to be discussed later in this chapter.

Ventricular size is another primary marker of structural change in the CNS with aging. Albert and Moss (1988) reviewed the literature on computerized tomography (CT) and aging to assess the age-related changes in ventricular size. They concluded that among studies that employed adequate numbers of subjects across the adult age span, ventricular size increases with age. There is less conclusive evidence, however, regarding the age of onset for increased ventricular size. The age of onset ranged from 40 to 60; the variability is thought to be attributable to methodological flaws in the studies such as including subjects who were not optimally healthy. Albert and Moss (1988) point out that when strict inclusion criteria are employed, significant increases in ventricular size begin between the sixth and seventh decade of life and continue thereafter. Indeed, ventricular size has been shown to increase significantly into the ninth decade of life (Earnest, Heaton, Wilkinson, & Manke, 1979). Other markers of structural change with age include increased sulcal size and gyral atrophy, both of which begin around the fourth decade, but have little relationship to increased ventricular size (Albert & Moss, 1988; Powers, 1994).

NEUROPSYCHOLOGICAL CHANGES WITH AGING

Much has been written regarding neuropsychological changes with aging (Albert & Moss, 1988; Goldstein, 1983; Goldstein & Nussbaum, 1997; Kaszniak, 1987; La Rue, 1992; Lovell & Nussbaum, 1994; Nussbaum, 1996). There are multiple theories of the neuropsychological profile related to normal aging, and a critical review of this literature highlights the changes in major conceptual perspectives over time. Wechsler (1955) molded early thinking on cognitive loss with age with the argument that human abilities demonstrate decline beginning in the second decade of life. This finding was generally thought to be confounded by the cross-sectional methodology employed by Wechsler. Cross-sectional research designs involve simultaneous comparison of different people varying in age. This comparison may involve subjects divided into young and old samples, or groups may be defined by decade or other unit of time. In contrast, longitudinal research designs study the same individuals over a period of time.

The field of aging and theory of brain–behavior relations across the life span vary considerably with particular research methodologies and design. Unfortunately, the two major research designs described above are flawed. Cross-sectional methodology, as used in the Wechsler (1955) study, has problems with matching groups, but a more serious problem of cohort effects (Schaie, 1965). Goldstein and Nussbaum (1997) note that a cohort or generational effect is related to, if not produced by, an individual's time of birth. For example, differences in cognitive performance of subjects between 30 and 70 years of age may not be a function of age, but rather due to educational or nutritional practices between the times these two age groups were in school. Longitudinal studies suffer from the problems of subject attrition and practice effects related to repeated use of the same measurement procedures.

Cohort effects have altered the traditional theoretical perspective of the aging process and cognitive change. Indeed, the original viewpoint of Wechsler (1955) that intelligence peaked during the second decade and gradually declined thereafter was eventually interpreted as reflecting the influence of a substantial cohort effect. However, Wechsler asserted that specific cognitive abilities are more vulnerable to decline with aging while others resist deterioration. Instruments that reflected the former were called "don't hold" tests while those measuring the latter were called "hold" tests.

The cohort effect was clearly demonstrated by Schaie's (1983) finding that consecutive cohorts of individuals show cognitive declines in particular domains of intellect at successively later time periods of the life span. In order to combat the inherent problems related to both cross-sectional and longitudinal research methodologies, Schaie and Parham (1977) developed a sequential design methodology that involves sequential data collection on cross-sectional and longitudinal samples. The finding that particular cognitive skills are more vulnerable to age-related decline was exemplified from studies employing both the Primary Mental Abilities Test (PMA) (Schaie & Parham, 1977) and the Wechsler Adult Intelligence Scale (WAIS) (Wechsler, 1955).

Using the sequential design methodology, Schaie and Parham (1977) found the word fluency factor to demonstrate an age-related decline for those ages 46 to 53. In contrast, the spatial factor and composite intelligence quotient evinced changes in the latter part of the sixth decade, while remaining factors including verbal relations, reasoning, and number concepts demonstrated decline by the middle of the seventh decade. In general, age-related declines occurred on all subtests by the sixth decade, but significant variability in performance across subtests of the PMA remained and could not be substantially accounted for until the seventh decade of life.

Likewise, performance of older adults on the WAIS has resulted in similar variability. Specifically, the Performance subtests (i.e., Picture Completion, Picture Arrangement, Block Design, Object Assembly, Digit Symbol) demonstrate greater age-related decline than Verbal subtests (Information, Vocabulary, Digit Span, Similarities, Comprehension, Arithmetic). Indeed, while deterioration on Performance subtests of the WAIS occurred in the sixth decade, skills measured by the Verbal subtests did not change until the eighth decade of life (Botwinick, 1977). This pattern of selective vulnerability to intellectual decline with age was originally described by Wechsler (1955).

Horn and Catell (1967) introduced *crystallized* and *fluid* intelligence to describe the two types of intellectual capacities measured by the WAIS, and thought to be differentially vulnerable to change with advanced age. Crystallized intelligence refers to well-learned, acquired academic-based knowledge that is resistant to extinction once learned. It is thought to be developed by the second decade of life with stabilization thereafter. Horn and Catell (1967) argued that the Verbal subtests of the WAIS represented crystallized intelligence, a theory similar to that of Wechsler who described the Verbal subtests as "hold" tests. Both characterize the fact that particular cognitive skills, most notably overlearned information acquired early in life, are resistant to change with advancing age. Fluid intelligence refers more to the capacity to deal adaptively with novel information and problem-solving. Fluid intelligence is thought to be biologically based, determined by the integrity of the CNS, and less by developmental acquisition. Indeed, Horn and Catell (1967) asserted that loss of fluid intelligence is related to degree of damage to the CNS (i.e., head injury, infection, stroke, etc.). The Performance subtests of the WAIS are thought to characterize abilities of fluid intelligence. Horn and Cattell (1967) argued that while fluid intelligence declines with advancing age, crystallized intelligence remains relatively stable.

The finding that older adults perform worse on tasks that measure nonverbal intelligence compared to verbally mediated cognitive operations raised speculation that with aging, there is a specific deterioration of the right cerebral hemisphere, labeled the *right hemisphere syndrome* (Goldstein, 1983; Goldstein & Shelly, 1981; Schaie & Schaie, 1977). However, La Rue (1992) points out that there are several problems with the right hemisphere hypothesis. First, studies supporting this hypothesis have relied on data derived from the WAIS and the Halstead–Reitan batteries. These two neuropsychological batteries do not adequately measure memory and new learning, and the WAIS does not provide clear interhemispheric differences. Indeed, individuals with bilateral brain injury often display a Verbal–Performance discrepancy (Matarazzo, 1972). There is also speculation that older individuals manifest greater difficulty with tasks of fluid intelligence because these types of cognitive tasks demand more effort. With advanced age, cognitive operations that demand more effortful processing are thought to be most vulnerable to change (Craik & McDowd, 1987; Hasher and Zacks, 1979). Hence, older adults may not perform well on tasks of nonverbal intelligence, thought to reflect a right hemisphere dysfunction, because of a diminished attentional capacity system that compromises their ability to carry out effort-demanding mental operations. Finally, La Rue (1992) asserts that there is minimal neurobiological evidence to corroborate a right hemisphere vulnerability with aging.

An alternative hypothesis is that the frontal lobe is particularly vulnerable to the aging process and that cognitive changes associated with advanced age can be described as a *frontal lobe syndrome* (Craik, Morris, Morris, & Loewen, 1990; Kobari, Stirling, & Ichijo, 1990; Mittenberg, Seidenberg, O'Leary, & DiGiulio, 1989; Schacter, Kaszniak, Kihlstrom, & Valdiserri, 1991; Squire, 1987; Uchiyama, Mitrushina, D'Elia, Satz, & Matthews, 1994). There is also support for frontal lobe vulnerability with aging from the neurobiological literature. Squire (1987) noted that relative to other brain regions, the prefrontal cortex suffers approximately 15% to 20% neuronal loss with old age and represents an area vulnerable to the aging process. Recent studies employing magnetic resonance imaging (MRI) technology with normal older adults (Awad *et al.*, 1986; Coffey, Figiel, Djang, & Weiner, 1990; Gerard & Weisberg, 1986; Lechner *et al.*, 1987; Steingart *et al.*, 1987; Sullivan, Pary, Telang, Rifai, & Zubenko, 1990; Zubenko *et al.*, 1990) have demonstrated that approximately 25% to 40% of older adults without dementia demonstrate *leukoencephalopathy*, defined as changes in deep white matter or subcortical gray matter. From review of this literature, anterior cortical regions and subcortical regions, intimately connected to the frontal cortex, are frequently cited as areas vulnerable to changes with aging. The relation between age-related cognitive changes, thought to represent frontal systems dysfunction, and reported neurobiological vulnerability of the frontal cortex is not yet clear.

Indeed, Schacter and colleagues (1991) cautioned that frontal lobe dysfunction remains a broad term encompassing multiple functional and cognitive systems. The relation of frontal lobe dysfunction to specific memory and other cognitive processes remains equivocal and in need of continued investigation. Some research has argued against frontal lobe cognitive change with age (Boone, Miller, Lesser, Hill, & D'Elia, 1990). Similarly, some have argued that cognitive profiles of normal elderly are heterogeneous and do not easily relate to specific regions of cortical dysfunction (Doraiswamy *et al.*, 1991; Valdois, Yves, Poissant, Ska, & Dehaut, 1990).

At present, it is well established that cognitive change does occur with advanced age. The frontal cortex hypothesis appears to have more support than the right hemisphere hypothesis for explaining the pattern of cognitive change manifested by older adults. In general, however, a coherent understanding of the neuropsychological changes associated with aging has not evolved. Change in cognitive functioning and ability is expected with advancing age, but heterogeneity of performance occurs, particularly with aging into the eighth and ninth decades of life. As such, global categorization of the "typical" pattern of cognitive change with aging is made difficult. Research indicates that with aging, specific areas of cognitive functioning decline while other areas remain preserved, even into the eighth decade of life.

Advanced age alone most likely does not totally explain the cognitive changes. Education and other demographic variables (Albert & Moss, 1988), as well as comorbid physical and mental disease (Goldstein, 1983; Hoyer, 1990), probably contribute substantially to cognitive changes in late life. Indeed, Goldstein (1983) proposed that no substantial decline in intellectual function occurs with advanced age except in those older adults who acquire CNS disease. However, Albert and Moss (1988) argued that age-related declines in intellectual processing are manifested even among the most healthy elderly. Methodological flaws most likely underscore the confusion in the literature. Studies do not adequately describe samples. Studies of normal aging often employ older adults with subtle neurological or cardiac illness, cross-sectional designs provide different results than longitudinal data, and older subjects with exceptionally high education levels are often included, making generalization of findings difficult. The challenge for neuropsychology is to better delineate brain–behavior changes that accompany normal aging and attempt to define the covariates of these changes. In addition, neuropsychology will benefit from a more holistic approach to the study of aging and do well to refrain from complete adherence to the disease-based medical model. Indeed, a major challenge for neuropsychology is to develop a model of brain–behavior relations for normal aging. The "abnormal" cannot be clearly understood until the "normal" is completely appreciated.

NEUROPSYCHOLOGICAL ASSESSMENT OF THE OLDER INDIVIDUAL

Neuropsychological assessment remains one of the few procedures that provide information about the functional integrity of the brain. As such, the neuropsychological examination is critical for differential diagnosis, prognostic accuracy, and delineation of cognitive and behavioral strengths and weaknesses. The purpose of neuropsychological examination of the older individual is to provide enough clinical information to answer the referral question. At the same time, the clinician must be conscious of the potential limitations of the older individual and must select neuropsychological instruments accordingly.

It is important to note that there is not a definitive neuropsychological battery or examination approach when evaluating an older individual. Lovell and Nussbaum (1994) have suggested a stepwise approach to the neuropsychological assessment of older individuals. Four steps are described: (1) the clinical interview that includes specific cognitive screening procedures (i.e., Folstein Mini-Mental State Examination) (Folstein, Folstein, & McHugh, 1975), (2) selection of neuropsychological instruments based on the interview and referral question, (3) interpretation of the examination results, and (4) specification of treatment and rehabilitation recommendations.

The stepwise process of the neuropsychological examination provides preliminary data from the cognitive screen that can then promote logical decisions regarding instrument selection, generate hypotheses to be tested, and permit an efficient means for longitudinal assessment. For example, individuals who perform poorly on the cognitive screen and cannot answer questions appropriately during the interview may not require additional assessment to make necessary clinical decisions. In contrast, those older individuals who have *some* difficulty on the cognitive screen will likely require additional assessment to better delineate diagnostic, prognostic, and treatment decisions. Selection of instruments to be included in the more extensive neuropsychological assessment should be based on the individual's behavior during the interview and performance on the cognitive screen. The clinician who conducts clinical examinations of older individuals needs to be sensitive to their potential sensory deficits, slowed cognitive processing, unfamiliarity with cognitive assessment, need for enhanced stimuli, need for frequent breaks secondary to reduced stamina, and need for questions to be repeated more often than for younger adults.

FIXED VERSUS FLEXIBLE NEUROPSYCHOLOGICAL BATTERIES

Clinical neuropsychology studies brain–behavior relations. The methodology involved in studying these relationships, however, varies (Kane, 1991). Within clinical neuropsychology, clinicians and researchers have generally adopted either a *fixed* battery approach or a *flexible* battery approach to cognitive assessment. The appropriateness and efficacy of these battery approaches to assessment of the older individual have been reviewed elsewhere (La Rue, 1992; Lovell & Nussbaum, 1994). These reviews can be summarized by the following:

1. Flexible batteries to neuropsychological assessment are most appropriate for older individuals because of their shorter administration time, economical favorability, modifiability to situation and individual needs, attention to qualitative aspects, and identification of aspects of brain disease beyond simply presence or absence of organicity.
2. The major fixed batteries appear to be most useful for research purposes. For clinical purposes, fixed batteries such as the Halstead–Reitan Neuropsychological Test Battery (HRNB) and the Luria–Nebraska (see Incagnoli, Goldstein, & Golden, 1986) generally are too time-consuming, lack ade-

quate normative data, and tend to overestimate the presence of organicity in older individuals. La Rue (1992) asserts, however, that particular subtests from the HRNB can be useful with neuropsychological assessment of the older adult as part of a larger flexible battery.

3. The Luria–Nebraska Neuropsychological Battery is somewhat more favorable than the HRNB for older adults because it requires a shorter administration time and promotes both a quantitative and qualitative analysis of test performance. A major drawback of the Luria–Nebraska is the paucity of empirical research on its use with older populations. Some research has documented good dissociation between healthy elderly and brain-injured older adults on all scales of the Luria–Nebraska (McInnes *et al.*, 1983). In addition, a short form of the Luria–Nebraska has recently been developed with demonstrated efficacy in differentiating older depressed individuals from those suffering progressive dementia (McCue, Goldstein, & Shelly, 1989).

4. Recent advancements have been made with regard to normative data for the HRNB for older individuals (Heaton, Grant, & Matthews, 1991).

5. While both types of battery approaches to neuropsychological assessment have advantages and disadvantages, the flexible battery has been proposed as the best approach for clinicians examining older individuals (Lovell & Nussbaum, 1994).

Specific selection of tests to include in a flexible neuropsychological battery should be based on the referral question, initial impression of the individual's cognitive strengths and weaknesses during the interview and cognitive screen, validity of the particular test, and available normative data. La Rue (1992) and Lovell and Nussbaum (1994) have reviewed the many different neuropsychological procedures available for assessment of the older adult. Issues of validity and relationship to particular brain regions have been highlighted for the clinician. La Rue (1992) has proposed a brief battery for geropsychological assessment specifically for the older adult. The battery includes tests that measure multiple domains of cognitive function and requires approximately 90 minutes to administer. In addition, La Rue (1992) provides information to guide the clinician about when the brief battery needs to be extended. The reader is encouraged to review this brief battery as a model for neuropsychological examination of the older adult.

BRIEF OLDER ADULT NEUROPSYCHOLOGICAL BATTERY

The present author proposes a brief older adult neuropsychological battery that includes: (1) the clinical interview with cognitive screen and (2) the neuropsychological assessment.

CLINICAL INTERVIEW WITH COGNITIVE SCREEN

The clinical interview is the earliest formal interaction with the older individual and therefore represents an opportunity to develop rapport, to educate the individual about the purpose of the examination, and to permit the older adult time to raise questions and concerns regarding the examination. A thorough interview should provide demographic and background information (e.g., age, education, marital status, handedness, occupational history, hobbies, etc.); medical and psychiatric history including current and past medication use, surgical interventions, hospitalizations, present and past drug use/abuse, need for mental health services; and an idea of the individual's general awareness of his or her condition or need for assessment. Additional information to be obtained from the interview includes observation of the individual's appearance, motor skills, appreciation of the individual's use of language and comprehension, mood and affect, psychotic behavior, and ability to interact with another in a formal setting. The clinician can gather important information about the individual in order to develop preliminary working hypotheses that assist in selection of assessment instruments for the brief neuropsychological battery.

An integral part of the clinical interview is the completion of a standard cognitive screen. The Folstein Mini-Mental State Examination (MMSE) (Folstein *et al.*, 1975) is a widely used mental status questionnaire that assesses orientation, mental calculation, memory, language, and visuoconstructional abilities. Early research using the MMSE revealed a cutoff score of 23 out of 30 to indicate the presence of cognitive impairment. The MMSE is correlated with several variables such as low education, poor health status, advanced age, and socioculture (Anthony, La Resche, Niaz, Von Korff, & Folstein, 1982; Escobar *et al.*,1986; Folstein, Anthony, Parhard, Duffy, & Gruenberg, 1985), and the validity of the scale is reduced when such variables are present. Clinical interpretation, therefore, should be made with caution when an individual presents with one or more of these variables.

It is also important to note that the MMSE was not designed to make differential diagnoses, but instead to indicate the need for further assessment. As such, it represents an ideal cognitive screen for older individuals, particularly if the clinician considers all potentially confounding variables to minimize the risk for false-positive errors.

Another relatively new cognitive screen is the Neurobehavioral Cognitive Status Examination (NCSE) (Kiernan, Mueller, Langston, & Van Dyke, 1987). An advantage of the NCSE is that it provides measurement of various cognitive domains without the expense of extended administration time. The NCSE is lacking in empirically based investigations with older adults, and until reliability and validity studies are published, the utility of the instrument for detecting neurological impairment remains unknown. Recent research indicates that the NCSE has limited efficacy with geropsychiatric patients (Engelhart, Eisenstein, & Meininger, 1994). In contrast, the NCSE was found to be more effective than the MMSE in differentiating individuals with organic brain syndrome from those with mood disorders (Osato, La Rue, & Yang, 1989). La Rue (1992) has recommended that the NCSE currently be employed with another, more established mental status questionnaire to guard against misinterpretation of results.

The quantitative and qualitative data generated from the interview and cognitive screen provide the basis for decisions regarding the need for and type of brief neuropsychological examination. Indeed, the interview and cognitive screen may indicate no need for further assessment and instead may only warrant logical treatment interventions or behavioral modifications. While this is certainly a reasonable occurrence, it would seem that neuropsychologists are unlikely to resist the need for additional data generated from the brief neuropsychological assessment. This probably reflects the nature of the neuropsychologist's training that warrants caution with interpretation until enough data is generated. (This then leads to the second component of the battery: the neuropsychological assessment.)

PAUL DAVID
NUSSBAUM

The purpose of the neuropsychological assessment is to provide more extensive data concerning the integrity of the individual's central nervous system. Six general domains of cognitive skills are assessed: general intellect, attention, memory, executive functioning, visuoconstruction, and language. Mood is also assessed. The assessment is designed to be (1) brief, because of the potential for the older individual to fatigue; (2) thorough, to measure multiple areas of cognitive functioning; (3) specific, to match the needs of the particular case; and (4) efficient, to reduce redundancy. While no flexible neuropsychological battery has been championed for older adults, there has been some discussion of particular tests to be included in an evaluation (Albert & Moss, 1988) and one proposed brief battery for the geriatric patient (La Rue, 1992). The purpose for proposing the following neuropsychological assessment for the older individual is to encourage critical thinking of test selection, to maximize validity and efficiency of the process, and to increase ability to answer the referral question.

The following brief neuropsychological assessment is proposed as one model for use in assessment of older adults (Table 1). This assessment requires approximately 2 hours and provides the neuropsychologist with information on the six cognitive domains outlined above. Mood is also measured. Test selection for the older adult battery was based on several factors including instrument validity, ease of administration, adequate normative data, and relationship to a particular cognitive domain and brain region. The assessment will be presented by cognitive domain, and specific instruments will include information on available normative data.

DESCRIPTION OF NEUROPSYCHOLOGICAL ASSESSMENT

The neuropsychological assessment (Table 1) is conducted after completion of the interview and cognitive screen. Relevant demographic information is secured from the interview as well as pertinent behavioral observations. The MMSE deter-

TABLE 1. BRIEF OLDER ADULT NEUROPSYCHOLOGICAL BATTERY

Cognitive domain	Tests
General intellect	Wechsler Adult Intelligence Scale-Revised (WAIS-R)—Information Similarities, Object Assembly, Block Design
Attention	Wechsler Memory Scale-Revised Attention and Concentration Index
Memory	Wechsler Memory Scale-Revised Logical Memory I and II
	Hopkins Verbal Learning Test
	Rey Complex Figure Test with 20-minute delay
Executive functions	Wisconsin Card Sorting Test-64 cards
	Trail Making Test Parts A and B
	Verbal Fluency
	Similarities subtest from WAIS-R
Visuoconstruction	Clock Drawing
	Intersecting Pentagons from Mini-Mental State Examination
	Block Design and Object Assembly from WAIS-R
Fine motor skill	Finger Tapping Test
Mood	Geriatric Depression Scale

[a]References for norms for each of the tests in the battery are given in the text.

mines the type of battery to be used. A score of 18 or higher on the MMSE warrants selection of the "high battery," respecting that the individual most likely maintains the ability to complete the examination. For those who score below 18 on the MMSE, the "low battery" is administered. This low battery includes the Mattis Dementia Rating Scale (DRS) (Mattis, 1976), given to generate an understanding of the severity of the cognitive impairment while appreciating the possibility that the individual's ability to complete the examination is compromised. Regardless of whether the high or low battery is used, fine motor skill and mood are assessed.

The high battery is presented first. The battery is divided into six different cognitive domains with some tests selected because they generate information important to more than one cognitive domain. These cognitive domains include *general intellect, attention, memory, executive functioning, language,* and *visuoconstruction. Fine motor skill* and *mood* are also assessed. The MMSE completed during the first step of the overall evaluation provides information on *orientation, simple language processing,* and *simple visuomotor skills.* Social appropriateness, mood, affect, awareness, and judgment can be examined grossly during the clinical interview.

General Intellect. General intellect is estimated using a truncated version of the Wechsler Adult Intelligence Scale-Revised (WAIS-R) (Wechsler, 1981). An attempt should be made to reasonably measure verbal and nonverbal intelligence. For the older adult battery, this is accomplished by completion of the Information, Vocabulary, and Similarities portions of the Verbal subtest and the Block Design and Object Assembly portions of the Performance subtest. While there is some inter-test correlation, performance on these subtests provides the neuropsychologist with a measure of general intellect and hemispheric integrity.

Each of the subtest raw scores is transformed into both scaled scores and age-appropriate scaled scores using the standard manual of the WAIS-R. Comparison of the individual's age-appropriate scaled scores on Information and Vocabulary—the "hold tests"—to an estimate of premorbid intelligence (Nelson, 1982; Wilson *et al.,* 1978) can provide some insight into whether the individual's current performance on subtests purported to measure intelligence in fact represents a decline in cognitive ability (i.e., dementia). Estimation of premorbid intelligence is a useful measure, particularly for the neuropsychological assessment of older adults where the question of a decline in cognitive functioning is important. However, caution is warranted with regard to interpretation of particular indices of premorbid intelligence, and the statistical properties of the index should be clearly understood (see Stebbins, Wilson, Gilley, Bernard, & Fox, 1990).

With regard to hemispheric integrity, the neuropsychologist can suggest primary dysfunction of either the left or right hemisphere based on a direct comparison of the individual's age-adjusted scaled scores on the Verbal and Performance subtests from the WAIS-R. It is important to consider the fact that the Performance subtests from the WAIS-R are timed while the Verbal subtests are not. Permitting the older adult to continue working beyond the standard time limit can rule out performance failure due to slowed performance alone.

The WAIS-R (Wechsler, 1981) provides normative data for individuals through age 74 only. Mayo's Older Americans Normative Studies (Ivnik *et al.,* 1992b) provides WAIS-R normative data for individuals ages 56 to 97.

Attention and Concentration. Attention and concentration are important cognitive functions to measure because older adults frequently suffer acute confusional syndrome that severely limits the individual's capacity to process information (Albert, 1988). Within the brief older adult neuropsychological battery, attention/concentration is measured in two ways. First, attentional processes are observed during the clinical interview. Second, a more formal assessment of attention is conducted using the Attention and Concentration Index score from the Wechsler Memory Scale-Revised (WMS-R) (Wechsler, 1987). The clinical interview can provide information about the individual's ability to follow instructions, ability to remain on task, vulnerability to distraction, and level of arousal. The WMS-R Attention and Concentration Index is composed of three subtests that include Digit Span forward and backward, Visual Memory Span forward and backward, and Mental Control. The index is constructed to have a mean of 100 and a standard deviation of 15, making it directly comparable to other WMS-R indices and WAIS-R intelligence quotients. Indeed, of the five different index scores derived for the WMS-R, the Attention and Concentration Index demonstrates the highest test–retest reliability coefficient ($r = .90$) and lowest standard error of measurement (4.93) for individuals in the age range 70–74.

The WMS-R provides normative data on individuals through age 74. Mayo's Older Americans Normative Studies (Ivnik *et al.*, 1992c) provides normative data for the WMS-R Attention and Concentration Index for individuals ages 56 to 94.

VERBAL MEMORY. The neuropsychological examination of older individuals must include an assessment of memory to provide the information necessary for differential diagnosis. For example, Alzheimer's disease frequently needs to be differentiated from the dementia syndrome of depression or other subcortical-based diseases. Memory is an extremely complicated cognitive process that cannot thoroughly evaluated using current neuropsychological assessment tools. Nonetheless, from a clinical perspective, neuropsychological assessment can provide information about an individual's ability to form new memories, ability to retrieve old memories, and any hemispheric advantage in processing verbal versus visual information. Qualitative information can also be obtained about the individuals's strategy for learning new information, effort exerted in learning new information, and any mood disorder that might be confounding the individual's performance on tests of explicit memory. The neuropsychologist is encouraged to use both a free recall and recognition paradigm in the assessment of memory to generate information relating to the integrity of the medial temporal lobe and subcortical frontal systems. In general, impaired free recall and recognition memory performance, in the context of intact arousal, suggests an inability to form new memories and is consistent with medial temporal lobe pathology. In contrast, impaired free recall but preserved recognition memory performance suggests a retrieval deficit with implications for subcortical-frontal systems dysfunction. Finally, an effort should be made to measure visual memory using the Rey Complex Figure Test (Rey, 1941) with 20-minute delay recall.

The Logical Memory subtest from the WMS-R is one measure of verbal free recall set in a prose or thematic format. Interestingly, the theme of this memory task (as well as other related memory paragraphs within neuropsychology) is weighted heavily with a negative or morbid tone. It is unknown if the emotional valence of such memory tests confounds the performance of particular individuals. The Logical Memory subtest from the WMS-R also provides an opportunity to determine the individual's ability to use strategies and task-inherent schemas (e.g., big cities generally have police who protect and help people). No recognition procedure has yet been developed for this verbal memory task.

The WMS-R Logical Memory subtest provides normative data for older adults through the age of 74. More recent normative data have been provided for older individuals through age 94 (Ivnik *et al.*, 1992c).

Free verbal recall is a particular type of memory assessment that provides information about the individual's ability to form new memories. Several methods of free recall exist and include the California Verbal Learning Test (CVLT) (Delis, Kramer, Kaplan, & Ober, 1987), Rey Auditory Verbal Learning Test (AVLT) (Rey, 1964), and more recently, the Hopkins Verbal Learning Test (HVLT) (Brandt, 1991). Each of these three memory tasks involves the presentation of a list of words in which the individual is asked to recall as many of the presented words as possible. The list of words is then presented again and the same recall task is conducted for as many as five trials (the HVLT uses three trials). Using this assessment procedure, learning can be measured across repeated trials and a learning curve can be produced.

For the older adult, the AVLT and CVLT have appeal because of their validity and available normative data. Indeed, norms for the AVLT are now available for older adults through age 97 (Ivnik *et al.*, 1992a). The CVLT has norms for older adults through age 80 (Delis *et al.*, 1987). The HVLT was designed specifically for older adults and derived some normative data for individuals through age 70. The HVLT has the advantage of providing a brief assessment of verbal learning because it employs three consecutive trials of 12 words each compared to the AVLT (five trials of 15 words) and the CVLT (five trials of six words each). The HVLT also uses a recognition memory procedure, conducted immediately after the third free recall trial, to provide information on the integrity of both encoding and retrieval properties of the individual's memory. Qualitative features of the individual's performance can also be appreciated using the HVLT: perseverative responses, use of semantic clustering, and number of false-positive versus false-negative errors.

The HVLT has one published study that provides norms for older adults through age 70 (Brandt, 1991).

The AVLT is a well-established instrument for the measurement of verbal learning. The recent publication of norms for individuals through the ninth decade (Ivnik *et al.*, 1992a) makes this tool attractive, particularly when brevity is not a necessary condition. The test requires more time than the HVLT, but provides more information such as a measure of interference and delayed recall.

The CVLT is a relatively new measurement of verbal learning that has well-established psychometric properties. It provides an enormous amount of clinical data to understand the different properties of the individual's declarative memory. Normative data are available for adults throughout the eighth decade, differentiated by gender (Delis *et al.*, 1987). The CVLT, like the AVLT, provides a more thorough assessment of verbal learning compared to the HVLT, but it also requires more time and is generally a more complicated tool to administer. A shortened form of the AVLT and CVLT would be beneficial with regard to assessment of verbal learning of the older adult.

VISUAL MEMORY. Visual memory should also be assessed as part of the brief older adult neuropsychological battery. Visual memory provides a measure of the integrity of the nondominant—typically right—hemisphere and serves as a contrast to the measurement of verbal memory and therefore the integrity of the dominant—typically left—hemisphere. At present, clinical neuropsychology is deficient in the number of valid assessment instruments that measure nonverbal memory. For example, while it may be assumed that the visual reproduction subtest from the WMS-R measures visual memory, principal components analyses conducted on the standardization sample yielded a general memory factor and an attention concentration factor (Wechsler, 1987). In addition, a recent factor analytic study using immediate memory subtests from the WMS-R revealed a general memory factor and an attention factor. Separate factors were not found for verbal memory and visual memory using immediate recall subtests of the WMS-R. A more recent study cautions interpretation of the visual reproduction subtest as a measure of memory functioning in patients with Alzheimer's disease due to potential confounds of visual perceptual and constructional properties of the test (Haut, Weber, Wilhelm, Keefover, & Rankin, 1994).

The Rey–Osterrieth Complex Figure test (ROCF) (Osterrieth, 1944; Rey, 1941) appears to provide a measure of nonverbal memory that is not easily performed using verbal cues. It also provides measurement of other generally nondominant cognitive functions such as perceptual and organizational skills. Inclusion of a 20- to 30-minute delayed recall component to the test appears to provide a sensitive measure of memory deficits (Loring, Martin, & Meador, 1990). A recent study demonstrated generally good psychometric properties of the ROCF in an elderly sample (Berry, Allen, & Schmitt, 1991). Using factor analysis, Berry and colleagues (1991) found the ROCF to load on a visuospatial perceptual/memory dimension. Normative studies of the ROCF are few in number and are difficult to interpret because of the variability of test administration across studies. For example, some studies do not employ an immediate and delay recall period while others use a 20- or 30-minute interval for delay recall. Interestingly, a recent study of the ROCF and older adults found no difference in recall using four different delay recall intervals (Berry & Carpenter, 1992). Some normative data are available using the ROCF with elderly populations (Boone, Lesser, Gutierrez-Hill, Berman, & D'Elia, 1993; Chiulli, Yeo, Haaland, & Garry, 1989; Kolb & Wishaw, 1985), and it is apparent that performance on the ROCF correlates with age, but more research is needed to assist the clinician with reliable interpretation of this test.

Normative data for older adults and the ROCF are available throughout the eighth decade of life. Interpretation of the normative data must be made with knowledge of the exact procedure used to derive the norms (see Boone *et al.*, 1993; Chiulli *et al.*, 1989; Kolb & Wishaw, 1985).

EXECUTIVE FUNCTIONING. Executive functions have been described to include goal formation, planning, carrying out goal-directed plans, and effective performance (Lezak, 1983). Each is a highly complex and multifaceted set of behaviors that should be assessed when a thorough measurement of executive skills is warranted. The executive functions are thought to reflect the general integrity of the frontal lobe. For this reason, they represent an important cognitive domain when examining an older adult. From the review provided above, the frontal lobe appears to be a primary area of vulnerability with aging, and an understanding of normal changes in executive functions from disease-induced deficiencies is necessary.

A complete assessment of executive functions clearly is not possible (and most likely unnecessary) as part of a brief neuropsychological assessment of the older adult. Three tests are proposed to provide preliminary information regarding the integrity of the executive system for the older adult. These include a measure of verbal abstraction using the Similarities subtest from the WAIS-R (already administered as part of general intellect), cognitive flexibility as measured by Trail Making Test Parts A and B (Reitan & Wolfson, 1985), and planning and novel problem-solving using the modified version of the Wisconsin Card Sorting Test (WCST) (Robinson, Kester, Saykin, Kaplan, & Gur, 1991).

The WCST is a complex, novel, problem-solving test that requires an extended period of time to administer, particularly with neurologically impaired individuals. To address this issue, several modifications to the standard WCST have been developed (Haaland, Vranes, Goodwin, & Garry, 1987; Nelson, 1976). The method espoused by Haaland and colleagues (1987) employs 64 cards (WCST-64) instead of the standard 128 and uses the same administration criteria as the standard WCST (Heaton, 1981). Measurement of the number of categories completed and number of perseverative responses is maintained. Research on the psychometric properties of the WCST-64 has demonstrated preliminary support for its use as a substitute for the standard WCST (Axelrod, Woodard, & Henry, 1992; Robinson *et al.*, 1991). Indeed, Robinson and colleagues (1991) have asserted that the WCST-64 is an acceptable alternative when administration of the full WCST is not possible. Axelrod and colleagues (1992) cautioned that the WCST-64 requires normative studies to assess the sensitivity of the different abbreviated WCST scoring variables. An additional potential problem is a reduction in the reliability of the WCST due to a reduction in the number of test items used with the WCST-64.

Some normative data for the WCST-64 and older adults are available (Axelrod, Jiron, & Henry, 1993; Haaland *et al.*, 1987), but more research is needed in this area. Axelrod and colleagues (1993) presented normative data using the WCST-64 for healthy individuals ages 20 to 90. Results indicate age-related changes on nearly all variables of the WCST-64, a finding in support of previous research (Axelrod, Woodard, & Henry, 1992).

The similarities subtest from the WAIS-R has published norms for individuals through age 74 (Wechsler, 1981) and through age 97 (Ivnik *et al.*, 1992b). Trail Making Test Parts A and B have published norms for the older adult (Davies, 1968; Spreen & Strauss, 1991). The modified version of the WCST has available norms for the older adult (Axelrod *et al.*, 1992; Axelrod *et al.*, 1993; Haaland *et al.*, 1987).

LANGUAGE. The brief older adult neuropsychological battery employs two measures of language functions: object naming and word fluency. While the clinical interview can provide important information pertaining to the individual's language function, a more formalized assessment using valid instruments is recommended. The brief neuropsychological battery proposed employs the Boston Naming Test (BNT) (Kaplan, Goodglass, & Weintraub, 1983) as a measure of object naming and the Controlled Oral Word Association Test (Benton & Hamsher, 1983) as a measure of word fluency.

The BNT measures the individual's ability to name 60 pictured items spontaneously, using phonemic prompts and semantic prompts. Normative data are available for adults ages 18 to 59 (Kaplan *et al.*, 1983) and more recently for older adults through age 85 (Van Gorp, Satz, Kiersch, & Henry, 1986). It is important to note that age and education correlate with performance on the BNT (Van Gorp *et al.*, 1986). Although age effects have been reported, they do not typically occur until after the seventh decade of life (Albert & Moss, 1988; Spreen & Strauss, 1991).

Word fluency is measured using the Controlled Oral Word Association Test (Benton & Hamsher, 1983) also known as the "F-A-S test of verbal fluency." The test measures the individual's ability to spontaneously produce words pertaining to a specific phonemic category (F-A-S) or semantic category (animals) under timed constraints. The test is thought to reflect both preservation of word knowledge and ability to self-initiate verbal output. As such, the test has been found to relate to both left frontal (Parks *et al.*, 1988) and bifrontal cortex (Miceli, Caltagirone, Gainotti, Masullo, & Silveri, 1981) function. Normative data exist for word fluency in healthy older adults through age 75 (Read, 1987). Age and education are related to performance on the word association test.

Normative data for the BNT are available for adults (Kaplan *et al.*, 1983) and for older adults through age 85 (Van Gorp *et al.*, 1986). The clinician is alerted to the effect of education on BNT performance. Normative data for word fluency are available for older adults through age 75 using both phonemic prompts (F-A-S) and semantic prompts (animal naming) (Read, 1987).

VISUOCONSTRUCTION. Visuoconstruction is broadly assessed using the Block Design and Object Assembly subtests from the WAIS-R (Wechsler, 1987), copy of the intersecting pentagons from the MMSE (Folstein *et al.*, 1975), and the Clock Drawing Test (Goodglass & Kaplan, 1972). Each of these tests purports to measure a broad range of cognitive skills thought to relate primarily to the nondominant (typically right) hemisphere. Use of these procedures provides enough information for the clinician to make an informed decision regarding the integrity of the nondominant hemisphere.

The performance of older individuals on the intersecting pentagons test is interpreted on a bivariate pass/fail scale, primarily at the discretion of the clinician. To this extent, no normative data are available, and caution is warranted in overinterpretation. The MMSE is proposed as part of the interview and thus can provide preliminary information on the potential of a visuoconstruction problem. This can then be confirmed with the WAIS-R subtests and the Clock Drawing Test.

The Block Design and Object Assembly subtests from the WAIS-R are also employed as part of the *general intellect* section of the battery and therefore do not require any additional time or effort to administer. Both subtests measure aspects of visuoconstruction and "nonverbal intelligence," although both also correlate with the Verbal Intelligence Quotient from the WAIS-R, suggesting that these tasks are not pure measures of nonverbal intelligence. Fortunately, new normative data are available for assessment of older adults.

The Clock Drawing Test has acquired a surge in popularity recently because of its brief and easy administration and the wealth of clinical information it produces. It is a measure of visual analytic ability correlating with other measures of visuoconstructional and visuoperceptual skills (Freedman *et al.*, 1994). Different procedures and scoring criteria exist for the Clock Drawing Test, each with normative data for the older adult (Freedman *et al.*, 1994). The older adult brief neuropsychological battery employs the Clock Drawing assessment and scoring procedure of Rouleau, Salmon, Butters, Kennedy, & McGuire (1992), after this procedure was found to demonstrate the highest inter-relater reliability among adult and older adult neuropsychiatric inpatients (Nussbaum, Fields, & Starratt, 1992).

Normative data for the Block Design and Object Assembly subtests of the WAIS-R are available for older adults through age 97 (Ivnik *et al.*, 1992b). Normative data are available for the Clock Drawing Test with older adults (Freedman *et al.*, 1994), but the clinician is cautioned to employ the appropriate norms for specific administrative and scoring procedures.

FINE MOTOR SKILL. Fine motor skill is included as part of the brief older adult neuropsychological battery. Fine motor speed is measured using the Finger Oscillation Task or Finger Tapping Test (Reitan, 1969). This task provides a gross measure of fine motor speed bilaterally and thus enables the clinician to generate additional interhemispheric comparisons. Age, gender, education, and hand dominance correlate with performance on the Finger Oscillation Task (Bornstein, 1985). Administrative aspects of the test can be found elsewhere (Spreen & Strauss, 1991).

Normative data for the Finger Tapping Test are sparse, particularly for older adults. Some norms are available based on healthy individuals from the general population and grouped by education, age, and handedness (Bornstein, 1985).

MOOD. Mood is an important domain given the fact that mood can affect cognitive performance in older individuals (Nussbaum, Kaszniak, Allender, & Rapcsak, 1995). There are many scales available to assess mood, but few have been designed specifically for older adults. The Geriatric Depression Scale (GDS) is a brief, 30-item inventory with good validity (Yesavage *et al.*, 1983). The GDS also has demonstrated utility in assessment of mood in individuals with cognitive impairment (Nussbaum & Sauer, 1993). Research on administration of the GDS has found that reading the items of the scale to individuals with progressive dementia does not sacrifice validity of the scale (Parmelee & Katz, 1990). For the brief older adult neuropsychological battery, the 30 items of the GDS are read to the individual with "yes–no" responses recorded by the clinician.

Normative data for the GDS are available (Yesavage *et al.*, 1983). Many studies using the GDS with different patient populations have also been conducted and provide important clinical information on pertinent issues such as differential diagnosis (see Nussbaum & Sauer, 1993).

The results from the neuropsychological assessment can be used for clinical decisions regarding diagnosis, prognosis, treatment, and placement issues. The neuropsychologist may decide that additional testing is required to answer a specific concern derived from the individual's performance on the brief older adult neuropsychological battery. In general, however, the brief battery is meant to provide sufficient clinical information on the individual's neurobehavioral status in a minimal amount of time.

The brief older adult neuropsychological battery is designed to assess multiple domains of cognitive functioning while attempting to address brain–behavior changes that occur with advancing age. As reviewed above, the prefrontal, medial temporal, and subcortical regions of the brain appear to be relatively vulnerable to the aging process. The fact that physiological and structural changes occur in these brain regions with advancing age implies that neurobehavioral and cognitive changes might also be expected. The literature on neuropsychological changes with age appears to have some preliminary support for this speculation. It is important, therefore, that the neuropsychologist know about the expected neurobehavioral and cognitive changes that occur with age in order to delineate normal aging from disease. The brief older adult neuropsychological battery is designed to assess multiple areas of cognitive functions while maintaining particular focus on the domains of memory (medial temporal and prefrontal brain regions), verbal abstraction, concept formation, cognitive flexibility, and verbal fluency (prefrontal and subcortical-frontal neural systems).

PAUL DAVID
NUSSBAUM

Clinical neuropsychology is at a critical point in its evolution, having experienced rapid growth and popularity during the past 15 years. With regard to the future direction for the field of neuropsychology and aging, several suggestions might be considered. First, it is clear that clinical neuropsychology needs to formalize a specialization in geriatrics to meet the demands of the demographic revolution occurring in our country today. This specialization could entail a structured graduate curriculum, internships, fellowships, and board certification in geriatrics, all designed to train neuropsychologists and produce experts in the area of aging. Second, neuropsychology will need to develop assessment tools that have predictive validity of functional status, particularly for older adults, where capacity for independent living is often of critical importance. Third, neuropsychologists should assume a role in the creation of new settings of care for the older adult, most of which will probably not reflect traditional, medical, disease-based models of care. The community-based continuums of care that presently exist represent a valid alternative to hospital-based models. Neuropsychology should begin to position itself as a viable and necessary discipline within these new settings of care. Training programs in geriatric neuropsychology can be located in these new settings, thereby providing a natural bridge for neuropsychology to develop a healthy and productive existence. A wellness-based theme of care for the aged can be championed by neuropsychology as a more useful model of care. Indeed, all neuropsychology graduate students should be expert in normal aging prior to study of neuropathological states in the elderly. The challenges noted above represent the frontier for neuropsychology and aging.

More traditional academic endeavors will likely focus on the relationship between the neuroanatomical, neurochemical, and neuropsychological changes with age. A major thrust of research is likely to occur in the area of normal aging, something that is desperately needed. This will include a better understanding of the heterogeneity of cognitive performance with advanced years and should result in publication of needed normative data. Preventative measures of health care will take on a more significant role, and integration of sophisticated computerized technology to promote such care will become reality. Identification of preclinical markers of progressive and chronic diseases suffered by the elderly will become standard and serve to guide treatment. Finally, the clinical landscape of eldercare will likely reflect a managed care model, at least into the near future. As such, neuropsychology will need to generate outcomes-based clinical research to justify its existence. A new conceptualization and vision for neuropsychology and aging would appear to be needed given the points noted above. The current state of the art is not suitable to meet the increasing demands of a rapidly expanding aging population.

CONCLUSION

This chapter presented a historical review of the literature on age-related cognitive change, paying particular attention to the relationship between changes in neuroanatomy and cognition. The frontal lobe theory of aging, which emphasizes the prefrontal cortical system's particular vulnerability to the effects of aging, is currently receiving much attention and critical analysis. The neuroanatomical substrates of aging are not yet clearly defined. Neuropsychology can contribute much to this area of study and represents a major player in understanding brain–behavior relations with advancing age. Indeed, cognitive and behavioral measures can be selected and developed based on continued clarification of the primary regions of neuronal loss with normal aging.

This chapter also reviewed the current status of neuropsychological assessment of the older individual. The advantage of using a brief, flexible neuropsychological battery was advanced, and a brief older adult neuropsychological battery was proposed for use by clinicians. For each cognitive domain assessed, particular tests were recommended and available normative data were presented. It is clear that more information is needed from fewer assessment tools, particularly with the older adult and in the current managed care environment. The brief older adult neuropsychological battery is designed to permit clinicians the flexibility of choosing specific tests and employing tests that provide information pertinent to more than one domain of interest.

A major conclusion from this chapter is that the status quo of clinical neuropsychology is not adequate to meet the demands of the older adult population. Major conceptual shifts and progressive ideas of care are needed for neuropsychology to survive as a clinical representative in the emerging eldercare system. Several ideas were described to better position neuropsychology as a leading provider of care for the older adult. Clinical neuropsychology has an opportunity to contribute to the field of neurosciences and aging. Indeed, geriatric neuropsychology has an opportunity to emerge as a formalized subspecialization that will thrive with the expanding population of older adults in the United States.

With a demographic revolution occurring in the United States, major societal changes will be needed to meet the rapidly growing needs of those 65 years of age and older. Indeed, by the year 2010, 76 million individuals who comprise the baby boom generation (those born between 1945 and 1964) will reach age 65. There will be so many adults within this age group and older that the use of age 65 as the "elderly threshold" will likely no longer apply. The political, medical, economical, and family landscape of this country will certainly be affected. Neuropsychology must now prepare with articulation of a clear vision for providing clinical services to older adults, training students for specialization in geriatrics, and promoting a wellness-based philosophy of aging. Indeed, neuropsychology can provide leadership to the country in changing the negative attitudes of aging to a positive, productive, and wellness-based philosophy. These are the challenges for clinical neuropsychology.

References

Albert, M. S. (1988). Acute confusional state. In M. S. Albert & M. B. Moss (Eds.), *Geriatric neuropsychology* (pp. 100–114). New York: Guilford Press.

Albert, M. S., & Moss, M. B. (Eds.). (1988). *Geriatric neuropsychology*. New York: Guilford Press.

Anthony, J. C., La Resche, L., Niaz, U., Von Korff, M., & Folstein, M. (1982). Limits of the Mini-Mental State as a screening test for dementia and delirium among hospital patients. *Psychological Medicine, 12,* 397–408.

Awad, I. A., Spetzler, R. F., Hodak, J. A., Awad, C. A., & Carey, R. (1986). Incidental subcortical lesions identified on magnetic resonance imaging in the elderly: Correlation with age and cerebrovascular risk factors. *Stroke, 17,* 1084–1089.

Axelrod, B. N., Jiron, C. C., & Henry, R. R. (1993). Performance of adults ages 20 to 90 on the abbreviated Wisconsin Card Sorting Test. *The Clinical Neuropsychologist, 7,* 205–209.

Axelrod, B. N., Woodard, J. L., & Henry, R. R. (1992). Analysis of an abbreviated form of the Wisconsin Card Sorting Test. *The Clinical Neuropsychologist, 6*, 27–31.

Benton, A. L., & Hamsher, K. (1983). *Multilingual aphasia examination.* Iowa City: AJA Associates.

Berry, D. T. R., Allen, R. S., & Schmitt, F. A. (1991). Rey–Osterrieth Complex Figure: Psychometric characteristics in a geriatric sample. *The Clinical Neuropsychologist, 5*, 143–153.

Berry, D. T. R., & Carpenter, G. S. (1992). Effect of four different delay periods on recall of the Rey–Osterrieth Complex Figure by older persons. *The Clinical Neuropsychologist, 6*, 80–84.

Boone, K. B., Lesser, I. M., Gutierrez-Hill, E., Berman, N. G., & D'Elia, L. F. (1993). Rey–Osterrieth Complex Figure performance in healthy, older adults: Relationship to age, education, sex, and IQ. *The Clinical Neuropsychologist, 7*, 22–28.

Boone, K. B., Miller, B. L., Lesser, I. M., Hill, E., & D'Elia, L. D. (1990). Performance on frontal lobe tests in healthy, older individuals. *Developmental Neuropsychology, 6*, 216–223.

Bornstein, R. A. (1985). Normative data on selected neuropsychological measures from a nonclinical sample. *Journal of Clinical Psychology, 41*, 651–659.

Botwinick, J. (1977). Intellectual abilities. In J. E. Birren & K. W. Schaie (Eds.), *Handbook of the psychology of aging* (pp. 580–605). New York: Springer.

Brandt, J. (1991). The Hopkins Verbal Learning Test: Development of a new memory test with six equivalent forms. *The Clinical Neuropsychologist, 5*, 125–144.

Chiulli, S. J., Yeo, R. A., Haaland, K. Y., & Garry, P. J. (1989). Complex figure copy and recall in the elderly. Paper presented to the International Neuropsychological Society, Vancouver, Canada.

Coffey, C. E., & Figiel, G. S. (1991). Neuropsychiatric significance of subcortical encephalomalacia. In B. J. Carroll & J. E. Barrett (Eds.), *Psychopathology and the brain* (pp. 243–264). New York: Raven Press.

Coffey, C. E., Figiel, G. S., Djang, W. T., & Weiner, R. D. (1990). Subcortical hyperintensity on magnetic resonance imaging: A comparison of normal and depressed elderly subjects. *American Journal of Psychiatry, 147*, 187–189.

Craik, F. I. M., & McDowd, J. M. (1987). Age differences in recall and recognition. *Journal of Experimental Psychology: Learning, Memory, and Cognition, 13*, 474–479.

Craik, F. I. M., Morris, L. W., Morris, R. G., & Loewen, E. R. (1990). Relations between source amnesia and frontal lobe functioning in older adults. *Psychology and Aging, 5*, 148–151.

Davies, A. (1968). The influence of age on trail making test performances. *Journal of Clinical Psychology, 24*, 96–98.

Dekaban, A. S., & Sadowsky, B. S. (1978). Changes in brain weights during the span of human life: Relation of brain weights to body height and weight. *Annals of Neurology, 4*, 345–357.

Delis, D. C., Kramer, J. H., Kaplan, E., & Ober, B. A. (1987). *The California Verbal Learning Test, research edition.* New York: Psychological Corporation.

Doraiswamy, P. M., Figiel, G. S., Husain, M. M., McDonald, W. M., Shah, S. A., Boyko, O. B., Ellinwood, E. H., & Krishnan, K. R. R. (1991). Aging of the human corpus callosum: Magnetic resonance imaging in normal volunteers. *The Journal of Neuropsychiatry and Clinical Neurosciences, 3*, 392–397.

Earnest, M. P., Heaton, R. K., Wilkinson, W. E., & Manke, W. F. (1979). Cortical atrophy, ventricular enlargement and intellectual impairment in the aged. *Neurology, 29*, 1138–1143.

Engelhart, C., Eisenstein, N., & Meininger, J. (1994). Psychometric properties of the Neurobehavioral Cognitive Status Exam. *The Clinical Neuropsychologist, 8*, 405–415.

Escobar, J. I., Burnam, A., Karno, M., Forsythe, A., Landsverk, J., & Golding, J. M. (1986). Use of the Mini-Mental State Examination in a community population of mixed ethnicity. *Journal of Nervous and Mental Disease, 174*, 607–614.

Folstein, M. F., Anthony, J. C., Parhard, I., Duffy, B., & Gruenberg, E. M. (1985). The meaning of cognitive impairment in the elderly. *Journal of the American Geriatrics Society, 33*, 228–235.

Folstein, M. F., Folstein, S. E., & McHugh, P. R. (1975). Mini-mental state: A practical method of grading the cognitive state of patients for the clinician. *Journal of Psychiatric Research, 12*, 189–198.

Freedman, M., Leach, L., Kaplan, E., Winocur, G., Shulman, K. I., & Delis, D. C. (1994). *Clock drawing: a neuropsychological analysis.* New York: Oxford University Press.

Gerard, G., & Weisberg, L. A. (1986). MRI periventricular lesions in adults. *Neurology, 36*, 998–1001.

Goldstein, G. (1983). Normal aging and the concept of dementia. In C. J. Golden & P. J. Vicente (Eds.), *Foundations of clinical neuropsychology* (pp. 249–271). New York: Plenum Press.

Goldstein, G., & Nussbaum, P. D. (1997). The neuropsychology of aging. In J. G. Beaumont & J. Sergent (Eds.), *Blackwell dictionary of neuropsychology.* Oxford: Blackwell Publishers.

Goldstein, G., & Shelly, C. (1981). Does the right hemisphere age more rapidly than the left? *Journal of Clinical Neuropsychology, 3,* 65–78.

Golomb, J., De Leon, M. J., Kluger, A., George, A. E., Tarshish, C., & Ferris, S. H. (1993). Hippocampal atrophy in normal aging. *Archives of Neurology, 50,* 967–973.

Goodglass, H., & Kaplan, E. (1972). *The assessment of aphasia and related disorders.* Philadelphia: Lea & Febiger.

Haaland, K. Y., Vranes, L. F., Goodwin, J. S., & Garry, P. J. (1987). Wisconsin card sort test in a healthy elderly population. *Journal of Gerontology, 42,* 345–346.

Hasher, L., & Zacks, R. T. (1979). Automatic and effortful processes in memory. *Journal of Experimental Psychology, General, 108,* 356–358.

Haug, H., Barmwater, V., Eggers, R., Fischer, D., Kuhl, S., & Sass, N. L. (1983). Anatomical changes in aging brain: Morphometric analysis of the human prosencephalon. In J. Cervos-Navarro & H. I. Sarkander (Eds.), *Neuropharmacology* (pp. 1–12). New York: Raven Press.

Haut, M. W., Weber, A. M., Wihelm, K. L., Keefover, R. W., & Rankin, E. D. (1994). The visual reproduction subtest as a measure of visual perceptual/constructional functioning in dementia of the Alzheimer's type. *The Clinical Neuropsychologist, 8,* 187–192.

Heaton, R. K. (1981). *A manual for the Wisconsin Card Sorting Test.* Odessa, FL: Psychological Assessment Services.

Heaton, R. K., Grant, I., & Matthews, C. G. (1991). *Comprehensive norms for an expanded Halstead–Reitan battery.* Odessa, FL: Psychological Assessment Resources.

Horn, J. L., & Catell, R. B. (1967). Age differences in fluid and crystallized intelligence. *Acta Psychobiologica, 26,* 107–129.

Hoyer, S. (1990). Brain glucose and energy metabolism during normal aging. *Aging, 2,* 245–258.

Incagnoli, T., Goldstein, G., & Golden, C. J. (1986). *Clinical application of neuropsychological test batteries.* New York: Plenum Press.

Ivnik, R. J., Malec, J. F., Smith, G. E., Tangalos, E. G., Peterson, R. C., Kokmen, E., & Kurland, L. T. (1992a). Mayo's older Americans normative studies: Updated AVLT norms for ages 56–97. *The Clinical Neuropsychologist, 6,* 83–104.

Ivnik, R. J., Malec, J. F., Smith, G. E., Tangalos, E. G., Peterson, R. C., Kokmen, E., & Kurland, L. T. (1992b). Mayo's older Americans normative studies: WAIS-R norms for ages 56–97. *The Clinical Neuropsychologist, 6,* 1–30.

Ivnik, R. J., Malec, J. F., Smith, G. E., Tangalos, E. G., Peterson, R. C., Kokmen, E., & Kurland, L. T. (1992c). Mayo's older Americans normative studies: WMS-R norms for ages 56–94. *The Clinical Neuropsychologist, 6,* 49–82.

Kane, R. (1991). Standardized and flexible batteries in neuropsychology: An assessment update. *Neuropsychology Review, 2,* 281–339.

Kaplan, E. F., Goodglass, H., & Weintraub, S. (1983). *The Boston Naming Test* (2nd ed.). Philadelphia: Lea & Febiger.

Kaszniak, A. W. (1987). Neuropsychological consultation to geriatricians: Issues in assessment of memory complaints. *Clinical Neuropsychologist, 1,* 35–46.

Katzman, R., & Rowe, J. W. (1992). *Principles of geriatric neurology.* Philadelphia: F. A. Davis Company.

Kiernan, R. J., Mueller, J., Langston, J. W., & Van Dyke, C. (1987). The neurobehavioral cognitive status examination: A brief but quantitative approach to cognitive assessment. *Annals of Internal Medicine, 107,* 481–485.

Kobari, M., Stirling, M., & Ichijo, M. (1990). Leukoaraiosis, cerebral atrophy, and cerebral perfusion in normal aging. *Archives of Neurology, 47,* 161–165.

Kolb, B., & Wishaw, I. Q. (1990). *Fundamentals of human neuropsychology* (3rd ed.). New York: W. H. Freeman and Company.

La Rue, A. (1992). *Aging and neuropsychological assessment.* New York: Plenum Press.

Lechner, H., Schmidt, R., Bertha, G., Justich, E., Offenbacher, H., & Schneider, G. (1987). Nuclear magnetic resonance image of white matter lesions and risk factors for stroke in normal individuals. *Stroke, 19,* 263–265.

Lezak, M. (1983). *Neuropsychological assessment* (2nd ed.). New York: Oxford University Press.

Lim, K. O., Zipursky, R. B., Murphy, G. M., & Pfefferbaum, A. (1990). In vivo quantification of the limbic system using MRI: Effects of normal aging. *Psychiatry Research: Neuroimaging, 35,* 15–26.

Loring, D. W., Martin, R. C., & Meador, K. J. (1990). Psychometric construction of the Rey–Osterrieth Complex Figure. *Archives of Clinical Neuropsychology, 5,* 1–14.

Lovell, M. R., & Nussbaum, P. D. (1994). Neuropsychological assessment. In C. E. Coffey & J. L.

Cummings (Eds.), *Textbook of geriatric neuropsychiatry* (pp. 129–144). Washington, DC: American Psychiatric Press.

Matarazzo, J. D. (1972). *Wechsler's measurement and appraisal of adult intelligence* (5th ed.). Baltimore, MD: Williams and Wilkins.

Mattis, S. (1976). Mental status examination for organic mental syndrome in the elderly patient. In L. Bellack & T. B. Karasu (Eds.), *Geriatric psychiatry* (pp. 79–121). New York: Grune & Stratton.

McCue, M., Goldstein, G., & Shelly, C. (1989). The application of a short form of the Luria Nebraska Neuropsychological Battery to discriminate between dementia and depression in the elderly. *International Journal of Clinical Neuropsychology, 11,* 21–29.

McInnes, W. D., Gillen, R. W., Golden, C. J., Graber, B., Cole, J. K., Uhl, H. S. M., & Greenhouse, A. H. (1983). Aging and performance on the Luria-Nebraska Neuropsychological Battery. *International Journal of Neuroscience, 19,* 179–180.

Miceli, G., Caltagirone, C., Gainotti, G., Masullo, C., & Silveri, M. C. (1981). Neuropsychological correlates of localized cerebral lesions in non-aphasic brain damaged patients. *Journal of Clinical Neuropsychology, 3,* 53–63.

Miller, A. K. H., Alston, R. L., & Corsellis, J. A. N. (1980). Variations with age in the volumes of grey and white matter in the cerebral hemispheres of man: Measurements with an image analyser. *Neuropathology and Applied Neurobiology, 6,* 119–132.

Miller, A. K. H., & Corsellis, J. A. N. (1977). Evidence for a secular increase in human brain weight during the past century. *Annals of Human Biology, 4,* 253–257.

Mittenberg, W., Seidenberg, M., O'Leary, D. S., & Digiulio, D. V. (1989). Changes in cerebral functioning associated with normal aging. *Journal of Clinical and Experimental Neuropsychology, 11,* 918–932.

Nelson, H. E. (1976). A modified card sorting test sensitive to frontal lobe defects. *Cortex, 12,* 313–324.

Nelson, H. E. (1982). *National Adult Reading Test manual.* Windsor, U.K.: NFER-Nelson Publishing Company.

Nussbaum, P. D. (1994). Pseudodementia: A slow death. *Neuropsychology Review, 4,* 71–90.

Nussbaum, P. D. (1996). Aging: Issues in health and neuropsychological functioning. In A. J. Goreczny (Ed.), *Handbook of recent advances in behavioral medicine.* New York: Plenum Press.

Nussbaum, P. D., Fields, R. B., & Starratt, C. (1992, February). *Comparison of three scoring procedures for the clock drawing.* Paper presented at the International Neuropsychological Society, San Diego, CA.

Nussbaum, P. D., Kaszniak, A. W., Allender, J., & Rapcsak, S. (1995). Cognitive decline and depression in the elderly: A follow-up study. *The Clinical Neuropsychologist, 9,* 101–111.

Nussbaum, P. D., & Sauer, L. (1993). Self-report of depression in elderly with and without progressive cognitive deterioration. *Clinical Gerontologist, 13,* 69–80.

Osato, S., La Rue, A., & Yang, J. (1989, November). *Screening for cognitive deficits in older psychiatric patients.* Paper presented at the annual meeting of the Gerontological Society of America, Minneapolis, MN.

Osterrieth, P. A. (1944). Le test de copie d'une figure complex: Contribution a l'etude de la perception et de la memoire. *Archives de Psychologie, 30,* 286–356.

Parmelee, P. A., & Katz, I. R. (1990). Geriatric Depression Scale. *Journal of the American Geriatrics Society, 38,* 1379–1380.

Powers, R. E. (1994). Neurobiology of aging. In C. E. Coffey & J. L. Cummings (Eds.), *Textbook of geriatric neuropsychiatry* (pp. 35–70). Washington, DC: American Psychiatric Press.

Read, D. E. (1987). *Neuropsychological assessment of memory in early dementia: Normative data for a new battery of memory tests.* Unpublished manuscript.

Reitan, R. M. (1969). *Manual for administration of neuropsychological test batteries for adults and children.* Unpublished manuscript.

Reitan, R. M., & Wolfson, D. (1985). *The Halstead–Reitan Neuropsychological Test Battery.* Tucson, AZ: Neuropsychology Press.

Rey, A. (1941). L'examen psychologique dans les cas d'encephalopathie traumatique. *Archives de Psychologie, 28,* 286–340.

Rey, A. (1964). *L'examen clinique en psychologie.* Paris: Presses Universitaires de France.

Robinson, L. J., Kester, D. B., Saykin, A. J., Kaplan, E. F., & Gur, R. C. (1991). Comparison of two short forms of the Wisconsin Card Sorting Test. *Archives of Clinical Neuropsychology, 6,* 27–33.

Rouleau, I., Salmon, D. P., Butters, N., Kennedy, C., & McGuire, K. (1992). Quantitative and qualitative analyses of clock drawings in Alzheimer's and Huntington's disease. *Brain and Cognition, 18,* 70–87.

Schacter, D. L., Kaszniak, A. W., Kihlstrom, J. F., & Valdiserri, M. (1991). The relation between source memory and aging. *Psychology and Aging, 6,* 559–568.

Schaie, K. W. (1965). A general model for the study of developmental problems. *Psychological Bulletin, 64,* 92–107.

Schaie, K. W. (1983). *Longitudinal studies of adult psychological development.* New York: Guilford.

Schaie, K. W., & Parham, I. (1977). Cohort-sequential analyses of adult intellectual development. *Developmental Psychology, 13,* 649–653.

Schaie, K. W., & Schaie, J. P. (1977). Clinical assessment and aging. In J. E. Birren & K. W. Schaie (Eds.), *The psychology of aging* (pp. 692–723). New York: Van Nostrand Reinhold.

Spreen, O., & Strauss, E. (1991). *A compendium of neuropsychological tests.* New York: Oxford University Press.

Squire, L. R. (1987). *Memory and brain.* New York: Oxford University Press.

Stebbins, G. T., Wilson, R. S., Gilley, D. W., Bernard, B. A., & Fox, J. H. (1990). Use of the National Adult Reading Test to estimate premorbid IQ in dementia. *The Clinical Neuropsychologist, 4,* 18–24.

Steingart, A., Hachinski, V. C., Lau, C., Fox, A. J., Diaz, F., Cape, R., Lee, D., Inzitari, D., & Merskey, H. (1987). Cognitive and neurologic findings in subjects with diffuse white matter lucencies on computed tomographic scan. *Archives of Neurology, 44,* 32–35.

Sullivan, P., Pary, R., Telang, F., Rifai, A. H., & Zubenko, G. S. (1990). Risk factors for white matter changes detected by magnetic resonance imaging in the elderly. *Stroke, 21,* 1424–1428.

Uchiyama, C. L., Mitrushina, M. N., D'Elia, L. F., Satz, P., & Matthews, A. (1994). Frontal lobe functioning in geriatric and non-geriatric samples: An argument for multimodal analyses. *Archives of Clinical Neuropsychology, 9,* 215–228.

Valdois, S., Yves, J., Poissant, A., Ska, B., & Dehaut, F. (1990). Heterogeneity in the cognitive profiles of the elderly. *Journal of Clinical and Experimental Neuropsychology, 12,* 587–598.

Van Gorp, W. G., Satz, P., Kiersch, M. E., & Henry, R. (1986). Normative data on the Boston Naming Test for a group of normal older adults. *Journal of Clinical and Experimental Neuropsychology, 8,* 702–705.

Wechsler, D. (1955). *Manual for the Wechsler Intelligence Scale.* New York: The Psychological Corporation.

Wechsler, D. (1981). *Wechsler Adult Intelligence Scale-Revised.* New York: Psychological Corporation.

Wechsler, D. (1987). *Wechsler Memory Scale-Revised. Manual.* New York: Psychological Corporation.

Wilson, R. S., Rosenbaum, G., Brown, G., Rourke, D., Whitman, D., & Grisell, J. (1978). An index of premorbid intelligence. *Journal of Consulting and Clinical Psychology, 46,* 1554–1555.

Yesavage, J. A., Brink, T. L., Rose, T. L., Lum, D., Huang, V., Adley, M., & Lerier, V. O. (1983). Development and screening of a geriatric depression rating scale: A preliminary report. *Journal of Psychiatric Research, 7,* 37–49.

Zubenko, G. S., Sullivan, P., Nelson, J. P., Belle, S. H., Huff, J., & Wolf, G. L. (1990). Brain imaging abnormalities in mental disorders of late life. *Archives of Neurology, 47,* 1107–1111.

Clinical Considerations

II

6

Evaluation of High-Functioning Autism

Don J. Siegel

Autism as a Pervasive Developmental Disorder

The presence of a developmental disorder is considered when an individual's abilities are not consistent with developmental or chronological age expectancy. In effect, there may be a significant departure from normal development in the onset and progression of speech, language, motor, cognitive, social, or self-help abilities. A wide range of severity, diverse symptom patterns, and functioning levels characterize individuals with developmental disorders. The course of development of these functions may be significantly delayed and/or erratic compared with normal variation in development. Developmental disorders are to be distinguished from acquired disorders which are related secondarily to a known medical condition, illness, trauma, or other causative agent. The causes of the developmental disorders are diverse and may be associated with chromosomal, metabolic, perinatal, pregnancy, or postnatal factors. Although often identified in the period of childhood and adolescence, impairments are usually chronic, are lifelong, and affect aspects of educational, social, and vocational functioning.

Kanner (1943) provided a clinical description of eight boys and three girls below the age of 11 years, each of whom demonstrated specific cognitive and communication delays and deficits, with peculiar social and behavioral characteristics including an inability to form emotional connections with others, which constituted a previously unreported syndrome. Kanner used the term *inborn autistic disturbances of affective contact* to describe the quality of the innate deficits in social and interpersonal functioning of these children. These children were unable to relate in the ordinary way to people and situations, overly attached to objects, socially inaccessible, and devoid of social awareness. People were disregarded or recognized as objects, as intruding on routines and efforts to maintain sameness or

Don J. Siegel   Western Psychiatric Institute and Clinic, University of Pittsburgh Medical Center, Pittsburgh, Pennsylvania 15213.

Neuropsychology, edited by Goldstein *et al.* Plenum Press, New York, 1998.

aloneness, and as interfering with the pursuit of special activities and preoccupations. One child did not acknowledge the presence of adults or children in the room, would draw on picture books which other children were using, and allowed his toys to be taken without responding. Another child catalogued issues of *Time* magazine by publication date, and though he also enjoyed going to the cinema and was able to recall the order in which he saw movies, did not derive pleasure from the characters, action, or story line. Parents described these children as being oblivious to surroundings, appearing to be hypnotized, preferring to be left alone, and acting as if others were not present. Disturbances in communication and language development were consistently observed which included the absence of gestures and inflected speech, repetitive and parroted speech patterns, a literal interpretation of word meanings, a reversal of personal pronouns, a mechanical use of phrases, and speech which did not have any purposeful communicative intent. Intact memory for rotely learned facts was apparent at an early age, as the children recited verses from the Bible or contents of the index page from an encyclopedia, and identified pictures of the presidents. Such odd behavior as spinning of objects or reflecting light, performing stereotyped movements, engaging in ritualistic play activities, insisting upon routines which if interrupted would result in tantrums, demonstrating atypical responses to touch, motion, and sound, and unusual fears (e.g., of an egg beater, vacuum cleaner, mechanical toys, a playground slide) were evident. Although the etiology of such autistic deficits was initially attributed to psychogenic factors, such as parenting style and its impact on psychosocial development, onset is no longer considered within this context, for autism is now generally accepted to be a neurobiological disorder involving early brain development.

Autism is one of the pervasive developmental disorders. As such, its manifestations encompass broad aspects of development including social, language, and cognitive abilities, with a range of severity possible within each primary deficit area, and a constellation of clinical features and impairments which are deviant with respect to an individual's developmental level. The Diagnostic and Statistical Manual of Mental Disorders–Fourth Edition (DSM-IV) (American Psychiatric Association [APA], 1994) criteria for autism specify the presence of delay or abnormal functioning in three areas: qualitative impairments of reciprocal social interaction, qualitative impairments in communication and language, and a restricted, repetitive pattern of behavior, interests, and activities. With regard to reciprocal social interaction, possible deficits include impairment in the use of nonverbal behavior to regulate social exchange (e.g., eye contact, facial expression), a failure to form peer relationships consistent with developmental expectancy or spontaneously share interests or achievements with others, and a lack of social or emotional reciprocity (e.g., an adherence to a social script, an inability to take turns or adopt the perspective of others). In the sphere of communication there may be a lack of language development, language delay, or regression with loss of language. Specific clinical features associated with emerging language in autism include echolalia, pronominal reversal, errors in verb tense, use of neologisms, and the production of speech which is perseverative, containing stereotyped content and phrases. There is typically failure to appreciate the social aspects of language (e.g., shifts in content, cues to begin or end conversation) with an associated deficit in the ability to initiate or sustain discourse with others. Included among the list of impairments in communication is a lack of imaginative, pretend, and symbolic play skills appropriate for the child's developmental level. Toys and games are typically used in a mechanical or inappropriate fashion, limited by the features of the objects, while play occurs without involvement of others. Lastly, deficits in the third area include a stereotyped and narrow range of interests which appear as preoccupations; a resistance to change and inflexible adherence to routines; stereotyped, repetitive motor movements; and attachments or preoccupation with the parts and sensory aspects of objects.

Assessment of Individuals with Autism: The High-Functioning Subgroup

Autism is a low-incidence disorder that is estimated to occur in two to five cases per 10,000 and is more common in males (APA, 1994). Establishing a diagnosis of autism is complicated in part by a pattern of delay in achieving certain milestones, with normal development occurring in other aspects of functioning. Diagnosis is also confused by the minority of cases where there is normal development for the first year or two and then regression in functioning. Delay or abnormal functioning in at least one of the three major deficit areas must occur, however, prior to age 3 years. Autism in individuals at the lower end of the ability distribution can be confounded by the appearance of symptoms associated with mental retardation (e.g., restricted play and poor social skills, self-stimulation, echolalic speech) which obscure performance and adaptive levels. Some of the presenting symptoms for which children are referred can be confused with developmental and behavioral disorders other than mental retardation, including deafness, emotional disturbance, a speech and language disorder such as aphasia, or hyperactivity.

Autism is a rather heterogeneous disorder with functioning levels extending from severely retarded to intellectually gifted. In contrast to the confound of mental retardation at early developmental levels, diagnosis of autism in non-retarded, high-functioning individuals poses other dilemmas, as deficits are likely to be qualitative, such as in understanding the communicative intent of others, rather than simply involving delay. For example, there must be documentation that language and behavior are deviant from developmental expectations rather than simply absent or deficient. The presentation of certain symptoms may also be specific to age, developmental factors, and the natural course of the disorder, as there is typically some improvement in functioning beyond the age of 5 years. While certain criteria may be met in a developmental perspective, enduring and consistent qualitative differences in social and communication abilities need to be present for diagnosis.

Although level of general intellectual function is not diagnostic of autism, with the disorder occurring across a wide range of intellectual ability, an estimated 66% to 75% of autistic individuals demonstrate Full Scale IQ scores equal to or less than 70, with approximately 50% of autistic individuals obtaining IQ scores below 50 (Smalley, Asarnow, & Spence, 1988). A sizable proportion, about 25% to 30% of autistic individuals, however, demonstrate intelligence quotients greater than 70, and thus comprise a substantial subgroup of the disorder. An epidemiological survey in the state of Utah reported that 34% of autistic individuals demonstrated IQ scores above 70, with a greater proportion of autistic males than females (6.3:1) having IQ scores in this range (Ritvo et al., 1989). In a case series study of consecutive referrals, autistic males were nearly nine times more likely than females to have Full Scale IQ scores greater than 70 (Volkmar, Szatmari, & Sparrow, 1993). In effect, not only is there a greater proportion of males than females with autism, sex differences in prevalence are also found at higher IQ levels.

While there are no existing criteria nor a consensus definition for high functioning autism (HFA), with reference made to various scores on verbal and nonverbal intelligence tests as being the cognitive standard, it has been recommended that study of autistic individuals without severe developmental delay and intellectual retardation be pursued. In this manner the influence of retardation can be controlled and core profiles common to all individuals with the disorder delineated. This approach has been adopted in some research settings, where the study of the subgroup of high-functioning autistic individuals has indeed proven advantageous for identifying a characteristic neuropsychologic profile of the disorder.

Different behavioral characteristics are found for low- (i.e., IQ below 70) and high-functioning autistic individuals (Freeman *et al.*, 1981), while level of language development and intellectual ability best characterize the degree of impairment for any particular autistic individual. Indeed, IQ and language function have proven to be important prognostic variables, for while the course of the disorder is relatively stable, chronic, and lifelong, outcome studies in autism have demonstrated that level of intelligence and the ability to use and comprehend language invariably predict future adaptive behavior and academic, social, and vocational adjustment (Venter, Lord, & Schopler, 1992).

An account of the social histories and adaptation of children initially diagnosed with autism in childhood, who were in their second and third decade of life, revealed that despite certain accomplishments, these high-functioning adults with autism continued to be limited by the fundamental deficits of the disorder (Kanner, Rodriguez, & Ashenden, 1972). Social relationships were transacted awkwardly, few genuine friendships had developed, and except for attending club meetings where a special interest or hobby was held in common, they remained socially isolated. In addition, the pursuit of special interests and an adherence to routine were rigidly maintained. These high-functioning autistic individuals were acutely aware of their limitations, however, and endeavored to conform and comply with social conventions and obligations in their everyday life. A certain degree of self-sufficiency had been achieved as these individuals drove cars, played instruments, learned foreign languages, obtained advanced educational degrees, and had purchased a house or apartment. Among the various positions they held were bank teller, duplicating machine operator, laboratory worker, accountant, and general office worker. Employers characteristically described them as being dependable, meticulous, and trustworthy. In a retrospective study of high-functioning, verbal autistic adolescents and adults, there was an improvement in aspects of communication and social behavior over time, although ritualistic, repetitive behavior appeared to persist in intensity (Piven, Harper, Palmer, & Arndt, 1996).

THE PROFILE OF COGNITIVE ABILITIES AND DEFICITS IN AUTISM

Neuropsychological studies in autism have taken divergent approaches in efforts to explain the clinical features of the disorder. Some investigations have focused either on specific, singular aspects of cognitive function which are considered to be primary and are then used to explain the various clinical features of the disorder. Study of a single neuropsychologic deficit has included sensory perception, memory, auditory information processing, conceptual reasoning, executive function, and aspects of attention. An alternative approach has been to examine for the presence of a pattern of multiple abilities and deficits across cognitive domains and sensory modalities which characterize autism (Goodman, 1989). Indeed, the use of comprehensive neuropsychological test batteries has delineated a specific pattern of abilities and deficits in autism which has been distinguished from that found in other developmental disorders with similar learning, social, and communication deficits. Although this section considers how individual aspects of cognitive function manifest in HFA, these collective findings and those from studies which simultaneously examined for a pattern of abilities and deficits indicate that a multiple primary deficit model may best characterize the disorder. This pattern of cognitive functions appears to be delineated by a sparing of visuospatial processing ability while encompassing deficits in other neuropsychologic domains that involve the processing of complex information (Minshew, Goldstein, & Siegel, 1997). Thus in HFA individuals, skilled motor, complex language, complex memory, and reasoning abilities have been found to be deficient, while basic attention, sensory perception, simple memory, simple language, and visual–spatial abilities are considered to be intact.

ATTENTION

One aspect of cognitive function which has been used to explain the clinical features of autism is attention. Courchesne *et al.* (1994) propose multiple deficits in the control mechanisms for attention, specifically focusing attention, selective attention, and shifting attention, as the underlying disorder in autism. Impairments in attention have been proposed to be fundamental to autism. In everyday life, input from social, language, and sensorimotor sources varies continuously and needs to be processed efficiently. Attentional deficits may thus interfere with the processing of complex information necessary for problem-solving and engaging in purposeful behavior. An experimental laboratory apparatus has been used to assess behavioral and neurophysiological responses to attention tasks and involves detecting the appearance of a target stimulus (e.g., a light or tone) and responding (i.e., by pressing a button). In the standard reaction time paradigm, subjects maintain attention and focus on stimuli in one sensory modality while ignoring the appearance of stimuli in the other modality. In the modality shift condition subjects are required to rapidly disengage, shift, and reengage their focus of attention from one modality to another in response to target stimuli. Stimuli have been used which alternate in presentation between auditory and visual modalities, color and form, or in visual–spatial location. False-alarm rates, correct target detection, and reaction time are recorded as measures of attentional functions. Using this methodology, Courchesne and colleagues (1994) reported that while autistic subjects were able to perform as well as controls and detect targets in the focus-attention experiment, they had difficulty responding rapidly and accurately in the shift-attention experiments.

The performance of high-functioning autistic and normal control subjects on neuropsychological tests selected to assess multiple elements of attention has failed, however, to distinguish between the groups (Minshew *et al.*, 1997). Subjects performed a variety of attention tasks including recalling and sequencing numbers, crossing out a target letter or number on a page with rows of either letters or numbers, or responding to the appearance of a target stimulus on a computer screen by pressing the keyboard or mouse. A possible explanation for the finding of an impairment in certain aspects of attention in the laboratory is that such studies actually involved a rather complex contingency paradigm which placed demands on decision-making and working memory which are impaired in autism.

SENSORY PERCEPTUAL ABILITIES

Investigation of possible left hemisphere dysfunction in autism was spurred by clinical and experimental findings. Administration of a dichotic listening task and evidence of an uneven cognitive profile highlighted by impaired sequencing ability and language skills in the presence of intact visual–spatial abilities initially suggested localized dysfunction of the cerebral hemispheres in autism (Dawson, 1983; Hoffman & Prior, 1982). Further assessment of sensory perception in autism relied on an assortment of procedures, scales, and instruments to reveal possible differences in the processing of sensory input which could be related to the right and left cerebral hemispheres. Data available from tests sensitive to left hemisphere dysfunction have not, however, supported a distinction for unilateral deficits in autism (Minshew & Rattan, 1992; Sussman & Lewandowski, 1990), for sensory-perceptual abilities of autistic individuals have been found to be intact bilaterally (Rumsey & Hamburger, 1988).

VISUAL–SPATIAL ABILITIES

Intact and superior visual–spatial ability is considered to be characteristic of autism. Evidence of intact visual–spatial ability has been repeatedly found on Wechsler Intelligence Scale subtests which involve manipulating colored blocks to construct a design printed on a card (Block Design) and the assembly of puzzles that do not have a frame (Object Assembly), with autistics generally demonstrating a higher Performance that Verbal or Full Scale IQ. When required to use perceptual organizational ability to copy a complex figure, HFA children have also performed as well as a normal control group in accurately reproducing the design (Prior & Hoffman, 1990). Clinical case reports of HFA individuals have reported the presence of certain savant abilities such as putting puzzle pieces together, explaining mathematical symbols, performing calculus derivatives, and rapidly performing mental calculations, which have also testified to intact and relatively enhanced visual–spatial abilities in autism. In a recent study of the profile of cognitive abilities in autism (Minshew *et al.*, 1997), visual–spatial ability was found to be intact as scores from Block Design and Object Assembly subtests were higher for the autistic group than for controls. These subtests, in combination with results from a task which required identifying the missing detail in a series of pictures of common objects and copying a complex line drawing, failed to adequately classify HFA and control subjects by diagnostic group.

Information about the particular strategy used by autistic individuals to perform visual–spatial tasks, however, is not conclusive. Block Design can be performed by using either an analytic (separating discrete elements from the whole) or holistic (i.e., global pattern recognition) strategy, and it is unclear which problem-solving approach is relied upon in autism. Using a figure-ground test, autistic children were able to successfully locate a target figure (e.g., a triangle) within a larger form (e.g., a carriage) (Shah & Frith, 1983). Examining the strategies used by subjects in this study, it was hypothesized that autistic individuals can recognize the arrangement of elements within a stimulus pattern, but may have relative difficulty visualizing pictorial stimuli when mental manipulation, rotation, twisting, or inversion of the stimulus material is involved, which is perhaps a relatively more complex task. The apparent integrity of visual–spatial abilities in autism in the context of other impaired abilities requires further investigation. It has been hypothesized that brain mechanisms for processing visual–spatial information are different from those involved with complex information processing in other cognitive domains and are spared in autism (Minshew *et al.*, 1997).

MOTOR ABILITIES

Motor incoordination as well as difficulty imitating motor acts, motor praxis, pantomime, and executing goal-directed motor behavior is found in autism (Leary & Hill, 1996; Smith & Bryson, 1994). Use of case history accounts to obtain information about developmental milestones has revealed that autistic individuals were delayed in the acquisition of motor skills and organized motor movements for drawing, dancing, handwriting, performing rhythm activities, and using utensils (DeMyer, Hingten, & Jackson, 1981). A recent study has further distinguished the motor abilities in autism by a dissociation between performing simple motor movements and a skilled sequence of movements (Minshew *et al.*, 1997). On a test of elementary motor skills involving speed and agility, a group of HFA individuals produced the same number of taps as normal controls for either hand. In contrast, tasks with greater information processing demands imposed on motor speed and planning required more time to complete for the group of HFA individuals than for normal controls.

MEMORY

A remarkable rote memory in autism was originally described by Kanner (1943) in his clinical account of young autistic children as they were able to remember poems, prayers, songs, pictures, and discrete facts. Subsequently, a series of neuropsychologic studies which used pictures, written words, and spoken word lists to delineate preserved and impaired aspects of memory in groups of autistic individuals proposed similarities between autism and the amnesic syndrome. In autism, memory impairments were reported to include deficits in free recall and recognition memory, but intact cued recall (Boucher & Warrington, 1976), and reduced primacy but normal recency effect (Boucher, 1981a). Impairments were also evident when retrieval cues had to be encoded at input, while there was intact memory with external cuing (Boucher, 1981b). Further study of rote memory processes on such tasks as the recall of digits presented auditorily or visually and for blocks of different shape and size has found recall under immediate and delayed conditions to be intact in autism (Hermelin & O'Connor, 1975; Prior & Chen, 1976; Rumsey & Hamburger, 1988).

In a study which performed a comprehensive assessment of memory using the California Verbal Learning Test (CVLT) (Delis, Kramer, Kaplan, & Ober, 1987), 27 of 33 test scores did not discriminate between HFA and normal control groups and thus failed to provide evidence of an amnesic disorder or poor memory capability in autism (Minshew & Goldstein, 1993). A few CVLT scores involving the formation of cognitive strategies for organizing information, such as categorizing items into fruits, spices, clothing, and tools, distinguished the groups. Although statistically significant differences were not found, controls did perform better than autistics on 30 of the CVLT scores. This pattern of performance was interpreted as suggesting some inefficiency in verbal memory for the autistics. Recent studies have further refuted the presence of an amnesic disorder in autism. Instead, an impairment in higher-order memory abilities which have a substantial information processing component, with deficits in the use of organizing strategies that support memory, has been reported in HFA (Minshew, Goldstein, & Siegel, 1996). Thus, rather than having difficulty learning new information, the memory difficulty in autism seems to involve storage and retrieval of complex sets of information which can be supported by the use of organizing strategies.

ACADEMIC SKILLS

Hyperlexia, an advanced ability to read words beyond one's mental or chronological age in the absence of any formal instruction and despite severe language delays, is often found in HFA. High-functioning autistic individuals typically demonstrate intact word analysis skills and perform competently and automatically when recognizing letters, establishing letter–sound correspondences, and decoding meaningful or pseudo-words. A dissociation between mechanical reading and the ability to acquire meaning from text has been found in HFA individuals (Rispens & Van Berckelaer, 1991; Whitehouse & Harris, 1984), for despite adequate decoding skills, knowledge of word meaning, an ample fund of information, and fluency in oral reading, reading comprehension is characteristically deficient. Recent investigation of the psychoeducational profile in HFA has confirmed a dissociation between intact procedural and mechanical skills and deficient analytic and interpretive skills (Minshew, Goldstein, Taylor, & Siegel, 1994). Autistic individuals performed relatively well and sometimes better than matched controls on some academic tests, demonstrating an intact ability to recognize letters and read single words, with relatively superior performance to controls in applying phonological skills to read pseudo-words. In contrast, the HFA group exhibited relative weaknesses in demonstrating interpretive and comprehension skills, as they performed more poorly than controls on measures of passage comprehension and following complex oral instructions.

COMMUNICATION AND LANGUAGE

Disturbances in the development of speech and language were emphasized by Kanner (1943) in his description of the syndrome of autism. While 8 of the 11 children described had developed the ability to speak and possessed an extensive vocabulary, the speech of these children was characterized by the absence of sentences which were spontaneously generated, echolalic speech, pronominal reversals, a literal understanding of language, and a lack of reciprocal communication. In severe cases of autism the communication deficits can involve mutism or the absence of speech, while for other autistic individuals there may a course of delayed onset of language, arrested development, or significant regression subsequent to normally emerging language skills. Although the development of language functions in autism can vary, certain intact linguistic abilities in the presence of specific impairments in speech and language appear to be universal.

High-functioning autistic individuals possess basic communicative competence while demonstrating selective linguistic deficits. In HFA there is fluent narrative discourse, intact word knowledge, defining abilities, and grammar, with specific linguistic impairment in the prosodic, semantic, and pragmatic aspects of language. Speech is often meaningful as sentences are produced with correct grammar and syntax. While the speech of high-functioning autistic individuals is comprehensible, the content of conversation is often ritualistic and perseverative, with a focus on topics of special, idiosyncratic interest, elaborated with the use of odd phrases and word choices, and vocalized with peculiar prosodic patterns. Furthermore, deficits in the pragmatic aspects of communication extend beyond unusual vocal inflection and melodic patterns; impairments in the meaning appended to speech by eye contact and the use and recognition of gestural and facial cues are also present. Reciprocal communication is further impaired by the inability of the autistic individual to adopt another's perspective, accurately interpret affective meanings of speech, understand cause and effect or temporal relationships, develop inferences about intervening events, or alternate and exchange speaker–listener roles in conversation. Comprehension of colloquialisms, figures of speech, humor, and jokes is likely to be absent or limited. The HFA individual is thus unable to comprehend subtle messages or the communicative intent of others, with language interpreted literally.

A comprehensive battery of language tests designed to assess communicative competence and contrast simple and complex linguistic abilities was used to evaluate HFA individuals (i.e., Verbal and Full Scale IQ of at least 70) and normal control subjects (Minshew, Goldstein, & Siegel, 1995). The test battery included measures of simple language skills, such as fluency and semantic knowledge, as well as language tests that assessed complex, interpretive, and inferential aspects of language. The mechanical and procedural aspects of language were found to be intact while deficits were present in interpretive skills involving the comprehension of metaphorical or figurative aspects of language.

Abstraction, Reasoning, and Problem-Solving

An impairment in conceptual problem-solving ability in autism has been demonstrated with the Wisconsin Card Sorting Test (WCST) (Heaton, 1981), as HFA men perseverated in responding and were unable to formulate rules or concepts necessary for performing the task (Rumsey, 1985). High-functioning autistic men also performed more poorly than normal controls on verbal and nonverbal reasoning and problem-solving tasks that involved identifying the improbably element of spoken statements, developing an inference about events in a brief story, recognizing an illogical aspect of a picture, and formulating a logical plan to perform a search activity (Rumsey & Hamburger, 1988, 1990). Study of the reasoning ability in HFA children has included problem-solving tasks that require learning, drawing, and recalling the solution to a maze, discovering a rule to complete a sorting task, and copying and remembering a complex design (Prior & Hoffman, 1990). The ability of these children to plan strategies, benefit from feedback about performance, and organize and categorize information was found to be impaired relative to normal controls. High-functioning autistic children required more time and committed more errors in solving the maze, and despite repeatedly making similar errors, were unable to generate novel solutions and persisted in employing ineffective strategies. As HFA children were able to accurately copy the complex design, but experienced difficulty reproducing the figure from memory, the presence of a deficit in organizing the material for efficient storage and retrieval was hypothesized. Further evidence of a deficit in conceptualizing information for verbal problem-solving was reported using the Twenty Questions Procedure (Mosher & Hornsby, 1996). A normal control group was able to formulate questions that conceptually grouped, ordered, and sorted pictures of common objects on a grid, thereby systematically reducing the number of possible answers. In contrast, a group of HFA individuals relied on a guessing procedure that involved random naming of the objects and was unable to form concepts or generate alternative strategies to solve the problem (Minshew, Siegel, Goldstein, & Weldy, 1994). Continued study of the abstraction and reasoning deficit in autism with a variety of instruments has further distinguished between intact performance on tests involving attribute identification or simple rule learning ability, and deficits on concept formation tests which require a self-initiated problem-solving strategy (Goldstein, Siegel, & Minshew, 1995).

EXECUTIVE FUNCTIONS

Executive function abilities involve the use of mental models or internal representations to guide behavior and perform such mental activities as planning, searching, flexibly employing strategies, and inhibiting responses in order to solve problems and attain a goal. An impairment in executive functions has been proposed as a primary deficit in autism (Ozonoff, 1995a). The WCST and Tower of Hanoi (TOH) (Borys, Spitz, & Dorans, 1982) task are among the tests frequently selected to demonstrate executive function deficits in HFA individuals, and have revealed that such impairment persists over a lengthy developmental period (Ozonoff & McEvoy, 1994). High-functioning autistic men have been found to respond to the WCST by making more errors and more perseverative errors than normal controls (Ozonoff, Pennington, & Rogers, 1991; Prior & Hoffman, 1990; Rumsey, 1985). Autistic subjects have been found to be less capable than controls in efficiently solving the TOH task across trials (Ozonoff & McEvoy, 1994; Ozonoff, Pennington, & Rogers, 1991).

INTELLECTUAL FUNCTIONING

The Arthur Point Scale, Seguin Form Board, nonlanguage items from the Stanford–Binet and Merrill–Palmer tests, and other procedures were originally used to assess the intelligence of autistic children (Kanner, 1943). It was reported that these autistic children had "good cognitive potentialities" as they demonstrated such skills as superior word knowledge, recall of remote events and complex patterns and sequences, as well as rote memory of poems and of names. The profile of IQ test scores in autism has been examined to delineate neurocognitive aspects of the disorder and enhance diagnostic accuracy. A characteristic uneven pattern of intelligence test scores has been found in samples of retarded and nonretarded individuals with autism. Using the age-appropriate version of the Wechsler Intelligence Scales, the typical pattern of scores is of an elevated Performance IQ (PIQ) relative to Verbal IQ (VIQ). This profile has been interpreted as being consistent with known patterns of neuropsychological functioning in autism, with the presence of significant language deficits with enhanced visual–spatial and perceptual organization ability accounting for a PIQ greater than VIQ. While the difference between Performance and Verbal IQ scores has been reported to be as large as 29 points, it has also been as small as 1 point, with some studies failing to find any PIQ–VIQ difference. A reversal in the direction of the difference has also been found, with VIQ as many as 9 points higher than PIQ. Inspection of the pattern of subtest scaled scores has indicated a predilection for the highest scaled score to occur on Block Design and the lowest on Comprehension. Repeatedly, Digit Span has been found to be the highest verbal subtest score while Picture Arrangement or Coding/Digit Symbol is the lowest nonverbal subtest score. Deficits in language comprehension, social awareness, and the knowledge of social conventions and their application to pragmatic problem-solving situations have been considered to explain poor scores on Comprehension and Picture Arrangement subtests. Intact rote memory processes are considered to explain the occurrence of high Digit Span scores.

Methodological and psychometric considerations limit the generalizability of such findings to the diagnosis of HFA. In several of these studies sample size was small, the scores of children and adults were combined, and sampling occurred principally from the lower end of the distribution of intelligence scores. Furthermore, the pattern appears to lack sufficient specificity (since it is also observed in normal individuals and other disorders such as the language impaired) and sensitivity (since studies have shown that the magnitude of the VIQ–PIQ difference is small or can occur in the opposite direction).

Indeed, in the largest sample of high-functioning autistic children (n = 45) and adults (n = 36) with Wechsler Full Scale IQ and VIQ ≥ 70 studied to date, the proposed prototypic PIQ > VIQ profile was not found (Siegel, Minshew, & Goldstein, 1996). Furthermore, in examining individual cases there were no significant PIQ–VIQ differences for the majority of subjects. With regard to the profile of subtest scores, results were consistent with previous findings, as each age cohort obtained highest scores on Block Design and lowest scores on Comprehension subtests. Inspection of the profile of scaled scores for individual cases, however, failed to reveal this presumed prototypic pattern in all cases. Although Wechsler IQ scale profiles have been considered by some clinicians and researchers to be characteristic of autistic disorder, caution should be exercised regarding the general applicability of this profile. Furthermore, examination of the profile of IQ test scores for high-functioning individuals with autism suggests that any unique pattern in IQ test scores may be ability-dependent, with VIQ–PIQ differences being smallest or absent in individuals with average-range ability (Rumsey, 1992).

DISTINGUISHING HIGH-FUNCTIONING AUTISM FROM OTHER DEVELOPMENTAL DISORDERS

The DSM-IV (APA, 1994) provides information about developmental course and symptoms to enable the clinician to distinguish between autism and four other categories of pervasive developmental disorder. High-functioning autism may be readily differentiated from Rett's Disorder, as the latter is a degenerative disorder, associated with severe or profound mental retardation, and has been diagnosed only in females. Rett's Disorder is also distinguished by its characteristic pattern of head growth deceleration, stereotyped hand movements, poor coordination of gait or trunk movements, and severe physical impairments which ultimately involve an inability to ambulate or perform purposeful hand skills. Childhood Disintegrative Disorder is distinguished by at least 2 years of apparently normal development, with subsequent marked regression in language, social, play, motor skills, adaptive behavior, or bowel–bladder control by age 3 or 4 years, and the usual presence of moderate to severe mental retardation and loss of speech. Differentiating a third pervasive developmental disorder, Asperger's Syndrome, from autism is discussed below. Finally, a category of Pervasive Developmental Disorder Not Otherwise Specified is coded if impairments in reciprocal social interaction, verbal and non-verbal communication, and stereotyped patterns of behavior, interests, and activities are present but the criteria for autism or a specific pervasive developmental disorder cannot be satisfied.

HIGH-FUNCTIONING AUTISM AND ASPERGER'S SYNDROME

At about the time Kanner (1943) was describing children with autistic disturbance, Asperger provided an account of male children and adolescents with *autistic psychopathy* who demonstrated clinical features that were remarkably similar to autism (i.e., problems related to social interaction, a restricted range of interests, motor control and coordination problems), as well as some differences in presentation (i.e., speech was not delayed, motor deficits were more common, onset occurred somewhat later, and the condition was initially found only in boys). Asperger's case reports provided information about the clinical characteristics of these individuals, which included but were not limited to the following features: (1) normal speech development but with pedantic content, invented words, poor comprehension of humor, and errors in the use of personal pronouns; (2) poor comprehension of the nonverbal aspects of speech, monotonous vocal intonation, limited use of gestures to accompany speech, and a lack of facial expression; (3) an inability to employ social conventions to engage in reciprocal social interactions; (4) an interest in repetitive activities and a resistance to change; (5) impaired motor coordination, stereotyped motor movements, and clumsiness, with odd posture and gait; and (6) an excellent rote memory, particularly for special interests and isolated topics for which facts were learned but with little comprehension (Wing, 1981).

For many years there was little agreement about a definition and criteria for the disorder, which promoted questions and controversy about the validity of Asperger's syndrome (AS) and its relationship to autism (Klin, 1994). Subsequent to an extensive international field trial, AS was included as one of the Pervasive Developmental Disorders in the DSM-IV (APA, 1994). Asperger's syndrome has been sometimes described as HFA because social disabilities similar to autism typically occur in the presence of normal-range intellectual ability, yet this allusion is misleading and inaccurate. Included among those clinical features of AS that are shared with autism are impairments in social interaction including nonverbal behavior which regulates social interaction, and a restricted, repetitive, stereotyped pattern of behavior, interests, and activities. Although Asperger's syndrome shares several clinical features with autism, it is distinguished by the absence of any impairment or delay in early language development, rather than language status later in life; in cognitive development; or in age-appropriate self-help skills, adaptive behavior, and curiosity about the environment (APA, 1994). Average to above-average levels of intellectual ability can be found in both AS and HFA individuals.

In a study that was conducted prior to the development of DSM-IV diagnostic criteria for AS, a comparison of groups of AS and HFA children attempted to identify possible neurocognitive patterns associated with each disorder (Szatmari, Tuff, Finlayson, & Bartolucci, 1990). In the absence of any significant between-group differences, it was concluded that the neurocognitive profiles for the disorders were similar. A standardized assessment of motor impairment involving manual dexterity, ball skills, and balance did not reveal any significant differences between HFA and AS children, thus suggesting that the groups cannot be differentiated with regard to deficits in motor movements or coordination (Manjiviona & Prior, 1995).

Small groups of HFA and AS individuals, of comparable gender, race, FSIQ, and chronological age, have been distinguished by their performance on certain cognitive and neuropsychological measures (Ozonoff, Rogers, & Pennington, 1991). Consistent with the prototypic intelligence test profiles associated with each disorder, individuals with AS demonstrated a higher VIQ than those with HFA, while the HFA group demonstrated a significant 25-point discrepancy between PIQ (101) and VIQ (76). A substantially elevated VIQ expected for the AS group was not found (PIQ of 95, VIQ of 92). Both HFA and AS groups performed poorly on executive function tasks (i.e., WCST, TOH) relative to a control group. The AS group performed better than the HFA group on tasks that required making judgments about mental and physical events and attributions about the mental states of others (i.e., theory of mind problems), and on verbal memory scores (i.e., long-term storage and consistent long-term retrieval). As performance on neuropsychological tests discriminated between the groups and correctly classified a significant proportion of subjects, it was concluded that AS can be distinguished as a subtype of autism.

Employing stringent diagnostic criteria in an effort to maximize possible differences between samples of AS and HFA individuals, a retrospective investigation was conducted of clinical records to distinguish neuropsychological profiles for each group (Klin, Volkmar, Sparrow, Cicchetti, & Rourke, 1995). Large VIQ–PIQ differences were found in the AS group (24 points) but not in the HFA group (2 points). This pattern is the converse of the one found by Ozonoff, Rogers, and Pennington (1991), who reported large differences in the HFA group (25 points) but not in the AS group (3 points). For the HFA group in each study PIQ was greater than VIQ, while there was no consistent pattern regarding the relative value of VIQ and PIQ in AS subjects. In each study, VIQ for the AS group was higher (i.e., 14 and 16 points) than the VIQ for the HFA group.

In the study by Klin *et al.* (1995), raters reviewed and evaluated observational and neuropsychological test data from records to determine the presence of 22 criteria, including neurocognitive assets and deficits used for diagnosis of Nonverbal Learning Disabilities (NLD) syndrome (Rourke, 1989). Half of the neuropsychological assets and deficits significantly discriminated between the HFA and AS groups, yet some were also found frequently in each group. A standard was established which required that at least 5 of 7 possible assets associated with NLD, in combination, with at least 10 of 15 possible neuropsychological deficits, be detected to establish the presence of the NLD profile in any subject. Applying this requirement, 18 of 22 AS individuals satisfied criteria for NLD while only 1 of 19 from the HFA group demonstrated the NLD profile. Findings were considered to suggest that the neuropsychological profiles for HFA and AS are different, that the neuropsychological profile for HFA diverges from that found in NLD, while some neuropsychological assets and deficits of AS and NLD appear to be shared.

AUTISM, LEARNING, AND LANGUAGE DISABILITY

Autism and learning disability (LD) are neurodevelopmental disorders that emerge early in life and occur disproportionately in males. Each disorder is chronic in nature and affects learning, vocational, and social adjustment. An uneven profile of skills is common to each disorder, involving cognitive, language, social, and interpersonal functioning and an assortment of behavioral difficulties related to attention, impulsivity, and organization. Well-defined intact and impaired abilities are also evident in both disorders. Each disorder encompasses a range of intellectual functioning, from borderline to gifted-range ability. Autism and LD are distinguished from each other, however, by numerous qualitative differences in deficient academic, language, communication, and social skills. Furthermore, the neuropsychological deficits found for LD may be mild, restricted, and specific, while those related to autism are characteristically severe, broad, and pervasive.

Perhaps the most striking difference between the psychoeducational profiles in autistics and in individuals with LD is in performance of reading skills. Typically, there is intact word recognition and sometimes a hyperlexic ability for HFA individuals to read single words. Despite a remarkably developed word recognition and oral reading ability, HFA individuals have deficits in understanding and acquiring meaning from text. In contrast to this academic profile, individuals with dyslexia encounter difficulty with the phonological aspects of reading and writing, such as recognizing, identifying, and sequencing letters and sounds, or using decoding skills and word recognition ability to fluently read words, but are nevertheless successful in comprehending text.

Evidence for complementary deficits in the profile of reading skills of HFA and dyslexia is found in a comparative study of children matched for reading age (Frith & Snowling, 1983). Autistic children performed better than dyslexic children in accurately reading nonsense words but experienced relative difficulty comprehending connected discourse and completing sentences which required semantic knowledge. While the failure to read words in isolation in dyslexia has often been characterized as *word blindness*, the term *sentence blindness* was assigned by the authors to refer to problems at the level of reading comprehension experienced by the HFA individuals.

Differing and complementary patterns of cognitive functioning for these neurodevelopmental disorders are also found in a comparison of the academic and neuropsychological test performance of adult autistic and dyslexic males with average range Verbal and Full Scale IQ scores (Rumsey & Hamburger, 1990). There were no differences between the groups on measures of perceptual organization and visual–spatial skills, simple language abilities including word fluency, tactile perception, and motor speed and coordination. The autistic group obtained significantly higher scores on academic skills tests and for the immediate auditory recall and sequencing of a series of digits. The group of dyslexic individuals demonstrated relatively superior verbal and nonverbal reasoning and problem-solving ability, and surpassed the HFA group in the recall of complex visual and verbal information for drawing a design and reciting details from a story after a delay.

With regard to motor, sensory, and perceptual abilities, motor coordination and performance problems are found in LD and HFA, while auditory and visual perceptual processing abilities are characteristically impaired in individuals with dyslexia. In HFA deficits in the imitation of motor acts, motor praxis, and the coordination of complex motor movements are often found. In HFA visual–spatial abilities are intact and may be superior to those of normal individuals.

Despite admonitions to disregard subtest profiles from the Wechsler Intelligence Scales in the diagnosis of clinical disorders including autism (Siegel, Minshew, & Goldstein, 1996) and LD (Kavale & Forness, 1985), the use of various profile analysis techniques has remained a popular practice for clinicians and researchers alike. Examiners are cautioned about the validity of methods used to interpret IQ test profiles for the purpose of diagnosis. If reliance on such approaches is inevitable, then reference to normative IQ profile types and consideration of base rates of occurrence are recommended to help evaluate the uniqueness of any profile, derived scatter index, or difference score. While the presence of a specific pathognomonic profile in autism is not supported, an uneven profile of intellectual abilities has been associated with groups of individuals with HFA.

Some similarities and differences have been reported in the pattern of intellectual ability associated with samples of LD and HFA individuals. In LD a prototypical pattern of intellectual strengths and weaknesses has been defined by using reorganized groupings of Wechsler scale subtests (Bannatyne, 1974). The ability profile which has emerged in LD corresponds to Spatial Ability > Verbal Conceptualization > Sequential Ability > Acquired Knowledge. The combination of subtests with highest scores in LD samples, Spatial Ability, includes Block Design, the subtest on which autistic individuals characteristically perform best. Sequential Ability, a grouping of subtests on which LD individuals typically perform poorly, includes Coding which is performed poorly by autistic individuals, but also Digit Span and Arithmetic which are usually the highest Verbal Scale subtest scores in autism. The presence of deficits on Arithmetic, Coding, Information, and Digit Span (ACID) subtests has also been considered in samples of LD individuals (Greenblatt, Mattis, & Trad, 1991). The ACID profile in LD contrasts with the Wechsler subtest profile found for HFA. Arithmetic and Digit Span subtests are usually performed satisfactorily but Comprehension is repeatedly the lowest subtest score obtained by HFA individuals. Significant differences between age-matched groups of HFA and dysphasic children were reported for WISC-R Comprehension, Vocabulary, Block Design, and Object Assembly subtests (Lincoln, Courchesne, Kilman, Elmasian, & Allen, 1988). Such differences are consistent with the prototypical subtest profile found for each disorder.

Distinctions between the language deficit found in autism and in specific language disorders have been demonstrated using retrospective review of clinical reports and interviews assessing developmental skills and stages of language acquisition, language samples obtained from play and interview assessments, and formal language testing. In autism there is a significant delay in the emergence of language skills with enduring impairments and qualitative deficits in language function. Deficits in the semantic and pragmatic aspects of language including difficulty with prosody, echolalia, and understanding and interpreting gestures and facial expression are characteristic of the disorder. In contrast, learning disabled individuals may experience problems with the articulation of speech, word-finding, use of syntax and grammar in oral expression, and written constructions. Unlike individuals with a motor or sensory aphasia, the autistic individual does not experience difficulty with the production of speech or fluency when semantic category cues are provided, discrimination and comprehension of speech sounds, or word naming and retrieval. The use of language for social purposes, while significantly impaired in autistic individuals, can be contrasted with the language deficit in some learning disabled individuals who effectively use social aspects of language but instead may have difficulty with syntax and grammatical constructions.

A study which used cognitive and language tests to demonstrate a distinction between the language abilities of HFA children and of those with developmental receptive language disorder correctly classified 40 of 42 subjects as being either autistic or dysphasic (Bartak, Rutter, & Cox, 1977). Conversational language samples have also been used to examine differences between HFA children and those with a diagnosis of receptive language disorder by analyzing various indices of pragmatic impairment (Eales, 1993). The HFA group was more impaired on global measures of pragmatic functions; their behavior was characterized by inappropriate utterances, failure to produce an utterance in a context where it could be expected, impairments in communicative intention, and the presence of stereotyped language.

AUTISM AND THE NONVERBAL LEARNING DISABILITY SYNDROME

A particular subtype of learning disability defined by a cluster of neuropsychological assets and deficits involving nonverbal aspects of functioning and presumed right hemisphere dysfunction, known as Nonverbal Learning Disability (NLD) syndrome (Rourke, 1989), provides some engaging comparisons with HFA. Descriptions of the clinical and neuropsychological features of NLD syndrome are also consistent with aspects of developmental learning disability of the right hemisphere (Weintraub & Mesulam, 1983). Weintraub and Mesulam's (1983) patients demonstrated average-range intellectual ability with VIQ typically higher than PIQ, a history of delay in achieving developmental milestones, visuospatial deficits, academic deficits in mathematics and spelling with intact reading ability, shyness and other interpersonal or social difficulties, an inability to establish eye contact, a disturbance in vocal prosody, and impaired use of facial and body gestures. Further review of the neuropsychological profile in NLD, considering similarities and differences with HFA and AS, seems relevant.

Several characteristic neuropsychological aspects of NLD can also be found in autism. These features include the performance of simple motor skills and the rote acquisition of material. Individuals with NLD or HFA also demonstrate intact auditory memory, the capacity to select and sustain attention for simple, repetitive verbal information, and superior word decoding and spelling skills. Neuropsychological deficits are also shared by the disorders, including difficulty performing complex psychomotor acts and processing novel information, and difficulty with concept formation, reasoning, strategy generation, and hypothesis testing. In the domain of academic functioning, reading comprehension skill is impaired relative to word recognition ability, with progressive difficulty for learning subject matter that requires processing complex information. Certain aspects of social and adaptive behavior are also common to both disorders including difficulty responding to new and complex situations; impaired social judgment; difficulty with problem-solving and perception, such as interpreting nonverbal aspects of communication; and hyperactivity in childhood. Children with developmental learning disability of the right hemisphere demonstrate deficits in the interpretation and expression of affect and other interpersonal skills which appear similar to HFA. In both of these disorders communication deficits involve an inability to understand incongruities and humor as well as the functional aspects of language such as pragmatics and prosody.

Despite some similar neuropsychological assets and deficits, the neuropsychological profiles of HFA and NLD diverge in several respects. Although basic language skills initially appear to be delayed in NLD, they ultimately develop rapidly. Deficits in visual–spatial organization ability, bilateral psychomotor coordination, and tactile and visual perception which are present in NLD are not found in HFA.

With regard to academic functioning, there are deficiencies in mechanical arithmetic and mathematics in NLD which are not characteristic of autism. Some characteristic assets of NLD which are not necessarily neuropsychological features of autism include strong receptive language skills and handwriting which is initially poor, but eventually becomes refined. A sharp contrast is drawn between HFA and NLD individuals in the use of language, as NLD individuals sometimes produce speech profusely, relying on language to engage others, acquire information, and minimize anxiety.

Study of reading, mathematics, and spelling skills in HFA has demonstrated similarities and differences with the academic profile proposed for NLD (Minshew, Goldstein, *et al.*, 1994). Other findings from small groups of HFA and NLD children have confirmed similarities and differences in the neuropsychological profiles of these disorders. A group of HFA individuals obtained higher scores than an NLD group on tests of visual–spatial, constructional, tactile–kinesthetic, and psychomotor tasks and lower scores on measures of verbal ability (Casey, Enright, & Gragg, 1996). The NLD group obtained a higher VIQ than PIQ, while the HFA group demonstrated the opposite pattern for VIQ and PIQ scores. Examining the profile of intelligence scale subtests, the highest scaled scores for the NLD group were on Similarities, Vocabulary, and Information, and the lowest scores were on Block Design and Object Assembly. In contrast to this profile and consistent with the classical pattern of IQ subtest profiles reported in autism, scaled scores for Block Design and Object Assembly were highest while Comprehension was the lowest subtest score obtained by the HFA group. Each group demonstrated a strength in word recognition and written spelling, with an unexpected finding of a weakness in performing computational arithmetic in the HFA group.

DESIGNING A NEUROPSYCHOLOGICAL TEST BATTERY

The purposes of assessment include diagnostic and prognostic decision-making, ascertaining current functioning levels, identifying asset and deficit patterns, obtaining information for planning educational and therapeutic interventions, and evaluating progress. Information about both the level and pattern of performance is important as this will indicate if functioning is within normal variation or atypical, as well as reveal relative strengths and weaknesses across various domains of functioning. The neuropsychological profile of autism requires an assessment of both simple and complex abilities across multiple domains of function, rather than a primary deficit in a single domain. Academic achievement tests, however, typically yield composite, cluster scores or quotients that summarize subject performance in a general academic domain or skill area. Reliance on such global scores can be misleading and fail to accurately portray strengths and deficits in HFA. Considering that the neurocognitive basis of autism involves a deficit in complex information processing, rather than a deficit in a single domain such as attention or memory, tests need to be selected which evaluate skills across domains and assess simple, mechanical, and procedural as well as complex, analytic, and interpretive abilities.

Intellectual, chronological, and developmental age factors are an essential context for developing a test battery and evaluating test results. For example, the performance of preschool-age HFA children on intelligence tests has been found to be relatively stable, for while there is some decline in IQ test scores compared with lower-functioning autistic children, a shift in range of intellectual functioning is less likely to occur with age (Lord & Schopler, 1989). Furthermore, there is evidence for some age-related changes in cognitive function in both normal and autistic individuals. For example, developmental changes in the performance by normal children and adolescents on tests of frontal lobe functioning have been described (Levin *et al.*, 1991). Age differences in performance of autistic individuals on tests of academic skills have also been delineated. In matched samples of autistic males and normal controls which were split at a median age of 14 years, some academic skills were found to remain at or above average for the autistic (e.g., single word reading), some skills did not attain average levels (e.g., comprehending complex oral direction), and other skills were intact during early school years, but were not maintained and declined in the group of older autistic subjects (i.e., reading comprehension) (Goldstein, Minshew, & Siegel, 1994). Finally, clinicians must remember that the natural course of the disorder is for improvement. Language and cognitive impairments are most severe in early childhood, while there may be minimal to substantial improvement for about half of all autistic individuals after the preschool years.

ASSESSING INTELLECTUAL FUNCTIONING

Administration of the age-appropriate version of the Wechsler Intelligence Scales can provide normative information about general level of intellectual function, detect possible mental retardation, ascertain patterns of verbal and nonverbal abilities, and generate hypotheses about cognitive strengths and weakness. Intelligence test scores should not be relied on exclusively in diagnostic decision-making, however, as various patterns and range in performance can be found in HFA (Siegel, Minshew, & Goldstein, 1996). Several limitations regarding the use of intelligence test scores in assessing HFA are relevant. First, the emergence of the prototypical IQ profile in autism appears to be severity-dependent and is likely to be absent in HFA. Secondly, Verbal IQ scores will not provide information about early language development, be sensitive to the nature of the language deficits in autism, or provide sufficient information for distinguishing between Asperger's Disorder and HFA. Thirdly, the Comprehension subtest will not adequately assess the nature of the social deficits in autism. Furthermore, while IQ test scores can characterize general intellectual level in HFA, these scores can be misleading, as adaptive functioning is significantly impaired relative to intellectual ability.

TESTING ACADEMIC SKILLS

The discrepancy between reading recognition and comprehension skills in autism requires that measures of word analysis and recognition as well as word and passage comprehension be administered. Included among well-standardized and normed tests of reading which can be used to evaluate diverse reading skills are the Kaufman Test of Educational Achievement Reading Decoding and Reading Comprehension subtests (Kaufman & Kaufman, 1985); Woodcock Reading Mastery Tests–Revised Word Recognition, Word Attack, and Passage Comprehension subtests (Woodcock, 1987); and the corresponding subtests from the Woodcock–Johnson Psycho-Educational Battery–Revised (WJ-R) (Woodcock & Johnson, 1989). To examine for intact mechanical skills in the presence of deficient analytic and interpretive abilities in the domain of mathematics, an assessment of both computational arithmetic and mathematical problem-solving abilities is necessary. The KeyMath

Revised (Connolly, 1988) or subtests from the WJ-R Tests of Achievement can provide a comprehensive assessment of mathematical abilities, including basic quantitative concepts, numerical operations, and mathematical applications, and help to distinguish between HFA and NLD.

ASSESSING MULTIPLE ASPECTS OF LANGUAGE

The development and pattern of linguistic competence found in autism require that a comprehensive evaluation of language functions be conducted, in order to establish diagnosis and provide information for planning interventions. This evaluation should include historical information provided by a caregiver about early language development, discourse samples, and the administration of standardized measures to assess current linguistic functions. Neuropsychological study of HFA has demonstrated a dissociation between intact formal language skills and impaired higher-order language comprehension and interpretive abilities. Evidence of possibly intact linguistic abilities should be documented with measures of simple, mechanical aspects of language. For example, verbal production and fluency tests can be administered. Word knowledge can be assessed with confrontation naming and picture vocabulary tests or the Vocabulary subtest from the Wechsler Intelligence Scales, while subtests from the Clinical Evaluation of Language Fundamentals–Revised (Semel, Wiig, & Secord, 1987) can be used to evaluate receptive and expressive language, syntax, semantics, and use of grammatical constructions.

Despite the presence of intact fluency, grammar, and semantic knowledge in HFA, as assessment needs to be conducted of complex aspects of language, including comprehension and interpretive abilities. The Token Test (Boller & Vignolo, 1966) assesses comprehension of oral directions with knowledge of relational terms, verbs, and descriptions of color, size, and shape. The Oral Directions subtest from the Detroit Tests of Learning Aptitude-2 (DTLA-2; Hammill, 1985) provides similar information about the ability to comprehend verbal information and the extent to which working memory is involved in executing complex oral directions. Information about the ability to comprehend metaphorical and figurative aspects of language, which is characteristically deficient in HFA, can be obtained from the Test of Language Competence–Expanded Edition (Wiig & Secord, 1989).

In the domain of written language, an assessment needs to encompass a variety of skills in isolation, including the ability to spell words from dictation; knowledge of mechanics such as grammar, style, tense, usage, and rules of capitalization and punctuation; as well as the ability to employ these skills in expository writing. The WJ-R Tests of Achievement can provide information about the simple, mechanical aspects of these skills, as there are subtests for motoric output (Handwriting), written spelling, punctuation, capitalization, and usage (Dictation, Proofing), and for generating connected discourse (Writing Samples, Writing Fluency) in a structured format. A sample of written language is also important for assessing whether knowledge of the formal rules of writing can be applied automatically in composition writing and if coherent, organized prose expressing a unified theme can be produced in a relatively unstructured context.

ASSESSING ATTENTIONAL PROCESSES

A comprehensive model of attention has proposed the coordination of four component processes: focus-execute, sustain, encode, and shift (Mirsky, Anthony,

Duncan, Ahearn, & Kellam, 1991). Performance on certain tests has been reported to reflect these different elements of attention and thus provides one approach for the assessment of these processes. The Stroop, Letter Cancellation, and Trail Making tests load on the focus–execute element of attention, Continuous Performance Test (CPT) on sustain, Digit Span on encode, and the WCST on the shift component. Number and letter cancellation tasks will also assess simple aspects of attention, while vigilance can be evaluated further with complex tests of attention. For example, a version of the CPT allows the examiner to individually modify and customize paradigm parameters including target stimuli, number of trails, inter-stimulus interval, and display time (Conners & MHS staff, 1994). Information about the kind of behavioral attentional deficits that affect classroom learning and adaptive functioning at home and in the community needs to be collected by using instruments such as the Conners' Teacher and Parent Rating Scales (Conners, 1990).

TESTING OF VISUAL–SPATIAL ABILITIES

Evidence of intact visual–spatial ability can be obtained from the Performance Scale subtests of the Wechsler Intelligence Scales and Block Design subtest in particular, as well as the copy score from various tests of visual–motor perception and production. Scores can be compared with normative data for the presence and accuracy of those elements of the figure which are reproduced and the strategy employed for completing the design. The rendering of figures from normed motor copying tasks should be relatively easy for HFA individuals to execute and should provide further evidence of intact visual–spatial abilities.

TESTING OF SENSORY AND MOTOR FUNCTIONS

Simple sensory functions are anticipated to be intact in HFA while skilled motor performance and processing of complex sensory information are expected to be impaired. Activities which require the examinee to produce gestures, imitate movements or vocal production, and respond to commands by pantomiming the use of common objects (e.g., hammering a nail) or performing symbolic gestures (e.g., hitchhiking) can be used to assess motor coordination deficits and apraxia. Instruments for assessing sensory function in autism include the Luria–Nebraska Neuropsychological Battery–Children's Revision (Golden, Purisch, & Hammeke, 1985) and tests from the Halstead–Reitan Neuropsychological Test Battery (Reitan & Wolfson, 1993). A standardized evaluation of motor impairment in children between the ages of 5 and 12 that has been successfully administered with HFA individuals is the Test of Motor Impairment–Henderson Revision (TOMI-H) (Manjiviona & Prior, 1995). The TOMI-H provides normative information for the assessment of motor functioning at four age periods, by using an increasingly difficult set of eight tasks which involve such abilities as coordinating both hands for a task, aiming and catching a ball, and maintaining balance under different conditions. The Bar Game, which employs a half-black, half-white rod and red and blue grooved discs, was designed as an experimental measure to examine the grip position of the dominant hand used by autistic children to place the rod in each disc (Hughes, 1996). When the starting position of the rod is varied, a subject would be disposed to use either an overhand or underhand grip to place the rod in the disc, with the planning ability of the autistic subjects assessed by determining if transfers were performed and the comfortable, overhand grip employed. On this goal-directed motor task autistic subjects demonstrated difficulty in planning and executing the motor sequences.

ASSESSMENT OF MEMORY

Episodic and semantic memory, and immediate and delayed recall of simple information, are typically intact in HFA, while deficits are found for retrieval of complex information requiring self-initiation of an organizing strategy. While the CVLT (Delis *et al.*, 1987) comprehensively assesses memory functions across various conditions including recognition, immediate and delayed recall, and free and cued recall, with scores for rate of learning, errors of intrusion, and repetition, and evidence of proactive and retroactive interference, deficits in most aspects of memory assessed with the CVLT are not necessarily found in HFA individuals. Instead, an assessment of working memory needs to be conducted to detect possible memory impairments. The DTLA-2 Oral Directions subtest is an example of a task which imposes significant demands on working memory. The Oral Directions subtest requires the examinee to listen to a series of increasingly complex instructions and after a brief delay, draw lines in a prescribed sequence to connect, circle, underline, or mark a set of pictures which contain common objects. The Rey Complex Figure (Osterrieth, 1944) delayed recall score will also provide information about the ability to process and retrieve complex visuospatial information, while the Wechsler Memory Scale Story Recall (Wechsler, 1987) can assess the recall of complex semantic information that requires an organizing strategy.

ASSESSING REASONING, PROBLEM-SOLVING, AND EXECUTIVE FUNCTIONS

Several tests and procedures have been used effectively to delineate an impairment in conceptual thinking and verbal and nonverbal problem-solving in HFA individuals. The Twenty Questions Procedure is a modification of the popular parlor game which requires the subject to detect a preselected picture from a grid of 42 common objects by asking the smallest number of questions which can be answered "yes" or "no" by the examiner. Questions formulated by subjects are classified by type, depending on whether (1) a feature, function, or location held in common by several objects is referenced ("Is it an animal?"), thereby eliminating other pictures (i.e., a constraint-seeking question); (2) an attribute is selected which refers exclusively to one picture ("Is it the thing that shines in the sky?") (i.e., a pseudoconstraint question); or (3) the object is simply named ("Is it the sun?") (i.e., a hypothesis scanning question). Picture Absurdities from the Stanford–Binet Intelligence Scale–Fourth Edition (Thorndike, Hagen, & Sattler, 1986) and the Test of Problem Solving (Zachman, Jorgensen, Huisingh, & Barrett, 1984) are measures of social awareness, reasoning, and problem-solving that use pictorial and verbal formats, while the solution of matrices is an example of a nonverbal conceptual reasoning task that requires examinees to evaluate configural and spatial relationships and select a design from several available choices to complete a pattern.

The WCST and TOH have revealed relatively enduring deficits in executive function abilities of HFA individuals (Ozonoff & McEvoy, 1994). The WCST assesses deficits in forming and maintaining concepts once established, recognizing shifts in category, and abandoning old response sets and flexibly initiating new classification procedures in response to feedback and changing perceptual stimu-

li. The cognitive processing demands of the WCST depict it as an attribute identification task (i.e., form, color, number) which is sensitive to perseverative responses (Perrine, 1993). This is a relevant distinction in assessing HFA, as differences can be found in the performance of HFA individuals on reasoning tests
that require simple rule-learning and those that require self-initiated concept
formation or strategy formation. A difficulty in shifting cognitive set was proposed, however, to explain the greater frequency of perseverative errors by
HFA individuals than normal controls on the WCST (Ozonoff, Pennington, &
Rogers, 1991). Use of the WCST in a sample of HFA individuals and normal
controls has in some instances failed to distinguish these groups (Goldstein *et al.*,
1995). The absence of impaired WCST performance in some groups of HFA
individuals may be attributed to the finding that measures of abstraction and
reasoning are correlated with general level of intellectual ability. The TOH,
which requires forming an internal representation of the intermediate position
(goal state) of the problem, has also been used for demonstrating executive
function ability in HFA individuals (Ozonoff, Pennington, & Rogers 1991). Different measures have been used to assess executive function in very young children with autism, with deficits also found in planning ability and flexibly shifting strategies (McEvoy, Rogers, & Pennington, 1993).

OUTCOMES, ISSUES, AND FUTURE DIRECTIONS

The precision of neuropsychological tests to assess cognitive impairments in
HFA needs improvement. First, assessment procedures need to be developed
which are sensitive and specific to autism and other pervasive developmental disorders. Second, tests are needed which evaluate unitary rather than multiple cognitive processes and functions, demonstrate adequate discriminant validity, and thus
yield pure measures of abilities in HFA individuals. Thirdly, neuropsychological
tests are needed which are sufficiently demanding and complex to distinguish
information processing abilities in HFA individuals from those in normal controls
of comparable intellectual level. Finally, tests need to be selected which impose
minimal demands on the linguistic abilities of HFA individuals.

A proliferation of computer-administered neuropsychological tests, including
versions of the WCST (Heaton & PAR Staff, 1993), CPT (Conners & MHS Staff,
1994), and TOH (Davis, Bajszar, & Squire, 1994), has introduced questions about
the reliability and validity of these clinical evaluation procedures. Reliance on a
computerized assessment format is intuitively appealing for use in evaluating individuals with pervasive developmental disorders because of the decreased demands
on social interaction and because of the inherent similarities to playing video games
which are typically enjoyed by HFA individuals. Further evaluation of the validity
of such procedures with HFA individuals is warranted, however, considering the
finding that a computerized and a standard administration of the WCST in groups
of HFA individuals were not equivalent (Ozonoff, 1995b).

Neuropsychological tests used for delineating the ability and deficit pattern in
HFA should lead directly to planning efficacious educational and vocational interventions. For example, knowledge of deficits in neuromotor development, gesture
imitation, and the coordination of complex motor activities which would impede
autistic children from acquiring sign language skills can guide educational planning. The pattern of academic abilities in HFA also has implications for educational

intervention, considering the divergent profile of reading abilities in autism and how this psychoeducational pattern is distinct from that of dyslexia and NLD. The dyslexic individual may demonstrate impaired visual and auditory perceptual processing abilities required for phonetic analysis and word recognition, but without concomitant deficits in word or passage comprehension. Specific implications emerge for teaching reading skills to autistic students, for unlike instruction designed for the dyslexic individual, there does not need to be training for phonological deficits such as sound–symbol relationships, remediation of phonetic or structural analysis skills, or teaching the orthographic features of words, syllabication, or blending. Rather than improving phonological processing by employing visual, auditory, kinesthetic–tactile, or combined approaches to teach reading, instruction for the autistic individual should be directed at developing comprehension skills by introducing strategies which can be used to interpret text. Unlike psychoeducational interventions planned for NLD individuals which employ highly systematic and concrete procedures, but rely considerably on the use of verbal information to explain mechanical arithmetic, instruction for HFA individuals should minimally involve linguistic information or the processing of language (Siegel, Goldstein, & Minshew, 1996).

As autism is a disorder encompassing a continuum of severity, there is some advantage for continuing study of those individuals with autism uncomplicated by mental retardation. Research findings have been successful in delineating the nature of the cognitive deficits in HFA, and have provided a profile of cognitive functions which is useful in supporting differential diagnosis. An expanded recognition of the group of HFA individuals has also occurred in clinic, school, and vocational training settings, which has generated an increase in the referral of individuals with pervasive developmental disorders and HFA in particular. Conducting a comprehensive assessment using standardized procedures and instruments has simultaneously improved diagnostic accuracy and succeeded in reliably distinguishing those with HFA from lower-functioning autistic individuals and from individuals with other developmental disorders or mental disorders such as schizophrenia. Continued reliance on measures that assess both simple and complex abilities, across cognitive domains, is crucial for delineating a valid profile of performance in HFA. Analysis of the methods used by HFA individuals in successfully performing tasks, as well as the particular strategies which are deficient or absent for accomplishing other problems, will provide information to further elaborate this profile of functioning and guide intervention.

REFERENCES

American Psychiatric Association. (1994). *Diagnostic and statistical manual of mental disorders* (4th ed.). Washington, DC: Author.

Bannatyne, A. (1974). Diagnosis: A note on recategorization of the WISC scaled scores. *Journal of Learning Disabilities, 7,* 272–274.

Bartak, L., Rutter, N., & Cox, A. (1977). A comparative study of infantile autism and specific developmental receptive language disorders. III. Discriminant function analysis. *Journal of Autism and Childhood Schizophrenia, 7,* 383–396.

Boller, F., & Vignolo, L. A. (1966). Latent sensory aphasia in hemisphere-damaged patients. An experimental study with the Token Test. *Brain, 89,* 132–133.

Borys, S. V., Spitz, H. W., & Dorans, B. A. (1982). Tower of Hanoi performance of retarded young adults and non-retarded children as a function of solution length and goal state. *Journal of Experimental Child Psychology, 33,* 87–110.

Boucher, J. (1981a). Memory for recent events in autistic children. *Journal of Autism and Developmental Disorders, 11,* 293–302.

Boucher, J. (1981b). Immediate free recall in early childhood autism: Another point of behavioural similarity with the amnesic syndrome. *British Journal of Psychology, 72,* 211–215.

Boucher, J., & Warrington, E. K. (1976). Memory deficits in early infantile autism: Some similarities to the amnesic syndrome. *British Journal of Psychiatry, 67,* 73–87.

Casey, J. E., Enright, C. A., & Gragg, M. M. (1996). High-functioning autism and the nonverbal learning disabilities syndrome: A comparison of neuropsychological and academic achievement functioning. *Journal of the International Neuropsychological Society, 2,* 40.

Conners, C. K. (1990). *Manual for Conners' Rating Scales.* Toronto: MHS.

Conners, C. K., & MHS Staff. (1994). *CPT Conners' Continuous Performance Test computer program.* Toronto: MHS.

Connolly, A. J. (1988). *Manual KeyMath Revised: A diagnostic inventory of essential mathematics.* Circle Pines, MN: American Guidance Service.

Courchesne, E., Lincoln, A. J., Townsend, J. P., James, H. E., Akshoomoff, N. A., Saitoh, A., Yeung-Courchesne, R., Egaas, B., Press, G. A., Haas, R. H., Murakami, J. W., & Schreibman, L. (1994). A new finding: Impairment in shifting attention in autistic and cerebellar patients. In S. H. Broman & J. Grafman (Eds.), *Atypical deficits in developmental disorders: Implications for brain function* (pp. 101–317). Hillsdale, NJ: Lawrence Erlbaum Associates.

Davis, H., Bajszar, G., & Squire, L. R. (1994). *Colorado Neuropsychology Tests Version 2.0c.* Colorado Springs, CO: CNT.

Dawson, G. (1983). Lateralized brain dysfunction in autism: Evidence from the Halstead–Reitan neuropsychological battery. *Journal of Autism and Developmental Disorders, 13,* 269–286.

Delis, D. C., Kramer, J. H., Kaplan, E., & Ober, B. A. (1987). *California Verbal Learning Test: Adult version.* San Antonio, TX: Psychological Corporation.

DeMyer, M. K., Hingten, J. N., & Jackson, R. K. (1981). Infantile autism reviewed: A decade of research. *Schizophrenia Bulletin, 7,* 388–451.

Eales, M. J. (1993). Pragmatic impairments in adults with childhood diagnoses of autism or developmental receptive language disorder. *Journal of Autism and Developmental Disorders, 23,* 593–617.

Freeman, B. J., Ritvo, E. R., Schroth, P. C., Tonick, I., Guthrie, D., & Wake, L. (1981). Behavioral characteristics of high- and low-IQ autistic children. *American Journal of Psychiatry, 138,* 25–29.

Frith, U., & Snowling, M. (1983). Reading for meaning and reading for sound in autistic and dyslexic children. *British Journal of Developmental Psychology, 1,* 329–342.

Golden, C. J., Purisch, A. D., & Hammeke, T. A. (1985). *Luria–Nebraska Neuropsychological Battery.* Los Angeles: Western Psychological Services.

Goldstein, G., Minshew, N. J., & Siegel, D. J. (1994). Age differences in academic achievement in high-functioning autistic individuals. *Journal of Clinical and Experimental Neuropsychology, 16,* 617–680.

Goldstein, G., Siegel, D. J., & Minshew, N. J. (1995). Abstraction and problem solving in autism: Further characterization of the fundamental deficit. *Archives of Clinical Neuropsychology, 10,* 335.

Goodman, R. (1989). Infantile autism: A syndrome of multiple primary deficits? *Journal of Autism and Developmental Disorders, 19,* 409–424.

Greenblatt, E., Mattis, S., & Trad, P. V. (1991). The ACID pattern and the freedom of distractibility factor in a child psychiatric population. *Developmental Neuropsychology, 7,* 121–130.

Hammill, D. D. (1985). *Detroit Tests of Learning Aptitude-2.* Austin, TX: Pro-Ed.

Heaton, R. K. (1981). *Wisconsin Card Sorting Test (WCST).* Odessa, FL: Psychological Assessment Resources.

Heaton, R. K., & PAR Staff (1993). *WCST: Computer Version-2 Research edition.* Odessa, FL: Psychological Assessment Resources.

Hermelin, B., & O'Connor, N. (1975). The recall of digits by normal, deaf and autistic children. *British Journal of Psychology, 66,* 203–209.

Hoffman, W. L., & Prior, M. (1982). Neuropsychological dimension of autism in children: A test of the hemispheric dysfunction hypothesis. *Journal of Clinical Neuropsychology, 4,* 27–41.

Hughes, C. (1996). Brief report: Planning problems in autism at the level of motor control. *Journal of Autism and Developmental Disorders, 26,* 99–107.

Kanner, L. (1943). Autistic disturbances of effective contact. *Nervous Child, 2,* 217–250.

Kanner, L., Rodriguez, A., & Ashenden, B. (1972). How far can autistic children go in matters of social adaptation? *Journal of Autism and Developmental Disorders, 2,* 9–33.

Kaufman, A. S., & Kaufman, N. L. (1985). *Manual for the Kaufman Test of Educational Achievement-Comprehensive Form (K-TEA).* Circle Pines, MN: American Guidance Service.

Kavale, K., & Forness, S. (1985). Learning disability: A pseudoscience. In *The science of learning disabilities* (pp. 5–38). San Diego, CA: College-Hill.

Klin, A. (1994). Asperger syndrome. *Child and Adolescent Psychiatric Clinics of North America, 3,* 131–148.

Klin, A., Volkmar, F. R., Sparrow, S. S., Cicchetti, D. V., & Rourke, B. P. (1995). Validity and neuropsychological characterization of Asperger syndrome: Convergence with nonverbal learning disabilities syndrome. *Journal of Child Psychology, Psychiatry, and Allied Disciplines, 36,* 1127–1140.

Leary, M. R., & Hill, D. A. (1996). Moving on: Autism and movement disturbance. *Mental Retardation, 34,* 39–53.

Levin, H. S., Culhane, K. A., Hartmann, J., Evankovich, K., Mattson, A. J., Harward, H., Ringholz, G., Ewing-Cobbs, L., & Fletcher, J. M. (1991). Developmental changes in performance on tests of purported frontal lobe functioning. *Developmental Neuropsychology, 7,* 377–395.

Lincoln, A. J., Courchesne, E., Kilman, B. A., Elmasian, R., & Allen, M. (1988). A study of intellectual abilities in high-functioning people with autism. *Journal of Autism and Developmental Disorders, 18,* 505–524.

Lord, C., & Schopler, E. (1989). The role of age at assessment, developmental level, and test in the stability of intelligence scores in young autistic children. *Journal of Autism and Developmental Disorders, 19,* 483–499.

Manjiviona, J., & Prior, M. (1995). Comparison of Asperger syndrome and high-functioning autistic children on a test of motor impairment. *Journal of Autism and Developmental Disorders, 25,* 23–39.

McEvoy, R. E., Rogers, S. J., & Pennington, B. F. (1993). Executive function and social communication deficits in young autistic children. *Journal of Child Psychology and Psychiatry and Allied Disciplines, 34,* 563–578.

Minshew, N. J., & Goldstein, G. (1993). Is autism an amnesic disorder? Evidence from the California Verbal Learning Test. *Neuropsychology, 7,* 1–8.

Minshew, N. J., Goldstein, G., & Siegel, D. J. (1995). Speech and language in high-functioning autistic individuals. *Neuropsychology, 9,* 255–261.

Minshew, N. J., Goldstein, G., & Siegel, D. J. (1996). Complex memory impairments in autism: Evidence of cortical dysfunction. *Annals of Neurology, 40,* 333.

Minshew, N. J., Goldstein, G., & Siegel, D. J. (1997). Neuropsychologic functioning in autism: Profile of a complex information processing disorder. *Journal of the International Neuropsychological Society, 3,* 303–316.

Minshew, N. J., Goldstein, G., Taylor, H. G., & Siegel, D. J. (1994). Academic achievement in high functioning autistic individuals. *Journal of Clinical and Experimental Neuropsychology, 16,* 261–270.

Minshew, N. J., & Rattan, A. I. (1992). The clinical syndrome of autism. In S. J. Segalowitz & I. Rapin (Eds.), *Handbook of neuropsychology: Vol. 7. Child neuropsychology* (pp. 65–89). Amsterdam: Elsevier.

Minshew, N. J., Siegel, D. J., Goldstein, G., & Weldy, S. (1994). Verbal problem solving in high functioning autistic individuals. *Archives of Clinical Neuropsychology, 9,* 31–40.

Mirsky, A. F., Anthony, B. J., Duncan, C. C., Ahearn, M. B., & Kellam, S. G. (1991). Analysis of the elements of attention: A neuropsychological approach. *Neuropsychology Review, 2,* 109–145.

Mosher, F. A., & Hornsby, J. R. (1966). On asking questions. In J. S. Bruner, R. R. Olver, & P. M. Greenfield (Eds.), *Studies in cognitive growth* (pp. 86–102). New York: John Wiley & Sons.

Osterrieth, P. A. (1944). Le test de copie d'une figure complexe. *Archives de Psychologie, 30,* 206–356.

Ozonoff, S. (1995a). Executive functions in autism. In E. Schopler & G. B. Mesibov (Eds.), *Learning and cognition in autism* (pp. 199–219). New York: Plenum Press.

Ozonoff, S. (1995b). Reliability and validity of the Wisconsin Card Sorting Test in studies of autism. *Neuropsychology, 9,* 491–500.

Ozonoff, S., & McEvoy, R. (1994). A longitudinal study of executive function and theory of mind development in autism. *Development and Psychopathology, 6,* 415–431.

Ozonoff, S., Pennington, B. F., & Rogers, S. J. (1991). Executive function deficits in high-functioning autistic individuals: Relationship to theory of mind. *Journal of Child Psychology and Psychiatry and Allied Disciplines, 32,* 1081–1105.

Ozonoff, S., Rogers, S. J., & Pennington, B. F. (1991). Asperger's syndrome: Evidence of an empirical distinction from high-functioning autism. *Journal of Child Psychology and Psychiatry, 32,* 1107–1122.

Perrine, K. (1993). Differential aspects of conceptual processing in the Category test and Wisconsin Card Sorting Test. *Journal of Clinical and Experimental Neuropsychology, 15,* 461–473.

Piven, J., Harper, J., Palmer, P., & Arndt, S. (1996). Course of behavioral change in autism: A retrospective study of high-IQ adolescents and adults. *Journal of the American Academy of Child and Adolescent Psychiatry, 35,* 523–529.

Prior, M. R., & Chen, C. S. (1976). Short-term and serial memory in autistic, retarded, and normal children. *Journal of Autism and Developmental Disorders, 6,* 121–131.

Prior, M., & Hoffman, W. (1990). Brief report: Neuropsychology testing of autistic children through an exploration with frontal lobe tests. *Journal of Autism and Developmental Disorders, 4,* 581–590.

Reitan, R. M., & Wolfson, D. (1993). *The Halstead–Reitan Neuropsychological Test Battery: Theory and clinical interpretation.* Tucson, AZ: Neuropsychology Press.

Rispens, J., & Van Berckelaer, I. A. (1991). Hyperlexia: Definition and criterion. In J. M. Joshi (Ed.), *Written language disorders* (pp. 143–163). Netherlands: Kluwer Academic Publishers.

Ritvo, E. R., Freeman, B. J., Pingree, C., Mason-Brothers, A., Jorde, L., Jenson, W. R., McMahon, W. M., Petersen, P. B., Mo, A., & Ritvo, A. (1989). The UCLA-University of Utah epidemiologic survey of autism: Prevalence. *American Journal of Psychiatry, 146,* 194–199.

Rourke, B. P. (1989). The NLD syndrome: Characteristics, dynamics, and manifestations. In *Nonverbal learning disabilities: The syndrome and the model* (pp. 80–110). New York: Guilford.

Rumsey, J. M. (1985). Conceptual problem-solving in highly verbal, nonretarded autistic men. *Journal of Autism and Developmental Disorders, 15,* 23–36.

Rumsey, J. M. (1992). Neuropsychological studies of high-level autism. In E. Schopler & G. B. Mesibov (Eds.), *High-functioning individuals with autism* (pp. 41–64). New York: Plenum Press.

Rumsey, J. M., & Hamburger, S. D. (1988). Neuropsychological findings in high-functioning men with infantile autism, residual state. *Journal of Clinical and Experimental Neuropsychology, 10,* 201–221.

Rumsey, J. M., & Hamburger, S. D. (1990). Neuropsychological divergence of high-level autism and severe dyslexia. *Journal of Autism and Developmental Disorders, 20,* 155–169.

Semel, E., Wiig, E. H., & Secord, W. (1987). *Clinical evaluation of language fundamentals-Revised administration manual.* San Antonio, TX: Psychological Corporation.

Shah, A., & Frith, U. (1983). An islet ability in autistic children: A research note. *Journal of Child Psychology and Psychiatry and Allied Disciplines, 24,* 613–620.

Siegel, D. J., Goldstein, G., & Minshew, N. J. (1996). Designing instruction for the high functioning autistic individual. *Journal of Developmental and Physical Disabilities, 8,* 1–19.

Siegel, D. J., Minshew, N. J., & Goldstein, G. (1996). Wechsler IQ profiles in diagnosis of high-functioning autism. *Journal of Autism and Developmental Disorders, 26,* 389–406.

Smalley, S. L., Asarnow, R. F., & Spence, M. A. (1988). Autism and genetics: A decade of research. *Archives of General Psychiatry, 45,* 953–961.

Smith, I. M., & Bryson, S. E. (1994). Imitation and action in autism: A critical review. *Psychological Bulletin, 116,* 259–273.

Sussman, K., & Lewandowski, L. (1990). Left-hemisphere dysfunction in autism: What are we measuring? *Archives of Clinical Neuropsychology, 5,* 137–146.

Szatmari, P., Tuff, L., Finlayson, M. A., & Bartolucci, G. (1990). Asperger's syndrome and autism: Neurocognitive aspects. *Journal of Autism and Developmental Disorders, 29,* 130–136.

Thorndike, R. L., Hagen, E. P., & Sattler, J. M. (1986). *Technical manual for the Stanford-Binet Intelligence Scale: Fourth Edition.* Chicago, IL: Riverside.

Venter, A., Lord, C., & Schopler, E. (1992). A follow-up study of high-functioning autistic children. *Journal of Child Psychology and Psychiatry and Allied Disciplines, 3,* 489–507.

Volkmar, F. R., Szatmari, P., & Sparrow, S. S. (1993). Sex differences in pervasive developmental disorders. *Journal of Autism and Developmental Disorders, 23,* 579–591.

Wechsler, D. (1987). *Wechsler Memory Scale-Revised manual.* San Antonio, TX: Psychological Corporation.

Weintraub, S., & Mesulam, M. M. (1983). Developmental learning disabilities of the right hemisphere: Emotional, interpersonal, and cognitive components. *Archives of Neurology, 40,* 463–468.

Whitehouse, D., & Harris, J. C. (1984). Hyperlexia in infantile autism. *Journal of Autism and Developmental Disorders, 14,* 281–289.

Wiig, E. H., & Secord, W. (1989). *Test of Language Competence-Expanded edition.* San Antonio, TX: Psychological Corporation.

Wing, L. (1981). Asperger's syndrome: A clinical account. *Psychological Medicine, 11,* 115–129.

Woodcock, R. W. (1987). *Woodcock Ready Mastery Tests-Revised.* Circle Pines, MN: American Guidance Service.

Woodcock, R. W., & Johnson, M. B. (1989). *Examiner's manual Woodcock–Johnson Psycho-Educational Battery-Revised Tests of Achievement.* Allen, TX: DLM Teaching Resources.

Zachman, L., Jorgensen, C., Huisingh, R., & Barrett, M. (1984). *Examiner's manual Test of Problem Solving TOPS.* East Moline, IL: LinguiSystems.

7

Evaluation of Head Trauma

Randy J. Smith, Jeffrey T. Barth, Robert Diamond, and Anthony J. Giuliano

Head trauma is a widespread, major public health problem often characterized as a silent epidemic. More than 2 million head injuries occur each year in the United States, about one every 15 seconds (National Institute of Neurological Disorders and Stroke [NINDS], 1989; National Head Injury Foundation [NHIF], 1993). Most head injuries (50%–86%) are classified as mild. Overall incidence rates range from 132 to 367 per 100,000, with an average incidence of 200 per 100,000 across population-based studies in the 1970s and 1980s (Sorenson & Kraus, 1991). This rate is comparable to that estimated for other countries (Gronwall, Wrightson, & Waddell, 1990).

The incidence of head injury in a population is difficult to measure due to diagnosis and classification issues, and these data may underestimate the actual occurrence because some victims of milder forms of head injury do not seek medical assistance (Torner & Shootman, 1996). Nevertheless, 500,000 (25%) of the annual head trauma cases in the United States are severe enough to require hospitalization for greater than 24 hours. In terms of mortality, an average of 39,400 persons died from head injury in each of several recent years, reflecting 2% of all deaths and 26% of injury-related fatalities (Sosin, Sacks, & Smith, 1989; Sosin, Nelson, & Sacks, 1992). In fact, head trauma ranks as the leading cause of death and disability among young Americans (NHIF, 1993). Penetrating head injury (PHI) in particular is associated with considerable morality (50%). The average per person costs of head injury of $115,300 translated into a staggering $37.8 billion total in 1985 which, no doubt, has since grown (Max, MacKenzie, & Rice, 1991). Treatment for survivors of severe head injury has an estimated lifetime cost of over $4 million, much of which is paid for by public funds (NINDS, 1989). The magni-

Randy J. Smith   Department of Psychology, Naval Medical Center, Portsmouth, Virginia 23708. Jeffrey T. Barth and Robert Diamond   Department of Psychiatric Medicine, University of Virginia Medical School, Charlottesville, Virginia 22908.   Anthony J. Giuliano   University of Hartford, West Hartford, Connecticut 06117; and Memory Disorders Clinic, Southwestern Vermont Medical Center, Bennington, Vermont 05201.

Neuropsychology, edited by Goldstein *et al.* Plenum Press, New York, 1998.

tude of the problem is further evidenced by the formation of a federal interagency head injury task force (Goldstein, 1990) and establishment of traumatic brain injury registries in a number of states (Harrison & Dijkers, 1992).

Motor vehicle accidents cause most head injuries (50%–60%), followed in order by falls (20%–30%), assaults (7%–45%, depending on the population studied), and trauma sustained in sports, recreational, and work-related accidents (Kraus, 1993; Sorenson & Kraus, 1991). These various causes generally produce closed head injuries (CHI) through blunt force or nonimpact mechanisms. PHIs resulting from military combat, suicide attempts, or crime-related gunshot wounds make up less than 10% of all head injury cases (Kampen & Grafman, 1989).

The risk of head trauma in general is highest for single young males ages 16 to 25, with slightly less risk for children and the elderly (Sorenson & Kraus, 1991). Overall, males are twice as likely as females to sustain a head injury, and they tend to be more severe (Dikmen, Machamer, Winn, & Temkin, 1995). Acute alcohol use, particularly in young males, is the primary predisposing factor across all levels of head injury severity. Most victims have positive blood alcohol levels and as many as 50% meet legal standards for intoxication at the time of injury, with over 33% diagnosed as chronic alcohol abusers (e.g., Dikmen, Donovan, Loberg, Machamer, & Temkin, 1993; Kraus, Morgenstern, Fife, Conroy, & Nourjah, 1989; Kreutzer, Doherty, Harris, & Zasler, 1990; O'Shanick, Scott, & Peterson, 1984; Rimel & Jane, 1983; Sparadeo, Strauss, & Barth, 1990; Wong, Dornan, Schentag, Ip, & Keating, 1993).

Increasing numbers of victims survive their head injuries due to life-saving advances in acute neurosurgical management and modern trauma center accessibility. This remarkable progress has also come with its own costs, namely, a corresponding increase in morbidity. Kraus (1993) used a number of variables to estimate a population-based disability rate and suggested that nearly 83,000 persons in the United States in 1990 had various degrees of residual cognitive and behavioral deficits which negatively impact or preclude participation in educational, vocational, and/or social activities. The neuropsychological consequences of head trauma are particularly costly for the many young adult victims entering their prime productive years.

Largely because of decreased mortality rates, increased morbidity, and greater recognition that a notable proportion of those suffering mild head trauma result in some type of disability, neuropsychologists are seeing growing numbers of head-injured patients for evaluation. Undoubtedly, similar factors are prompting an increased referral rate to other health care professionals who, depending on the setting, may or may not have experience with head trauma evaluation by clinical neuropsychologists. While awareness of head injury has greatly increased across health care disciplines in recent years, this chapter is intended as a brief introduction to neuropsychological concepts and methods in the evaluation and management of head trauma in adults. Clearly, the scope and nature of the problem requires a multidisciplinary approach, of which neuropsychological evaluations may be one useful component.

MECHANISMS OF HEAD TRAUMA

A comprehensive description of the putative mechanisms underlying head injury and its consequences is complex and would necessarily include such variables as injury mechanism (i.e., shear strain, compressive, tensile, penetrating), severity (i.e., mild, moderate, severe), neuropathology (i.e., diffuse, focal, hypoxic-ischemic, edema), primary versus secondary events, age, premorbid status (i.e., education, level of adjustment), and other medical and psychosocial factors. More specifically, appreciation of the number and kind of interacting events that take place within the brain (and other areas of the body) when head trauma occurs continues to grow with developments in basic neuroscience and applied clinical research. For example, recent findings indicate that in addition to structural damage to the brain, neurochemical mechanisms at the cellular level (such as increased neurotransmitter flow causing excitotoxicity, or excessive excitation of brain cells) contribute to the pathophysiology of head trauma and possibly to specific neurocognitive sequelae such as attention and memory deficits (Dixon, Taft, & Hayes, 1993; Hayes, Jenkins, & Lyeth, 1992). While the pathogenesis of head trauma requires further specification, these major mechanisms and their effects on neural tissue, some of which are unique, require understanding before the neurobehavioral consequences of a particular head injury can be appreciated.

There are two major types of head injuries: penetrating and closed. Both penetrating and nonpenetrating injuries involve *primary mechanisms* (those that occur at impact) and *secondary mechanisms* (a cascade of later effects), some of which are produced by the primary effects (Pang, 1989). In Figure 1, the various primary and secondary head injury mechanisms are presented, and organized according to whether the damage is focal or diffuse.

Penetrating head injuries are open brain wounds produced when the skull is

Type of Mechanism

	Primary	Secondary
Focal	Laceration Contusion Depressed skull fracture Cavitation from PHI[1]	Hematoma Hygroma Localized edema Focal hypoxia-ischemia Herniation
Diffuse	Acceleration/deceleration and rotation producing diffuse axonal injury and hemorrhage	Increased ICP[2] Hydrocephalus Generalized edema Diffuse hypoxia-ischemia Neurochemical changes Herniation

Nature of Damage

[1]PHI = Penetrating head injury.
[2]ICP = Intracranial pressure.

Figure 1. Head injury mechanisms organized by type and nature of damage.

perforated by an external intruding object such as a bullet, knife blade, or exploding shell fragment. In PHI, *focal primary* damage results from formation of a cavity (cavitation) as the intruding object enters the brain and destroys tissue along the wound track. *Diffuse primary* damage throughout the brain also occurs in the form of hemorrhaging and tissue disruption produced by resultant shock waves. The amount of damage depends primarily on the velocity of the penetrating object in that faster missiles impart more kinetic energy and cause more cavitation of brain tissue. Due to its low incidence and high mortality, neuropsychologists typically see few individuals with PHI.

Of more relevance to most neuropsychologists' practices are the primary mechanisms of CHI, which may also involve *focal primary* injury at the location of impact (variously called impact injury, impression trauma, or coup lesion) in the form of a laceration, contusion, or depressed skull fracture often caused by contact with a blunt object (Miller, 1991). The most significant and *diffuse primary* mechanism in CHI involves rapid *acceleration/deceleration* and *rotation* of the movable brain within the skull. These dynamics are often the central mechanisms of CHI because of the widespread, devastating effects produced when the brain repeatedly impacts the irregular, interior surface of the skull as it lags behind the skull during acceleration, continues to oscillate during deceleration, or continues to plunge forward after the skull comes to an abrupt stop. Further damage occurs as a result of shear strain as the brain rotates (i.e., swirls) on its axis about the brain stem and is transiently stretched and distorted.

Acceleration and rotation force injuries of variable severity can occur in the absence of a direct blow to the head such as when a stationary or decelerating motor vehicle is rear-ended, producing a whiplash effect, or a moving vehicle collides with a stationary object and the belted occupant does not come into contact with an immovable surface. Such nonimpact brain injury may be especially important in appreciating one form of mild head injury that sometimes results in neuropsychological impairment (Radanov, Stefano, Schnidrig, Sturzenegger, & Augustiny, 1993; Sweeney, 1992).

These primary acceleration/deceleration and rotational forces produce mechanical stress that has several immediate consequences. First, *focal* structural damage in the form of lacerations (wounds or cuts in the brain parenchyma) and contusions (bruises) can occur to specific locations at the site of impact or where the brain abrades against the inner skull. Rotational forces may produce focal *contrecoup* injury, or damage to the brain opposite the site of impact, although the concept of contrecoup has been the source of some debate (Joseph, 1990; Katz, 1992). Damage most commonly occurs to brain tissue on the inferior surface of the frontal lobes (orbitofrontal cortex) where the bony surface of the skull is rough and irregular, and to the inferior anterior tips of the temporal lobes (Katz, 1992; Pang, 1989). Goldman-Rakic (1993) has observed that the frontal lobes are the most frequently damaged site in head trauma. The skull's frontal bony structures (anterior and middle cranial fossa) are sometimes alluded to as the "dashboard" of the brain because of their tendency to cause damage to adjacent cortex (Varney & Menefee, 1993).

Second, acceleration/deceleration mechanisms cause *diffuse* damage in the form of shear strain, or microscopic stretching, compressing, and tearing of axons and disruption of cell bodies, particularly in the superior frontal gyrus and isthmus of the temporal lobe, and vascular perturbations such as traumatic hemorrhages of the subarachnoid space or in subcortical white matter, especially over the convexity and in the frontal, temporal, callosal, and periaqueductal regions (Adams, Doyle,

Graham, Lawrence, & McLellan, 1986; Adams, Graham, Murray, & Scott, 1982). *Diffuse axonal injury* likely occurs at all levels of head injury severity, and the concept of diffuse axonal injury is central to understanding the neurological impact of head trauma, including mild head injury (Williams, Levin, & Eisenberg, 1990). For example, diffuse axonal injury may explain some of those instances in which patients with mild head injury report and may demonstrate neuropsychological deficits on testing in the absence of documented structural damage on neuroimaging techniques such as head computed tomography (CT) and magnetic resonance imaging (MRI) scans.

Although initial studies focused on diffuse axonal injury as an immediate structural phenomenon, more recent investigations have found evidence of delayed axonal injury. Consequently, diffuse axonal injury has been reconceptualized as an evolving process reflecting the interaction of biomechanical and biochemical events (Dixon *et al.*, 1993). Diffuse axonal injury may also be accompanied by alteration of other aspects of neuronal function, such as those pertaining to the cytoskeleton and neurotransmitter release (Dixon *et al.*, 1993).

The primary mechanisms of CHI either result in or are associated with a pathophysiological cascade of *secondary events,* which are common to both CHI and PHI and which contribute to further neurologic dysfunction. These secondary mechanisms can be focal or diffuse, intracranial or systemic in origin, and can involve vascular, respiratory, or metabolic systems (i.e., critical blood, oxygen, and glucose supplies to vital brain centers). *Focal* secondary mechanisms include hematomas (e.g., subdural or epidural hemorrhages, which displace or herniate brain structures), hygromas, localized edema or swelling of brain tissue around a contusion or laceration, and focal hypoxia-ischemia.

Diffuse secondary insults include increased intracranial pressure, generalized edema sometimes leading to herniation of brain tissue, diffuse hypoxia-ischemia due to more widespread vascular and respiratory compromise, various alterations of brain metabolism and neurotransmitter function, and sometimes epileptic seizures (which have a higher incidence in PHI; e.g., Grafman & Salazar, 1987; Kampen & Grafman, 1989). Important neurochemical alterations include excitotoxicity as well as oxygen free radical toxicity, which experimental evidence has shown may be an important mediator of traumatic brain injury (Zasler, 1992). Because head injuries rarely occur in isolation from injury to other parts of the body (and because certain brain and brain stem centers control critical respiratory and cardiovascular functions), there are a number of secondary insults of *systemic* origin which can add to the intracranial mechanisms: arterial hypotension, hypoxemia, hypercarbia, and hyponatremia (e.g., Miller, 1991). For example, arterial hypotension is usually caused by blood loss from bodily injury, while hypoxemia may result from pulmonary trauma (Miller, 1991; Pang, 1989).

An important point is that many of these mechanisms, whether primary or secondary, focal or diffuse, intracranial or systemic in origin, mutually influence and interact with one another. For example, cerebral edema may precipitate a rise in intracranial pressure which may aggravate the edema and, in turn, may lead to herniation of brain tissue. The final outcome of the various head trauma mechanisms may be necrosis of brain tissue on the one hand, or alterations in cellular processes on the other, both of which disrupt normal brain functioning and can lead to clinical changes in a patient's neurocognitive and behavioral status. Of course, there are a number of demographic, psychosocial, and other factors which may interact with the various head trauma mechanisms and affect neuropsychological test results. These factors include age, education, physical pain, and psychiatric status (e.g., Rimel, Giordani, Barth, Boll, & Jane, 1981) and will be discussed later.

Neuropsychological Assessment along the Continuum of Recovery

Neurobehavioral recovery from head injury is an ongoing process, yet it is possible to identify discrete, sequential phases early in recovery and to conceptualize later stages somewhat more arbitrarily for convenience. Different approaches to and methods of neuropsychological assessment are appropriate at each stage of recovery and depend on initial severity of head trauma. In Figure 2, the role of neuropsychological evaluation at each stage is depicted on a hypothetical recovery curve. Jones (1992) divides the recovery process into acute and chronic phases, whereas Stuss and Buckle (1992) and Williams (1992) envision early (acute), middle, and late recovery stages. Gronwall (1989) further divides the acute phase into the periods of unconsciousness, posttraumatic amnesia, and early posttraumatic period.

Although head trauma victims may display any combination of neuropsychological deficits, and later in recovery may show specific patterns of focal impairment, the general sequence of recovery in severe head trauma is marked by a relatively orderly process involving a predictable sequence of events (e.g., impaired consciousness followed by period of posttraumatic amnesia) (Gronwall, 1989; Jones, 1992). Although the sequence is constant, the phases of recovery may vary in length across different severities of head injury. The earliest acute phase typically involves a period of *altered* or *impaired consciousness* ranging from seconds in the most mild injuries to weeks or sometimes months in severe head trauma. There

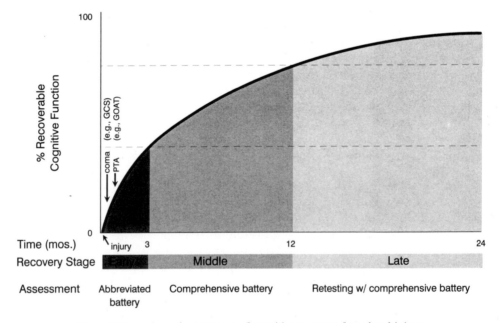

Figure 2. Stages in and assessment of cognitive recovery from head injury.

are, of course, degrees of impaired arousal ranging from obtundation and stupor to brief but complete loss of consciousness through to coma or prolonged psychological unresponsiveness. In addition, occasionally reports of alterations in consciousness (i.e., being "dazed," seeing "stars," confused) are not indicative of brain injury at all (Rizzo & Tranel, 1996).

Regardless of extent, impaired consciousness immediately following moderate of severe head trauma is generally attributed to the effects of mechanical stress on the upper brain stem (midbrain) often due to rapid acceleration/deceleration and/or rotational forces. The rotational mechanism stretches the midbrain reticular activating centers that mediate arousal and consciousness. Shear strain of the ascending fiber pathways of the reticular activating system also serves to disconnect the midbrain arousal centers from their targets in the cerebrum. Because these ascending reticular activating pathways project throughout the brain, widespread shear strain in the white matter of both cerebral hemispheres can also disrupt consciousness. Prolonged and deeper levels of unconsciousness or deterioration to a comatose state are less common, usually associated with more severe traumatic brain injury, and tend to be caused by diffuse axonal injury (e.g., Taylor, 1992), presumably of a more pervasive nature, and/or extra-axial or intraparenchymal hemorrhages. Head trauma victims can also lapse into coma shortly after their injuries when midbrain centers responsible for alertness are subjected to pressure by downward herniation of brain tissue produced by cerebral edema or other mass-occupying lesions.

The Glasgow Coma Scale (GCS; Teasdale & Jennett, 1974) is the most widely used instrument to grade the initial severity of head trauma and for early coma assessment. The GCS is a well-validated measure that objectively quantifies level of cerebral responsiveness by grading the patient's best eye opening, verbal, and motor responses on a 15-point scale. Coma is defined as no eye opening, no motor response to command, and no intelligible verbal response. Coma depth is typically related to variations in motor responsiveness (Capruso & Levin, 1996). Initial GCS scores of 13–15 (in conjunction with unconsciousness for less than 20 minutes, no focal neurological deficit or complications, and no structural abnormalities on neuroimaging) are typically associated with mild head injury; 9–12 with moderate injury; and 8 or less with severe trauma (e.g., Gronwall, 1989). Increased intracranial pressure must be ruled out in patients with initial GCS scores of less than 7 (Feler & Watridge, 1992).

An alternative measure of level of arousal, responsiveness, and neurobehavioral recovery most often used in rehabilitation settings is the Ranchos Los Amigos Scale of Cognitive Function (Malkmus, Booth, & Kodimer, 1980). Head injury patients are assigned to one of eight levels from Level I (no response to pain, touch, light, or sound) to Level VIII (exhibits purposeful, appropriate behavior). The degree and quality of confusion is appraised in Levels IV to VI. Other scales that are sometimes used include the Grady Coma Scale (Cooper *et al.*, 1979; Fleischer, Payne, & Tindall, 1976) and the Ransohoff Coma Scale (Ransohoff & Fleischer, 1975).

Accurate measurement of depth and duration of coma is important because it is the basis for classifying head injury severity and is a useful predictor of neurobehavioral outcome (Katz, 1992; Levin, Gary, Eisenberg *et al.*, 1990). Longer and deeper comas tend to be associated with more serious head injuries and lower acute GCS scores (Wilson, Teasdale, Hadley, Wiedmann, & Lang, 1993). Neuropsychologists monitor acute and serial GCS scores to predict coma duration and neurobehavioral outcome. Of particular interest to neuropsychologists is the finding that acute and early GCS scores are related to neuropsychological status at various postinjury intervals (Alexandre, Colombo, Nertempi, & Benedetti, 1983; Dikmen, Machamer, Temkin, & McLean, 1990; Levin *et al.*, 1990). Less reliable is the relationship between longer coma durations and cognitive outcome (Capruso & Levin, 1996).

The next acute subphase is the period of *posttraumatic amnesia* (PTA), which refers to a transitional phase of disorientation marked by a loss of the capacity to encode and recall new information or ongoing events (Capruso & Levin, 1996; Long & Schmitter, 1992). The term *anterograde amnesia* is used to refer to the inability to learn new information or register what happened after the traumatic event, though memory fragments or "islands" of intact memory are sometimes observed (Capruso & Levin, 1996). The period of PTA is clinically defined as the interval between emergence from coma and restoration of continuous memory and orientation (Russell, 1971), while anterograde amnesia generally encompasses the duration of coma (by definition), the PTA interval, and ongoing memory impairments.

In reality, the PTA interval after coma is a multidimensional phenomenon (Saneda & Corrigan, 1992) characterized not only by amnesia but also by confusion, disorientation, inattention, fluctuating arousal, and behavioral disturbances such as agitation. It is likely that multiple head injury mechanisms (most notably diffuse axonal injury to cerebral white matter) contribute to the brain dysfunction responsible for the appearance of these various PTA sequelae. It can be argued that these mechanisms produce an acute confusional state or delirium (referred to as PTA) whose various symptomatology is secondary to a fundamental disturbance in arousal and attention (cf. Mesulam, 1985). In any case, while PTA is variable in length, it follows a course of progressive improvement during recovery. The degrees of memory impairment range from initially poor storage after emergence from coma to patchy, then more continuous recall (Gronwall, 1989). The PTA ends when an individual is finally able to provide a clear and consecutive account of what is taking place around him or her (Stuss & Buckle, 1992). Nonintoxicated, uncomplicated mild head injury patients generally clear very quickly and, even though it may take several weeks, most survivors of more significant head trauma clear early in acute rehabilitation as they recover from the more pronounced effects of diffuse axonal injury (Saneda & Corrigan, 1992).

During the period of PTA, patients may also suffer from *retrograde amnesia,* or an inability to recall recent preinjury events. Extent of retrograde amnesia is somewhat variable, but typically diminishes over the course of PTA, and the modal head injury patient is left with a small residual amnestic window of less than 30 minutes before the traumatic event (High, Levin, & Gary, 1990).

Although the end of the PTA period marks an important clinical milestone in recovery, it is sometimes difficult to assess accurately because (1) transient pockets or "islands" or postinjury event recall can be mistaken for permanent return of continuous memory (Brooks, 1989); Gronwall & Wrightson, 1980; (2) efficient memory in uninjured normals may not be perfectly continuous; and (3) memory deficits are common in head-injured patients who have clearly recovered well beyond the PTA period (Gronwall, 1989). These difficulties are compounded by the fact that clinical assessment of PTA duration is often retrospective and thus depends on the patient's self-report; primary medical records are often unavailable or incomplete and thus fail to clarify the issue.

Measurement of PTA and its resolution became standardized with the development of the Galveston Orientation and Amnesia Test (GOAT; Levin, O'Donnell, & Grossman, 1979), which allows orientation (person, place, time), PTA duration, and retrograde amnesia to be assessed serially over the course of recovery. Total GOAT scores of 76–100 are considered unimpaired, scores of 66–75 define the borderline impaired range, and scores below 66 fall in the impaired range. Most recently, GOAT scores of 75 or above for 2 consecutive days have been proposed as a criterion to mark the end of the PTA period and to signify testability with more comprehensive neuropsychological methods (Levin, High, & Eisenberg, 1988). High and colleagues (1990) used the GOAT serially to investigate recovery of orientation in a sample of CHI patients. They found a sequential ordering of orientation, with the most common sequence being return of awareness of person first, followed by place, and then time. Most patients exhibited backward displacement of the date (gave an earlier rather than a future date), particularly those with, among other characteristics, longer durations of PTA. The dual assessment of disorientation and PTA is desirable because the two may not be equivalent (Crovitz, 1987; Gronwall, 1989; Gronwall & Wrightson, 1980). Use of the GOAT overcomes some, but not all, of the problems inherent in assessing PTA duration (Crovitz, 1987; Gronwall, 1989; Stuss & Buckle, 1992).

A recently proposed alternative to the GOAT is the Orientation Group Monitoring System (OGMS; Mysiw, Corrigan, Carpenter, & Chock, 1990; Saneda & Corrigan, 1992). The OGMS is a reliable observational system applicable within the context of daily therapeutic orientation groups. Preliminary research by its authors has shown that the OGMS is similar to the GOAT in its ability to monitor PTA, with each having relative advantages depending on the aim of assessment and setting characteristics.

Estimation of PTA duration has practical value in charting recovery, both in terms of its usefulness for day-to-day clinical management and intervention, and in terms of signaling when further neuropsychological assessment may be appropriate. Disoriented patients with marked attentional and memory deficits are typically not yet testable using conventional neuropsychological measures (Gronwall, 1989; Williams, 1992). Assessment of PTA duration is important prognostically because, like coma, it has been shown to be a broad predictor of head injury severity and outcome (e.g., Russell, 1971). For example, patients with less than 14 days of PTA typically demonstrate a good recovery whereas the opposite holds for those with PTA durations over 2–4 weeks (Katz, 1992; Levin *et al.*, 1979). Across studies, longer periods of PTA have been associated with more significant and diffuse neuropathology, higher risk for changes in personality, and less favorable vocational and other outcomes (Capruso & Levin, 1996).

The third and final acute subphase of recovery is the early posttraumatic period (Gronwall, 1989), which blends into what can simply be called early recovery. With the end of the PTA period, the patient is oriented, better able to focus and sustain attention and remember ongoing experiences. The more generalized disruption of cognitive functions from the acute trauma has resolved sufficiently to reveal specific neuropsychological deficits. Essentially, cortical functioning has recovered to the point where formal neuropsychological testing using a battery of more demanding measures is now possible, and in fact testability has been used to define the early stage of recovery (e.g., Gronwall, 1989). The point at which individual patients become testable varies considerably with the severity of the injury, although Williams (1992) has noted that most neuropsychological assessments are typically carried out at least within 12 weeks of the head injury. Importantly, prediction of longer term neurobehavioral outcome appears reasonably sound when based on assessments of level of recovery at 3 to 6 months (e.g., Choi *et al.*, 1994; Capruso & Levin, 1996).

During the early stage of recovery, neuropsychological test batteries may be abbreviated and tailored to the patient's specific level of functioning and ability to participate secondary to endurance and fatigue. Measures may also be selected based on deficits expected to be most prominent at this stage, particularly those involving information processing capacity, attention and concentration, and memory functioning, or related to known focal lesions. When possible, and particularly as the patient moves into the middle and late stages of recovery, administration of a comprehensive neuropsychological test battery is advisable since head-injured patients may manifest practically any neuropsychological deficit during their recovery depending on the nature and severity of the neurologic impairment. Findings are then used to guide rehabilitation, psychosocial, educational, and vocational interventions and plans.

To some extent, division of recovery stages after the discrete early subphases is somewhat arbitrary and heuristic. Stuss and Buckle (1992) divide middle and late recovery at the first 12-month point. Whereas the purpose of assessment during the acute inpatient stages may have been to help estimate severity and testability and to assist with acute management and discharge decisions, the purpose of comprehensive neuropsychological assessment during the middle phase is to develop an integrated explanation of behavior, combining hypotheses about altered brain function and descriptions of neurobehavioral strengths and weaknesses with an understanding of an individual's personality, current circumstances, and social relationships (cf. LaRue, 1994). That is, a comprehensive assessment elucidates which cognitive abilities have been spared or impaired by the head injury, and to what degree. Both focal disturbances and diffuse impairments can and should be identified at this time (Stuss & Buckle, 1992). This information is used to (1) establish a baseline against which the results of future neuropsychological assessments can be compared as recovery is monitored over time; (2) make recommendations for further neuromedical, pharmacological, psychological, and other allied health evaluations and treatment approaches; (3) help plan for and guide a program of physical, cognitive, and vocational rehabilitation; (4) make educational recommendations if the individual is in school; (5) provide feedback to the patient and his or her family and recommendations for neurobehavioral management; (6) assist in medicolegal decision-making (e.g., disability determinations, civil litigation).

The purpose of neuropsychological assessment during late recovery is to document continued recovery of function against the earlier baseline findings (and perhaps also to estimate when a plateau may have been reached). This information is then used to make any necessary adjustments to initial recommendations concerning school, vocational, or everyday functioning. Also, the extent of psychosocial consequences of the head trauma becomes clearer during late recovery (Stuss & Buckle, 1992). It is important to assess psychosocial outcome at this time for a number of reasons, not the least of which is the possibility that it may reflect personality or executive control disturbances due to frontal-subcortical system dysfunction, and to guide more specific neurobehavioral management, pharmacological, psychological, and/or family interventions.

To understand how neuropsychologists approach the assessment of persons who have sustained a head injury, it is necessary to establish a context derived from extant empirical findings. The neuropsychological consequences of head injury include cognitive, emotional, and behavioral sequelae that can be conceptualized in terms of outcome at a particular point in time or in terms of recovery as a process which occurs over time, both of which must take into account acute injury characteristics and various other premorbid and postinjury factors. Head-injured patients vary considerably in the kind and degree of neuropsychological deficits they may manifest, and neuropsychological impairments after head trauma are common. They can display virtually any cognitive deficit, including impairments in attention/concentration, information processing, learning and memory, executive functions, language, motor speed and coordination, visuospatial organizational skills, sensory/perceptual functions, intelligence, and personality/emotional functioning, as well as a variety of difficult to document postconcussive symptoms (i.e., headaches, fatigue, dizziness). The most frequently observed deficits tend to involve impairments of attention, memory, and information processing ability and speed. Again, however, the extent to which additional deficits are present depends on numerous factors, including the severity of the injury, time elapsed since the trauma, location of the lesion, and various premorbid characteristics such as age, education, number of previous head injuries, and general health.

Generally speaking, persons who have sustained moderate to severe CHI manifest impairments in numerous cognitive domains when assessed relatively early in recovery (e.g., 1 or 3 months; Dikmen *et al.*, 1990). In contrast, earlier work examining neuropsychological sequelae in mild head injury found that although mild injuries can cause cognitive impairments, they tend to do so selectively. Some researchers have also found a trimodal distribution at 3 months postinjury (Barth *et al.*, 1983). Using the Halstead–Reitan Neuropsychological Test Battery (HRB) as well as other measures, about one third of the subjects in the Barth and colleagues' (1983) study showed no or very mild impairment, another third showed mild to mild-moderate deficits, and the final third fell in the severely impaired range. Later investigations that controlled for pertinent background variables likely to affect neuropsychological test results (e.g., polysubstance abuse, previous head injury, pending litigation) found that most healthier mild head injury patients have few if any cognitive problems after 3 months (e.g., Dikmen, McLean, & Temkin, 1986; Levin *et al.*, 1987). A general conclusion that can be drawn from the literature is that most survivors of moderate to severe head trauma will initially demonstrate significant neuropsychological deficits across a range of functions, whereas mild head injury is associated with a smaller number of patients showing a more limited range of deficits, with attention and memory problems most frequently documented followed by rapid improvement.

Turning to the recovery process, or positive change in cognition over time, the general trend is one of rapid early improvement followed by a slower rate of change. Those head injury patients with initial cognitive impairment tend to show the most improvement within the first 3–6 months; cognitive recovery curves tend to reach a relative asymptote by 6–12 months (Capruso & Levin, 1996), with recovery extending over longer periods in some cases. Neuropsychological recovery is a curvilinear rather than linear process (Jones, 1992), although few studies have examined recovery past 1 year or longer intervals (Dikmen *et al.*, 1990). Levin *et al.* (1990) investigated outcome after severe head injury at the 1-year point and found that impairments of memory and information processing persisted but that language and visuospatial abilities recovered compared to controls. Dikmen *et al.* (1990) reported that after 2 years, their moderate-severe CHI subjects remained impaired in a number of neuropsychological functions relative to a control group. Dikmen and colleagues also noted, however, that recovery in the second year may be more modest, selective, dependent on severity, and somewhat obscured by various methodological issues and confounds.

Clinical estimation of recovery curves for individual patients should take into account a number of premorbid, psychosocial, and ecologically relevant factors in addition to acute head injury characteristics. These various factors, which were alluded to earlier, are summarized in Table 1. Some of the factor (e.g., general health) are presented as dichotomous variables but of course actually lie on continua and should be so regarded. The factors included in Table 1 illustrate the importance of both premorbid and current functioning in recovery from head trauma, though more complex, multivariate and longitudinal studies will be required to specify the nature of the relationships among variables over time and their relative importance to specific outcomes.

Alcohol use at the time of injury, regardless of a history of abuse or dependence, can result in poor outcome (Sparadeo *et al.*, 1990). Medical researchers are well aware of and concerned about the relationship between substance abuse and traumatic brain injury (cf. *The Journal of Head Trauma Rehabilitation*, 1990, 5). The substantial numbers of head injury victims with undiagnosed, untreated alcohol or drug problems, or alcohol in their systems at the time of injury, has clear implications for neuropsychological assessment after head trauma (Solomon & Malloy, 1992).

Other findings regarding neuropsychological outcome are that it can be predicted by some neurological variables such as coma length (Dikmen *et al.*, 1990), and that it can in turn predict other major dimensions of outcome such as later cognitive functioning, success in rehabilitation, and return to work or school (Cre-

TABLE 1. FACTORS THAT AFFECT INDIVIDUAL RECOVERY CURVES IN HEAD INJURY

Factor	Recovery curve	
	Probable good/rapid	Possible poor/slower
Severity of head injury	Mild to moderate	Severe
Age	Young adult	Older adult (or young child)
Premorbid general health	Good	Poor
Premorbid intellectual and cognitive abilities	High	Low
Previous head injury or other neurologic disorder	None	Present
History of substance abuse or use at time of injury	None	Present
Psychiatric history or current mood disorder	None	Present
Physical pain disorder	Absent/mild	More severe
Support systems (family, vocational, economic)	Strong	Weak
Persisting postconcussive syndrome	No	Yes
Head injury information provided at hospital	Yes	No
Pending litigation	No	Yes

peau & Scherzer, 1993; Dikmen *et al.*, 1994; Karzmark, 1992; Ruff *et al.*, 1993; Ryan, Sautter, Capps, Meneese, & Barth, 1992).

As with CHI, the outcome for survivors of PHI can range from poor to good, although PHI in general tends to be associated with more significant trauma and therefore worse overall outcome. Neuropsychological deficits associated with PHI encompass the full range as they do for CHI, although as discussed later, unique patterns of focal cognitive impairment may result from PHI due to the idiosyncratic nature of this form of trauma.

Comprehensive Neuropsychological Evaluation

Comprehensive neuropsychological evaluation of the head-injured patient consists of a history, clinical interview, and, as appropriate to level of recovery, bedside evaluation or administration of a battery of formal tests. Feedback to the patient and others is also considered to be an integral part of the evaluation process. Although formal testing often consumes the majority of time in a neuropsychological evaluation, all four of these elements represent critical cornerstones of an appropriate examination in head trauma cases. Because testing takes the most time, and because it clearly represents a compact, efficient method of making multiple observations of behavior in a standardized manner, the other components sometimes receive less emphasis than they should. Yet, as discussed below (1) interpretation of test data is impossible without a good history (to place data in an appropriate context), (2) testing as a method of behavioral observation should be supplemented by the neuropsychologist's direct observations during the clinical interview; and (3) feedback to the patient is not only an ethical responsibility, but should be viewed as an extension of the evaluation process itself and as a clinical intervention.

A good history and clinical interview are essential precursors to formal neuropsychological testing. In fact, Phay (1992) has gone so far as to state that "the only value of neuropsychological testing in the hands of an experienced neuropsychologist is to support the conclusions already reached through interview and history taking" (p. 35). Although Phay intended his statement to be provocative, it does underscore the critical importance of these facets of the evaluation process. A skilled interview and history contribute to the development of hypotheses about altered brain function and neurobehavioral functioning and their interaction with current stressors and psychosocial circumstances. This information in turn facilitates accurate interpretation of neuropsychological test performance, identification of coexisting disorders (e.g., depression, substance abuse) requiring treatment, and estimation of a recovery curve applicable to the particular individual (see Table 1). For these various reasons, it should be apparent that the neuropsychologist must go beyond formal test data and obtain other forms of information to evaluate and manage head injury adequately (Jarvis & Barth, 1994).

The History

The history and clinical interview represent overlapping but not identical functions. The history is obtained from both interviews of the patient and other informants as well as from a review of available medical and/or school and vocational records. Depending on the setting, available records may consist of the hospital chart, referral letter, office notes of the referring clinic, and reports of relevant specialists such as neurosurgeons, neurologists, neuroradiologists, neurophthalmologists, and speech/language pathologists. Thorough review of all medical documents is crucial but not sufficient, as vital portions of the history are obtained through the clinical interview.

The history encompasses both history of the presenting problem and past history. In taking the history of the head injury, neuropsychologists attempt to understand as fully as possible all relevant characteristics of the injury event and its natural history (e.g., timing and mechanisms of injury, duration of unconsciousness, extent of retrograde and posttraumatic amnesia, acute and subsequent neurodiagnostic exam results, GCS scores, length of hospitalization, postinjury cognitive and behavioral changes, current status). In this regard, Alexander (1995) provides a particularly useful approach to mild traumatic brain injury. More broadly, an effort is made to understand the person's level of functioning in all key roles of present everyday life. Neuropsychologists use this information to begin formulating the nature of the head injury and its likely neuropsychological consequences.

The medical records, especially EMT and emergency room reports, may clarify the nature of the head injury, its acute characteristics, and other physical trauma (e.g., pulmonary damage) which have implications for neuropsychological diagnosis and prognosis. A detailed history as obtained through the interview can elucidate current sequelae, and a careful inquiry into possible changes in all areas of cognitive, behavioral, and emotional functioning should be carried out. Sometimes, head injury patients are unaware of certain changes, cannot remember them due to a memory disorder, or for various reasons may be motivated to distort certain facts. Therefore, it is essential that family members or other reliable informants be included in the interview process whenever possible. For example, sometimes only the family of the head trauma survivor may notice alterations in personality or emotional functioning which might suggest frontal lobe dysfunction or reactive depression. Finally, in taking the immediate history, determining current substance use, including prescribed medications (e.g., antiepileptic medication), is critically important as these factors may well influence neuropsychological functioning and therefore performance on tests.

Past history should include an early history (prenatal and early development, including possible birth injury due to forceps delivery or anoxia), prior neurologic difficulties (including previous head injuries and illnesses such as meningitis), general medical history, school achievement, work history (including any military service), family/marital background, psychiatric history (including outpatient treatment and hospitalizations), and substance abuse or dependence problems. The past history in each of these areas should be detailed. For example, an educational history of simply the number of years completed is insufficient to help estimate the head-injured patient's premorbid intellectual functioning or understand rival explanations of impaired test performance. The neuropsychologist also inquires about grades earned, possible learning disabilities, need for special education services, grade repetitions, performance on standardized achievement tests, and academic strengths and weaknesses. It may turn out that a low score on measures of calculation ability is due more to a premorbid weakness in this area rather than to the individual's head injury.

Overall, there are several ways in which the history of the head injury and the past history are useful in neuropsychological evaluation. First, through the history, the extent and severity of the head injury, along with its functional consequences and the patient's adaptation to it, can begin to be appreciated. Second, potential areas of deficit can be identified to facilitate test selection. Third, information necessary for estimating overall premorbid functioning, especially premorbid intellectual level, is generated. Measuring premorbid intelligence is important in formulating what level of performance is expected on neuropsychological tests. Estimates of premorbid intellectual abilities constitute one example of the fourth use of the history, which is to identify alternative factors which may explain deficient test performance or individual behavior. Finally, the history may reveal coexisting or secondary disorders (e.g., depression, substance abuse) warranting recommendations for further assessment and intervention. For further information on history taking or methods for estimating premorbid intellectual functioning, the reader is referred to Filskov and Leli (1981), Lezak (1995), Jarvis and Barth (1994), and Phay (1992).

Clinical Interview

Most head trauma specialists include interviews as part of their evaluations in order to obtain information about patients' functioning in their areas of expertise. Neuropsychologists conduct clinical interviews to assess essential aspects of the head-injured patient's neurocognitive status which are not accessible through standardized neuropsychological tests.

For example, *awareness* is one such aspect. A number of neurologically impaired patients are partially or completely unaware of their deficits or the implications of their deficits. Unawareness (commonly referred to as anosognosia) also occurs following head injury depending on the type and location of the lesion. Anosognosia for cognitive impairment (e.g., memory deficits), emotional and behavioral disturbances are sometimes seen in patients with moderate and severe head injuries. Neuropsychologists should routinely include assessment of awareness in the clinical interview in order to distinguish between problems with awareness and genuine absence of deficits and because of its obvious implications for rehabilitation and environmental competence. It remains extremely difficult to help patients develop compensatory strategies if they do not have some awareness of their deficits. Unawareness is associated with reduced motivation for treatment, overestimation of improvement, and other behavioral problems (McGlynn & Schacter, 1989).

In addition to awareness, neuropsychologists assess a number of other functions in the clinical interview, including observations of the patient's appearance, level of alertness, ability to attend, emotional state (mood, affect), fluency of conversational speech, coherence of thought processes, and degree of preservation of social skills. Such mental status observations alert the neuropsychologist to neurologically based difficulties related to the head injury, some of which may warrant further study through formal testing. In addition, mental status findings can indicate factors that might compromise the validity of the test data. A dramatic example is the patient who appeared fatigued upon arrival for a morning evaluation. Questioning about his tired appearance revealed that he had arrived at the neuropsychological laboratory in Virginia shortly after having flown from New Mexico on an overnight flight that included a significant layover! At the same time, this information also suggested that some further assessment of planning, judgment, and other executive control functions was indicated.

Some of the clinical observations made during the interview are designed to supplement information obtained from formal test procedures. For example, qualitative clinical observations of slowness can help the neuropsychologist to appreciate the pervasiveness and severity of a deficit in information processing speed in a manner not possible by quantitative assessment alone.

Given that no standardized neuropsychological tests have yet proven exclusively sensitive to frontal lobe executive dysfunction in head-injured patients, the history and clinical interview must be utilized. In fact, these facets of the neuropsychological evaluation process may be the best suited for the task, and allow the neuropsychologist to capitalize on his or her expertise in the clinical observation of behavior. For example, neuropsychologists may also conduct various simple but revealing neurobehavioral tests (or "bedside methods") to elicit neurologically based difficulties. For instance, some patients, usually those who have sustained damage to the dominant cerebral hemisphere, may be apraxic or unable to perform certain skilled learned movements (e.g., blowing out a match or waving goodbye). Single and serial limb movements may reveal signs of apraxia and help to localize lesions and appreciate their severity. Other neurobehavioral tests include asking head-injured patients to squeeze one's hand when the word "red" is spoken but not when they hear "green." Such "go–no go" paradigms are a means of assessing possible impairment in motor regulation, a possible sign of dysfunction in orbital/inhibitory systems. Malloy and Richardson (1994) provide a more detailed description of bedside and psychometric methods for conducting a thorough assessment of functions subserved by frontal lobe systems.

One promising means by which the clinician can organize, structure, and systematize the assessment of frontal lobe executive functions is through use of the Profile of Executive Control System (PRO-EX), developed by Braswell *et al.* (1993). The PRO-EX is essentially a comprehensive data gathering tool for behaviors reflecting executive control functions. The patient's best level of functioning is described in each of the following seven key areas: goal selection, planning/sequencing, initiation, execution, time sense, awareness of deficits, and self-monitoring. The PRO-EX is versatile in that it can be used either as the basis for a structured interview with head trauma patients suspected of executive control dysfunction, or as an observational rating scale if the opportunity exists to observe the patient in a setting outside of the clinician's office. The format of the PRO-EX also allows for repeated measurements. Although not a standardized neuropsychological test, the neuropsychologist's clinical observation skills can be organized by the PRO-EX in a manner that may enhance the reliability and validity of the assessment of frontal lobe executive functioning.

FORMAL NEUROPSYCHOLOGICAL TESTING

Due to the complexity of brain–behavior relationships and the variety of cognitive and behavioral disturbances that follow head injury, batteries of such tests are virtually always used today rather than single tests (Jarvis & Barth, 1994; Reitan & Wolfson, 1993). In selecting an assessment strategy, neuropsychologists are essentially faced with one of three choices: using one of the established, fixed batteries available in the field, using such batteries with modification, or composing their own battery of specific test procedures based on the referral question and other individualized considerations (Lezak, 1995). The first choice reflects selection of a standardized test battery, whereas the second and third choices represent use of a flexible battery. Most practitioners have identified the best features of several different approaches, and thus integrate selected tests from fixed batteries with methods designed to examine specific impairments identified by clinical history and research as relevant (Jones, 1996).

STANDARDIZED VERSUS FLEXIBLE BATTERIES. There is a clear distinction between fixed or standardized batteries, in which all patients take all tests, and flexible or customized batteries, which reflect test selection by the individual neuropsychologist geared to the individual patient (Kane, 1991). These two approaches are rooted in different scientific traditions and conceptual assumptions (e.g., Lezak, 1995). In addition, each approach has its variations as well as respective advantages and disadvantages. A unique variation on the flexible approach is the process approach.

STANDARDIZED NEUROPSYCHOLOGICAL BATTERIES. Neuropsychologists who use standardized batteries typically employ either the HRB (Reitan & Wolfson, 1986, 1993) or the Luria–Nebraska Neuropsychological Battery (LNB; Golden, 1981; Golden, Hammeke, Purisch, & Moses, 1984; Golden & Maruish, 1986). The quantitatively oriented HRB is the product of an empirical actuarial approach, whereas the LNB is an outgrowth of a more clinical hypothesis-testing tradition and has a quantitative and qualitative focus. Nevertheless, both batteries are composed of an established set of tests collectively intended as a reasonably comprehensive assessment of neurocognitive functions susceptible to disruption from neurologic impairment, including that sustained after head injury.

The Halstead–Reitan Neuropsychological Test Battery. As the more senior of the two standardized batteries, the HRB is more commonly used as well validated. The battery today consists of eight tests which measure functional abilities across attentional, linguistic, conceptual, sensory, motor, and spatial domains, and which represent varying levels of task complexity and specificity of cerebral functioning. The basic battery consists of the Category Test, Trail Making, Speech–Sounds Perception, Seashore Rhythm, Tactual Performance Test, Finger Oscillation, Aphasia Screening, and Sensory-Perceptual Examination. These tests are augmented by two allied procedures—the Wechsler Adult Intelligence Scale-Revised (WAIS-R; Wechsler, 1981), the most widely used test battery to assess intellectual functioning, and the Minnesota Multiphasic Personality Inventory-2 (MMPI-2; Graham, 1990)— which incorporate intellectual and personality functioning into the HRB assessment. The specifics of the HRB tests and allied procedures are discussed in detail elsewhere (e.g., Jarvis & Barth, 1994; Reitan & Wolfson, 1993).

Along with individual tests scores from the basic HRB battery, an overall Impairment Index is calculated. The individual scores and the Impairment Index are corrected for the effects of demographic variables (age, gender, education) empirically known to affect the scores (Heaton, Grant, & Matthews, 1991). A summary neuropsychological deficit score, which is more comprehensive than the impairment index, can also be generated (Reitan & Wolfson, 1993).

The data from the HRB are used to answer a series of questions regarding the presence, severity, progressiveness, lateralization, location, and nature of the cerebral impairment. The neuropsychologist also generates a description of the patient's cognitive and behavioral strengths and weaknesses as they relate to daily living skills, treatment, and rehabilitation. To increase the likelihood that valid inferences about brain–behavior relationships are formulated in addressing these questions, four methods of interpretation are applied. These methods include analyses of the level of performance, pattern of performance, left–right differences between the two sides of the body, and presence of pathognomonic signs (Reitan & Wolfson, 1993).

The HRB has been subjected to extensive validation across a variety of neuropathological conditions (see Reitan & Wolfson, 1986, 1993, for reviews), and its psychometric properties are well established. Depending on the specific study, factor analyses generally have yielded four factors reflecting varying combinations of impairment in verbal, spatial, sensory, motor, and attentional abilities (Kane, 1991). Using data from CHI subjects, Thomas and Trexler (1982) reported a four-factor solution with the following factor labels: global cognitive, visual–motor performance, memory, and motor speed. The HRB was also shown to be sensitive to the effects of head trauma in a series of investigations by Dikmen, Reitan, and their colleagues (Dikmen *et al.,* 1986; Dikmen & Reitan, 1976, 1977a, 1977b, 1978; Dikmen, Reitan, & Temkin, 1983; Reitan & Wolfson, 1988).

The strengths of the HRB lie in (1) its comprehensiveness for assessing a wide range of neuropsychological functions subserved by the brain, including those for which impairment or strength may not be evident by clinical history, (2) its inclusion of both general and specific measures of brain functioning, (3) the sound psychometric approach taken in its development, including standardization based on a large number of cases, (4) recent provision of comprehensive norms which take into account key demographic variables, and (5) its well-established validity over a wide variety of neurologic conditions including head trauma.

Limitations of the HRB include (1) its lack of a theoretical basis (although this is refuted by some [e.g., Russell, 1981]); (2) its questionable ability based on factor analytic studies to furnish a refined breakdown of cognitive impairments (Kane, 1991); (3) criticism that it is unresponsive to specific clinical issues and does not adequately assess some crucial deficits in head injury (e.g., memory); (4) its very lengthy administration time of 5 to 6 hours (especially when supplemented by intellectual, verbal memory, and personality tests); (5) its expense and lack of portability if the original test versions and equipment are used; and (6) the quantitative emphasis on scores with less consideration of how the scores were achieved (i.e., it does not always define meaningfully qualitative aspects of a patient's deficits which in turn can limit appreciation of the fundamental nature of the impairment and issues relevant to person–environment fit).

The Luria–Nebraska Neuropsychological Battery. The LNB is an alternative fixed battery based on Luria's (1966, 1973) conceptualization of cortical functions, as well as later investigations by Christensen (1975). Developed more recently than the HRB, the LNB (Golden, Hammeke, & Purisch, 1978) has gained increased acceptance as a valid set of standardized procedures for assessing brain–behavior relationships (Golden & Maruish, 1986). The LNB consists of 11 broad ability scales supplemented by two lateralizing scales, a pathognomonic scale, and various localization and factor scales, all of which are based on administration of 269 items. The 11 basic clinical scales sample abilities in the following areas: motor, rhythm, tactile, visual, receptive speech, expressive speech, reading, writing, arithmetic, memory, and intellectual processes. Each scale yields a total score which is converted to a standard *T* score adjusted for the effects of age and education. Two parallel forms of the LNB have been developed, and Form I has been modified in several respects to augment its clinical sensitivity (Golden, Purisch, & Hammeke, 1985; cf. Kane,

1991). As with the HRB, neuropsychologists use various inferential schemes (e.g., pattern analysis of the clinical scales) to interpret the results in order to lateralize, localize, and determine the severity of the head-injured patient's lesion(s).

Ample validation studies now exist which establish that the LNB can reliably discriminate various neurological groups from controls (e.g., see Golden & Maruish, 1986). Golden and Maruish (1986) also reviewed evidence in support of the battery's ability to lateralize lesions to either the right of left cerebral hemisphere. However, Goldstein, Shelly, McCue, and Kane (1987), using cluster analysis, found that the LNB failed to lateralize for left, right, or diffuse brain damage. Those investigators consequently expressed concern over the battery's sensitivity to cerebral dysfunction associated with head injury. In a review of several factor analytic studies, Kane (1991) expressed his concern that the LNB may lack adequate factorial diversity, and concluded that further investigation is needed of the battery's capacity to localize lesions and provide clinical information of sufficient specificity. Further work that specifically addresses the validity of the LNB for head trauma populations is clearly indicated.

Strengths of the LNB are considered to be (1) like the HRB, the comprehensiveness and standardization of the assessment, (2) its theory-driven development, (3) the breakdown of functions into their most basic components, which allows for a combined quantitative/qualitative approach to evaluation and in turn a meaningful definition of the nature of specific deficits, (4) the relatively short administration time of 2 to 3 hours, (5) its relatively low cost, and (6) the self-contained assessments of memory and intelligence, which obviate the requirement for additional allied procedures in these areas. Limitations that are often noted include (1) its heavy concentration on specific measures to the exclusion of tests at the highest levels of central processing (overall reasoning and concept formation), (2) its considerable verbal demands either make patients with intact language functions appear "normal" or those whose language abilities are deficient for whatever reason appear very impaired (Lezak, 1995; p. 720), (3) the need for further exploration of the battery's psychometric properties (validity), and (4) its less extensive history of use and consequently smaller data base and opportunity for multicenter research. Lezak (1995) has advised that "given its many psychometric defects, the examiner must be extremely cautious about drawing conclusions based on the scores and indices of this battery as presently constituted" (p. 722).

A major strength of the large battery approach is its comprehensiveness, which allows (1) unsuspected deficits in head injury to be uncovered (2) unrecognized patient strengths to be defined, (3) large normative data bases to be developed which facilitate both clinical research and the standardization process, and (4) a broad foundation and mastery experience for neuropsychologists in training (Kane, 1991). Limitations of the two fixed batteries concern their (1) reduced effectiveness in focusing on a particular patient's deficits if administered in full or in isolation from other measures, (2) constraints in assessing some of the key deficits in head injury unless other measures are included, (3) relative time and cost inefficiency if the full battery is routinely given in an inflexible fashion (Kane, 1991), and (4) incompletely established ecological validity, or ability to make valid predictions about everyday functioning (e.g., Chelune & Moehle, 1986; Heinrichs, 1990).

FLEXIBLE NEUROPSYCHOLOGICAL BATTERIES. The alternative to the fixed HRB or LNB is the flexible battery in which the neuropsychologist constructs a customized battery of specific tests or modifies a basic battery based on individual patient issues and history. Test selection is guided by information available about the patient from medical records, deficits suggested by the clinical interview and testing process itself, and the research literature on the neuropsychology of head injuries (Brooks, 1989).

Lezak's Hypothesis-Testing Approach. Lezak's (1983) individualized examination constitutes one approach to flexible battery testing:

> In an individualized examination, the examiner rarely knows exactly which tests he is going to give before he beings working with the patient. He usually starts with a basic battery that touches upon the major dimensions of intellectual behavior and makes many of his choices as he proceeds. The patient's strengths and limitations and specific handicaps will determine how he uses the tests in the battery, which he must discard, and which require modifications to suit the patient's capabilities. As he raises and tests hypotheses regarding possible diagnoses, areas of intellectual dysfunction, and psychosocial or emotional contributions to the total behavioral picture, he may need to go outside the basic battery and use techniques relevant to *this* patient at *this* time. (p. 107)

Lezak's (1983) flexible examination represents a hypothesis-testing approach which she characterizes as falling at the middle of the quantitative–qualitative interpretative continuum. Both Lezak (1983; 1995) and Spreen and Strauss (1991) provide complete compendia of available neuropsychological tests that can be included in flexible batteries. These compendia describe each test and its administration, scoring, psychometric properties, and normative data.

The Boston Process Approach. Kaplan's (1988, 1990) Boston Process Approach represents a variation on the flexible approach, although it is clearly distinguishable from a flexible battery if the term battery connotes a group of standardized tests administered and scored in a standardized manner. In the process approach, a core battery of procedures are administered; however, the administration and scoring of some of the measures are modified. These modifications are linked to the focus of the process approach, which is on appraisal of the qualitative nature and effectiveness of the behavior and strategies that the patient uses in attempting to solve the problems posed by the test procedures, and which relates the behavior to experimental neuropsychological and cognitive neuroscience conceptual frameworks (Milberg, Hebben, & Kaplan, 1996). Essentially, the qualitative problem-solving behavior of the head-injured patient is carefully observed and analyzed during testing. Kaplan and colleagues believe that this emphasis on process (the "how" along with the "whether") yields more useful information about underlying brain–behavior relations than a sole focus on the outcome of problem–solving from a global right–wrong perspective (the "whether" only). However, the process approach is not strictly qualitative in nature (Kane, 1991). The qualitative observations made during process assessment are scored and quantified, although sometimes using expanded or refined scoring schemes.

Just as each of the two fixed batteries has respective merits and drawbacks, there are advantages and disadvantages to the use of flexible batteries. Strengths of flexible and process approaches in general center on their (1) ability to provide a focused, refined evaluation of a specific patient's deficits; (2) increased sensitivity to neuropsychological sequelae of head injury because the examination can be tailored to this problem; (3) readily modifiable nature such that new tests can be easily incorporated which allows the field to advance (Kane, 1991); and (4) relative cost effectiveness and time efficiency.

Limitations of flexible batteries involve their (1) inability to be standardized for psychometric and normative purposes, (2) potential for missing unanticipated deficits and strengths in a particular head-injured patient, and (3) incompletely established ecological validity.

Because flexible batteries are by definition unstandardized and vary, their overall validity as a set of procedures cannot be determined, nor can they be compared empirically with fixed batteries (cf. Kane, 1991). The psychometric properties of specific measures making up such batteries can, of course, be appraised and are reviewed in a number of sources (e.g., Lezak, 1995; Spreen & Strauss, 1991).

In actual practice, it is quite likely that many neuropsychologists with an overall preference for one of the two existing standardized batteries do, in fact, flexibly adapt their assessment not only to head trauma, but also to the individual patient. In addition, many experienced neuropsychologists utilize some degree of clinical hypothesis-testing and are sensitive to qualitative aspects of patient performance on tests, although perhaps not in the exact manner as Lezak or Kaplan and her colleagues suggest. An obvious implication is that batteries alone do not determine the effectiveness of the assessment. The expertise and experience of the examining neuropsychologist are of critical importance to the evaluation.

Is There an Ideal Head Injury Battery?

In view of the various considerations stated earlier, there may be no *single* suitable battery. Rather, it may be most appropriate to think in terms of specific *features* that allow any battery to be characterized as suitable. The battery should (1) be used or supervised by a neuropsychologist with expertise and experience in head trauma, (2) be reasonably comprehensive in terms of the range of cognitive functions assessed, and (3) include specific and detailed measures of common head injury deficits. In other words, in addition to the neuropsychologist's experience level, the battery would be comprehensive to ensure adequate screening of the broad range of neurocognitive functions and focused to ensure adequate assessment of the most common deficits in head injury. A comprehensive assessment is critical because, even though there are common head injury sequelae, no neurocognitive functions are free from potential disruption.

Table 2 presents essential abilities and processes that should be assessed in any

TABLE 2. ABILITIES AND BEHAVIORS TO BE ASSESSED
IN ANY HEAD INJURY EVALUATION

Orientation
Attention/concentration
Language functions
Intellectual functions
Reasoning, concept formation, problem-solving, mental flexibility
Executive functions
Learning and memory
Information processing capacity and speed
Basic sensory-perceptual processes
Visuospatial, visuoperceptual, and visuoconstructive abilities
Psychomotor functions
Emotional/personality status
Substance abuse

neuropsychological evaluation of head trauma in order to ensure both comprehensiveness and specificity. Cognitive functions are represented at both a general and specific level. In addition, specificity of processes most susceptible to head trauma is reflected by the inclusion of attention, learning/memory, information processing ability, psychomotor speed, and personality, and executive control functions.

Of course, the areas noted in Table 2 are general in nature and risk obscuring the obvious complexity of the constructs. The neuropsychologist's understanding and conceptual appreciation of the constructs guides the assessment strategy at the level of instrument selection. For example, attention is regarded as a nonunitary process whose assessment must necessarily be multifaceted. Neuropsychologists often administer multiple attentional measures which variously tap immediate versus sustained concentration, auditory versus visual attention, and attentional capacity under slower versus more rapid processing conditions.

In addition, neuropsychologists are often faced with a choice of measures for each dimension of each major ability area, which may be equally desirable based on the available literature supporting their general psychometric properties or validity for head trauma populations. For example, to assess one component of learning/memory, namely auditory verbal learning, the Rey Auditory Verbal Learning Test (Lezak, 1983), Buschke Selective Reminding Test (Buschke, 1973), California Verbal Learning Test (Delis, Kramer, Kaplan, & Ober, 1987), or the paradigm included in the Memory Assessment Scales (Williams, 1990) could be used. Despite the fact that each of these measures has respective strengths and weaknesses, a case could probably be made that all are reasonably sensitive to auditory verbal learning deficits in head injury. Frequently, the choice is dictated by personal preference, experience with a particular instrument, and/or clinical research issues.

Table 3 presents the basic head injury battery used at the outpatient assessment laboratory at the University of Virginia Health Sciences Center. Although the practice of neuropsychology at this laboratory originally grew out of the Halstead–Reitan orientation, the current approach to the evaluation of head trauma is best

TABLE 3. BASIC HEAD INJURY BATTERY AT THE UNIVERSITY
OF VIRGINIA

Galveston Orientation and Amnesia Test
Trail Making Test[a]
Seashore Rhythm Test[a]
Paced Auditory Serial Addition Test-Revised
Aphasia Screening Examination[a]
Controlled Oral Word Association Test
Wechsler Adult Intelligence Scale-Revised
Halstead Category Test[a] and/or Wisconsin Card Sorting Test
Buschke Selective Reminding Test
Wechsler Memory Scale-Revised (Logical Memory, Visual Reproduction)
Sensory-Perceptual Examination[a]
Rey Complex Figure Test
Hooper Visual Organization Test
Finger Oscillation Test[a]
Lafayette Grooved Pegboard
Academic achievement battery (Wide Range Achievement Test-3)
Minnesota Multiphasic Personality Inventory-2
Beck Depression Inventory

---

[a]From the Halstead–Reitan Neuropsychological Test Battery.

described as a flexible one. Like Lezak (1983), a basic battery based on recent research findings regarding head injury is used and modified according to information contained in the consult and medical record, hypotheses suggested by the clinical interview, as well as the patient's behavior during testing. The basic battery presented in Table 3 reflects an effort to operationalize the abilities set forth in Table 2, again in terms of both comprehensiveness and specificity for head trauma.

Assessment of some cognitive processes cuts across specific measures listed in Table 3. Returning to the concept of attention as an example, at least seven tests and subtests provide information on various facets of attention/concentration: the Digit Span, Arithmetic, Digit Symbol, and Picture Completion subtests from the WAIS-R, Trail Making Test (Reitan & Wolfson, 1993), Seashore Rhythm Test (Reitan & Wolfson, 1993), and Paced Auditory Serial Addition Test-Revised (Brittain, LaMarche, Reeder, Roth, & Boll, 1991; Gronwall, 1977). Of course, some of these instruments tap components other than attention, but depending on the pattern of findings, often shed light on attentional processes. In addition, other measures (e.g., memory) may also reflect impairment in attention.

In a flexible approach, the basic battery presented in Table 3 must sometimes be modified. For example, although the Halstead Category Test is typically administered as a measure of higher-level abstraction and concept formation, new problem-solving, and mental flexibility, there are instances when a simpler measure of nonverbal reasoning ability (e.g., Raven's Standard or Coloured Progressive Matrices; Raven, 1938) is substituted. Such a decision might be based on clinical judgment that the patient would perform poorly not only due to his or her head injury, but also because of low premorbid functioning in this ability area. An individual with a history of limited sociocultural opportunities and only four years of formal education might warrant substitution of the Raven's for the Category Test. There are also assessment situations (e.g., a question of cognitive perseveration in a head-injured patient with suspected frontal lobe impairment) in which the assessment of concept formation may be expanded to include both the Category Test and the Wisconsin Card Sorting Test (Heaton, 1981; Perrine, 1993).

Another example of the need for flexibility in head injury assessment concerns a patient's emotional/personality status. Although the Depression scale from the MMPI-2 is reasonably sensitive to depression, which is relatively common after head trauma, there are instances in which a more comprehensive assessment of depression is desirable based on information obtained from the history and clinical interview. In such cases, an instrument such as the Beck Depression Inventory (Beck, 1987) may be added or substituted for the MMPI-2. Exner's Rorschach Comprehensive System (Exner, 1993) might also be included to further elucidate psychopathology such as preexisting or comorbid states that might be complicating recovery from a head injury.

As previously noted, no current standardized neuropsychological tests appear adequate for a comprehensive assessment of executive control dysfunction due to common frontal lobe damage after head injury. The importance of a careful history, clinical interview of both patient and informants, addition of specific methods to assess certain frontal-lobe functions (see Malloy & Richardson, 1994), and perhaps use of a structured data gathering tool such as the PRO-EX described earlier, should be seriously considered.

In addition, despite its inclusion as a core area of inquiry in Table 2, no specific instrument for the assessment of alcohol and drug abuse is listed in Table 3. An appropriate assessment of substance abuse is best undertaken through a structured interview format, rather than through standardized tests such as the MacAndrews Scale from the MMPI (MacAndrew, 1965; Lee, 1992) or Michigan Alcoholism Screening Test (MAST; Selzer, 1971; Lee, 1992). These instruments, including short forms of the MAST (Zung, 1984), have been shown to have questionable validity (Lee, 1992). Although widely used, the MAST is known to yield a high rate of false-negatives in that individuals with alcohol problems often go undetected (as a result of denial or lying) on this face-valid, self-report questionnaire.

Appropriate assessment of substance abuse, as with frontal lobe dysfunction, falls back on the clinical skills of the neuropsychologist in conjunction with procedures other than standardized neuropsychological tests. A structured interview, such as the Substance Use Disorder Diagnostic Schedule (SUDDS; Harrison & Hoffman, 1989), is highly recommended for the all too frequent alcohol problem that may have precipitated the patient's head injury. Elevated blood alcohol levels are often recorded during emergent care of the head-injured patient (especially after a motor vehicle accident), but the neuropsychologist may be the only health care professional involved in the chain of follow-up care with the specific skills and opportunity to make a competent diagnosis or much needed treatment recommendations.

FEEDBACK TO THE PATIENT AND OTHERS

Feedback by the neuropsychologist to the head-injured patient and his or her family is an integral part of the evaluation process (Gass & Brown, 1992; Pope, 1992). There are several reasons why feedback is important in the context of neuropsychological evaluation of head trauma. First, in contrast to some laboratory and consultation services (e.g., neuroradiology), neuropsychological test data involve actual behavior. Patients, family members, educators, and/or employers can therefore more easily understand the test results, which may have clear implications for educational or vocational goals, everyday functioning, decision making, and environmental strategies. Neuropsychologists as behavioral experts are best suited to interpret and communicate evaluation findings compared to most other service providers.

Second, as consumers of neuropsychological services, head-injured individuals, like other patients, have a right to be informed in plain language of the assessment results and their implications (Gass & Brown, 1992). This right to receive and corresponding responsibility to provide interpretive feedback derives from established guidelines for professional conduct to which neuropsychologists are obligated to adhere (American Psychological Association, 1987, 1992). The feedback process provides ample opportunity for neuropsychologists to address questions and concerns patients may have about the test procedures, diagnostic impressions, and recommendations. When construed as a dynamic, interactive process, assessment often continues during the feedback session (Pope, 1992), which can further increase understanding of the cognitive or behavioral consequences suffered by the head-injured patient. A final ethical/professional issue particularly relevant to feedback in head injury cases concerns the neuropsychologist's responsibility to prepare the patient for the possibility that the test findings could eventually be requested or subpoenaed in civil actions such as a disability evaluation, worker compensation claim, or tort litigation, even though it may not have been part of the referral question (Pope, 1992).

Third, the feedback session has definite value in providing emotional support and practical guidance for decision-making to the head-injured person and family members. Along with the patient, families are often thrown into a state of uncertainty, distress, and readjustment as they struggle to cope with a general lack of understanding of head injury, its functional consequences, and prognosis; helplessness in new roles as caregivers to a family member now with special needs; and altered behaviors in the patient, some of which may be frustrating, inappropriate, embarrassing, or dangerous (Gass & Brown, 1992).

Through a sensitive feedback session, the neuropsychologist can provide supportive and educative counseling which clarifies common neuropsychological deficits after head injury and outlines typical recovery curves. In addition, common tendencies of families either to overprotect or to deny the extent of the patient's deficits (which may increase potential for failure experiences) can be addressed (Gass & Brown, 1992). These feedback functions help foster realistic expectations, accurate understanding, and better emotional well-being for patient and family. They represent a valuable and often critical contribution by the neuropsychologist in an often fragmented health care delivery system in which primary providers may have little time to spend with individual patients (Pope, 1992). The exact manner in which neuropsychologists provide feedback varies with the specific setting (e.g., acute neurosurgical versus rehabilitation).

Neuropsychological Evaluation of Penetrating versus Closed Head Injury

Kampen and Grafman (1989) assert that a different neuropsychological assessment strategy is needed for PHI patients because focal deficits prevail over the diffuse cognitive dysfunction more often seen after CHI. In addition, because of the idiosyncratic nature of PHI wounds (which cross structural and vascular boundaries), the resulting focal deficits may be themselves idiosyncratic compared to CHI and may not correspond to those typically seen in other focal neurological syndromes such as stroke. For these reasons, Kampen and Grafman (1989) advocate that assessments need to be especially broad-based in nature, and that evaluation of specific deficit areas should be detailed and multifactorial, even though in both instances conventional neuropsychological tests may be used. The implication is that neuropsychologists may expect to find, and should slant their assessments toward, specific, subtle, or unique patterns of focal cognitive dysfunction.

There is, of course, overlap in the neuropsychological evaluation of PHI and CHI. Because memory impairment is common after PHI, memory testing should be as thorough in PHI as it is in CHI. In addition, a complete examination of frontal lobe functioning (emotional, motor, and executive functions) should routinely be included because the frontal lobes are often the point of entry in PHI. Such frontal lobe assessment, as stated earlier, is particularly challenging since few, if any, standard neuropsychological test procedures are exclusively sensitive to functions subserved by this specific neuroanatomic area and its systems.

Neuropsychological Evaluation of Head Injury in Children

Neuropsychological evaluation of head injury in children is both similar to and distinct from evaluation of head injury in adults. Children can exhibit any of the behavioral sequelae of head injury seen in adults. In addition, the stages along the continuum of recovery (e.g., impaired consciousness, posttraumatic amnesia) are all seen in children, although children tend to progress more rapidly through the initial stages than do adults.

The neuropsychological evaluation of head injury in children should be comprehensive and sensitive to all of the processes and abilities listed in Table 2. When evaluating older children and adolescents, routine administration of academic achievement tests (e.g., Wechsler Individual Achievement Test, Psychological Corporation, 1992; Peabody Individual Achievement Test-Revised, Markwardt, 1989; Woodcock–Johnson Psychoeducational Battery-Revised, Woodcock & Johnson, 1989; Wide Range Achievement Test-3, Wilkinson, 1993), as well as a measure of general intellectual functioning (e.g., Wechsler Intelligence Scale for Children-Third Edition [WISC-III], Wechsler, 1991), is important. The appropriate age range versions of the HRB Category Test, Trail Making Test, Aphasia Screening Examination, Sensory-Perceptual Examination, and Finger Oscillation Test (Reitan & Wolfson, 1993) can be all administered. In addition, the Buschke Selective Reminding Test (Clodfelter, Dickson, Newton Wilkes, & Johnson, 1987; Morgan, 1982), California Verbal Learning Test-Children's Version (Delis, Kramer, Kaplan, & Ober, 1994), Wide Range Assessment of Memory and Learning (Sheslow & Adams, 1990), Test of Learning and Memory (Reynolds & Bigler, 1994), Rey Complex Figure, and Lafayette Grooved Pegboard can be used as measures of memory, constructional ability, and motor coordination, respectively. These procedures may be supplemented with other instruments as dictated by the individual case. With age-appropriate adolescents, the MMPI-Adolescent Edition (Butcher *et al.*, 1992) can provide useful information about personality functioning.

Neuropsychological assessment of head injury in younger children is complicated by all of the factors that limit the reliability and validity of any psychological assessment of younger children. Most notably, these factors include the increased significance of behavioral variability and rapport, and the importance of maintaining an appropriate test taking set. Moreover, traditional neuropsychological procedures are often less appropriate with younger children due to the relative plasticity of the developing brain. For example, lateralization of cerebral function is not evident under the age of 6, and is not well established until approximately 9 or 10 years of age.

Accordingly, assessment of head injury in younger children is much less consistent than work with older children and adolescents. Nevertheless, routine administration of academic achievement tests and a measure of general intellectual functioning is indicated. Usually, the WISC-III is appropriate, although alternative procedures such as the Differential Abilities Test (Elliot, 1990) or the Kaufman Assessment Battery for Children (Kaufman & Kaufman, 1983) may be substituted. Beyond this, test selection depends on individual situations consistent with a flexible battery approach. The Motor scale of the McCarthy Scales of Children's Abilities (McCarthy, 1972), the G-F-W Auditory Selective Attention Test (Goldman, Fristoe, & Woodcock, 1974), and the copy-only format of the Rey Complex Figure are useful procedures for assessing the educationally relevant factors of fine and gross motor skills, auditory attention, and graphomotor skills, respectively.

In evaluating children with head injury, it is necessary to consider two interrelated aspects of development: the internal neurocognitive development of brain function (Allison, 1992; Hynd & Willis, 1987; Kolb & Fantie, 1989) and the external changes in development demands and expectations of the environment (Rudel,

Holmes, & Pardes, 1988; Ylvisaker, Hartwick, & Stevens, 1991). In adults, the typical course for a head injury is for initial impairment and subsequent improvement of function over an 18- to 24-month period, with the potential for permanent residual sequelae. Although this pattern may be seen in children, especially in cases involving focal and unilateral insult, it is crucial to consider the possibility that childhood head injury may lead to manifestation of difficulties at a later age. Such "delayed deficit" may be attributed either to damage to areas necessary for the emergence of more developmentally mature behaviors or to interference with the course of neurodevelopment itself, thus resulting in greater cognitive impairment in the developing brain than in the mature brain (Capruso & Levin, 1996). Rourke, Fisk, and Strang (1986) have suggested that the emergence of executive control dysfunction in individuals with a history of childhood head injury may reflect an example of delayed deficit.

Evaluation of head injury in children also requires consideration of family factors (e.g., socioeconomic disadvantage), dynamics (e.g., marital conflict), and the emotional response (e.g., guilt, anger) of the child's caretakers. Two common responses of caretakers to a child's head injury are denial and overprotection. These factors can have significant effects on a child's optimal recovery and psychosocial outcome. Premorbid learning and developmental disabilities, behavioral disorders, and poor academic achievement are risk factors for head injury and complicate outcome.

Due to development and psychosocial factors, the evaluation of head injury in children frequently evolves into a long-term relationship with the child and the family. For evaluations in the early recovery period, it is important to stress the need for flexibility. An early return of the child to school (perhaps with partial day attendance) may be encouraged, but it will be important to caution against resumption of preinjury academic expectations. In most cases, the return to school is framed as allowing reintroduction of the learning set and fostering of age-appropriate peer interactions. Specific academic recommendations, such as the possible need for long-term special education services, are generally best postponed until approximately 9 to 12 months following head injury, postponement of specific recommendations allows the caretakers and the child time to come to terms with the head injury.

In summary, perhaps even more so than with adults, the neuropsychological evaluation of head injury in children involves more than specific procedures and tests. That is, the evaluation and clinical consultation often represent a long-term process that requires ongoing consideration of neuropsychological, developmental, educational, and psychosocial factors.

FORENSIC NEUROPSYCHOLOGICAL EVALUATION

Neuropsychological assessment of head injury within a medicolegal context involves unique issues and considerations. Ideally, a forensic head trauma evaluation and a comprehensive nonforensic (clinical) head injury assessment should be equivalent. However, typical legal issues such as disability determination and personal injury litigation demand more attention to level and quality of premorbid functioning, details of medical and psychosocial history, and specific facts regarding the accident and acute injury characteristics. All of this information is critical if the neuropsychological assessment and neurocognitive outcome are to be placed in the appropriate context with regard to what the injury costs the individual and family. The neuropsychologist should be in a position to address these issues through an evaluation of all of the relevant circumstances and their relationship to the head injury assessment data.

The medical and psychosocial history can be complied from several sources, including direct review of medical, employment, and school records, which the attorney (if requesting the assessment) should provide (Barth, Ryan, Schear, & Puente, 1992). Interviews with family members, school personnel, employers, and other treating physicians and health care professionals can augment the record review, patient interview, and behavioral observations.

Because moderate to severe head injuries may follow an 18- to 24-month recovery curve, with the most notable improvements occurring within the first 3 to 6 months and slower subsequent progression, 3-month, 1-year, and 2-year assessments using comprehensive neuropsychological assessment techniques with alternative forms are recommended (see Table 3). Most instances of single, uncomplicated mild head injury show an expected course of recovery which is more rapid and complete, though in some cases protracted and complicated recovery has been documented. Serial evaluations allow documentation of deficits and expected improvements in cognitive functions over time. It is furthermore important to consider and address the effects of chronic pain, reactive anxiety and depression, fatigue, frustration, and/or personality changes on test performance over time (see Table 1).

When litigation is involved in head trauma assessment, the neuropsychologist must be sensitive to and consider the possibility of malingering or exaggerated impairment. The *Diagnostic and Statistical Manual of Mental Disorders Fourth Edition* (American Psychiatric Association, 1994) defines malingering as "intentional production of false or grossly exaggerated physical or psychological symptoms, motivated by external incentives such as avoiding military duty, avoiding work, obtaining financial compensation, evading criminal prosecution, or obtaining drugs" (p. 683). Malingering is not a diagnosis, but rather is considered a condition *not* attributable to a mental disorder but a focus of clinical attention or treatment. Since the base rate of frank malingering is unknown, Binder (1992) has advocated that "the only sensible strategy is to consider the possibility of malingering in each and every patient who has any monetary or external incentives for faking bad on a neuropsychological examination. Any other tactic is indefensible" (p. 355). A combination of strategies involving both interview and testing methods has been recommended in this regard. When assessing the head-injured individual who is involved in litigation, it is important to reassure the patient beforehand that the neuropsychological test procedures are sensitive enough to assess adequately the neurocognitive deficits associated with their injury, at the same time warning them that the tests will also provide an index of their test-taking efforts which will necessarily be documented in the neuropsychological report (Barth *et al.*, 1992). Formal testing of level of motivation and potential malingering through the use of multiple methods, including symptom validity assessment and recognition of test performance patterns associated with poor effort, should also be considered when addressing the issue of malingering or exaggerated impairment (e.g., Binder & Willis, 1991; Guilmette, Hart, & Giuliano, 1993; Mittenberg, Azrin, Millsaps, & Heilbronner, 1993).

As Barth *et al.* (1992) have stated, it is important to "recognize . . . that there remains today considerable controversy about the accurate determination of malingering and that an obvious spectrum of motivational issues should be considered if malingering is suspected" (p. 538). Issues such as exaggeration of symptoms, non-deliberate distortion, Ganser's syndrome, factitious disorders, Munchausen's syndrome, somatoform or conversion disorders, and posttraumatic stress disorder also require consideration when motivation and malingering are addressed in neuropsychological assessment.

COMPUTERIZED NEUROPSYCHOLOGICAL EVALUATION OF HEAD TRAUMA

The past decade has seen increased reliance on online computer administration of neuropsychological tests (Kane & Kay, 1992), including adaptations or analogues of many measures found in conventional head injury batteries (e.g., finger oscillation, reaction time, symbol digit, digit span, trail making, auditory rhythm, letter cancellation, selective reminding, and auditory verbal learning tasks). Reports have begun to appear which either contrast standard and computer versions of existing tests, or compare the new analogues with their traditional counterparts. For example, preliminary findings suggest equivalence between the standard and computerized version of the Halstead Category Test (Choca & Morris, 1992). On the other hand, in comparing analogue measures of memory and attention and concentration with traditional tests, psychometric properties were either better or worse depending on the ability area assessed (Youngjohn, Larrabee, & Crook, 1992).

Considerable work remains to be done in (1) comparing standard versus computerized versions of established tests, (2) contrasting analogue and conventional measures, and (3) validating and standardizing the computer tests themselves. Few of the computerized tests or batteries have yet to be validated adequately for use with head-injured populations, although initial studies are in progress (Kane & Kay, 1992). The proponents of computerized neuropsychological testing cite its various advantages such as standard administration, increased scoring accuracy, cost efficiency, and data base management (Kane & Kay, 1992). As better validation and marketing are achieved, a gain in the popularity of computerized approaches to the neuropsychological evaluation of head trauma can be expected. However, the continued presence of well-trained professional staff to make expert observations of human behavior during testing, and to conduct supplemental mental status and neurobehavioral testing, will be critical in the foreseeable future. Moreover, ease of administration and scoring in computerized test batteries should never outweigh professional standards in the use of these instruments (Division 40 task force report on computer-assisted neuropsychological evaluation, 1987).

SUMMARY AND FUTURE DIRECTIONS

Over the past two decades, clinical neuropsychology has added a new dimension to the clinical evaluation of head trauma by extending and refining standardized methods for examining common cognitive and behavioral sequelae which rely on empirically derived normative data. Using these methods, research investigations have been able to study and specify differential recovery patterns, natural history, and neurobehavioral outcomes over time, which are more precisely referable to head injury severity, age, and other factors. Reasonable approaches to neuropsychological assessment of head-injured patients can be based on either fixed battery or flexible strategies as long as the evaluation retains both comprehensiveness and specificity for common neuropsychological deficits and individual case circumstances and features. In either case, it is imperative that neuropsychologists supplement standardized tests with a careful history and clinical interview to assess potential unawareness of deficits, executive control dysfunction, adjustment problems, and substance abuse, and provide psychoeducational and supportive feedback to the patient and family after the assessment to facilitate postinjury adaptation and recovery.

There are a number of ways to improve the contribution of clinical neuropsychology to the evaluation and treatment of head trauma. Clearly, the ecological validity of its standardized measures continues to warrant further study, as does development and validation of both computerized and brief test batteries for specific diagnostic or prognostic questions. Given the multidimensional nature of most neuropsychological constructs (i.e., attention, memory, executive functions), identification of which aspects of these general constructs are more often disturbed by head trauma represents an important research focus. Clarification of the essential characteristics of mild head injury, and of persisting effects of mild head injury over periods of several years, will likely contribute to some reduction of the controversy that surrounds these issues. In addition, research on neuropsychological outcome is still in a state of relative infancy, particularly regarding later stages of recovery in moderate and severe injuries, long-term outcomes of a representative group of persons with mild head injury, the effects of multiple mild head injuries, and efficacy of neuroprotective drugs and standardized treatment protocols for postconcussive syndrome. Such work will allow neuropsychological knowledge and practice to keep pace with advances in the basic, cognitive, and applied clinical neurosciences related to head injury, and will augment the data base on what remains a significant public health problem.

REFERENCES

Adams, J. H., Doyle, D., Graham, D. I., Lawrence, A. E., & McLellan, D. R. (1986). Gliding contusions in nonmissile head injury in humans. *Archives of Pathology and Laboratory Medicine, 110,* 485–488.

Adams, J. H., Graham, D. I., Murray, L. S., & Scott, G. (1982). Diffuse axonal injury due to nonmissile head injury in humans: An analysis of 45 cases. *Annals of Neurology, 12,* 557–563.

Alexander, M. P. (1995). Mild traumatic brain injury: Pathophysiology, natural history, and clinical management. *Neurology, 45,* 1253–1260.

Alexandre, A., Colombo, F., Nertempi, P., & Benedetti, A. (1983). Cognitive outcome and early indices of severity of head injury. *Journal of Neurosurgery, 59,* 751–761.

Allison, M. (1992). The effects of neurologic injury on the maturing brain: New developments. *Headlines, 3,* 2–11.

American Psychiatric Association. 1994. *Diagnostic and statistical manual of mental disorders* (4th ed.). Washington, DC: Author.

American Psychological Association. (1987). *General guidelines for providers of psychological services.* Washington, DC: Author.

American Psychological Association. (1992). Ethical principles of psychologists and code of conduct. *American Psychologist, 47,*1597–1611.

Barth, J. T., Macciocchi, S. N., Giordani, B., Rimel, R., Jane, J. A., & Boll, T. J. (1983). Neuropsychological sequelae of minor head injury. *Neurosurgery, 13,* 529–533.

Barth, J. T., Ryan, T. V., Shear, J. M., & Puente, A. E. (1992). Forensic assessment and expert testimony in neuropsychology. In S. Hanson & D. Tucher (Eds.), *Neuropsychological assessment: Diagnostic and* clinical applications. *Physical medicine and rehabilitation: State of the art review* (pp. 531–546). Philadelphia: Hanley and Belfus.

Beck, A. T. (1987). *Beck Depression Inventory: Manual.* San Antonio, TX: Psychological Corporation.

Beers, S. R. (1992). Cognitive effects of mild head injury in children and adolescents. *Neuropsychology Review, 3,* 281–320.

Binder, L. M. (1992). Deception and malingering. In A. E. Puente & R. J. McCaffrey (Eds.), *Handbook of neuropsychological assessment: A psychosocial perspective* (pp. 353–374). New York: Plenum.

Binder, L. M., & Willis, S. C. (1991). Assessment of motivation after financially compensable minor head injury. *Psychological Assessment, 3,* 175–181.

Braswell, D., Hartry, A., Hoornbeek, S., Johansen, A., Johnson, L., Schultz, J., & Sohlberg, M. M. (1993). *Manual for the Profile of Executive Control System (PRO-EX).* Puyallup, WA: Association for Neuropsychological Research and Development.

Brittain, J. L., LaMarche, J. A., Reeder, K. P., Roth, D. L., & Boll, T. J. (1991). Effects of age and IQ on Paced Auditory Serial Addition Task (PASAT) performance. *Clinical Neuropsychologist, 5,* 163–175.

Brooks, N. (1989). Closed head trauma: Assessing the common cognitive problems. In M. D. Lezak (Ed.), *Assessment of the behavioral consequences of head trauma* (pp. 61–85). New York: Alan R. Liss.

Buschke, H. (1973). Selective reminding for analysis of memory and learning. *Journal of Verbal Learning and Verbal Behavior, 12,* 543–550.

Butcher, J. N., Williams, C. L., Graham, J. R., Archer, R. P., Tellegen, A., Ben-Porath, Y. S., & Kaemmer, B. (1992). *MMPI-A: Minnesota Multiphasic Personality Inventory-Adolescent Edition: Manual for administration, scoring, and interpretation.* Minneapolis, MN: University of Minnesota Press.

Capruso, D. X., & Levin, H. S. (1996). Neurobehavioral outcome of head trauma. In R. W. Evans (Ed.), *Neurology and trauma* (pp. 201–221). Philadelphia: W. B. Saunders Co.

Chelune, G. J., & Moehle, K. A. (1986). Neuropsychological assessment and everyday functioning. In D. Wedding, A. M. Horton, Jr., & J. Webster (Eds.), *The neuropsychology handbook: Behavioral and clinical perspectives* (pp. 489–525). New York: Springer.

Choca, J., & Morris, J. (1992). Administering the Category Test by computer: Equivalence of results. *The Clinical Neuropsychologist, 6,* 9–15.

Choi, S. C., Barnes, T. Y., Bullock, R., Germanson, T. A., Marmarou, A., & Young, H. F. (1994). Temporal profile of outcomes in severe head injury. *Journal of Neurosurgery, 81,* 169–173.

Christensen, A. L. (1975). *Luria's neuropsychological investigation.* New York: Spectrum.

Clodfelter, C. J., Dickson, A. L., Newton Wilkes, C., & Johnson, R. B. (1987). Alternate forms of selective reminding for children. *The Clinical Neuropsychologist, 1,* 243–249.

Cooper, P. R., Moody, S., Clark, W. K., Kirkpatrick, J., Maravilla, K., Gould, A. L., & Drane, W. (1979). Dexamethasone and severe head injury: A prospective double blind study. *Journal of Neurosurgery, 51,* 307–316.

Crepeau, F., & Scherzer, P. (1993). Predictors and indicators of work status after traumatic brain injury: A meta-analysis. *Neuropsychological Rehabilitation, 3,* 5–35.

Crovitz, H. F. (1987). Techniques to investigate posttraumatic and retrograde amnesia after head injury. In H. S. Levin, J. Grafman, & H. M. Eisenberg (Eds.), *Neurobehavioral recovery from head injury* (pp. 330–340). New York: Oxford University Press.

Delis, D. C., Kramer, J. H., Kaplan, E., & Ober, B. A. (1994). *California Verbal Learning Test—Children's Version.* San Antonio, TX: Psychological Corporation.

Dikmen, S. S., Donovan, D. M., Loberg, T., Machamer, J. E., & Temkin, N. R. (1993). Alcohol use and its effects on neuropsychological outcome in head injury. *Neuropsychology, 7,* 296–305.

Dikmen, S., Machamer, J., Temkin, N., & McLean, A. (1990). Neuropsychological recovery in patients with moderate to severe head injury: 2 year follow-up. *Journal of Clinical and Experimental Neuropsychology, 12,* 507–519.

Dikmen, S. S., Machamer, J. E., Winn, H. R., & Temkin, N. R. (1995). Neuropsychological outcome at 1 year post–head injury. *Neuropsychology, 9,* 80–90.

Dikmen, S., McLean, A., & Temkin, N. (1986). Neuropsychological and psychosocial consequences of minor head injury. *Journal of Neurology, Neurosurgery, and Psychiatry, 49,* 1227–1232.

Dikmen, S., & Reitan, R. M. (1976). Psychological deficits and recovery of functions after head injury. *Transactions of the American Neurological Association, 101,* 72–77.

Dikmen, S., & Reitan, R. M. (1977a). Emotional sequelae of head injury. *Annals of Neurology, 2,* 492–494.

Dikmen, S., & Reitan, R. M. (1977b). MMPI correlates of adaptive ability deficits in patients with brain lesions. *Journal of Nervous and Mental Disease, 165,* 247–254.

Dikmen, S., & Reitan, R. M. (1978). Neuropsychological performance in post-traumatic epilepsy. *Epilepsia, 19,* 177–183.

Dikmen, S., Reitan, R. M., & Temkin, N. R. (1983). Neuropsychological recovery in head injury. *Archives of Neurology, 40,* 333–338.

Dikmen, S. S., Temkin, N. R., Machamer, J. E., Holubkov, A. L., Fraser, R. T., & Winn, H. R. (1994). Employment following traumatic head injuries. *Archives of Neurology, 51,* 177–186.

Division 40 task force report on computer-assisted neuropsychological evaluation (1987). *The Clinical Neuropsychologist, 2,* 161–184.

Dixon, C. E., Taft, W. C., & Hayes, R. L. (1993). Mechanisms of mild traumatic brain injury. *Journal of Head Trauma Rehabilitation, 8,* 1–12.

Elliot, C. C. (1990). *Differential Abilities Scales.* San Antonio, TX: Psychological Corporation.

Exner, J. E., Jr. (1993). *The Rorschach: A comprehensive system: Vol. 1. Basic foundations* (3rd ed). New York: Wiley.

Feler, C. A., & Watridge, C. B. (1992). Initial management of head trauma. In C. J. Long & L. K. Ross (Eds.), *Handbook of head trauma: Acute care to recovery* (pp. 19–31). New York: Plenum.

Filskov, S. B., & Leli, D. A. (1981). Assessment of the individual in neuropsychological practice. In S. B. Filskov & T. J. Boll (Eds.), *Handbook of clinical neuropsychology* (Vol. 1, pp. 545–576). New York: Wiley.

Fleischer, A. S., Payne, N. S., & Tindall, G. T. (1976). Continuous monitoring of intracranial pressure in severe closed head injury without mass lesions. *Surgical Neurology, 6,* 31–34.

Gass, C. S., & Brown, M. C. (1992). Neuropsychological test feedback to patients with brain dysfunction. *Psychological Assessment, 4,* 272–277.

Golden, C. J. (1981). A standardized version of Luria's neuropsychological tests: A quantitative and qualitative approach to neuropsychological evaluation. In S. B. Filskov & T. J. Boll (Eds.), *Handbook of clinical neuropsychology* (Vol. 1, pp. 608–642). New York: Wiley.

Golden, C. J., Hammeke, T. A., & Purisch, A. D. (1978). Diagnostic validity of a standardized neuropsychological battery derived from Luria's neuropsychological tests. *Journal of Consulting and Clinical Psychology, 46,* 1258–1265.

Golden, C. J., Hammeke, T. A., Purisch, A. D., & Moses, J. A. (1984). *A manual for the administration and interpretation of the Luria–Nebraska Neuropsychological Battery.* Los Angeles: Western Psychological Services.

Golden, C. J., & Maruish, M. (1986). The Luria–Nebraska Neuropsychological Battery. In D. Wedding, A. M. Horton, Jr., & J. Webster (Eds.), *The neuropsychology handbook: Behavioral and clinical perspectives* (pp. 161–193). New York: Springer.

Golden, C. J., Purisch, A. D., & Hammeke, T. A. (1985). *Luria–Nebraska Neuropsychological Battery: Forms I and II.* Los Angeles: Western Psychological Services.

Goldman, R., Fristoe, M., & Woodcock, R. W. (1974). *G-F-W Auditory Selective Attention Test.* Circle Pines, MN: American Guidance Service.

Goldman-Rakic, P. S. (1993). Specification of higher cortical functions. *Journal of Head Trauma Rehabilitation, 8,* 13–23.

Goldstein, G., Shelly, C., McCue, M., & Kane, R. L. (1987). Classification with the Luria–Nebraska Neuropsychological Battery: An application of cluster and ipsative profile analysis. *Archives of Clinical Neuropsychology, 2,* 215–235.

Goldstein, M. (1990). Traumatic brain injury: A silent epidemic. *Annals of Neurology, 27,* 327.

Grafman, J., & Salazar, A. (1987). Methodological considerations relevant to the comparison of recovery from penetrating and closed head injuries. In H. S. Levin, J. Grafman, & H. M. Eisenberg (Eds.), *Neurobehavioral recovery from head injury* (pp. 43–54). New York: Oxford University Press.

Graham, J. R. (1990). *MMPI-2: Assessing personality and psychopathology.* New York: Oxford University Press.

Gronwall, D. M. A. (1977). Paced Auditory Serial Addition Task: A measure of recovery from concussion. *Perceptual and Motor Skills, 44,* 367–373.

Gronwall, D. (1989). Behavioral assessment during the acute stages of traumatic brain injury. In M. D. Lezak (Ed.), *Assessment of the behavioral consequences of head trauma* (pp. 19–36). New York: Alan R. Liss.

Gronwall, D., & Wrightson, P. (1980). Duration of post-traumatic amnesia after mild head injury. *Journal of Clinical Neuropsychology, 2,* 51–60.

Gronwall, D., Wrightson, P., & Waddell, P. (1990). *Head injury: The facts.* New York: Oxford University Press.

Guilmette, T. J., Hart, K. J., & Giuliano, A. J. (1993). Malingering detection: The use of a forced-choice method in identifying organic versus simulated memory impairment. *The Clinical Neuropsychologist, 7,* 59–69.

Harrison, C. L. & Dijkers, M. (1992). Traumatic brain injury registries in the United States: An overview. *Brain Injury, 6,* 203–212.

Harrison, P. A., & Hoffman, N. G. (1989). *SUDDS Substance Use Disorder Diagnostic Schedule manual.* St. Paul, MN: Ramsey Clinic.

Hayes, R. L., Jenkins, L. W., & Lyeth, B. G. (1992). Neurochemical aspects of head injury: Role of excitatory neurotransmission. *Journal of Head Trauma Rehabilitation, 7,* 16–28.

Heaton, R. K. (1981). *Wisconsin Card Sorting Test manual.* Odessa, FL: Psychological Assessment Resources.

Heaton, R. K., Grant, I., & Matthews, C. G. (1991). *Comprehensive norms for an expanded Halstead–Reitan battery.* Odessa, FL: Psychological Assessment Resources.

Heinrichs, R. W. (1990). Current and emergent applications of neuropsychological assessment: Problems of validity and utility. *Professional Psychology: Research and Practice, 21,* 171–176.

High, W. M., Jr., Levin, H. S., & Gary, H. E., Jr. (1990). Recovery of orientation following closed head injury. *Journal of Clinical and Experimental Neuropsychology, 12,* 703–714.

Hynd, G. W., & Willis, W. G. (1987). *Pediatric neuropsychology.* Orlando, FL: Grune & Stratton.

Jarvis, P. E., & Barth, J. T. (1994). *The Halstead–Reitan Neuropsychological Battery: A guide to interpretation and clinical applications.* Odessa, FL: Psychological Assessment Resources.

Jones, C. L. (1992). Recovery from head trauma: A curvilinear process? In C. J. Long & L. K. Ross (Eds.), *Handbook of head trauma: Acute care to recovery* (pp. 247–270). New York: Plenum.

Jones, R. D. (1996). Neuropsychological assessment of patients with traumatic brain injury: The Iowa–Benton approach. In M. Rizzo and D. Tranel (Eds.), *Head injury and postconcussive syndrome* (pp. 375–393). New York: Churchill Livingstone.

Joseph, R. (1990). *Neuropsychology, neuropsychiatry, and behavioral neurology.* New York: Plenum.

Kampen, D. L., & Grafman, J. (1989). Neuropsychological evaluation of penetrating head injury. In M. D. Lezak (Ed.), *Assessment of the behavioral consequences of head trauma* (pp. 49–60). New York: Alan R. Liss.

Kane, R. L. (1991). Standardized and flexible batteries in neuropsychology: An assessment update. *Neuropsychology Review, 2,* 281–339.

Kane, R. L., & Kay, G. C. (1992). Computerized assessment in neuropsychology: A review of tests and test batteries. *Neuropsychology Review, 3,* 1–117.

Kaplan, E. (1988). A process approach to neuropsychological assessment. In T. Boll & B. K. Bryant (Eds.), *Clinical neuropsychology and brain function: Research, measurement, and practice.* (pp. 129–167). Washington, DC: American Psychological Association.

Kaplan, E. (1990). The process approach to neuropsychological assessment of psychiatric patients. *Journal of Neuropsychology and Clinical Neurosciences, 2,* 72–87.

Karzmark, P. (1992). Prediction of long-term cognitive outcome of brain injury with neuropsychological, severity of injury, and demographic data. *Brain Injury, 6,* 213–217.

Katz, D. I. (1992). Neuropathology and neurobehavioral recovery from closed head injury. *Journal of Head Trauma Rehabilitation, 7,* 1–15.

Kaufman, A. S., & Kaufman, N. L. (1983). *K-ABC: Kaufman Assessment Battery for Children.* Circle Pines, MN: American Guidance Service.

Kolb, B., & Fantie, B. (1989). Development of the child's brain and behavior. In C. R. Reynolds & E. Fletcher-Janzen (Eds.), *Handbook of clinical neuropsychology* (pp. 17–41). New York: Plenum.

Kraus, J. F. (1993). Epidemiology of head injury. In P. R. Cooper (Ed.), *Head injury* (pp. 1–26). Baltimore: Williams & Wilkins.

Kraus, J. F., Morgenstern, H., Fife, D., Conroy, C., & Nourjah, P. (1989). Blood alcohol tests, prevalence of involvement, and outcomes following brain injury. *American Journal of Public Health, 79,* 294–299.

Kreutzer, J. S., Doherty, K. R., Harris, J. A., & Zasler, N. D. (1990). Alcohol use among persons with traumatic brain injury. *Journal of Head Trauma Rehabilitation, 5,* 9–20.

LaRue, A. (1992). *Aging and neuropsychological assessment.* New York: Plenum.

Lee, A. L. (1992). Diagnosing alcoholism: Toward a multisource approach. In C. E. Stout, J. L. Levitt, & D. H. Ruben (Eds.), *Handbook for assessing and treating addictive disorders.* New York: Greenwood Press.

Levin, H. S., Gary, H. E., Jr., Eisenberg, H. M., Ruff, R. M., Barth, J. T., Kreutzer, J., High, W. M., Jr., Portman, S., Foulkes, M. A., Jane, J. A., Marmarou, A., & Marshall, L. F. (1990). Neurobehavioral outcome one year after severe head injury: Experience of the traumatic coma data bank. *Journal of Neurosurgery, 73,* 699–709.

Levin, H. S., High, W. M., Jrs., & Eisenberg, H. M. (1988). Learning and forgetting during posttraumatic amnesia in head injured patients. *Journal of Neurology, Neurosurgery, and Psychiatry, 51,* 14–20.

Levin, H. S., Mattis, S., Ruff, R. M., Eisenberg, H. M., Marshall, L. F., Tabaddor, K., High, W. M., & Frankowski, R. A. (1987). Neurbehavioral outcome following minor head injury: A three center study. *Journal of Neurosurgery, 66,* 234–243.

Levin, H. S., O'Donnell, V. M., & Grossman, R. G. (1979). The Galveston Orientation and Amnesia Test: A practical scale to assess cognition after head injury. *Journal of Nervous and Mental Disease, 167,* 675–684.

Lezak, M. D. (1983). *Neuropsychological assessment* (2nd ed.). New York: Oxford University Press.

Lezak, M. D. (1995). *Neuropsychological assessment* (3rd ed.). New York: Oxford University Press.

Long, C. J., & Schmitter, M. E. (1992). Cognitive sequelae in closed head injury. In C. J. Long & L. K. Ross (Eds.), *Handbook of head trauma: Acute care to recovery* (pp. 107–122). New York: Plenum.

Luria, A. R. (1966). *Higher cortical functions in man.* New York: Basic Books.

Luria, A. R. (1973). *The working brain.* New York: Basic Books.

MacAndrew, C. (1965). The differentiation of male alcoholic out-patients from nonalcoholic psychiatric patients by means of the MMPI. *Quarterly Journal of Studies on Alcohol, 26,* 238–246.

Malkmus, D., Booth, B. J., & Kodimer, C. (1980). *Rehabilitation of the head injured adult: Comprehensive cognitive management.* Downey, CA: Professional Staff Association of the Ranchos Los Amigos Hospital.

Malloy, P. F. & Richardson, E. D. (1994). Assessment of frontal lobe functions. *The Journal of Clinical Neuropsychiatry, 6,* 399–410.

Markwardt, F. C. (1989). *Peabody Individual Achievement Test-Revised.* Circle Pines, MN: American Guidance Service.

Max, W., MacKenzie, E. J., & Rice, D. P. (1991). Head injuries: Costs and consequences. *Journal of Head Trauma Rehabilitation, 6,* 76–91.

McCarthy, D. (1972). *Manual for the McCarthy Scales of Children's Abilities.* San Antonio, TX: Psychological Corporation.

McGlynn, S. M., & Schacter, D. L. (1989). Unawareness of deficits in neuropsychological syndromes. *Journal of Clinical and Experimental Neuropsychology, 11,* 143–205.

Mesulam, M. (1985). Attention, confusional states, and neglect. In M. Mesulam (Ed.), *Principles of behavioral neurology* (pp. 125–168). Philadelphia: F. A. Davis.

Milberg, W. P., Hebben, N., & Kaplan, E. (1996). The Boston Process Approach to neuropsychological assessment. In I. Grant & K. M. Adams (Eds.), *Neuropsychological assessment of neuropsychiatric disorders* (pp. 58–80). New York: Oxford University Press.

Miller, J. D. (1991). Pathophysiology and management of head injury. *Neuropsychology, 5,* 235–261.

Mittenberg, N., Azrin, R., Millsaps, C., & Heilbronner, R. (1993). Identification of malingered head injury on the Wechsler Memory Scale-Revised. *Psychological Assessment, 5,* 34–40.

Morgan, S. F. (1982). Measuring long term memory, storage and retrieval in children. *Journal of Clinical Neuropsychology, 4,* 77–85.

Mysiw, W. J., Corrigan, J. D., Carpenter, D., & Chock, S. K. L. (1990). Prospective assessment of postraumatic amnesia: A comparison of the GOAT and the OGMS. *Journal of Head Trauma Rehabilitation, 5,* 65–72.

National Head Injury Foundation. (1993). *Fact sheet and pamphlet.* Washington, DC: Author.

National Institute of Neurological Disorders and Stroke. (1989). *Intergency head injury task force report.* Bethesda, MD: Author.

O'Shanick, G. J., Scott, R., & Peterson, L. G. (1984). Psychiatric referral after head trauma. *Psychiatric Medicine, 2,* 131–137.

Pang, D. (1989). Physics and pathophysiology of closed head injury. In M. D. Lezak (Ed.), *Assessment of the behavioral consequences of head trauma* (pp. 1–17). New York: Alan R. Liss.

Perrine, K. (1993). Differential aspects of conceptual processing in the Category Test and Wisconsin Card Sorting Test. *Journal of Clinical and Experimental Neuropsychology, 15,* 461–473.

Phay, A. J. (1992). Use of history in neuropsychological assessments. In C. J. Long & L. K. Ross (Eds.), *Handbook of head trauma: Acute care to recovery* (pp. 35–56). New York: Plenum.

Pope, K. S. (1992). Responsibilities in providing psychological test feedback to clients. *Psychological Assessment, 4,* 268–271.

Psychological Corporation. (1992). *Wechsler Individual Achievement Test: manual.* San Antonio, TX: Author.

Radanov, B. P., Stefano, G. D., Schnidrig, A., Sturzenegger, M., & Augustiny, K. F. (1993). Cognitive functioning after common whiplash: A controlled follow-up study. *Archives of Neurology, 50,* 87–91.

Ransohoff, J., & Fleischer, A. (1975). Head injuries. *Journal of the American Medical Association, 234,* 861–864.

Raven, J. C. (1938). *Progressive Matrices: A perceptual test of intelligence: Individual form.* London: H. K. Lewis.

Reitan, R. M., & Wolfson, D. (1986). The Halstead–Reitan Neuropsychological Test Battery. In D. Wedding, A. M. Horton, Jr., & J. Webster (Eds.), *The neuropsychology handbook: Behavioral and clinical perspectives* (pp. 134–160). New York: Springer.

Reitan, R. M., & Wolfson, D. (1988). *Traumatic brain injury: Vol. II. Recovery and rehabilitation.* Tucson, AZ: Neuropsychology Press.

Reitan, R. M., & Wolfson, D. (1993). *The Halstead–Reitan Neuropsychological Test Battery: Theory and clinical interpretation* (2nd ed.). Tucson, AZ: Neuropsychology Press.

Reynolds, C. R., & Bigler, E. D. (1994). *Test of Memory and Learning.* Austin, TX: Pro-Ed.

Rimel, R. W., Giordani, B., Barth, J. T., Boll, T. J., & Jane, J. A. (1981). Disability caused by minor head injury. *Neurosurgery, 9,* 221–228.

Rimel, R., & Jane, J. (1983). Characteristics of the head injured patient. In M. Rosenthal, E. R. Griffith, M. R. Bond, & J. D. Miller (Eds.), *Rehabilitation of the head injured adult.* Philadelphia: Davis.

Rizzo, H. & Tranel, D. (1996). Overview of head injury and postconcussive syndrome. In M. Rizzo & D. Tranel (Eds.), *Head injury and postconcussive syndrome* (pp. 1–18). New York: Churchill Livingstone.

Rogers, R. (Ed.), (1988). *Clinical assessment of malingering an deception.* New York: Guilford.

Rourke, B. P., Fisk, J. L., & Strang, J. D. (1986). *Neuropsychological assessment of children: A treatment-oriented approach.* New York: Guilford.

Rudel, R. G., Holmes, J. M., & Pardes, J. (1988). *Assessment of developmental learning disorders: A neuropsychological approach.* New York: Basic.

Ruff, R. M., Marshall, L. F., Crouch, J., Klauber, M. R., Levin, H. S., Barth, J., Kreutzer, J., Blunt, B. A., Foulkes, M. A., Eisenberg, H. M., Jane, J. A., & Marmarou, A. (1993). Predictors of outcome following severe head trauma: Follow-up data from the Traumatic Coma Data Bank. *Brain Injury, 7,* 101–111.

Russell, E. W. (1981). The psychometric foundation of clinical neuropsychology. In S. B. Filskov & T. J. Boll (Eds.), *Handbook of clinical neuropsychology* (Vol. 2, pp. 45–80). New York: Wiley.

Russell, W. R. (1971). *The traumatic amnesias.* New York: Oxford University Press.

Ryan, T. V., Sautter, S. W., Capps, C. F., Meneese, W., & Barth, J. T. (1992). Utilizing neuropsychological measures to predict vocational outcome in a head trauma population. *Brain Injury, 6,* 175–182.

Saneda, D. L., & Corrigan, J. D. (1992). Predicting clearing of post-traumatic amnesia following closed-head injury. *Brain Injury, 6,* 167–174.

Selzer, M. L. (1971). The Michigan Alcoholism Screening Test: The quest for a new diagnostic instrument. *American Journal of Psychiatry, 127,* 89–94.

Sheslow, D., & Adams, W. (1990). *Wide range assessment of memory and learning.* Wilmington, DE: Jastak Assessment Systems.

Solomon, D. A., & Malloy, P. F. (1992). Alcohol, head injury, and neuropsychological function. *Neuropsychology Review, 3,* 249–280.

Sorenson, S. B., & Kraus, J. F. (1991). Occurrence, severity, and outcomes of brain injury. *Journal of Head Trauma Rehabilitation, 6,* 1–10.

Sosin, D. M., Nelson, D. E., & Sacks, J. J. (1992). Head injury deaths: The enormity of firearms. *Journal of the American Medical Association, 268,* 791.

Sosin, D. M., Sacks, J. J., & Smith, S. M. (1989). Head injury-associated deaths in the United States from 1979 to 1986. *Journal of the American Medical Association, 262,* 2251–2255.

Sparadeo, F. R., Strauss, D., & Barth, J. T. (1990). The incidence, impact, and treatment of substance abuse in head trauma rehabilitation. *Journal of Head Trauma Rehabilitation, 5,* 1–8.

Spreen, O., & Strauss, E. (1991). *A compendium of neuropsychological tests: Administration, norms, and commentary.* New York: Oxford University Press.

Stuss, D. T., & Buckle, L. (1992). Traumatic brain injury: Neuropsychological deficits and evaluation at different stages of recovery and in different pathologic subtypes. *Journal of Head Trauma Rehabilitation, 7,* 40–49.

Sweeney, J. E. (1992). Nonimpact brain injury: Grounds for clinical study of the neuropsychological effects of acceleration forces. *The Clinical Neuropsychologist, 6,* 443–457.

Taylor, D. A. (1992). Traumatic brain injury: Outcome and predictors of outcome. In C. J. Long & L. K. Ross (Eds.), *Handbook of head trauma: Acute care to recovery* (pp. 293–306). New York: Plenum.

Teasdale, G., & Jennett, B. (1974). Assessment of coma and impaired consciousness: A practical scale. *Lancet, 2,* 81–84.

Thomas, J. D., & Trexler, L. E. (1982). Behavioral and cognitive deficits in cerebrovascular accident and closed head injury: Implications for cognitive rehabilitation. In L. E. Trexler (Ed.), *Cognitive rehabilitation: Conceptualization and intervention.* New York: Plenum.

Torner, J. C. & Shootman, M. (1996). Epidemiology of closed head injury. In M. Rizzo & D. Tranel (Eds.), *Head injury and postconcussive syndrome* (pp. 19–46). New York: Churchill Livingstone.

Varney, N. R., & Menefee, L. (1993). Psychosocial and executive deficits following closed head injury: Implications for orbital frontal cortex. *Journal of Head Trauma Rehabilitation, 8,* 32–44.

Wechsler, D. (1981). *Manual for the WAIS-R.* New York: Psychological Corporation.

Wechsler, D. (1991). *Wechsler Intelligence Scale for Children-Third Edition.* San Antonio, TX: Psychological Corporation.

Wilkinson, G. (1993). *The Wide Range Achievement Test-3: Administration manual.* Wilmington, DE: Wide Range, Inc.

Williams, D. H., Levin, H. S., & Eisenberg, H. M. (1990). Mild head injury classification. *Neurosurgery, 27,* 422–428.

Williams, J. M. (1990). *Memory Assessment Scales.* Odessa, FL: Psychological Assessment Resources.

Williams, J. M. (1992). Neuropsychological assessment of traumatic brain injury in the intensive care and acute care environment. In C. J. Long & L. K. Ross (Eds.), *Handbook of head trauma: Acute care to recovery* (pp. 271–292). New York: Plenum.

Wilson, J. T. L., Teasdale, G. M., Hadley, D. M., Wiedmann, K. D., & Lang, D. (1993). Posttraumatic amnesia: Still a valuable yardstick. *Journal of Neurology, Neurosurgery & Psychiatry, 56,* 198–201.

Wong, P. P., Dornan, J., Schentag, C. T., Ip, R., & Keating, A. M. (1993). Statistical profile of traumatic brain injury: A Canadian rehabilitation population. *Brain Injury, 7,* 283–294.

Woodcock, R. W., & Johnson, M. B. (1989). *Woodcock–Johnson Psycho-Educational Battery—Revised.* Allen, TX: DLM Teaching Resources.

Ylvisaker, M., Hartwick, P., & Stevens, M. (1991). School reentry following head injury: Managing the transition from hospital to school. *Journal of Head Trauma Rehabilitation, 6,* 10–22.

Youngjohn, J. R., Larrabee, G. J., & Crook, T. H., III. (1992). Test–retest reliability of computerized, everyday memory measures and traditional memory tests. *The Clinical Neuropsychologist, 6,* 276–286.

Zasler, N. D. (1992). Acute neurochemical alterations following traumatic brain injury: Research implications for clinical treatment. *Journal of Head Trauma Rehabilitation, 7,* 102–105.

Zung, B. J. (1984). Reliability and validity of the short Michigan Alcoholism Screening Test among psychiatric patients. *Journal of Clinical Psychology, 40,* 347–350.

8

Evaluation of Cerebrovascular Disease

C. DELLA MORA AND ROBERT A. BORNSTEIN

INTRODUCTION

Cerebrovascular disease (CVD) is broadly defined as any disruption of brain function arising from some pathological condition related to the blood vessels (Walsh, 1987). In lay terms, it is more commonly referred to as stroke and is one of the most common neurologic diseases of adult life (Adams & Victor, 1989). Stroke represents the third leading cause of death in the United States, after heart disease and cancer (American Heart Association, 1988), and accounts for at least 50% of the neurological disorders present in a general hospital setting (Adams & Victor, 1989). In general, there has been an increased awareness and understanding of stroke during the second half of the 20th century (Wiebe-Velazquez & Hachinski, 1991).

Traditionally, patients with CVD have provided a major source of clinical and theoretical information. Benton (1991) chronicles the role that stroke has played throughout the history of neuropsychology. From the 17th century up to the present, the behavioral effects observed in patients with CVD have assumed a major role in the development of our knowledge about brain–behavior relationships. Cerebrovascular disease shed light on certain deficits in cognitive processing and behavior, eventually making it possible to relate these deficits to specific abnormalities of cerebral circulation. Focal ischemic lesions, for example, have contributed to our understanding of various disorders including aphasia, apraxia, anomia, pure alexia, prosopagnosia, and neglect (Adams & Victor, 1989; Benton, 1991; Bornstein, 1986). Information on the lateralization and localization of higher cognitive functions, sensory-perceptual and motor functions, and emotional functions have also been influenced by the study of ischemic infarction (Hom, 1991).

C. DELLA MORA   Department of Psychiatry/Neuropsychology, The Ohio State University, Columbus, Ohio 43210.   ROBERT A. BORNSTEIN   Neuropsychology Program, The Ohio State University, Columbus, Ohio 43210.

Neuropsychology, edited by Goldstein *et al*. Plenum Press, New York, 1998.

Cerebrovascular disease is a heterogeneous entity which produces great variability in terms of the degree of behavioral deficit and potential for recovery. Many patients who survive their strokes are left with significant physical and/or psychological deficits which result in great personal and economic cost (Bornstein, 1986).

The primary focus of the present chapter is on the neuropsychological assessment of stroke. To lay the groundwork for assessment, however, the chapter begins with a brief overview of the epidemiology of stroke, the vascular supply to the brain, mechanisms of stroke, and neurovascular syndromes. More detailed discussions of these areas are available in references cited.

EPIDEMIOLOGY

As already noted, stroke represents one of the leading causes of death in the United States and North America in general. It is the primary cause of death in Japan and China, and the third leading cause of death in Europe as well. The vast majority of stroke deaths occur in the elderly, but stroke is also the third most common cause of death in middle-aged adults (American Heart Association, 1988). Nonetheless, the incidence of stroke does increase with advancing age (Wiebe-Velazquez & Hachinski, 1991). The most common risk factors of stroke include hypertension, cardiac disease, diabetes mellitus, and smoking (Bornstein & Kelly, 1991).

The last three decades have witnessed a progressive decline in stroke mortality. This decline can be attributed to a number of factors including improved medical care of stroke patients, identification and control of risk factors such as hypertension, decreased incidence of cerebral embolism from heart disease, and the advent of diagnostic tools such as angiography, magnetic resonance imaging (MRI), and computed tomography (CT) scans (Adams & Victor, 1989; American Heart Association, 1988; Wiebe-Velazquez & Hachinski, 1991). As a result of reduced mortality rates, there are now more people surviving with stroke-related neurobehavioral deficits.

There is a reciprocal relationship between the incidence of CVD and cardiovascular disease. Stroke patients are more likely to die from heart disease than from a subsequent stroke.

Estimates of mortality rates vary from 20% to 38% during the acute phase and up to 30 days after stroke (Joseph, 1990; Wiebe-Velazquez & Hachinski, 1991). Fifty percent of patients who survive the first month die over the course of the next 7 years. The longer the patient lives, the greater the likelihood that death will be due to myocardial infarction. Short-term mortality is predicted by such factors as impaired consciousness, leg weakness, and increasing age. Long-term mortality is predicted by low activity level, advanced age, male gender, heart disease, and hypertension. Mortality rates are higher for patients with hemorrhagic strokes than for patients with infarctions. Hyperglycemia and diabetes mellitus are associated with poor neurological recovery and higher short-term mortality because they accentuate ischemic damage. Oxygen metabolism correlates better with clinical status and functional recovery than does cerebral blood flow. However, the single most important determinant of prognosis is the severity of the stroke, regardless of its etiology or location (Wiebe-Velazquez & Hachinski, 1991).

The greatest degree of recovery from stroke often occurs during the first 30 days after stroke, although recovery continues for some time thereafter. Estimates suggest that although about 60% of major stroke patients are able to achieve total independence, only 10% of the initial survivors are able to return to their employment without gross or obvious disability. Of the remaining survivors, 40% demonstrate mild disability, 40% are severely disabled, and 10% require institutionalization. The 5-year cumulative risk of repeated stroke is 42% for males and 24% for females (Joseph, 1990).

CEREBRAL VASCULATURE

The brain comprises only 2% of the total body weight, but uses 15% to 20% of the total cardiac output. Approximately 1 liter of blood passes through the brain per minute. Blood supplies the brain with glucose and oxygen, while dispersing heat and metabolic products of cerebral activity (Liebman, 1988). When the brain is not adequately perfused with blood and is deprived of much oxygen, glucose, and other nutrients, various neurological, neuropsychiatric, and neuropsychological abnormalities may result, depending on the duration of vascular interruption and which portions of the brain are involved (Joseph, 1990). After 3 to 4 minutes of arterial deprivation, neurons begin to die, with those of the cerebral cortex being most susceptible and those of the brain stem being more resistant. Brain damage is irreversible after 5 minutes of deprivation.

The vascular supply of the brain is presented in detail elsewhere (Brodal, 1992; Joseph, 1990; Liebman, 1988). In brief, there are two primary arterial systems that supply blood to the brain: the vertebral arteries and the internal carotid arteries. The symmetric vertebral arteries enter the skull through the foramen magnum at the base of the skull and pass along the ventral surface of the medulla. At the level of the pons, the vertebral arteries join to form the basilar artery which then extends up to the midbrain and branches to form the posterior cerebral arteries (PCAs). The PCAs supply the posterior part of the cerebral hemispheres including the medial and ventral surface of the temporal lobes (including important memory structures such as the hippocampal complex) and the visual cortex of the occipital lobes. Branches of the PCAs nourish the subthalamic and some thalamic nuclei. Eventually, the basilar artery divides into the inferior and superior cerebellar arteries and the pontine branches.

The common carotid artery branches in the neck into internal and external divisions, forming the internal carotid arteries (ICA). The ICAs enter the skull through the foramen lacerum and sit next to the lateral border of the optic chiasma. Here, they branch into the anterior and middle cerebral arteries (ACAs and MCAs, respectively). The ACAs supplies the mesial surface of the brain from the anterior frontal region to the parieto-occipital fissure. Branches of the ACAs feed the anterior limb of the internal capsule, putamen, and head of the caudate nucleus. The MCAs are the largest of the cerebral arteries and supply most of the lateral surface of the brain, affecting all the lobes. Striate arteries branching off the MCAs supply the remaining parts of the basal ganglia and posterior limb of the internal capsule. The striate arteries are frequent sites of stroke, and are sometimes referred to as "arteries of stroke" (Liebman, 1988).

Anteriorly, the left and right ACAs are connected by the anterior communicating artery. Posteriorly, the MCAs and PCAs are linked by the posterior communicating artery. A ring is formed at the base of the brain between the vertebral arteries and internal carotid arteries, known as the "Circle of Willis." The Circle of

Willis represents a collateral or backup circulation system such that, if one artery becomes occluded, blood can potentially reach the deprived area by means of another blood vessel. The Circle of Willis is often the site of aneurysms which occur when blood pressure at a weakened spot of the vessel wall balloons out, pressing on adjacent structures or possibly bursting.

MECHANISMS OF STROKE

The cardinal pathogenic feature of cerebrovascular accidents is a sudden interruption in the supply of nutrients to the brain due to disruption in blood flow. This results in a lack of oxygen, nutrients, and/or impaired removal of metabolic products, thereby causing damage or death of neurons, glia, and vasculature. When cells die, their membranes burst and release lipids, fatty acids, and other substances that can produce systemic and local effects (Joseph, 1990).

Most cases of CVD are due to cerebral infarction or hemorrhage (Lezak, 1983; Walsh, 1987). Infarctions are the result of occlusions or obstructions of blood vessels, and can be thrombotic or embolic (see below) in nature. Hemorrhages often occur secondary to hypertension or a ruptured aneurysm or arteriovenous malformation (AVM). Each of the main vascular disorders is briefly described below.

INFARCTION

The most common source of obstruction of blood flow to the brain is the buildup of fatty deposits or atherosclerotic plaques within the artery walls (Lezak, 1983). Atherosclerosis is a noninflammatory degenerative disease that can result in arterial abnormalities throughout the body. Atherosclerotic development usually starts early in life. Onset is usually in childhood and remains silent, growing slowly for 20 or 30 years before becoming symptomatic. It appears to peak between the ages of 50 and 75 years, and affects twice the number of men as women (Joseph, 1990). Common sites of atherosclerotic plaques are the areas where the cerebral arteries branch and curve.

Atherosclerosis promotes the gradual narrowing of the major arteries which reduces the perfusion of brain tissue. Cerebral blood flow does not change substantially unless the degree of stenosis reaches approximately 90%. A critical level of stenosis leads to ischemia and ultimately cerebral infarction (Joseph, 1990). As an infarction evolves, local edema can lead to further reduction in tissue perfusion, thereby exacerbating the infarct. Thromboembolism accounts for two thirds of all the cases of cerebrovascular disability (Walsh, 1987).

During an ischemic attack, reduced blood flow usually extends beyond infarct boundaries such that not only are directly affected neural tissue and its function lost, but several neural areas related to the damaged areas are also depressed and their function is consequently compromised for a period. This may make the functional effects of a stroke appear more pervasive and profound (Kolb & Whishaw, 1990). Studies have demonstrated bilateral reduction of hemispheric blood flow in patients with unilateral cerebral infarction (Lavy, Melamed, & Portnoy, 1975; Meyer *et al.* 1970; Slater, Reivich, Goldberg, Banka, & Greenberg, 1977). This condition is referred to as *diaschisis* and was first described by Von Monakow (1914), who concluded based on clinical observation that following localized injury to the brain, temporary depression of function may occur in remote areas of the nervous system. The bilateral depression of hemispheric blood flow appears to be due to an overall reduction of cerebral metabolism (Lavy et al., 1975; Meyer et al., 1970). Depressed blood flow in the noninfarcted hemisphere may persist for as long as 3 weeks (Welch & Meyer, 1975) and is due to a composite of decreased neuronal stimulation modified by loss of autoregulation, release of vasoactive substances, cerebral edema, and other factors (Slater et al., 1977). After 2 to 3 weeks, blood flow on the healthy side increases while that on the diseased side remains relatively low. As ischemia subsides and blood flow is increased to depressed areas, there is a considerable degree of recovery.

THROMBOEMBOLIC STROKES

The source of vessel occlusion may exist at the actual site of occlusion or may originate at some point distant from the site. The former situation is known as a thrombosis and is caused by a localized buildup of blood particles and tissue overgrowth (thrombus). As the thrombus grows, it narrows the opening of the blood vessel, thereby reducing blood flow or closing the vessel off altogether. The thrombus develops gradually, with symptoms and signs usually following a stepwise progression. The gradual development of the thrombus may facilitate the development of collateral blood flow as a means of compensating for the obstruction in blood flow. The presence and integrity of collateral vessels influence the size of infarction and therefore can attenuate the severity of neurobehavioral deficits sustained by the patient.

Common causes of thrombotic strokes include arteriosclerosis, inflammatory disorders, hematologic disorders, and problems with cerebral circulation secondary to cardiac dysfunction. Blockage is most common in the bifurcations of the carotid arteries. Although any cerebral vessels may be affected, there is a greater likelihood of middle cerebral artery involvement.

With embolic strokes, obstruction is caused by an embolus which most often consists of fragments of atherosclerotic plaque or other debris that are carried in the bloodstream from elsewhere in the body such as the heart or the neck. The embolus becomes lodged within a cerebral blood vessel, creating an ischemic cerebral infarction. Emboli may have a variety of origins, but are most commonly secondary to cardiac disease. Onset is usually sudden, and the full course of the stroke is completed within minutes. Due to the abrupt onset, there is no time for the development of collateral blood flow. Hence, the severity of the stroke and resulting neurobehavioral sequelae tends to be greater. As with thrombotic strokes, most emboli have an affinity for the MCA system. In some cases, multiple emboli may occur, and are a likely source of transient ischemic attacks.

TRANSIENT ISCHEMIC ATTACKS (TIAs)

A major stroke may be preceded by one or more episodes of focal neurologic deficit which occur secondary to insufficient blood supply, are sudden in onset, and resolve completely within 24 hours with usually no residuum (Toole, 1990). For a long time, TIAs were considered inconsequential. Even today, there is considerable debate over the etiology, natural history, and management of these brief episodes of focal neurologic deficit. Although the seriousness of TIA continues to be underscored, TIAs represent the best means by which individuals at high risk for cerebral infarction can be identified, so that preventive interventions can be instituted.

Transient ischemic attacks are more common in white than in black or Asian individuals, which is likely due to the greater prevalence of atherosclerosis in white populations. They are twice as prevalent in males than in females, and most often occur in the 50–70 years age group (Toole, 1990). Patients who suffer TIAs usually have other associated diseases such as hypertension, ischemic heart disease, and diabetes.

Transient ischemic attacks usually affect distal arterial branches and are most commonly associated with thrombosis (Adams & Victor, 1989), although they may be related to minor embolizations or chronic hypertension as well. Approximately 90% of TIAs occur in the carotid distribution, 7% in the vertebrobasilar distribution, and 3% in both. Up to 40% of patients with TIAs eventually sustain a major stroke, most of which occur within the first 6 months of the initial episode. The greatest risk of stroke following TIAs is within the first year, and particularly the first 30 days; 20% occur within the first month and 50% occur within the first year (Joseph, 1990; Lezak, 1983; Toole, 1990).

The nature of the neurobehavioral disturbances experienced depends on whether the carotid system or vertebrobasilar system is involved (see Toole, 1990, for discussion). In many instances, it may be difficult to determine which type of TIA has occurred since they seldomly occur and/or have fully resolved by the time a medical examination takes place. Most often, data are based on information provided by patients and their families. Such information may be less than accurate due to the patient's condition at the time of the event, the family's emotional response and concern, and unfamiliarity with medical information. Furthermore, there are no laboratory tests which provide an objective measure of TIAs. Hence, classification of TIA is very subject to observer bias.

When an episode lasts more than 24 hours but fully resolves within 1 or 2 weeks, it is referred to as a Reversible Ischemic Neurologic Deficit (RIND) (Toole, 1990). RINDs are classified separately from infarctions in which there is permanent deficit. Both TIAs and RINDs may be part of a continuum since they are both precursors of infarction. Rigid boundaries of classification, however, have proven to be artificial and unsatisfactory. For example, a common belief is that TIAs are not associated with any actual neuronal death, but this may not be the case. An absence of findings may be due to the insensitivity of functional measures used and to the relatively silent location of the affected territory (Toole, 1990; Walsh, 1987). The availability of more sensitive measures may reveal the presence of an actual infarction. There is some suggestion that recurrent TIAs may result in more permanent deficits of higher cognitive function (Lezak, 1983).

HEMORRHAGE

Hemorrhages are highly variable with respect to rate of onset, size, and location. They can occur anywhere in the brain, although determining the lesion site is complicated by several factors such as vasospasm, edema, and disruption of neuronal input into brain regions that are remote from the hemorrhage (Brown, Spicer, & Malik, 1991). Common causes include traumatic brain injury, hypertension, rupture of an aneurysm or AVM, weakening of a segment of the vasculature secondary to thrombi or emboli, or vessel wall necrosis due to ischemia and occlusion (Bigler, 1988; Joseph, 1990). Warning signs are rare for hemorrhagic strokes.

Most often, patients complain of extremely painful headaches accompanied by nausea and vomiting and followed within hours by evidence of neurological dysfunction such as stiff neck and focal neurological signs (Lezak, 1983). Significant bleeding from a major blood vessel may lead to infarctions in other regions due to a compressive distortion of the brain. Hypertensive intracerebral hemorrhages typically affect small branches of blood vessels that supply deeper brain regions and tend to result in damage to structures such as the thalamus, basal ganglia, and brain stem. Hemorrhagic strokes have a mortality rate of 50%, which is considerably higher than that for ischemic strokes. It is estimated that 15% to 25% of CVD disorders are due to subarachnoid or intracerebral hemorrhages (Walsh, 1987). These hemorrhagic disorders are discussed in detail elsewhere (e.g., Adams & Victor, 1989; Joseph, 1990) and are briefly described below.

SUBARACHNOID HEMORRHAGE. Subarachnoid hemorrhage (SAH) occurs with greatest frequency after the age of 50. In younger individuals, it is usually secondary to congenital vascular abnormalities such as an angioma, aneurysm, or an AVM. It involves a sudden and abrupt onset of bleeding into the subarachnoid space. Immediate common sequelae are severe headache, vomiting, and severe backache and neck stiffness. Mild cases are characterized by severe headache and possible development of mild focal deficits. Confusion and irritability are common in moderate cases. Coma and eventual death are likely with severe cases. The mortality rate is over 50%; one third of cases die within the first 24 hours. Twenty-five percent of those who survive make a good recovery (Joseph, 1990).

INTRACEREBRAL HEMORRHAGE. Intracerebral hemorrhages are commonly the result of hypertension and associated degenerative changes in the vessel walls of penetrating arteries, making them susceptible to rupture. Onset may be sudden or very gradual. After a large hemorrhage, swelling of the affected hemisphere can disrupt the function of the opposite hemisphere as a result of midline shift. Coma and death ensue when vital brain stem areas are compressed. Intracerebral hemorrhages most often involve the internal capsule, corona radiata, frontal lobe, pons, thalamus, and putamen. It is estimated that less than 40% of survivors achieve good recovery, and 30% to 75% die within 30 days. In most cases there are significant and permanent neurological deficits. In more than 60% of cases, intracerebral hemorrhage is related to hypertensive CVD.

AVMs AND ANEURYSMS. As already noted, hemorrhages can also occur secondary to AVMs and aneurysms, particularly in younger persons. An AVM is "a congenital entanglement of blood vessels that forms an abnormal connection between arterial and venous circulations" (Brown et al., 1991, p. 202). They are rare in occurrence and evolve slowly, with fetal abnormalities arising as early as 3 weeks after conception. Congenital AVMs can alter and disrupt the normal development of cerebral organization, thus complicating behavioral inferences. AVMs are characterized by a variety of neuropathological features which likely account for the variability of neuropsychological impairment seen among AVM patients. Slow-onset lesions are often associated with less marked lateralized or focal neuropsychological findings.

An aneurysm is a balloon-shaped dilation of a blood vessel resulting from a weak vessel wall which can ultimately burst under pressure. Aneurysms have an incidence of about 2% in the population (Lezak, 1983) and can produce disruption of cerebral functioning by rupturing, shedding emboli, and producing mass effects and vasospasm (Brown *et al.*, 1991). Ruptured aneurysms involving the anterior cerebral and anterior communicating arteries result in a variety of cognitive and personality changes implicating frontal lobe dysfunction, and less frequently, callosal involvement.

NEUROBEHAVIORAL SEQUELAE OF STROKE

Infarction in the territory of a major cerebral artery has important implications regarding a patient's neurobehavioral function. Neuropsychological and behavioral deficits associated with infarction of the major cerebral arteries are discussed and summarized by Bigler (1988) and Funkenstein (1988). What follows is a brief overview of the neurobehavioral deficits resulting from major vessel infarction. In addition, the effects of affective disorders, particularly depression, are briefly discussed in terms of their impact on cognitive function and recovery.

ANTERIOR CEREBRAL ARTERY (ACA)

Infarction in the disruption of the ACA produces contralateral weakness or paralysis, usually affecting mostly the distal part of the lower extremity. Contralateral sensory deficits may be mild or absent. In the acute phase, patients may show abulia (lack of spontaneity in speech or motor activity), difficulties with focusing attention and concentration, and distractibility. There may be diminished speed and output of motor behavior. Impaired articulation of speech or a disturbance in motor inertia may be seen. There may be an initial period of mutism which eventually resolves into normal grammatical speech. Apraxia may be present and is likely to be more severe ipsilaterally. Difficulties with new learning and memory may be observed, but visuospatial and visuoperceptual abilities appear to be unaffected. Behaviorally, a frontal lobe syndrome may develop. With right hemisphere involvement, prosody may be affected. Prosody refers to the variation in stress, pitch, and rhythm of speech by which differences in meaning, intent, attitude, and feeling are conveyed (Joseph, 1990; Kolb & Whishaw, 1990).

MIDDLE CEREBRAL ARTERY (MCA)

The deficits associated with infarction in the MCA distribution depend on the extent of the territory involved and location of lesion. Infarcts involving the anterior portion of the MCA impact the frontal and anterior parietal lobes. In this case, varying combinations of contralateral paralysis (face and upper extremity affected more than lower extremity), contralateral sensory disturbances, and gaze palsy may be evident. Left hemispheric lesions may produce nonfluent aphasias including Broca's aphasia, motor aphasia, and transcortical motor aphasia, as well as verbal memory disturbances. Infarcts involving the posterior division of the MCA usually produce neuropsychological deficits without marked motor or sensory problems. However, a visual field cut is not uncommon. Left hemispheric lesions may produce Wernicke's, transcortical sensory, conduction, and amnestic aphasias. An infarction of the left angular gyrus may produce components of Gerstmann syndrome and alexia with agraphia. Lesions in the right hemisphere may result in impairments in spatial/nonverbal memory, topographic memory and orientation, constructional ability, as well as difficulties with prosody. Left-sided neglect, anosagnosia, and constructional and dressing apraxias are common with right hemispheric involvement. Consequences of bilateral infarcts in the parietal lobe areas include prosopagnosia, severe constructional difficulties, dressing apraxia, and Balint's syndrome. Balint's syndrome is an agnosic syndrome consisting of paralysis of eye fixation with inability to look voluntarily into and scan the peripheral field, optic ataxia (inability to precisely grasp or touch an object under visual guidance), and disturbance of visual attention involving neglect of the peripheral field (Adams & Victor, 1989; Kolb & Wishaw, 1990).

Posterior Cerebral Artery (PCA)

The hallmark of PCA involvement is a relatively permanent and dense quandrantanopsia or hemianopsia secondary to a loss of the blood supply to the calcarine cortex. Central vision is generally spared (unlike the visual deficits associated with the MCA) since the occipital pole is usually supplied by the middle cerebral cortex. Motor involvement is not common unless the proximal arteries are affected. Thalamic involvement raises the possibility of sensory disturbances and/or pain. Memory difficulties consisting of transient or permanent global amnesia as well as permanent short-term memory deficits may arise as a result of disrupted blood flow in the branches supplying the medial temporal lobe (hippocampus and amygdala). Left PCA strokes involving the left occipital lobe and splenium of the corpus callosum may produce alexia without agraphia. This classic neurobehavioral syndrome involves a disconnection between the posterior speech area and the visual association cortex. Hence, the patient is able to talk and to write to dictation, but is unable to identify words. Left hemispheric involvement may produce other aphasic difficulties as well. Right hemispheric lesions may result in constructional apraxia and impaired visuospatial functioning. Bilateral infarctions may produce cortical blindness and a persistent denial or visual problems known as Anton's syndrome. In some instances, bilateral PCA infarcts may produce a prolonged delirium with agitation and confusion.

Vertebrobasilar Arteries

Infarction in the distribution of the vertebrobasilar arteries can produce various combinations of ataxia of gait or limbs, dysarthria, dysphagia, weakness and sensory loss in one or more limbs, facial weakness and numbness, and diplopia. Bilateral lesions may produce a "locked-in" syndrome in which an individual is incapable of voluntary muscle movement below the eyes, but has preserved consciousness and thought. An infarction in the medial thalamus and mesencephalon typically results in prolonged sleepiness or akinetic mutism, and memory deficits.

Affective Function

In the patient who has sustained a stroke, there may occur affective changes which have diagnostic significance. Although this may be true with other neurologic, nonvascular disorders, the effects tend to be more localized and circumscribed with vascular disorders (Bigler, 1988).

Depression is the most common emotional disorder following stroke (Starkstein & Robinson, 1989). Based on a review of the research in this area by

Starkstein and Robinson, major and minor depressive disorders occur in 30% to 50% of stroke patients. Without treatment, major depression lasts more than 1 year and most minor depressions last more than 2 years, although both respond to tricyclic antidepressants. The etiology of post stroke depression (PSD) remains unknown, but serotonergic or noradrenergic dysfunction may play important roles. Methodological issues in PSD research include level of consciousness, presence of aphasia, diagnostic criteria for PSD, weaknesses in the assessment of the brain lesion itself, and the distant effects of lesions (diaschisis), and are discussed in detail elsewhere (Starkstein, Bolla, & Robinson, 1991; Starkstein & Robinson, 1989).

Lesion location is very important in the pathogenesis of PSD. Depression occurs with greater frequency in patients with left hemisphere lesions (cortical and subcortical) than in patients with comparable right hemisphere lesions (Starkstein *et al.*, 1991). Furthermore, depression is often associated with left frontal or left basal ganglia lesions and preexisting subcortical atrophy, and increases in severity the closer the lesion is to the frontal pole.

In aphasic patients, resultant affective disturbances may be associated with the psychological consequences of verbal and cognitive deficits, and may be a reaction to loss of function and an inability to communicate effectively. Profound or catastrophic depression more often results from left hemispheric strokes that are primarily cortically based. This catastrophic affect occurs less frequently, however, when a subcortical vascular lesion is identified.

Among aphasic patients, depression is more common in patients with nonfluent aphasias (usually following left frontal lobe lesions) than in patients with fluent or global aphasias (involving parietal and temporal lobe lesions). The association between nonfluent aphasia and depression may be due to lesions in similar locations which produce both the depression and aphasia. The aphasia itself does not appear to cause depression, but rather aphasia and depression seem to be parallel processes which often coexist, depending on lesion location.

In contrast to patients with left hemisphere lesions, patients with right hemispheric strokes, particularly right frontal lesions, tend to appear apathetic or inappropriately cheerful, with no affective presentation of depression. This suggests that no relationship exists between depression and right hemisphere dysfunction, however, there may be evidence to the contrary (Sinyor *et al.*, 1986). Patients with right hemispheric strokes may display deficits in emotional output associated with aprosodic voice patterns and dysfunctional gesturing. What appears to be flattened affect may be a disturbance in prosody. Such patients may report symptoms of dysphoria but demonstrate behavior and affect that are incongruous. Alternately, pathologic crying or laughing may be exhibited. It is possible that underlying depression may be masked by such inappropriate emotional presentation. Bigler (1988) suggests that if a right hemispheric lesion (particularly in the right inferior frontal region) can be documented, observed affective disturbance can usually be attributed to underlying organic deficit rather than to some functional or nonneurologic cause.

Secondary mania is a relatively rare consequence of stroke and is similar to mania without brain injury in terms of its presentation and response to treatment. Poststroke mania (PSM) is strongly associated with a right hemisphere lesion in a limbic-connected area and perhaps with a predisposing factor, such as genetic loading, subcortical atrophy, or seizure disorder (Starkstein & Robinson, 1989). As with PSD, the mechanism of PSM remains unknown.

Although the brain constitutes only a small percentage of one's total body size, it receives a disproportionately large amount of the body's total blood supply. Given such a large vascular demand, it is therefore not surprising that many vascular disorders lead to cognitive dysfunction (Bigler, 1988). Cerebrovascular disease produces a variety of complex disorders that, in turn, result in a vast array of neurobehavioral deficits. The cognitive impairments associated with vascular dementia result primarily from obstructions of many vessels with infarctions in corresponding areas of cerebral tissue (Cummings & Benson, 1992). The specific distribution of the vessels involved determines the clinical characteristics in an individual patient.

Vascular dementia is the second most common cause of chronic progressive intellectual decline (second to dementia of the Alzheimer type [DAT]) (Stuss & Cummings, 1990). Although vascular dementia represents a heterogeneous syndrome, more generally, it has an earlier age of onset and tends to affect men more than women. In addition, duration of survival after onset of mental status changes is approximately 20% lower in vascular dementia than in DAT.

Identification and diagnosis of vascular dementia is made on the basis of characteristic historical, clinical, and neuroimaging findings. Common historical features that facilitate the diagnosis of vascular dementia include an abrupt onset, stepwise deterioration, fluctuating course, increased confusion at night, previous hypertension, and a history of previous syndromes of CVD. The classic clinical syndrome consists of bilateral, asymmetric pyramidal and extrapyramidal signs. Some degree of limb rigidity, spasticity, hyperreflexia, extensor plantar responses, gait abnormality, and incontinence is often observed. Pseudobulbar palsy with extreme laughter or weeping and mild dysarthria are not common. Patients may be hypertensive and have abnormal electrocardiograms (ECGs) due to various cardiac abnormalities.

Neuropsychological deficits commonly associated with vascular dementia include impaired orientation, short-term memory, abstraction, and calculation. Language function may show deficits in writing, following complex commands, and grammatical complexity. Depression, psychosis, and changes in personality are common.

The extent and pattern of neuropsychological deficits are determined by the location, degree of ischemic tissue damage, and size of the obstructed vessels. Cortical dementias involve the obstruction of extracranial (carotid) and intracranial vessels or small vessels supplying the cortical surface. The pattern of resulting neurologic deficits depends on the cortical territory supplied by the involved vessel. When several focal infarctions have occurred, the resulting clinical picture commonly includes dementia. When dementia results from cortical infarction, the infarcts are multiple and bilateral (Cummings & Benson, 1992). An example of cortical dementia is the angular gyrus syndrome which, in its complete form, is characterized by fluent aphasia, alexia with agraphia, Gerstmann syndrome (dysgraphia, right–left disorientation, finger agnosia, and acalculia), and constructional disturbances. Clinically the syndrome closely resembles Alzheimer's disease, making it potentially difficult to distinguish between the two disorders (Cummings & Benson, 1992). Subcortical dementias result from the obstruction of arteriole-sized vessels irrigating deep gray and white matter structures, and are characterized by abulia-aspontaneity, dysarthria, incontinence, lower limb spasticity, gait abnormalities, bradyphrenia, cognitive deterioration, and mood alterations (Cummings & Mahler, 1991). Examples of subcortical dementias are lacunar state, Binswanger's disease, and thalamic dementia. Detailed discussions of cortical and subcortical vascular dementias are presented elsewhere (e.g., Funkenstein, 1988; Cummings & Benson, 1992; Cummings & Mahler, 1991; Stuss & Cummings, 1990).

ASSESSMENT

Stroke can produce a variety of disorders which affect all facets of an individual's life including cognitive, physical, emotional, family/social, and vocational areas. Neuropsychological assessment represents a bridge between the physiological changes associated with stroke and their effects on quality of life (Brown, Baird, & Shatz, 1986).

Cerebrovascular disease can produce a wide range of neurobehavioral pathology depending on the arterial distribution affected. For example, occlusion of the middle cerebral artery can result in a widespread primary lesion which affects blood flow in many related regions. In contrast, occlusion of distributions of smaller penetrating branches tend to produce smaller and more focal lesions. As a result, strokes can produce a broad range of deficits in higher cognitive function including intelligence, abstract reasoning, problem-solving, learning and memory, attention and concentration, and cognitive flexibility. Relatively specific deficits may relate to primary functions of the particular area of the brain, whereas nonspecific effects likely reflect the disruption of a number of neurobehavioral systems with which the lesioned area is linked. Hence, neuropsychological functions that are somewhat independent of location of brain damage or are dependent on the integrated function of various brain regions are also likely to be impaired. It follows that a primary focus on only discrete dysfunctions would increase the likelihood of overlooking the full extent of cognitive dysfunction. Hence, the patient with cerebrovascular disorder should always have a comprehensive neuropsychological evaluation which addresses multiple dimensions of function, thereby providing a stronger foundation on which to base treatment goals and recommendations regarding rehabilitation.

A thorough neuropsychological evaluation consists of current and background information, the neuropsychological examination itself, and feedback.

CURRENT AND BACKGROUND INFORMATION

Current and background information are important components of a comprehensive evaluation (Incagnoli, 1986). Although this information may not be used for test selection, it allows for the generation of hypotheses and provides a context in which to interpret neuropsychological data. In addition, it facilitates the development and provision of recommendations. Sources of patient information include medical records and a clinical interview of the patient and/or significant others. Phay, Gainer, & Goldstein (1986) review several aspects of the clinical interview and history-taking process and stress the importance of integrating this information with objective psychometric and laboratory data in order to formulate an optimal treatment plan.

NEUROPSYCHOLOGICAL EVALUATION

There are different approaches to neuropsychological assessment. Measurement of general brain function is usually based on a battery of neuropsychological tests which evaluate both general and specific aspects of brain function. Inferences about the functional integrity of the brain are drawn on the basis of a systematic evaluation of intellectual, cognitive, and sensorimotor abilities using standardized measures.

Neuropsychological batteries can be flexible or fixed in nature. A flexible approach involves selecting appropriate tests based on the examiner's thorough knowledge of the case history, familiarity with known features of the disorder in question, and the nature of the referral question. Advantages of this approach include a lack of redundancy in evaluation, a more in-depth analysis of particular functions, and the opportunity to integrate new research findings with tests reflecting such knowledge (Incagnoli, 1986). On the other hand, a flexible approach relies heavily on the skill of the examiner, thereby limiting the use of technicians and risking the possibility of overlooking some deficits. Selected flexible batteries are described by Lezak (1983) and by Benton and colleagues (Benton, Hamsher, Varney, & Spreen, 1983).

The fixed battery approach entails the use of a standard battery of tests which is administered in the same way to all individuals evaluated. Detailed discussions of fixed neuropsychological batteries are available elsewhere (see chapters by Tarter, Parsons, and Golden, in Incagnoli, 1986). In general, fixed batteries are comprehensive and therefore may be more likely to reveal subtle patterns of neuropsychological function and dysfunction that may be overlooked by less comprehensive methodologies (Hom, 1991). They are also more suitable for technician administration. However, disadvantages are related to redundancy of information, its poor ability to account for failure, and the possibility that other areas of function may not be explained. One of the most widely used standardized batteries, representing the fixed battery approach, is the Halstead–Reitan Neuropsychological Test Battery (Reitan & Wolfson, 1993).

Data can be evaluated along a quantitative/qualitative dimension. A quantitative system focuses on achievement and is therefore more concerned with whether or not a problem is solved correctly and if so, to what extent. This approach compares an individual's performance to a cutoff score which separates brain-damaged from non-brain-damaged performance. Here, psychological deficit is inferred on the basis of the level and pattern of performance, pathognomonic signs, and relative performance of the two sides of the body. In contrast, the qualitative approach relies more on the observations of the examiner to draw clinical inferences.

A qualitative approach examines the process by which a problem or task is solved or failed. According to Kaplan (1983), a process approach "informs us of the compensatory strategies, both adaptive and maladaptive, that an individual has developed to cope with his or her deficit(s). . . . [It] provides information concerning the individual's differential response to varying task demands as well as stimulus parameters that may induce a more or less effective response" (p. 155). A primary strength of this approach is its relevance to patients' daily lives and hence, its contribution to treatment planning and rehabilitation (Milberg, Hebben, & Kaplan, 1986).

There are suggestions that the complexity and variability of patient behavior are optimally addressed by a combination of both the qualitative and quantitative approaches (Lezak, 1983). Indeed, such an integrated approach would seem to be ideal for assessing stroke, given the broad range of resultant neurobehavioral dysfunction discussed earlier.

Stroke patients may present with problems such as hemiparesis, neglect, and depression which can interfere with their performance on a neuropsychological examination. Although these problems are not specific to stroke, they are common sequelae of stroke and represent potential confounds when interpreting test results. Hence, it will be important to take into account these and other potential confounds when considering the length, content, and format of evaluation to be used.

FEEDBACK

Feedback is an integral part of the assessment process. It provides the opportunity to relate the results of the neuropsychological examination to the stroke patient's daily function, and to make appropriate recommendations accordingly. It also provides a mechanism for informing and educating patients and their families about stroke and its impact on cognitive, emotional, and psychosocial function. Gass and Brown (1992) emphasize that the provision of feedback is of great importance in helping patients and families cope with the consequences of stroke or any brain injury.

TESTING CONSIDERATIONS

ASSESSMENT OF THE ELDERLY

Since the incidence of vascular disorders increases with age, age effects should be given special consideration in the evaluation of older patients with vascular disorders. The elderly often present with complex diagnostic problems, and symptoms often reflect a confluence of medical, personal, social, and sometimes psychiatric issues. Factors such as affective problems, medical disorders with central nervous system effects, and medications influence the neuropsychological performance of all persons, but particularly of geriatric patients. It is important for the clinical neuropsychologist to be familiar with both normal changes related to aging and changes associated with or characteristic of abnormal processes.

TIMING AND PURPOSE OF EVALUATION

The issue of when to assess is not unique to stroke. The timing of a neuropsychological evaluation is linked to the goals or purpose of the evaluation. In many cases, a one-time evaluation is performed to determine, for example, a patient's cognitive strengths and weaknesses. In some instances, it is important to perform an assessment as early as possible in the course of the patient's poststroke recovery in order to achieve a baseline. This is especially warranted when the goal is to monitor the patient's recovery. Baseline evaluations often may not be feasible, however, due to factors such as medical complications, patients' early discharge or transfer, death, or significant acute cognitive deficits which interfere with the administration and interpretation of test results. Serial or follow-up evaluations may be helpful in tracking changes in cognitive status over time. Serial assessments can help delineate the rate of change and determine whether the pathological process is degenerative or progressive. Identification of the pattern of decline over time can be of diagnostic and treatment value.

The nature of a neuropsychological assessment is also driven by the purpose of the assessment and by the patient's ability level. Common reasons for evaluating the patient with CVD include assessing relative cognitive strengths and weaknesses, clarifying the pattern of cognitive decline (e.g., sudden and abrupt vs. gradual, insidious), determining the effects of neuropathology, assisting in treatment planning, and guiding rehabilitative efforts and the return to daily life.

185

EVALUATION
OF CEREBRO-
VASCULAR
DISEASE

SUMMARY

Stroke can produce a broad range of neurobehavioral disorders which affect all facets of an individual's life. The widespread impact of stroke underscores the need for a thorough and comprehensive neuropsychological examination. Regardless of the assessment approach utilized (i.e., flexible vs. fixed battery, etc.), care must be taken to ensure that one is evaluating the effects of stroke while taking into account patient characteristics and behaviors that can affect neuropsychological performance. Factors such as age, fatigue, and depression, if overlooked, can confound the interpretation of test results. A comprehensive evaluation addresses multiple dimensions of function, yielding a profile of cognitive strengths and weaknesses which assists in directing treatment goals and rehabilitative efforts.

As a major source of clinical and theoretical information, stroke has played a significant role in the history of neuropsychology. However, despite the knowledge acquired to date, there remains much information that is unknown about stroke. For example, relatively few studies have examined the intellectual, cognitive, and neuropsychological deficits associated with stroke, and the effects of different types of strokes (i.e., occlusive vs. hemorrhagic) on neuropsychological function. In addition, we have a long way to go in furthering our understanding of the impact of stroke-related neuropsychological deficits on psychosocial and vocational adjustment, and adaptive daily functioning. Clearly, neuropsychological assessment will play a critical role in addressing these and many other questions related to the impact of stroke and care of the stroke patient.

REFERENCES

Adams, R. D., & Victor, M. (1989). *Principles of neurology* (4th ed.). New York: McGraw-Hill, Inc.

American Heart Association. (1988). *Facts about stroke*. [Brochure.] Dallas, Texas.

Benton, A. L. (1991). Cerebrovascular disease in the history of clinical neuropsychology. In R. A. Bornstein & G. Brown (Eds.), *Neurobehavioral aspects of cerebrovascular disease* (pp. 3–13). New York: Oxford University Press.

Benton, A. L., deS. Hamsher, K., Varney, N. R., & Spreen, O. (1983). *Contributions to neuropsychological assessment*. New York: Oxford University Press.

Bigler, E. D. (1988). *Diagnostic clinical neuropsychology*. Austin, TX: University of Texas Press.

Bornstein, R. A. (1986). Neuropsychological aspects of cerebrovascular disease and its treatment. In G. Goldstein and R. E. Tarter (Eds.), *Advances in clinical neuropsychology* (pp. 55–94). New York: Plenum.

Bornstein, R. A., & Kelly, M. P. (1991). Risk factors for stroke and neuropsychological performance. In R. A. Bornstein & G. Brown (Eds.), *Neurobehavioral aspects of cerebrovascular disease* (pp. 182–201). New York: Oxford University Press.

Brodal, P. (1992). *The central nervous system: Structure and function*. New York: Oxford University Press.

Brown, G. G., Baird, A. D., & Shatz, M. W. (1986). The effects of cerebrovascular disease and its treatment. In I. Grant & K. M. Adams (Eds.), *Neuropsychological assessment of neuropsychiatric disorders* (pp. 384–414). New York: Oxford University Press.

Brown, G. G., Spicer, K. G., & Malik, G. (1991). Neurobehavioral correlates of arteriovenous malformations and cerebral aneurysms. In R. A. Bornstein & G. Brown (Eds.), *Neurobehavioral aspects of cerebrovascular disease* (pp. 202–223). New York: Oxford University Press.

Cummings, J. L., & Benson, D. F. (1992). *Dementia: A clinical approach.* Boston: Butterworth-Heinemann.

Cummings, J. L., & Mahler, M. E. (1991). Cerebrovascular dementia. In R. A. Bornstein & G. Brown (Eds.), *Neurobehavioral aspects of cerebrovascular disease* (pp. 131–149). New York: Oxford University Press.

Franzen, M. D., & Rasmussen, P. R. (1990). Clinical neuropsychology and older populations. In A. M. Horton (Ed.), *Neuropsychology across the life-span: Assessment and treatment* (pp. 81–102). New York: Springer Publishing Company.

Funkenstein, H. H. (1988). Cerebrovascular disorders. In M. S. Albert & M. B. Moss (Eds.), *Geriatric neuropsychology* (pp. 179–207). New York: Guilford Press.

Gass, C., & Brown, M. C. (1992). Neuropsychological test feedback to patients with brain dysfunction. *Psychological Assessment, 4* (3), 272–277.

Hom, J. (1991). Contributions of the Halstead–Reitan battery in the neuropsychological investigation of stroke. In R. A. Bornstein & G. Brown (Eds.), *Neurobehavioral aspects of cerebrovascular disease* (pp. 165–181). New York: Oxford University Press.

Incagnoli, T. (1986). Current directions and future trends in clinical neuropsychology. In T. Incagnoli, G. Goldstein, & C. J. Golden (Eds.), *Clinical application of neuropsychological test batteries* (pp. 1–44). New York: Plenum.

Joseph, R. (1990). *Neuropsychology, neuropsychiatry, and behavioral neurology.* New York: Plenum Press.

Kaplan, E. (1983). Process and achievement revisited. In S. Warner & B. Kaplan (Eds.), *Toward a holistic developmental psychology* (pp. 143–156). Hillsdale, NJ: Lawrence Erlbaum Associates.

Kolb, B., & Whishaw, I. Q. (1990). *Fundamentals of human neuropsychology.* New York: W. H. Freeman and Company.

Lavy, S., Melamed, E., & Portnoy, Z. (1975). The effect of cerebral infarction on the regional cerebral blood flow of the contralateral hemisphere. *Stroke, 6,* 160–163.

Lezak, M. (1983). *Neuropsychological assessment.* New York: Oxford University Press.

Liebman, M. (1988). *Neuroanatomy made easy and understandable.* Rockville, MD: Aspen Publishers, Inc.

Meyer, J. S., Shinohara, Y., Kanda, T., Fukuuchi, Y., Ericsson, A. D., & Kok, N. K. (1970). Diaschisis resulting from acute unilateral cerebral infarction. *Archives of Neurology, 23,* 241–247.

Milberg, W. P., Hebben, N., & Kaplan, E. (1986). The Boston Process approach to neuropsychological assessment. In I. Grant & K. M. Adams (Eds.), *Neuropsychological assessment and neuropsychiatric disorders* (pp. 65–86). New York: Oxford University Publishing.

Phay, A., Gainer, C., & Goldstein, G. (1986). Clinical interviewing of the patient and history in neuropsychological assessment. In T. Incagnoli, G. Goldstein, & C. J., Golden (Eds.), *Clinical application of neuropsychological test batteries* (pp. 45–74). New York: Plenum.

Reitan, R. M., & Wolfson, D. (1993). *The Halstead–Reitan Neuropsychological Test Battery.* Tucson, AZ: Neuropsychology Press.

Sinyor, D., Jacques, P., Kaloupek, D., Becker, R., Goldenberg, M., & Coopersmith, H. (1986). Poststroke depression and lesion location. *Brain, 109,* 537–546.

Slater, R., Reivich, M., Goldberg, H., Banka, R., & Greenberg, J. (1977). Diaschisis with cerebral infarction. *Stroke, 8*(6), 684–690.

Starkstein, S. E., Bolla, K. I., & Robinson, R. G. (1991). Dementia syndrome of depression in patients with stroke. In R. A. Bornstein & G. Brown (Eds.), *Neurobehavioral aspects of cerebrovascular disease* (pp. 150–164). New York: Oxford University Press.

Starkstein, S. E., & Robinson, R. E. (1989). Affective disorders and cerebral vascular disease. *British Journal of Psychiatry, 154,* 170–182.

Stuss, D. T., & Cummings, J. L. (1990). Subcortical vascular dementias. In J. L. Cummings (Ed.), *Subcortical dementia* (pp. 145–163). New York: Oxford University Press.

Toole, J. F. (1990). *Cerebrovascular disorders.* New York: Raven Press.

Von Monakow, C. (1914). *Die Lokalisation im Groshirn und der Abbau der Funktion durch kortidale Herde* (pp. 26–34). Wiesbaden, Germany: JF Bergmann.

Walsh, K. (1987). *Neuropsychology: A clinical approach.* London: Churchill Livingstone.

Welch, K. M. A., & Meyer, J. S. (1975). Disordered cerebral metabolism after cerebral ischemia and infarction: Therapeutic implications. In J. S. Meyer (Ed.), *Modern concepts of cerebrovascular disease* (pp. 87–112). New York: Spectrum Publications, Inc.

Wiebe-Velazquez, S., & Hachinski, V. (1991). Overview of clinical issues in stroke. In R. A. Bornstein & G. Brown (Eds.), *Neurobehavioral aspects of cerebrovascular disease* (pp. 111–130). New York: Oxford University Press.

9

Evaluation of Demyelinating and Degenerative Disorders

Daniel N. Allen, David G. Sprenkel, Rock A. Heyman, Carol J. Schramke, and Nicole Englund Heffron

INTRODUCTION

The terms demyelinating and degenerative have been used to classify two groups of disorders that have distinct neuropathological features. The disorders that compose these categories, and particularly the degenerative category, are quite diverse. As the term implies, *degenerative* is used to classify a group of disorders that involve degeneration of neurons in the central nervous system (CNS). Degenerative disorders include Alzheimer's disease (AD), vascular dementia, Parkinson's disease, Huntington's chorea, progressive supranuclear palsy, and Pick's disease, among others. The etiology of many degenerative disorders is unknown and there are no identified antecedents to their onset. Rather, degenerative disorders begin insidiously and progress slowly, often over the course of many years. Typically, neuronal degeneration is bilateral. For many of these disorders, specific neuronal systems or neuroanatomical areas are affected more so than others. On the other hand, demyelinating disorders are grouped together because myelin destruction is their defining characteristic. Generally, the nerve cell itself is spared. Examples of demyelinating disorders include multiple sclerosis (MS), leukoencephalopathy, and acute disseminated encephalomyelitis. While the etiology of some demyelinating disorders is known, for many the etiology is not known. The course of demyelinating diseases is varied. In some cases the disease can begin insidiously and then progress either gradually or quickly. In other cases, there appears to be only one or two exacerbations during the entire course of the disorder and little or no cognitive or physical disability.

Daniel N. Allen, David G. Sprenkel, Carol J. Schramke, and Nicole Englund Heffron Psychology Service (116B), VA Pittsburgh Healthcare System, Highland Drive Division, Pittsburgh, Pennsylvania 15206-1297. Rock A. Heyman Department of Neurology, University of Pittsburgh School of Medicine, Pittsburgh, Pennsylvania 15261.

Neuropsychology, edited by Goldstein *et al.* Plenum Press, New York, 1998.

We have limited our discussion to MS and AD because they are the most common demyelinating and degenerative disorders, respectively. The neuropathology, symptomatology, and diagnosis of the other demyelinating and degenerative disorders are thoroughly discussed elsewhere (Adams & Victor, 1997; Samuels & Feske, 1996).

NEUROPATHOLOGY OF MULTIPLE SCLEROSIS AND ALZHEIMER'S DISEASE

Multiple sclerosis is a disease restricted to the CNS. Disease effects are multifocal and result in lesions called plaques. Plaques are predominantly in myelinated white matter regions. Plaques vary in size from less than 1 millimeter to several centimeters. As the myelin covering of groups of axons is disrupted, nerve messages are either slowed, distorted, or blocked. Nerve cell function becomes energy inefficient and is more easily affected by temperature. (The phenomenon of heat associated with recurrence of neurological dysfunction in patients with MS is known as Uthoff's phenomenon.) Only CNS myelin formed by oligodendrocytes is affected. Acute MS lesions may cause marked cognitive deficits and physical disabilities; however, significant recovery of function often occurs in 1 to 3 months. Recovery is postulated to occur because of improvement in membrane electrical functions, resolution of initial swelling, and to a lesser extent, repair of myelin. Steroid medications may speed this process in some patients but do not alter eventual outcome. Many MS attacks may be subclinical, undetected by the patient or physician. In early, milder forms of MS, axons often remain intact and excellent recovery of function frequently occurs. With more severe or repeated attacks, axons may be lost as well. Astrocytes may contribute to scar formation in plaques (Raine, 1990). Attacks of MS are scattered over time and occur when the body's own immune system attacks CNS myelin. Because of this, MS is considered an autoimmune disease (e.g., rheumatoid arthritis and systemic lupus erythematosus). The target of the attack may be myelin proteins such as myelin basic protein or proteolipid protein. Etiology of this process is unknown; however, activated white blood cells migrate from the bloodstream into the CNS and destroy myelin (Raine, 1990). In an experimental model of MS (experimental allergic encephalomyelitis), attacks may be triggered in susceptible animals by immune stimulation with myelin proteins. Current MS treatments are aimed at controlling immune mediated attacks. Multiple sclerosis may follow a variable, unpredictable course (Sibley, 1990). Most patients have intermittent attacks followed by significant recovery (relapsing–remitting type). The disease may remain this way. However, most patients have attacks leading to lasting disability which may gradually increase (secondary progressive type). A minority of patients (10%–15%) experience progressive disability from the onset (primary progressive type).

Alzheimer's disease contrasts with multiple sclerosis in many ways. Because the progressive degeneration of AD involves the nerve cell bodies themselves, gray matter of the cortex atrophies with overall decrease in brain size and compensatory enlargement of the whole ventricular system which contains spinal fluid. Degree of atrophy overlaps with normal aging changes. A more pronounced atrophy often involves posterior parietal, inferior temporal, hippocampal, and frontal regions. Individual patients may initially have lateralized or more focal changes. Certain regions are characteristically spared including the cerebellum and caudate nucleus. The more developed, multilayer cortical regions (neocortex) are most involved

189

EVALUATION OF
DEMYELINATING
AND
DEGENERATIVE
DISORDERS

with loss of many larger neurons. In addition to progressive cell loss, microscopic hallmarks of AD include senile amyloid plaques and neurofibrillary tangles (NFT). Amyloid plaques consist of a core of amyloid protein surrounded by enlarged degenerating nerve endings. These plaques occur frequently in the amygdala and the hippocampal and neocortical regions. They are a pathologic hallmark of AD but also may be seen at a lower frequency in normal aging. Neurofibrillary tangles are composed of collections of microscopic filaments within a neuron. A significant component of the filaments is formed from aggregations of an intracellular protein known as tau. These NFT often first occur in the hippocampus, particularly in the CA1 region, although they also become widespread in the neocortex. Although NFT occur in other degenerative diseases, involvement of the neocortex appears to be essential for development of AD (Bierer *et al.*, 1995). Astrocytes proliferate in affected areas and a variety of other nonspecific changes also occur.

In AD, the process most correlated with cognitive decline is loss of synaptic connections. This is associated with neurotransmitter changes in particular regions emphasizing the neurotransmitter acetylcholine. Two enzymes related to this system, choline acetyltransferase and acetylcholinesterase, decrease as does acetylcholine level itself. Changes are most pronounced in deep nuclear regions (medial septum, diagonal band of Broca, and nucleus basalis of Meynert) which provide input to the amygdala, hippocampus, and neocortex. Other neurotransmitter changes occur but are not as marked, although changes in the noradrenergic brain stem area called the locus ceruleus may be associated with depression. The etiology of AD remains unknown although certain genetic factors have been identified. An early onset of AD is associated with trisomy of chromosome 21 (Down's syndrome) and less strongly linked with certain blood lipoprotein patterns. Symptomatic treatment currently involves pharmacological attempts to boost function of the acetylcholine system.

Symptomatology of MS and AD

Exacerbations of MS may produce symptoms by disrupting any CNS function. Symptoms may be difficult to describe and may include loss of functions (e.g., weakness, numbness, incoordination, or blindness) or abnormal functions (e.g., spasms or pain). Some characteristic symptoms include unilateral vision loss (optic neuritis), double vision, urinary bladder dysfunction, or focal weakness or numbness. An electric-like shock termed Lhermitte's phenomenon may occur with neck flexion. Fatigue is a very common symptom of MS. While fatigue can be related to impaired sleep, impaired mobility, depression or medication, primary fatigue independent of these factors can be severe and disabling. The clinical course of AD is variable. Characteristic early abnormalities involve language, recent memory, attention, visual–spatial function, problem-solving, and calculations. One particular area of deficit may be emphasized in individual patients at first presentation. As the disease progresses, personality, thought processes, affect, and praxis may also be affected. Incontinence, impaired motor skills, and seizures are usually late signs. Social skills are relatively preserved. Sudden increases in symptoms may be precipitated by medical conditions (e.g., medication, infection) or environmental changes (e.g., loss of caretaker, new residence). This disease course may vary from 1 to 15 years, with most patients surviving 5 to 12 years after onset.

Cognitive deficits associated with MS and AD vary markedly. We have described MS and AD in the context of demyelinating versus degenerative categories of neurological disease. However, some investigators have suggested that there are distinct patterns of cognitive deficits that differentiate disorders that have primarily cortical versus subcortical pathology. This distinction has been controversial (Whitehouse, 1986), although some neuropsychological studies suggest that cortical and subcortical disorders do produce different cognitive profiles (Paulsen *et al.*, 1995). Others have provided in-depth reviews of cognitive and behavioral symptomatology associated with cortical and subcortical disorders (Cummings, 1990; Cummings & Benson, 1992; Rao, 1993). To summarize, cortical dementias, such as AD, are distinguished by degeneration of the cerebral cortex. Neurocognitive deficits noted in cortical dementias include impairment of intellect and memory. Memory deficits involve encoding and/or storage. Also, aphasia, agnosia, apraxia, and visuospatial and constructional deficits are common. Typically, personality, gait, and other motor functions are preserved until the late stages of the disease (Cummings & Benson, 1992). Subcortical disorders are also accompanied by memory impairment but this impairment appears to be retrieval-based (see Rao, 1993, for review). In addition, subcortical disorders cause decreased information processing speed, impaired visuospatial ability, deficient executive functions (e.g., planning ability, initiation and cognitive flexibility), and impaired sustained attention (Cummings & Benson, 1992; Rao, 1993). Also, prominent impairment of motor functions and mood disturbance are common. However, general intellectual functioning and language abilities remain grossly intact. We will consider cognitive deficits characteristic of MS and AD in greater detail when full neuropsychological evaluation is discussed later in this chapter.

NEUROPSYCHOLOGICAL ASSESSMENT

Assessment can entail short or intermediate cognitive screening or full neuropsychological evaluation. Full neuropsychological assessment is a time- and resource-intensive process that requires expertise often unavailable in clinical settings. Because of this, it is impractical to provide all patients with suspected or confirmed MS or AD with full neuropsychological evaluation. As a result, investigators and clinicians have developed brief and intermediate-length screening procedures. We have somewhat arbitrarily divided these instruments according to administration time; brief instruments take 10 minutes or less to complete while intermediate-length measures may take as long as 60 minutes. This is in contrast to full neuropsychological evaluation, which may take anywhere from 3 to 8 hours. Brief and intermediate-length screening instruments have the longest history of use with demented patients, particularly with AD patients. Their usefulness in patients with MS is less well established. In the following sections, we discuss the neuropsychological assessment of MS and AD according to the categories of (1) brief screening instruments, (2) intermediate screening procedures, and (3) full neuropsychological assessment.

BRIEF SCREENING METHODS

While brief screening instruments do not generally provide information helpful in making differential diagnoses, they are useful in determining presence or

191

EVALUATION OF
DEMYELINATING
AND
DEGENERATIVE
DISORDERS

absence of cognitive impairment and in making gross estimates of level of cognitive impairment. Typically, performance below designated cutoff scores for these short screening instruments indicates presence of significant cognitive impairment. We discuss the psychometric characteristics of these instruments, first for MS and then for AD.

BRIEF SCREENING METHODS FOR MS. For MS, initial data is available for two brief screening measures, the Mini-Mental State Examination (Folstein, Folstein, & McHugh, 1975) and the Cognitive Capacity Screening Examination (Jacobs, Bernhard, Delgado, & Strain, 1977).

Mini-Mental State Examination (MMSE). The MMSE (Folstein *et al.*, 1975) is the best known cognitive screening instrument. It consists of 19 questions that yield a score between 0 and 30. The MMSE items assess orientation, registration, attention, calculation, verbal memory, language, and visual construction. The standard cutoff score for impairment is 23 or below. However, the following cutoff scores can be used to provide a more precise classification of cognitive functioning: 24–30 = no cognitive impairment; 18–23 = mild cognitive impairment; 11–17 = moderate cognitive impairment; 0–10 = severe cognitive impairment. Few studies provide information on utility of the MMSE in detecting cognitive deficits in individuals with MS (Heaton, Thompson, Nelson, Filley, & Franklin, 1990; Rao, Leo, Bernardin, & Unverzagt, 1991; Swirsky-Sacchetti *et al.*, 1992).

Rao, Leo, Bernardin, & Unverzagt (1991) used the MMSE to classify 100 individuals with MS as cognitively impaired or unimpaired. The standard MMSE cutoff was not used to distinguish between cognitively impaired and cognitively unimpaired subjects. Rather, Rao and co-workers considered any score below the 5th percentile of the control group's average score as impaired. This strategy ensured that the optimal MMSE cutoff score for individuals with MS was identified. Using this strategy, the MMSE categorized 11% of individuals with MS as cognitively impaired. In contrast, full neuropsychological assessment indicated 43% were actually cognitively impaired. Thus, in their sample, the MMSE had excellent specificity (98%) but inadequate sensitivity (23%).

Heaton *et al.* (1990) also examined sensitivity and specificity of the MMSE in a group of 40 individuals with MS, 55% of whom exhibited cognitive impairment on the Halstead–Reitan Neuropsychological Test Battery. Using an MMSE cutoff score of less than 21, the MMSE failed to correctly classify any of the individuals as cognitively impaired. Using a cutoff score defined as one standard deviation below the mean, the MMSE classified 22.5% of the MS group and 12.2% of the control group (N = 90) as impaired. These classification percentages yield a specificity rate of 0.94 and a sensitivity rate of 0.36. Swirsky-Sacchetti *et al.* (1992) reported similar results. In 56 individuals with definite MS, sensitivity of the MMSE to cognitive impairment was low. Also, MMSE scores did not correlate well with total lesion area calculated from MRI scans. In contrast, high correlations were observed between standard neuropsychological tests and total lesion area.

Beatty and Goodkin (1990) attempted to modify the MMSE so that it would be more sensitive to cognitive deficits in patients with MS. They found that, when using a cutoff score of 27 and below, the MMSE was not sensitive enough to detect patients with MS who had dementia. A cutoff score of 28 misclassified too many (16%) nonimpaired control group members as impaired. Suggestions for modifying the MMSE included (1) replacing MMSE naming items (two items) with the

Boston Naming Test (60 items; Kaplan, Goodglass, & Weintraub, 1983); (2) changing the three verbal free recall items to seven items; and (3) including the Symbol Digit Modalities Test (Smith, 1982). In their initial study (Beatty and Goodkin, 1990), these modifications did not appear to substantially increase sensitivity of the MMSE. Further research is needed to determine the potential value of these modifications.

Cognitive Capacity Screening Examination (CCSE). The CCSE (Jacobs *et al.*, 1977) consists of 30 items and has a maximum score of 30. Items assess orientation, attention, concentration, registration, calculation, verbal concept formation, and verbal memory. Scores below 20 are indicative of gross cognitive impairment (Jacobs *et al.*, 1977). Heaton *et al.* (1990) used the CCSE with 40 individuals who had MS. Neuropsychological assessment indicated 55% were cognitively impaired. The standard CCSE cutoff score of less than 20 did not classify any individuals with MS as cognitively impaired. When CCSE scores that fell one standard deviation below the mean were used as the cutoff, the CCSE had a sensitivity of 0.41 and specificity of 0.94. In the MS group, CCSE scores correlated -0.70 with Halstead–Reitan Neuropsychological Test Battery impairment ratings.

BRIEF SCREENING METHODS FOR AD. A much larger body of literature addresses brief cognitive screening of individuals with dementias. Some of the more popular measures that we review include the Mini-Mental State Examination (Folstein *et al.*, 1975), the Short Portable Mental Status Questionnaire (Pfeiffer, 1975), the Blessed Information-Memory-Concentration test (Blessed, Tomlinson, & Roth, 1968), and the Short Orientation-Memory-Concentration test (Katzman *et al.*, 1983).

Mini-Mental State Examination (MMSE). The most extensively studied and widely used brief cognitive screening instrument is the MMSE (Folstein *et al.*, 1975). Studies examining the reliability and validity of the MMSE suggest that it has adequate reliability, with internal consistency estimates ranging from 0.77 to 0.92 (Holzer, Tischler, Leaf, & Myers, 1984; Morris *et al.*, 1989; Weiler, Chiriboga, & Black, 1994), inter-rater reliability of 0.97 (O'Connor *et al.*, 1989), and test–retest reliabilities ranging from 0.74 to 0.84 (Morris *et al.*, 1989; O'Connor *et al.*, 1989). There is also support for concurrent and construct validity of the MMSE (Morris *et al.*, 1989). Scores on the MMSE are significantly correlated with the Short Orientation-Memory-Concentration test ($r = -0.89$; Katzman *et al.*, 1983), Consortium to Establish a Registry for Alzheimer's Disease (CERAD) Word List Memory ($r = 0.85$), CERAD Word List Recall ($r = 0.80$), and CERAD Word List Recognition ($r = 0.74$) (see Morris *et al.*, 1989). Also, when the MMSE was administered longitudinally to individuals with AD, significant correlations were present between magnitude of decline in MMSE scores and measures of cerebral atrophy (cerebrospinal fluid volume increase on computerized tomography scans; Shear *et al.*, 1995). In mixed samples, MMSE specificity ranges from 0.92 to 1.0 and its sensitivity ranges from 0.70 to 0.86 (Baker, Robinson, & Stewart, 1993; Monsch, Foldi, *et al.*, 1995; O'Connor *et al.*, 1989).

Performance on the MMSE is significantly related to age and education for individuals with and without dementia (Magni, Binetti, Cappa, Bianchetti, & Trabucchi, 1995). Individuals with 8 or fewer years of formal education perform more poorly than those with more than 8 years of education (Anthony, LeResche, Niaz,

193

EVALUATION OF
DEMYELINATING
AND
DEGENERATIVE
DISORDERS

VonKorff, & Folstein, 1982). Age and education influence performance on the MMSE independent of other conditions, such as medical conditions (Launer, Dinkgreve, Jonker, Hooijer, & Lindeboom, 1993; Tombaugh, McDowell, Kristjansson, & Hubley, 1996). Because education does affect MMSE performance, some authors suggest that the MMSE should be adjusted so that a cutoff score of 20 or below is used for those with 8 years of education or less and a score of 25 or below for well-educated individuals (Galasko *et al.*, 1990; also see Monsch, Foldi, *et al.*, 1995). Marshall and Mungas (1995) developed a statistical technique that increases sensitivity of the MMSE by adjusting MMSE scores according to age and years of education. Other investigators have developed MMSE short forms and, in some cases, these short forms show similar levels of specificity and sensitivity when used to distinguish individuals without AD from those with mild AD (see Fillenbaum, Wilkinson, Welsh, & Mohs, 1994).

Short Portable Mental Status Questionnaire (SPMSQ). The SPMSQ (Pfeiffer, 1975) consists of 10 items that assess orientation, remote memory, and concentration. It has been used with community, medical, and psychiatric samples. Although the SPMSQ has adequate test–retest reliability ($r = 0.82$ to 0.83; Pfeiffer, 1975; Roccaforte, Burke, Bayer, & Wengel, 1994) when the standard cutoff score is used (four or more errors), its sensitivity to detect cognitive impairment may be limited. For medical inpatients, sensitivity increased from 0.76 to 0.86 to 1.0 and specificity decreased from 1.0 to 0.99 to 0.89 when cutoff scores of 4, 3, and 2 were used, respectively (Erkinjuntti, Sulkava, Wikström, & Autio, 1987). These results are consistent with studies of other patient groups (Erkinjuntti *et al.*, 1987; Roccaforte *et al.*, 1994). Based on these findings, some have advocated lowering the cutoff score to 3 for dementia and 2 for delirium in order to increase sensitivity (Erkinjuntti *et al.*, 1987). Others have found that the SPMSQ was not significantly related to clinical or neuropsychological diagnosis in a sample of psychiatric and neurological patients (Dalton, Pederson, Blom, & Holmes, 1987), reinforcing the notion that brief screening measures such as the SPMSQ do not contribute much to differential diagnosis, especially between different types of dementia. However, the SPMSQ is significantly correlated with MMSE scores ($r = 0.81$; Roccaforte *et al.*, 1994). Finally, Pfeiffer (1975) reported that SPMSQ scores were related to education and race, although this assertion has not been consistently supported (Erkinjuntti *et al.*, 1987).

Blessed Information-Memory-Concentration (IMC) Test. The IMC test (Blessed *et al.*, 1968) is composed of three of the four subscales that make up the Blessed Dementia Scale (Blessed *et al.*, 1968): the Information subscale, Memory subscale, and Concentration subscale. In the United States version, the three cognitive subscales contain 26 items that assess knowledge of basic personal information, orientation, memory (remote and recent), attention, and concentration (Fuld, 1978; Katzman *et al.*, 1983). For the United States version, scores range from 1 to 33. Higher scores indicate greater impairment. For the IMC test, investigators report test–retest reliabilities ranging from 0.82 to 0.90 (Thal, Grundman, & Golden, 1986) and an internal consistency of 0.94 (Weiler, Chiriboga, & Black, 1994). Salmon, Thal, Butters, and Heindel (1990) provided some support for the validity of the IMC test by reporting significant correlations between the IMC total score and scores on the MMSE ($r = -0.88$) and Mattis Dementia Rating Scale ($r = -0.79$; Mattis, 1988). It also appears that IMC test scores increase as AD progresses

(Salmon *et al.*, 1990; Stern *et al.*, 1992), although the IMC test does not appear to be sensitive to changes in cognitive functioning in severe dementia (Salmon *et al.*, 1990).

Short Orientation-Memory-Concentration (SOMC) Test. The SOMC (Katzman *et al.*, 1983) is a six-item version of the IMC test (Blessed *et al.*, 1968). Higher SOMC test scores indicate more severe impairment. In the initial validity study, Katzman and colleagues (1983) found that the SOMC test accounted for 88.6% to 92.6% of the variance in the IMC test, depending on the sample. The SOMC test has adequate test–retest reliability ($r = 0.77$; Fillenbaum, Heyman, Wilkinson, & Haynes, 1987) and high internal consistency ($r > 0.92$; Morris *et al.*, 1989). It correlated highly with the MMSE ($r = -0.89$), CERAD Word List Memory ($r = -0.84$), CERAD Word List Recall ($r = -0.84$), and CERAD Word List Recognition ($r = -0.77$) (Morris *et al.*, 1989; also see Fillenbaum *et al.*, 1987). In addition, SOMC test scores were significantly correlated ($r = 0.54$, $p < .001$) with temporal, parietal, and frontal cortex plaque counts of 38 autopsied patients (Katzman *et al.*, 1983). Finally, it discriminates between individuals with dementia and those without; its sensitivity is 0.87 and its specificity is 0.94 (Davous, Lamour, Debrand, & Rondot, 1987). Darvous and colleagues (1987) suggest that the SOMC test is as effective as the MMSE when used to screen for dementia.

SUMMARY. For individuals with MS, the CCSE appears to be somewhat more sensitive to cognitive impairment than the MMSE. However, because of the poor sensitivity of the MMSE and CCSE, most cases of cognitive impairment in individuals with MS would go undetected. Based on this, we recommend that clinicians and investigators use brief cognitive screening techniques cautiously with individuals who have MS, if at all. Certainly, based on MMSE or CCSE results in the "unimpaired" range one cannot conclude that patients with MS are deficit-free; to do so would be a great disservice to those individuals with MS who truly do have cognitive deficits.

Studies of individuals with AD are more conclusive. They indicate that the brief cognitive screening instruments reviewed have adequate sensitivity and specificity. Also, reliability and validity estimates are acceptable. These measures may be useful because they predict ADL capacity and rated level of care by caregivers for patients with AD (Blessed *et al.*, 1968; Folstein *et al.*, 1975; Weiler *et al.*, 1994). Also, few demonstrable practice effects are apparent for the MMSE and the IMC test when they are repeatedly administered to patients who have AD (Thal *et al.*, 1986). While there is support for use of any of the measures previously described, we recommend use of the MMSE (unless there are compelling reasons to the contrary) simply because the MMSE has the most support and validation data of any brief cognitive screening instrument.

INTERMEDIATE-LENGTH COGNITIVE SCREENING

INTERMEDIATE COGNITIVE SCREENING IN MS. Some authors have developed cognitive screening batteries specifically for patients with MS by compiling tests that appear relatively more sensitive to cognitive dysfunction in these individuals (Beatty *et al.*, 1995; Heaton *et al.*, 1990; Rao, Leo, Bernardin, & Unverzagt, 1991).

Brief Cognitive Battery (BCB). After extensive neuropsychological evaluation of 100 individuals with MS, Rao, Leo, Bernardin, and Unverzagt (1991) developed the BCB by combining test scores that were particularly sensitive to cognitive deficits. The test scores used in the BCB include (1) the consistent long-term retrieval score from the Selective Reminding Test (Buschke & Fuld, 1974); (2) the total recall score from the 7/24 Spatial Recall Test (Barbizet & Cany, 1968); (3) the total number of words generated on the Controlled Oral Word Association Test (Benton & Hamsher, 1976); and (4) the percentage correct on the hard form of the Paced Auditory Serial Addition Test (Gronwall, 1977). When failure of one or more of these four test scores was used as the cutoff, sensitivity was 0.90 and specificity was 0.77. When failure on two or more test scores was used, the sensitivity decreased to 0.71 and the specificity increased to 0.94. In contrast, the MMSE had a sensitivity of 0.23 and a specificity of 0.98 for the same individuals.

Neuropsychological Screening Battery (NSB). Heaton *et al.* (1990) also reported use of an intermediate-length cognitive screening battery, the NSB, with patients who had MS. The NSB is composed of 18 brief tests (for an in-depth description of these tests and the scores derived from them see Heaton *et al.*, 1990). The NSB takes approximately 45 minutes to administer and yields individual scores for each instrument. Two additional scores can also be calculated: a Total Summary Score (TSS), which includes tests of sensorimotor and cognitive abilities; and a Cognitive Summary Score (CSS), which includes only tests of cognitive abilities (no motor component). The CSS and TSS performed better than brief cognitive screens (MMSE and CCSE) in detecting cognitive impairment (Heaton *et al.*, 1990). Sensitivity and specificity of the TSS were 0.77 and 0.56, respectively. When the CSS was used, sensitivity decreased to 0.55 but specificity increased to 0.94. Because the sensitivity of the TSS and CSS are relatively low, Heaton *et al.* (1990) suggested that use of such summary scores may actually obscure detection of cognitive deficits in patients with MS. Examination of individual test scores may increase the NSB's sensitivity, although further research is needed to determine if this is actually the case.

Screening Examination for Cognitive Impairment (SEFCI). The SEFCI was recently developed by Beatty and co-workers (1995). The SEFCI includes a short list learning task (10 words, 3 trials, and 12-minute delayed recall), the Shipley Institute of Living Scale (SILS: Zachary, 1986), and the Symbol Digit Modalities Test (SDMT; Smith, 1982) oral response version. It takes approximately 30 minutes to complete. In their initial investigation, they examined 103 individuals with clinically definite MS and 32 controls. All subjects were administered a battery of neuropsychological tests and the tests that composed the SEFCI. When norms were derived based on the control group's performance, the SEFCI had a sensitivity of 0.86 and a specificity of 0.90. When published norms for the SILS and SDMT were used, the sensitivity decreased to 0.74 and the specificity increased to 0.91.

INTERMEDIATE COGNITIVE SCREENING IN AD. For individuals with AD, we review the Mattis Dementia Rating Scale (Mattis, 1988) and the Middlesex Elderly Assessment of Mental State (Golding, 1989). While other intermediate-length screening measures are available, we selected these for review because they were specifically designed to assess neurocognitive function of elderly individuals or individuals with dementia.

195

EVALUATION OF
DEMYELINATING
AND
DEGENERATIVE
DISORDERS

Mattis Dementia Rating Scale (DRS). The DRS (Mattis, 1988) is composed of 36 tasks. From these tasks, five subscale scores are calculated, including Attention, Initiation/ Perseveration, Construction, Conceptualization, and Memory. Scores range from 0 to 144 and individuals without cognitive impairment typically obtain scores between 140 and 144. Scores that fall two standard deviations below the mean (122 and below) indicate cognitive impairment. Split-half and test–retest reliability for the DRS are 0.90 (Gardner, Oliver-Muñoz, Fisher, & Empting, 1981) and 0.97 (Coblentz *et al.*, 1973), respectively. The DRS is excellent in detecting even mild cognitive impairment in elderly individuals (Monsch, Bondi, *et al.*, 1995) and the DRS subscales are able to distinguish between mild and moderate dementia (Mattis, 1988). Overall, the DRS is sensitive to changes in cognitive deterioration (Salmon *et al.*, 1990) and it is more sensitive to these changes than brief screening measures (e.g., MMSE). Some DRS subscales are also highly correlated with clinical ratings; ratings of AD patients' competency to make medical decisions are significantly correlated with scores on the Memory subscale ($r = 0.80$) and the Initiation/Perseveration subscale ($r = 0.90$) (Marson, Herfkens, Brooks, Ingram, & Harrell, 1995). Also, two studies indicate that the Attention subtest distinguishes patients with mild functional impairment from patients with severe functional impairment (Shay *et al.*, 1991). Finally, factor analytic studies indicate that the DRS is composed of three underlying factors: conceptualization/organization abilities, visuospatial abilities, and memory and orientation (Woodward, Salthouse, Godsall, & Green, 1996). These factors have shown strong correlations with neuropsychological tests that assess similar cognitive abilities (Woodward *et al.*, 1996).

Middlesex Elderly Assessment of Mental State (MEAMS). The MEAMS (Golding, 1989) is composed of 54 items that make up 12 subtests: Orientation, Name Learning, Naming, Comprehension, Remembering Pictures, Arithmetic, Spatial Construction, Fragmented Letter Perception, Unusual Views, Usual Views, Verbal Fluency, and Motor Perseveration. The MEAMS has a number of unique features including the ability to distinguish sensory from perceptual deficits (Unusual Views and Usual Views) and two equivalent forms. Scores of 10 to 12 are considered normal (range = 0–12), scores of 8 and 9 are indicative of borderline impairment, and scores of 7 and below indicate cognitive impairment. The MEAMS has an inter-rater reliability of 0.98 (Golding, 1989) and alternate form reliabilities ranging from 0.82 (patients with dementia) to 0.95 (normal controls; Golding, 1989; Powell, Brooker, & Papadopolous, 1993). Validity studies indicate it distinguishes between depression and dementia. Also, there is some initial evidence suggesting it can distinguish between (1) individuals with AD and those with multi-infarct dementia (Golding, 1989) and (2) individuals with right versus left hemisphere strokes (Shiel & Wilson, 1992). When used in an elderly psychiatric sample the MEAMS was more sensitive to mild cognitive impairment than the MMSE (Husband & Tarbuck, 1994). Calculations we performed on information presented in the test manual indicate the MEAMS has a sensitivity of 0.91 and a specificity of 0.95 when differentiating patients with depression from those with dementia. These estimates may increase when the MEAMS is used to distinguish between cognitively impaired and normal control groups.

SUMMARY

Intermediate-length screening batteries appear to hold more promise than brief screening techniques (e.g., MMSE and CCSE) for individuals with MS because they tend to have higher sensitivity and specificity. Of the three intermediate-length screening techniques, the SEFCI had the highest sensitivity and comparable specificity. Also, while the SEFCI takes about 30 minutes to complete, the SILS requires 20 minutes and is self-administered. The other two screening techniques (Heaton *et al.*, 1990; Rao, Bernardin, & Unverzagt, 1991) take more examiner time to administer completely but are not as sensitive to cognitive deficits. Because of this, we recommend use of the SEFCI for cognitive screening of individuals with MS.

When conducting cognitive screening with patients who have AD, the DRS should be strongly considered simply because there is a substantial body of litera-ture that supports its validity and reliability as well as its sensitivity when detecting even mild cognitive impairment associated with AD. The MEAMS is less well estab-lished but appears to hold some promise as a cognitive screening measure for use with elderly individuals. Additional research is needed to establish more thorough-ly the psychometric properties of the MEAMS, including whether or not it is as sensitive to cognitive deficits as the DRS. When compared to brief screening instru-ments (e.g., MMSE), future research may demonstrate that instruments that evalu-ate a broader range of cognitive abilities (e.g., DRS and MEAMS) are more sensitive to differences in dementias based on their etiologies and thus assist in differential diagnosis.

FULL NEUROPSYCHOLOGICAL EVALUATION

Full neuropsychological evaluation is warranted if, based on results of cogni-tive screening, cognitive deficits are apparent. Evaluation may also be warranted if significant subjective complaints of memory, attention, or other deficits are pre-sented by the client, a family member, or a health care provider. In the following sections, issues relevant to the assessment of MS and AD are presented. However, these discussions are not meant to be exhaustive. Rather, cognitive deficits common to each disorder are emphasized. Comprehensive discussions of neuropsychologi-cal evaluation often used neuropsychological tests and the most popular philoso-phies underlying neuropsychological evaluation have been proposed by Lezak (1995) and in Reitan and Wolfson (1993).

EVALUATION OF MS

GENERAL CONSIDERATIONS. There are a number of reasons why it is important to evaluate cognitive functioning in individuals with MS. Cognitive impairment may not be readily apparent in this group because the degree of cognitive impair-ment is not strongly associated with severity of physical impairment or duration of illness. Individuals with relatively little physical disability can be severely cognitively impaired (Beatty *et al.*, 1995; Peyser, Edwards, Poser, & Filskov, 1980; Rao, Leo, Bernardin, & Unverzagt, 1991; Ryan, Clark, Klonoff, Li, & Paty, 1996). Because of these and other factors, cognitive impairment often goes undetected (Peyser *et al.*, 1980). On the other hand, cognitive impairment does significantly impact social and vocational functioning, so that cognitively impaired individuals with MS are significantly less likely to have gainful employment, are less likely to participate in social activities, require greater physical assistance, exhibit more problems main-taining personal hygiene and activities of daily living, are less able to follow a simple

197

EVALUATION OF
DEMYELINATING
AND
DEGENERATIVE
DISORDERS

recipe when cooking, and have higher levels of overall functional impairment (Kessler, Cohen, Lauer, & Kausch, 1992; Rao, Leo, Ellington, *et al.*, 1991; Ryan *et al.*, 1996). Because cognitive deficits often go undetected even though they do negatively impact functioning, full neuropsychological evaluation can be invaluable because it allows for quantification of suspected or unsuspected cognitive deficits.

TYPE AND PREVALENCE OF COGNITIVE DEFICITS. Although cognitive impairment was initially believed to be infrequent in MS, recent studies suggest that it is actually quite prevalent. Prevalence rates of cognitive impairment range between 33% and 72% (Heaton, Nelson, Thompson, Burks, & Franklin, 1985; McIntosh-Michaelis *et al.*, 1991; Rao, Leo, Bernardin, & Unverzagt, 1991; Ryan *et al.*, 1996). Much of this discrepancy appears due to types of patients under investigation, duration of MS symptoms, course and progression of the disorder, and type, sensitivity, and number of neuropsychological tests administered. For example, individuals with chronic progressive MS exhibit more severe cognitive deficits than those with relapsing remitting MS (Beatty, Goodkin, Monson, & Beatty, 1986; Beatty, Goodkin, Monson, Beatty, & Hertsgaard, 1988; Beatty, Goodkin, Beatty, & Monson, 1989; Rao, Leo, Bernardin, & Unverzagt, 1991). Two studies have provided estimates of cognitive impairment in individuals with MS who live in the community and have minimal levels of disability (Rao, Leo, Bernardin, & Unverzagt, 1991; Ryan *et al.*, 1996). We review these studies because of their excellent methodologies and because they provide results that are applicable to most individuals with MS, namely, the community dwelling MS population.

Rao, Leo, Bernardin, & Unverzagt (1991) evaluated 100 individuals with MS living in the community (average age = 45.7 ± 11.3; years of education = 13.2 ± 2.4 years; predicted premorbid IQ = 106.8 ± 7.3) and 100 controls matched for age, education, and premorbid IQ (average age = 46.0 ± 11.6; years of education = 13.3 ± 2.0 years; predicted premorbid IQ = 106.5 ± 6.9). Subjects with MS received a formal diagnosis of MS an average of 9.5 years (SD = 9.0) prior to the study. Thirty-nine had relapsing-remitting MS, 19 had chronic-progressive MS, and 42 had chronic-stable MS. All subjects (including controls) underwent extensive screening to rule out conditions other than MS that could adversely influence cognitive functioning. Neuropsychological evaluation included tests of verbal intelligence, memory (immediate, recent, and remote), abstracting abilities, attention/concentration, language abilities, and visuospatial abilities. Performance on individual tests was considered impaired if the test score fell at or below the fifth percentile, with the fifth percentile determined by performance of normal controls. Individuals with MS performed in the impaired range on significantly more neuropsychological tests indices (4.64 ± 4.9) than did controls (1.13 ± 1.8). Using the criteria of four or more test indices failed, 48% of the MS group were categorized as cognitively impaired, as were 5% of controls. Of individuals with MS, 31% failed tests of recent memory ability, approximately 25% failed tests of sustained attention and verbal fluency, approximately 20% were impaired on tests of visuospatial perception and conceptual reasoning, and 15% to 21% had impaired Verbal IQ composite or verbal subtest scores.

Ryan *et al.* (1996) performed a less extensive evaluation of cognitive functioning in 177 subjects with MS (age = 36.2 ± 7.8 years; education = 13.5 ± 2.2 years) and in 89 matched controls (age = 35.2 ± 7.0 years; education = 13.9 ± 2.2 years). Subjects with MS were formally diagnosed with MS approximately 5.2 years prior to the study (this was calculated from information presented by Ryan *et al.*, 1996).

199

EVALUATION OF
DEMYELINATING
AND
DEGENERATIVE
DISORDERS

All had definite MS with a relapsing-remitting course. A standard disability rating scale, the Expanded Disability Status Scale (EDSS; Kurtzke, 1983), yielded scores of 2.0 ± 1.2 which indicated only minimal physical disability. Using a similar cutoff to Rao, Leo, Bernardin, & Unverzagt (1991), Ryan and co-workers (1996) found that their MS group failed significantly more tests than controls (tests failed = 1.76 ± 1.91 vs. 0.73 ± 1.08, respectively). Using this criteria, 33% of the MS group (n = 58) and 5% of controls (n = 5) were cognitively impaired. Cluster analyses were performed using scores from the Benton Visual Retention Test (VBR; Benton, 1974), Word Fluency letters F, A, and S (WF; Lezak, 1995), and Paired Associate Learning (PA; Wechsler, 1945). Of the six clusters resulting from the analysis, subjects with MS in two of the clusters exhibited either moderate impairment (n = 19) or severe impairment (n = 11) on all three tests. Of the remaining subjects with MS, nine were impaired on BVR and PA but not WF (cluster 3), eight were impaired on BVR but not on WF or PA (cluster 4), ten were impaired on WF and PA but not on BVR (cluster 5), and one subject was not impaired on any of the tests (cluster 6).

When test scores of subjects classified as unimpaired (MS and controls) were subjected to cluster analysis, 10 clusters were derived. Of these clusters, three had an overrepresentation of subjects with MS. Two of these three clusters were similar to the impaired groups' cluster 4 and cluster 5. Ryan et al. (1996) concluded that "there is no single pattern of cognitive impairment characteristic of MS" (p. 188). However, this conclusion should be qualified. First, over half (n = 30) of the MS group classified as cognitively impaired did exhibit moderate to severe impairment on all three test indices, that is, half of the subjects exhibited a similar global pattern of impairment differentiated only by severity of impairment. From the available information, it appears that this pattern of impairment is consistent with a typical subcortical dementia profile. Second, overall, Ryan's sample had a relatively short disease duration, was relatively young, had a substantial proportion of subjects with MS who did not exhibit any cognitive impairment (35%), and was composed entirely of individuals with relapsing-remitting MS. It is not surprising then that the whole group did not exhibit a cognitive profile characteristic of subcortical dementia and that significant variability was present from one individual to the next. In patients with mild cognitive impairment that probably does not meet criteria for dementia per se, it is more likely that a number of specific deficits will emerge on neuropsychological testing. Some of these deficits demonstrated on neuropsychological testing will have direct correlations to localized cerebral lesions while other cognitive deficits will not (Ryan et al., 1996). However, for some patients with discrete cognitive deficits, as lesion load increases, it is probable that a more global pattern of cognitive deficits will emerge that is similar to the characteristic pattern noted in subcortical dementia.

When comparing the two studies (Rao, Leo, Bernardin, & Unverzagt, 1991; Ryan et al., 1996), Rao and co-workers reported a higher level of cognitive impairment. However, because Rao's group was almost 10 years older than Ryan's group (45.7 ± 11.3 vs. 36.2 ± 7.8, respectively), had longer disease histories (formal diagnosis = 9.5 vs. 5.2 years, respectively), included individuals with chronic-progressive MS, and exhibited greater physical disability (EDSS score = 2.0 ± 1.2 vs. 4.1 ± 2.2), this probably represents a real difference.

Even though estimates of cognitive impairment reported by Rao, Leo, Bernardin, & Unverzagt (1991) and Ryan et al. (1996) are lower than some previous reports, the patterns of their results are similar to those reported by others (Beatty et al., 1989; Beatty et al., 1988; Grossman et al., 1994; Heaton et al., 1985; Kessler et *al.*, 1992; McIntosh-Michaelis *et al.*, 1991). Lower prevalence of cognitive impairment reported by Rao, Leo, Bernardin, and Unverzagt (1991) and Ryan *et al.* (1996) reflects the more general and less severely impaired populations they examined. Their inclusion/exclusion criteria undoubtedly excluded some severely impaired patients, which also decreased level of overall cognitive impairment in their samples. However, it is important to remember that their results reflect cognitive functioning of individuals with MS who live in the community. It is expected that these individuals would be less severely impaired than patients who were institutionalized or seeking evaluation for cognitive impairment from a health care provider. In general then, these studies indicate that individuals with MS exhibit a pattern of cognitive impairment characterized by deficits in recent memory (possibly retrieval-based), decreased information processing speed, impaired planning abilities and concept formation, impaired visuospatial processing, and impaired motor abilities.

CHOICE OF NEUROPSYCHOLOGICAL TESTS. While a number of neuropsychological tests or test batteries could be used to assess cognitive functioning in patients with MS, many tests and batteries have at least two disadvantages. First, they were not developed specifically for use with MS patients so their tasks are often inappropriate (e.g., heavily loaded with tasks that have substantial motor components). Second, they do not provide normative information specifically for MS patients. One solution to this problem is to use batteries developed specifically for individuals with MS, such as the battery used by Rao, Leo, Bernardin, and Unverzagt (1991). This battery takes into consideration the motor impairment that many patients with MS exhibit and that interfere with test performance, and the means and standard deviations for Rao's normal controls can be used to determine the test failure. Rao provides a detailed description of this battery elsewhere (Rao, Leo, Bernardin, & Unverzagt, 1991; Rao, Leo, Haughton, St. Aubin-Faubert, & Bernardin, 1989).

EVALUATION OF AD

GENERAL CONSIDERATIONS. A number of factors unique to assessment of elderly individuals with dementia are worth considering throughout the evaluation process. These arise from two general sources: (1) client characteristics and (2) instrument and test environment characteristics. In this volume, P. D. Nussbaum provides an excellent review of the area of geriatric assessment. Other reviews are also available (e.g., Storandt, 1994) that specifically address assessment of elderly individuals. For our purposes, suffice it to say that differences arising from age-related changes in physical health, cognitive functioning, patterns of medication use, amount and quality of education, and sociocultural experiences can significantly impact neuropsychological evaluation and test results. Therefore, differences between the young and old need to be considered when assessing elderly individuals.

Along these same lines, special consideration is required when assessing individuals with severe dementia because these patients have significantly decreased attention spans and it is often difficult to establish adequate rapport (Saxton, McGonigle-Gibson, Swihart, & Boller, 1993). While we focus most of our discussion on assessment of individuals with mild to moderate AD, assessment of severe dementia requires special consideration because many neuropsychological tests are not appropriate for evaluating severe dementia (see Nussbaum & Allen, 1997, for review). Two factors limit use of standard neuropsychological tests with patients who have severe dementia. First, instructions are often incomprehensible to these patients, and second, items that compose standard tests are often too difficult so that significant floor effects occur or they are insensitive to fluctuations in cognitive function (Salmon *et al.*, 1990).

Of those tests currently available for the assessment of severe dementia, the Severe Impairment Battery (SIB; Saxton *et al.*, 1993) has been the most thoroughly investigated and assesses a broader range of cognitive abilities than other instruments. It was specifically designed to assess individuals with severe dementia. It takes approximately 20 minutes to administer and it assesses attention, orientation, language, memory, visuospatial perception, and construction. There is no cutoff for normal performance and the SIB is sensitive to differences in levels of cognitive impairment in patients who score less than 10 on the MMSE. As such, the SIB is a useful tool for evaluating patients who are typically considered untestable. Even so, some patients categorized as being severely impaired (i.e., vegetative, nonverbal, and/or responsive only to pain stimuli) are not testable using neuropsychological evaluation procedures.

201

EVALUATION OF
DEMYELINATING
AND
DEGENERATIVE
DISORDERS

Type and Prevalence of Cognitive Deficits

The most extensive literature available regarding the neuropsychological assessment of individuals with AD is for those who have mild to moderate cognitive impairment. As previously mentioned, cognitive deficits associated with AD and other cortical disorders include intellectual impairment, encoding and/or storage-based memory deficits, aphasia, agnosia, apraxia, and visuospatial and constructional deficits. However, two areas of cognitive functioning, language and memory, show distinct patterns of deficits and thus may have potential use in differential diagnosis. Researchers have found that these areas are useful in distinguishing individuals with AD from normal controls (Hom, 1992; Binetti *et al.*, 1993) and from those with vascular dementia (Barr, Benedict, Tune, & Brandt, 1992).

MEMORY DEFICITS IN AD. Impaired memory functioning is often the first sign of a dementia and, in AD, the first sign of impaired memory is a decreased ability to learn and retain new information. Anterograde amnesia occurs early in AD and impairs memory for both verbal and nonverbal material (Wilson, Bacon, & Kaszniak, 1983). Information is not retained after brief delays and performance is impaired on both recall and recognition tasks (Wilson *et al.*, 1983). Thus, individuals with AD benefit very little from cueing during testing. Remote memory (i.e., retrograde memory) is also impaired in AD. Early in the disease course, however, remote memory loss is temporally graded (Beatty, Salmon, Butters, Heindel, & Granholm, 1987; Sager, Cohen, Sullivan, Corkin, & Growdon, 1988). That is, information from the distant past is more likely to be recalled than information from the more recent past. An area of memory functioning that appears to be relatively preserved in AD is procedural memory (Eslinger & Damasio, 1986).

LANGUAGE DYSFUNCTION IN AD. The first indication of language difficulties associated with AD is often word-finding problems (Huff, Corkin, & Growdon, 1986; Hart, Smith, & Swash, 1988). Semantic paraphasias (e.g., spoon for fork) and circumlocution are frequently encountered in language performance while phonemic paraphasias (e.g., spork for fork) are a relatively rare occurrence (Appell, Kertesz, & Fisman, 1982). Confrontational naming is often impaired in AD and marked by progressive deterioration (Hodges, Salmon, & Butters, 1991). Generative naming has also been found to be a sensitive marker for AD and useful in distinguishing those with the disease from normal controls (Storandt, Botwinick, Danziger, Berg, & Hughes, 1984).

DIFFERENTIAL DIAGNOSIS. Much of the testing research examining differential diagnosis of dementias has employed differences between scores on a variety of tests (e.g., Wechsler Memory Scale scores, Boston Naming Test scores, etc.) in an attempt to find tests that will make distinctions between groups. The results of these studies generally indicate that many neuropsychological tests do not effectively differentiate among different types of dementia (e.g., AD vs. MID) (Loring & Largen, 1985; Miller, 1981), although they are able to distinguish individuals with cognitive impairment from those with no impairment (Storandt & Hill, 1989; Eslinger, Damasio, Benton, & Van Allen, 1985). Given that different dementias may present themselves with different patterns of scores in one or more areas of functioning (e.g., memory and language) future research may show that neuropsychological data can provide differential diagnoses by examining patterns of scores rather than differences in individual test scores. An important step toward this end is the development of a battery of tests standardized on one or more patient groups with a known dementia. For example, a battery given to individuals with AD will theoretically reveal consistencies in their pattern of results. These patterns can then be used as a reference when administering this same battery to individuals with a dementia of unknown etiology. By comparing the pattern of scores from known diagnostic groups to unknown diagnostic groups, we may be better able to contribute to differential diagnosis using neuropsychological data. The development of the CERAD neuropsychological battery is a step in this direction and may prove helpful in differential diagnosis. As such, we will discuss it further during our consideration of the specific assessment methods for AD. At the present time, the best method of making a differential diagnosis of AD based upon neuropsychological data is by assessing a number of areas of cognitive functioning in greater breadth than is afforded by either the brief or intermediate scales outlined above.

CHOICE OF NEUROPSYCHOLOGICAL TESTS

Although many instruments can be used to assess the cognitive deficits present in AD, the CERAD neuropsychological assessment battery was developed by investigators across the United States as a brief but comprehensive battery for assessing the cognitive functioning of patients with a diagnosis of AD. It has two assessment batteries, a clinical and a neuropsychological battery. The Clinical Assessment Battery was "designed to furnish experienced clinicians with the minimum information necessary to make a confident diagnosis of probable AD" (Morris et al., 1989, p. 1160). In addition to two semistructured interviews, one with the patient and the other with an informant, the Clinical Assessment Battery consists of physical, neurological, and laboratory evaluations and a medical history and drug inventory. The Neuropsychological Assessment Battery consists of a battery of tests assessing areas of cognitive functioning that are typically impaired in AD. These areas include language functioning, memory, praxis, and general intellectual functioning. Specific tests include Verbal Fluency, Modified Boston Naming Test, MMSE, Word List Memory, Constructional Praxis, Word List Recall, and Word List Recognition. The authors report that the Neuropsychological Assessment Battery can be completed within 20 to 30 minutes.

203

EVALUATION OF
DEMYELINATING
AND
DEGENERATIVE
DISORDERS

The initial report on the CERAD was based on findings from 354 subjects with a clinical diagnosis of AD and 278 elderly subjects without dementia (Morris *et al.*, 1989). The subjects with AD were categorized as having either mild or moderate levels of dementia based upon the Clinical Dementia Rating Scale (Hughes, Berg, Danziger, Coben, & Martin, 1982; Berg, 1988). All subtests from the Neuropsychological Assessment Battery revealed significant differences among all groups (controls, mild AD, moderate AD). The control subjects performed better than the other groups and the moderate AD subjects showed the greatest impairment. While the differences noted between groups suggests that the battery may have discriminate validity, the authors note that the range of scores on many of the tests overlapped among the three groups. One-month test–retest correlations are also reported to have been significant for all measures, ranging from a high of 0.90 on the Boston Naming Test for the moderate AD group to a low of 0.36 for the control subjects on Word List Recognition (Morris *et al.*, 1989). The lower test–retest correlations for control subjects were attributed by the authors to possible ceiling effects of the tests, resulting in a restricted range of values.

Since the initial report of the CERAD the battery has undergone further investigation. The original investigation reported means and ranges of scores for control subjects, mild AD, and moderate AD (Morris *et al.*, 1989). More recently, there were norms published for 413 individuals without cognitive impairment (Welsh *et al.*, 1994). These norms are also based on age, education, and gender which have all been shown to affect test scores. While the CERAD has been shown to differentiate AD from normals as well as mild AD from moderate AD, to date there has been no research exploring the battery's ability to differentiate dementias based on etiology. Future research exploring differential diagnosis based upon pattern of neuropsychological scores may find that the CERAD makes a valuable contribution to this important area of neuropsychology.

Summary

In summary then, MS and AD have relatively distinct patterns of cognitive impairment. Individuals with MS tend to exhibit deficits in recent memory (possibly retrieval-based), decreased information processing speed, impaired planning abilities and concept formation, impaired visuospatial processing, and impaired motor abilities. On the other hand, individuals with AD often manifest more general intellectual impairment, encoding and/or storage-based memory deficits, aphasia, agnosia, apraxia, and visuospatial and constructional deficits. Because of the differing pattern of cognitive impairment and the age differences often present between individuals with MS and those with AD, different approaches to test administration and test selection are required. Fortunately, there are specific test batteries that have been developed for assessment of these two groups. These neuropsychological batteries can be supplemented with additional tests when necessary, or specific instruments taken from these more extensive batteries can be used when indicated.

Acknowledgments

This work was supported in part by a grant from the National Multiple Sclerosis Society to Daniel N. Allen, Ph.D.

REFERENCES

Adams, R. D., & Victor, M. (1997). *Principles of neurology* (6th ed.). New York: McGraw-Hill Information Services Company.

Anthony, J. C., LeResche, L., Niaz, U., VonKorff, M. R., & Folstein, M. F. (1982). Limits of the "Mini-Mental State" as a screening test for dementia and delirium among hospital patients. *Psychological Medicine, 12,* 397–408.

Appell, J., Kertesz, A., & Fisman, A. (1982). A study of language functioning in Alzheimer's patients. *Brain and Language, 17,* 73–91.

Baker, F. M., Robinson, B. H., & Stewart, B. (1993). Use of the Mini-Mental State Examination in African American elders. *Clinical Gerontologist, 14,* 5–13.

Barbizet, J., & Cany, E. (1968). Clinical and psychometrical study of a patient with memory disturbances. *International Journal of Neurology, 7,* 44–54.

Barr, A., Benedict, R., Tune, L., & Brandt, J. (1992). Neuropsychological differentiation of Alzheimer's disease from vascular dementia. *International Journal of Geriatric Psychiatry, 7,* 621–627.

Beatty, W. W., & Goodkin, D. E. (1990). Screening for cognitive impairment in multiple sclerosis: An evaluation of the Mini-Mental State Examination. *Archives of Neurology, 47,* 297–301.

Beatty, W. W., Goodkin, D. E., Beatty, P. A., & Monson, N. (1989). Frontal lobe dysfunction and memory impairment in patients with chronic progressive multiple sclerosis. *Brain and Cognition, 11,* 73–86.

Beatty, W. W., Goodkin, D. E., Monson, N., & Beatty, P. A. (1986). Cognitive disturbances in patients with relapsing remitting multiple sclerosis. *Archives of Neurology, 46,* 1113–1119.

Beatty, W. W., Goodkin, D. E., Monson, N., Beatty, P. A., & Hertsgaard, D. (1988). Anterograde and retrograde amnesia in patients with chronic progressive multiple sclerosis. *Archives of Neurology, 45,* 611–619.

Beatty, W. W., Paul, R. H., Wilbanks, S. L., Hames, K. A., Blanco, C. R., & Goodkin, D. E. (1995). Identifying multiple sclerosis patients with mild or global cognitive impairment using the Screening Examination for Cognitive Impairment. *Neurology, 45,* 718–723.

Beatty, W. W., Salmon, D. P., Butters, N., Heindel, W. G., & Granholm, E. P. (1987). Retrograde amnesia in patients with Alzheimer's disease or Huntington's disease. *Neurobiology of Aging, 9,* 181–186.

Benton, A. L. (1974). *The revised visual retention test* (4th ed.). New York: Psychological Corporation.

Benton, A. L., & Hamsher, K. de S. (1976). *Multilingual Aphasia Examination.* Iowa City, IA: University of Iowa.

Berg, L. (1988). Clinical Dementia Rating (CDR). *Psychopharmacology Bulletin, 24,* 637–639.

Bierer, L. M., Hof, P. R., Purohit, D. P., Carlin, L., Schmeidler, J., Davis, K. L., & Perl, D. P. (1995). Neocortical neurofibrillary tangles correlates with dementia severity in Alzheimer's disease. *Archives of Neurology, 52,* 81–88.

Binetti, G., Magni, E., Padovani, A., Cappa, S. F., Bianchetti, A., & Trabucchi, M. (1993). Neuropsychological heterogeneity in mild Alzheimer's disease. *Dementia, 4,* 321–326.

Blessed, G., Tomlinson, E., & Roth, M. (1968). The association between quantitative measures of dementia and senile change in the cerebral grey matter of elderly subjects. *British Journal of Psychiatric Medicine, 114,* 797–811.

Buschke, H., & Fuld, P. A. (1974). Evaluating storage, retention, and retrieval in disordered memory and learning. *Neurology, 24,* 1019–1025.

Coblentz, J. M., Mattis, S., Zingesser, L., Kasoff, S. S., Wisniewski, H. M., & Katzman, R. (1973). Presenile dementia: Clinical aspects and evaluation of cerebral spinal fluid dynamics. *Archives of Neurology, 25,* 299–308.

Cummings, J. L. (1990). *Subcortical dementias.* New York: Oxford University Press.

Cummings, J. L., & Benson, D. F. (1992). *Dementia: A clinical approach* (2nd ed.). Boston: Buttersworth-Heinemann.

Dalton, J. E., Pederson, S. L., Blom, B. E., & Holmes, N. R. (1987). Diagnostic errors using the Short

205

EVALUATION OF
DEMYELINATING
AND
DEGENERATIVE
DISORDERS

Portable Mental Status Questionnaire with a mixed clinical population. *Journal of Gerontology, 42,* 512–514.

Davous, P. L. Y., Lamour, Y., Debrand, E., & Rondot, P. (1987). A comparative evaluation of the short orientation memory concentration test of cognitive impairment. *Journal of Neurology, Neurosurgery, and Psychiatry, 50,* 1312–1317.

Erkinjuntti, T., Sulkava, R., Wikström, J., & Autio, L. (1987). Short Portable Mental Status Questionnaire as a screening test for dementia and delirium among the elderly. *Journal of the American Geriatrics Society, 35,* 412–416.

Eslinger, P. J., & Damasio, A. R. (1986). Preserved motor learning in Alzheimer's disease: Implications for anatomy and behavior. *Journal of Neurosciences, 6,* 3006–3009.

Eslinger, P. J., Damasio, A. R., Benton, A. L., & Van Allen, M. (1985). Neuropsychologic detection of abnormal mental decline in older persons. *Journal of the American Medical Association, 253,* 670–674.

Fillenbaum, G. G., Heyman, A., Wilkinson, W. E., & Haynes, C. S. (1987). Comparison of two screening tests in Alzheimer's disease: The correlation and reliability of the Mini-Mental State Examination and the Modified Blessed Test. *Archives of Neurology, 44,* 924–927.

Fillenbaum, G. G., Wilkinson, W. E., Welsh, K. A., & Mohs, R. C. (1994). Discrimination between stages of Alzheimer's disease with subsets of the mini-mental state examination items: An analysis of the consortium to establish a registry for Alzheimer's disease data. *Archives of Neurology, 51,* 916–921.

Folstein, M. F., Folstein, S. E., & McHugh, P. R. (1975). Mini-Mental State. A practical method for grading the cognitive state of patients for the clinician. *Journal of Psychiatric Research, 12,* 189–198.

Fuld, P. A. (1978). Psychological testing in the differential diagnosis of dementias. In R. Katzman, R. D. Terry, & K. L. Bick (Eds.), *Alzheimer's disease: Senile dementia and related disorders* (pp. 185–193). New York: Raven Press.

Galasko, D., Klauber, M. R., Hofstetter, C. R., Salmon, D. P., Lasker, B., & Thal, L. J. (1990). The Mini-Mental State examination in the early diagnosis of Alzheimer's disease. *Archives of Neurology, 47,* 49–52.

Gardner, R., Oliver-Muñoz, S., Fisher, L., & Empting, L. (1981). Mattis Dementia Rating Scale: Internal reliability study using a diffusely impaired population. *Journal of Clinical Neuropsychology, 3,* 271–275.

Golding, E. (1989). *MEAMS: The Middlesex Elderly Assessment of Mental State description and validation.* Fareham, England: Thames Valley Test Company.

Gronwall, D. M. A. (1977). Paced auditory serial-addition task: A measure of recovery from concussion. *Perceptual and Motor Skills, 44,* 367–373.

Grossman, M., Armstrong, C., Onishi, K., Thompson, H., Schaefer, B., Robinson, K., D'Esposito, M., Cohen, J., Brennan, D., Rostami, A., Gonzalez-Scarano, F., Kolson, D., Constantinescu, C., & Silberberg, D. (1994). Patterns of cognitive impairment in relapsing-remitting and chronic progressive multiple sclerosis. *Neuropsychiatry, Neuropsychology, and Behavioral Neurology, 7,* 194–210.

Hart, R. P., Smith, C. M., & Swash, M. (1988). Word fluency in patients with early dementia of Alzheimer type. *British Journal of Clinical Psychology, 27,* 115–124.

Heaton, R., Nelson, L., Thompson, D. S., Burks, J. S., & Franklin, G. M. (1985). Neuropsychological findings in relapsing-remitting and chronic-progressive multiple sclerosis. *Journal of Consulting and Clinical Psychology, 53,* 103–110.

Heaton, R., Thompson, L. L., Nelson, L. M., Filley, C. M., & Franklin, G. M. (1990). Brief and intermediate-length screening of neuropsychological impairment in multiple sclerosis. In S. A. Rao (Ed.), *Neurobehavioral aspects of multiple sclerosis* (pp. 149–160). New York: Oxford University Press.

Hodges, J. R., Salmon, D. P., & Butters, N. (1991). The nature of the naming deficit in Alzheimer's and Huntington's disease. *Brain, 114,* 1547–1558.

Holzer, C. E. III, Tischler, G. L., Leaf, P. J., & Myers, J. K. (1984). An epidemiologic assessment of cognitive impairment in a community population. In J. R. Greenley (Ed.), *Research in community mental health* (vol. 4, pp. 3–32). London, England: JAI Press.

Hom, J. (1992). General and specific cognitive dysfunctions in patients with Alzheimer's disease. *Archives of Clinical Neuropsychology, 7,* 121–133.

Huff, F. J., Corkin, S., & Growdon, J. H. (1986). Semantic impairment and anomia in Alzheimer's disease. *Brain and Language, 28,* 235–249.

Hughes, C. P., Berg, L., Danziger, W., Coben, L., & Martin, R. (1982). A new clinical scale for the staging of dementia. *British Journal of Psychiatry, 140,* 566–572.

Husband, H. J., & Tarbuck, A. F. (1994). Cognitive rating scales: A comparison of the Mini-Mental State Examination and the Middlesex Elderly Assessment of Mental State. *International Journal of Geriatric Psychiatry, 9,* 797–802.

Jacobs, J. W., Bernhard, M. R., Delgado, A., & Strain, J. J. (1977). Screening for organic mental syndromes in the medically ill. *Annals of Internal Medicine, 86,* 40–46.

Kaplan, E., Goodglass, H., & Weintraub, S. (1983). *Boston Naming Test.* Philadelphia: Lea & Febiger.

Katzman, R., Brown, T., Fuld, P., Peck, A., Schechter, R., & Schimmel, H. (1983). Validation of a short orientation memory concentration test of cognitive impairment. *American Journal of Psychiatry, 140,* 734–739.

Kessler, H. R., Cohen, R. A., Lauer, K., & Kausch, D. F. (1992). The relationship between disability and memory dysfunction in multiple sclerosis. *International Journal of Neuroscience, 62,* 17–34.

Kurtzke, J. F. (1983). Rating neurological impairment in multiple sclerosis: An expanded disability status scale (EDSS). *Neurology, 33,* 1444–1452.

Launer, L. J., Dinkgreve, M. A. H. M., Jonker, C., Hooijer, C., & Lindeboom, J. (1993). Are age and education independent correlates of the Mini-Mental State Exam. *Journal of Gerontology: Psychological Sciences, 48,* 271–277.

Lezak, M. D. (1995). *Neuropsychological assessment* (2nd ed.). New York: Oxford University Press.

Loring, D., & Largen, J. (1985). Neuropsychological patterns of presenile and senile dementia of the Alzheimer type. *Neuropsychologia, 23,* 351–357.

Magni, E., Binetti, G., Cappa, S., Bianchetti, A., & Trabucchi, M. (1995). Effect of age and education on performance on the Mini-Mental State Examination in a healthy older population and during the course of Alzheimer's disease. *Journal of the American Geriatrics Society, 43,* 942–943.

Marshall, S. C., & Mungas, D. (1995). Age and education correction for the Mini-Mental State Exam. *Journal of the International Neuropsychological Society, 1,* 166.

Marson, D., Herfkens, K., Brooks, A., Ingram, K., & Harrell, L. (1995). Relevance of dementia screening instruments to physicians' competency judgments in Alzheimer's disease. *Journal of the International Neuropsychological Society, 1,* 143.

Mattis, S. (1988). *DRS: Dementia Rating Scale professional manual.* Odessa, FL: Psychological Assessment Resources, Inc.

McIntosh-Michaelis, S. M., Roberts, M. H., Wilkinson, S. M., Diamond, I. D., McLellan, D. L., Martin, J. P., & Spackman, A. J. (1991). The prevalence of cognitive impairment in a community survey of multiple sclerosis. *British Journal of Clinical Psychology, 30,* 333–348.

Miller, E. (1981). The differential psychological evaluation. In N. E. Miller & G. D. Cohen (Eds.), *Clinical aspects of Alzheimer's disease and senile dementia, Aging* (Vol. 15, pp. 121–138). New York: Raven Press.

Monsch, A. U., Bondi, M. W., Salmon, D. P., Butters, N., Thal, L. J., Hansen, L. A., Wiederholt, W. C., Cahn, D. A., & Klauber, M. R. (1995). Clinical validity of the Mattis Dementia Rating Scale in detecting dementia of the Alzheimer type: A double cross-validation and application to a community-dwelling sample. *Archives of Neurology, 52,* 899–904.

Monsch, A. U., Goldi, N. S., Ermini-Funfschilling, D. E., Berres, M., Tayler, K. I., Seifritz, E., Stähelin, H. B., & Spiegel, R. (1995). Improving the diagnostic accuracy of the Mini-Mental State Examination. *Acta Neurologica Scandinavica, 92,* 145–150.

Morris, J. C., Heyman, A., Mohs, R. C., Hughes, J. P., van Belle, G., Fillenbaum, G., Mellits, E. D., Clark, S., & the CERAD investigators. (1989). The Consortium to Establish a Registry for Alzheimer's Disease (CERAD). Part I. Clinical and neuropsychological assessment of Alzheimer's disease. *Neurology, 39,* 1159–1164.

Nussbaum, P. D., & Allen, D. N. (1994). Recent developments in neuropsychological assessment of the elderly and individuals with severe dementia. In G. Goldstein & T. M. Incagnoli (Eds.), *Contemporary approaches to neuropsychological assessment* (pp. 277–324). New York: Plenum Press.

O'Connor, D. W., Pollitt, P. A., Hyde, J. B., Fellows, J. L., Miller, N. D., Brook, C. P. B., & Reiss, B. B. (1989). The reliability and validity of the Mini-Mental State in a British community survey. *Journal of Psychiatric Research, 23,* 87–96.

Paulsen, J. S., Butters, N., Sadek, J. R., Johnson, S. A., Salmon, D. P., Swerdlow, N. R., & Swenson, M. R. (1995). Distinct cognitive profiles of cortical and subcortical dementia in advanced illness. *Neurology, 45,* 951–956.

Peyser, J. M., Edwards, K. R., Poser, C. M., & Filskov, S. B. (1980). Cognitive function in patients with multiple sclerosis. *Archives of Neurology, 37,* 577–579.

Pfeiffer, E. (1975). Short Portable Mental Status Questionnaire for the assessment of organic brain deficit in elderly patients. *Journal of the American Geriatrics Society, 23,* 433–441.

Powell, T., Brooker, D. J. R., & Papadopolous, A. (1993). Test–retest reliability of the Middlesex Assessment of Mental State (MEAMS): A preliminary investigation in people with probable dementia. *British Journal of Clinical Psychology, 32,* 224–226.

207

EVALUATION OF
DEMYELINATING
AND
DEGENERATIVE
DISORDERS

Raine, C. S. (1990). Neuropathology. In S. A. Rao (Ed.), *Neurobehavioral aspects of multiple sclerosis* (pp. 15–36). New York: Oxford University Press.

Rao, S. M. (1993). White matter dementias. In R. W. Parks, R. F. Zec, & R. S. Wilson (Eds.), *Neuropsychology of Alzheimer's disease and other dementias* (pp. 438–456). New York: Oxford University Press.

Rao, S. M., Leo, G. J., Bernardin, L., & Unverzagt, F. (1991). Cognitive dysfunction in multiple sclerosis. I. Frequency, patterns, and prediction. *Neurology, 21,* 685–691.

Rao, S. M., Leo, G. J., Ellington, L., Nauertz, T., Bernardin, L., & Unverzagt, F. (1991). Cognitive dysfunction in multiple sclerosis. II. Impact on employment and social functioning. *Neurology, 21,* 692–696.

Rao, S. M., Leo, G. J., Haughton, V. M., St. Aubin-Faubert, P., & Bernardin, L. (1989). Correlation of magnetic resonance imaging with neuropsychological testing in multiple sclerosis. *Neurology, 39,* 161–166.

Reitan, R. M., & Wolfson, D. (1993). The Halstead–Reitan Neuropsychological Test Battery: Theory and clinical interpretation. Tucson, AZ: Neuropsychology Press.

Roccaforte, W. H., Burke, W. J., Bayer, B. L., & Wengel, S. P. (1994). Reliability and validity of the Short Portable Mental Status Questionnaire administered by telephone. *Journal of Geriatric Psychiatry and Neurology, 7,* 33–38.

Ryan, L., Clark, C. M., Klonoff, H., Li, D., & Paty, D. (1996). Patterns of cognitive impairment in relapsing-remitting multiple sclerosis and their relationship to neuropathology on magnetic resonance images. *Neuropsychology, 10,* 176–193.

Sager, H. J., Cohen, N. J., Sullivan, E. V., Corkin, S., & Growdon, J. H. (1988). Remote memory function in Alzheimer's disease and Parkinson's disease. *Brain, 111,* 941–959.

Salmon, D. P., Thal, L. J., Butters, N., & Heindel, W. C. (1990). Longitudinal evaluation of dementia of the Alzheimer type: A comparison of three standardized mental status examinations. *Neurology, 40,* 1225–1230.

Samuels, M. A., & Feske, S. (1996). *Office practice of neurology.* New York: Churchill Livingstone.

Saxton, J., McGonigle-Gibson, K., Swihart, A., & Boller, F. (1993). *The Severe Impairment Battery (SIB) manual.* Suffolk, England: Thames Valley Test Company.

Shay, K. A., Duke, L. W., Conboy, T., Harrell, L. E., Callaway, R., & Folks, D. G. (1991). The clinical validity of the Mattis Dementia Rating Scale in staging Alzheimer's dementia. *Journal of Geriatric Psychiatry and Neurology, 4,* 18–25.

Shear, P. K., Sullivan, E. V., Mathalon, D. H., Lim, K. O., Davis, L. F., Yesavage, J. A., Tinklenberg, J. R., & Pfefferbaum, A. (1995). Longitudinal volumetric computed tomographic analysis of regional cerebral brain changes in normal aging and Alzheimer's disease. *Archives of Neurology, 52,* 392–402.

Shiel, A., & Wilson, B. A. (1992). Performance of stroke patients on the Middlesex Elderly Assessment of Mental State. *Clinical Rehabilitation, 6,* 283–289.

Sibley, W. A. (1990). Diagnosis and course of multiple sclerosis. In S. M. Rao (Ed.), *Neurobehavioral aspects of multiple sclerosis* (pp. 5–14). New York: Oxford University Press.

Smith, A. (1982). *Symbol Digit Modalities Test: Manual.* Los Angeles: Western Psychological Services.

Stern, R. G., Mohs, R. C., Bierer, L. M., Silverman, J. M., Schmeidler, J., Davidson, M., & Davis, K. L. (1992). Deterioration on the Blessed test in Alzheimer's disease: Longitudinal data and their implications for clinical trials and identification of subtypes. *Psychiatry Research, 42,* 101–110.

Storandt, M. (1994). General principles of assessment of older adults. In M. Storandt & G. R. Van den Bos (Eds.), *Neuropsychological assessment of depression and dementia in older adults: A clinician's guide* (pp. 7–32). Washington, DC: American Psychological Association.

Storandt, M., Botwinick, J., Danziger, W. L., Berg, L., & Hughes, C. P. (1984). Psychometric differentiation and mild senile dementia of the Alzheimer's type. *Archives of Neurology, 41,* 497–499.

Storandt, M., & Hill, R. D. (1989). Very mild senile dementia of the Alzheimer's type, II. Psychometric test performance. *Archives of Neurology, 46,* 383–386.

Swirsky-Sacchetti, T., Field, H. L., Mitchell, D. R., Seward, J., Lublin, F. D., Knobler, R. L., & Gonzalez, C. F. (1992). The sensitivity of the Mini-Mental State Exam in the white matter dementia of multiple sclerosis. *Journal of Clinical Psychology, 48,* 779–786.

Thal, L. J., Grundman, M., & Golden, R. (1986). Alzheimer's disease: A correlational analysis of the Blessed Information-Memory-Concentration Test and the Mini-Mental State Exam. *Neurology, 36,* 262–264.

Tombaugh, T. N., McDowell, I., Kristjansson, B., & Hubley, A. M. (1996). Mini-Mental State Examination (MMSE) and the modified MMSE (3MS): A psychometric comparison and normative data. *Psychological Assessment, 8,* 48–59.

Wechsler, D. A. (1945). A standardized memory scale for clinical use. *Journal of Psychology, 19,* 87–95.

Weiler, P. G., Chiriboga, D. A., & Black, S. A. (1994). Comparison of mental status tests: Implications for Alzheimer's patients and their caregivers. *Journal of Gerontology, 49,* S44–S51.

Welsh, K. A., Butters, N., Mohs, R. C., Beekley, D., Edland, S., Fillenbaum, G., & Heyman, A. (1994). The Consortium to Establish a Registry for Alzheimer's Disease (CERAD). Part V. A normative study of the neuropsychological battery. *Neurology, 44,* 609–614.

Whitehouse, P. J. (1986). The concept of subcortical and cortical dementia: Another look. *Annals of Neurology, 19,* 1–6.

Wilson, R. S., Bacon, L. D., Fox, J. H., & Kaszniak, A. W. (1983). Primary memory and secondary memory in dementia of the Alzheimer's type. *Journal of Clinical Neuropsychology, 5,* 337–344.

Woodward, J. L., Salthouse, T. A., Godsall, R. E., & Green, R. C. (1996). Confirmatory factor analysis of Mattis Dementia Rating Scale in patients with Alzheimer's disease. *Psychological Assessment, 8,* 85–91.

Zachary, R. A. (1986). *Shipley Institute of Living Scale. Revised manual.* Los Angeles: Western Psychological Services.

10

Assessment following Neurotoxic Exposure

Lisa A. Morrow

INTRODUCTION

In the middle part of this century, toxicity assessment was determined most often by what is known as the LD_{50}—the lethal dose (LD) for 50% of test animals tested with a chemical agent. That is, a group of animals, typically rats, were administered various doses of chemicals and the dose that resulted in 50% dying after 14 days was estimated as the LD_{50}. As an example of the range of toxicity for various chemicals, alcohol has an approximate LD_{50} of 14,000 mg/kg, compared to DDT which has an LD_{50} of 100 mg/kg (Kamrin, 1988). However, what applies to rats does not necessarily apply to humans. Toxicity assessment has since expanded its definition of impairment by relying on other measures such as biochemical indices and, more and more, behavioral testing. What constitutes a "safe threshold" today is much different from what was considered safe just a few years ago. The work of Herbert Needleman on the behavioral effects of lead, particularly in young children, was instrumental in lowering permissible lead levels in gasoline and drinking water (Needleman *et al.*, 1979; Needleman, Schell, Bellinger, Leviton, & Allred, 1990). Increasingly, neuropsychology is coming to the forefront as one of the best ways to assess the toxic effects of chemicals.

Ordinarily it is the physical symptoms produced by overexposure to a neurotoxic agent that bring the patient to the attention of a physician or emergency room personnel. The most common exposures, in the home or on the job, are to metals such as lead and mercury, pesticides, carbon monoxide, and organic solvents. In terms of physical signs and symptoms, certain features may distinguish the toxins. With *heavy metal exposure,* for example to arsenic or lead, sensory neuropathy is often a sign of overexposure, with classic "wrist drop" associated with very high lead levels. *Mercury poisoning,* often associated with the "mad hatters" syndrome

LISA A. MORROW   Department of Psychiatry, Western Psychiatric Institute and Clinic, and University of Pittsburgh School of Medicine, Pittsburgh, Pennsylvania 15213.

Neuropsychology, edited by Goldstein *et al.* Plenum Press, New York, 1998.

because of symptoms displayed by workers in the felt-hat industry, produces "hatters shakes," "salivation," and "psychic irritability" (Weeden, 1989). An acute overexposure to *pesticides* is fairly easy to document, as serum cholinesterase activity is reduced and the patient presents with excessive lacrimation, sweating, intestinal cramps, diarrhea, fatigue, ataxia, and respiratory distress. The initial symptom of *carbon monoxide* (CO) exposure is headache, which can occur with an air concentration of 0.05%: A concentration greater than 0.1% may produce transient cortical blindness, mania, amnesia, convulsions, and death. In persons who survive CO poisoning, there have often been reports of delayed sequelae. That is, patients will appear to be neurologically intact for several days or weeks and then become confused and display parkinsonian symptoms, amnesia, mutism, and sometimes psychotic disturbances (Min, 1986). The acute presentation in a person exposed to *organic solvents*—either accidentally or from voluntary abuse—may vary depending on the solvent or solvent mixture, but typically consists of headaches, dizziness, nausea or vomiting, fatigue, and mental confusion. A differential diagnosis is difficult following solvent exposure as solvents have a very short half-life, and unless blood and urine measures are collected within several hours of exposure it is almost impossible to document the nature and extent of exposure.

While specific neurotoxins may have different presenting physical symptoms and complaints, the neuropsychiatric symptoms are quite similar. Most often the patient complains of lassitude, periods of confusion, difficulty concentrating, increased forgetfulness, clumsiness or motor incoordination, and mood changes. Though these symptoms are typically reported following acute exposure, persons with chronic exposure to levels within the permissible range also report similar cognitive and behavioral changes. There are numerous studies in the literature that find that chronic low-level exposure to lead, solvents, and other neurotoxins produces cognitive and personality changes (cf. Hartman, 1995, for an extensive review). Most health professionals do not question the neuropsychiatric symptoms following acute exposure to neurotoxic chemicals, but many assume that the symptoms will remit as soon as the exposure is terminated. This is not necessarily the case. Most follow-up studies of neurotoxin exposure find that patients with acute or chronic exposure to neurotoxins—whether solvents, lead, or CO—continue to report neuropsychiatric symptoms, and cognitive deficits may be apparent long after exposure ceases, and in some cases, may actually worsen (Antti-Poika, 1982; Choi, 1983; Edling, Ekberg, & Ahlborg, 1990; Gregersen, Klausen, & Uldal Elsnab, 1987; Juntunen, Antti-Poika, Tola, & Partanen, 1982; Min, 1986; Morrow, Ryan, Hodgson, & Robin, 1991; Needleman *et al.*, 1990; Smith & Brandon, 1970).

Several behavioral test batteries, as well as specific tests, have been developed for assessing neurobehavioral impairment following exposure to toxic chemicals. Neurophysiological and neuroimaging techniques have also been shown to provide additional indications of subtle changes in brain function following chemical exposure. This chapter reviews the most widely used neuropsychological test batteries and several specific tests designed to assess the effects of toxic chemicals. Current findings from neurophysiological and neuroimaging studies are also reviewed.

NEUROPSYCHOLOGICAL ASSESSMENT

Virtually all test batteries compiled to assess the effects of environmental or occupational neurotoxins have relied on standardized tests of neuropsychological function. Several of the batteries were designed to evaluate persons with varying cultural backgrounds or to quickly and economically evaluate large groups of individuals, such as factory workers. A drawback of many of the batteries is that the majority are what most neuropsychologists would view as brief screenings. While they generally tap several cognitive domains, such as learning or motor speed, brief screenings do not constitute a thorough neuropsychological evaluation. In addition, there is often a lack of published normative data, and to date, most batteries have not been validated on a wide range of exposed populations. With few exceptions, the batteries were not formulated to test for a single neurotoxin. This is not surprising given the fact that the neuropsychological deficits following neurotoxic exposure, whether the exposure is to lead, solvents, or pesticides, are quite similar—reductions in learning and memory, attentional deficits, poor visuospatial ability, slowed response time, and decreased mental flexibility. While we know certain toxins may have an affinity for a specific brain region or target a specific neurotransmitter (e.g., CO has been shown to damage the globus pallidus, and pesticides inhibit cholinesterase function) there have been no studies that have delineated chemically exposed groups (e.g., lead vs. solvents) based on neuropsychological findings.

THE WORLD HEALTH ORGANIZATION (WHO) NEUROBEHAVIORAL CORE TEST BATTERY

In 1983 the WHO, along with the National Institute for Occupational Safety and Health sponsored a workshop that resulted in the formulation of the Neurobehavioral Core Test Battery (NCTB). This battery was recommended as a "test instrument having widespread usefulness for identifying neurotoxic effects" (Cassitto, Camerino, Hanninen, & Anger, 1990, p. 203). The battery was conceived to be sensitive to changes in CNS function and applicable to a variety of cultures, as well as quick and economical to administer. There are seven tests included in the battery and they focus on visuoperceptual and visuomotor skills as well as mood state. Because the targeted populations are worldwide, tests with a major language component are not included (see Table 1).

Few clinical or epidemiological studies have been published using the NCTB in its totality. However, numerous studies have been conducted that use individual tests from the battery, and, for the most part, the tests tend to discriminate between exposed and unexposed groups (Anger *et al.*, 1993). An initial evaluation of the battery was done to assess the feasibility and validity across cultures (Cassitto *et al.*, 1990). A summary of the findings from testing done at European cities showed significant skewness for most tests and wide variation among the groups tested. Norms for over 2,300 persons unexposed to chemical substances from a variety of countries have been collected and tabulated by Anger *et al.* (1993). This study reported a wide range of mean scores across countries but fairly consistent within-country data. The authors suggest that the NCTB can be used for men and women with 8 to 12 years of education, but published normative data should not be substituted for appropriate control groups in future studies. As with any neuropsychological test battery, the feasibility of the NCTB for poorly educated populations is still in question, and as Anger and colleagues (1993) point out, there are a number of dominant cultural factors (e.g., socioeconomic status, living conditions) that must be dealt with when applying neuropsychological tests to both industrial and nonindustrial countries.

TABLE 1. NEUROPSYCHOLOGIAL TEST BATTERIES

World Health Organization (WHO) Neurobehavioral Core Test Battery
  Simple Reaction Time
  Digit Span
  Santa Ana Dexterity
  Digit Symbol
  Benton Visual Retention
  Pursuit Aiming
  Profile of Mood States
Milan Automated Neurobehavioral System
  Simple Reaction Time
  Digit Span
  Benton Visual Retention
  Digit Serial
  Symbol Digit
  Pursuit Aiming
  Santa Ana Dexterity
Neurobehavioral Evaluation System (NES)
  Finger Tapping
  Hand–Eye Coordination
  Simple Reaction Time
  Continuous Performance
  Symbol Digit
  Visual Retention
  Pattern Recognition
  Pattern Memory
  Digit Span
  Serial Digit Learning
  Paired Associate Learning
  Paired Associate Delayed Recall
  Mood Scales
  Memory Scanning
  Vocabulary
  Horizontal Addition
  Switching Attention

MILAN AUTOMATED NEUROBEHAVIORAL SYSTEM

The Milan Automated Neurobehavioral System (MANS) is a computerized version of six of the seven tests from the NCTB with the addition of one visual learning measure (Cassitto, Gilioli, & Camerino, 1989; see Table 1). Correlations between the standard NCTB administration and the MANS were computed and found to be in the moderate range (0.73 to 0.79). The authors of the MANS suggest that it can be used in various sociocultural settings with little adaptation and is well received by subjects. They report on several field studies, including a finding of a positive correlation between the computerized test scores and blood lead in smelter workers. Camerino (1991) administered the MANS to several groups with occupational exposures (e.g., lead, zinc, and manganese) and compared performance to a referent group. He found a higher prevalence of clinical impairment in the exposed group, defined as scores at least two standard deviations below those of the nonexposed group, and this impairment was related to "type and severity" of exposure.

Agnew, Schwartz, Bolla, Ford, and Bleecker (1991) compared the examiner-

Table 1. (Continued)

213

ASSESSMENT
FOLLOWING
NEUROTOXIC
EXPOSURE

Pittsburgh Occupational Exposures Test (POET)
    Verbal Associative Learning
    Delayed Verbal Recall
    Symbol-Digit Learning
    Delayed Symbol-Digit Recall
    Incidental Memory
    Recurring Words
    Wechsler Memory Scale Immediate Visual Reproductions
    Wechsler Memory Scale Delayed Visual Reproductions
    Embedded Figures Score
    Grooved Pegboard-Dominant
    Grooved Pegboard-Nondominant
    Trail Making-Part A
    Embedded Figures-Mean Time
    Wechsler Adult Intelligence Scale-Revised (WAIS-R) Information
    WAIS-R Similarities
    WAIS-R Picture Completion
    WAIS-R Block Design
    WAIS-R Digit Span
    WAIS-R Digit Symbol
    Trail Making Part B
Carbon Monoxide Neuropsychological Screening Battery (CONSB)
    General Orientation
    Digit Span
    Trail Making
    WAIS Digit Symbol
    Aphasia Screening
    WAIS Block Design

administered and the computer-administered version of four tests from the MANS in a group of men working at paint manufacturing plants. They found correlations to range between 0.45 and 0.77. Not surprisingly, the lowest correlation was seen on the visual memory test. In the computer version of the visual memory test the subject selects one of several designs shown on the screen, while in the examiner version the subject draws a design from memory. Age, vocabulary, dexterity, and solvent exposure were found to be significantly related to the difference between the computerized and examiner-administered testings.

Neurobehavioral Evaluation System (NES)

The NES battery is also computerized and was designed for epidemiological studies of populations at risk for developing neurobehavioral complications from workplace exposure (Baker, Letz, & Fidler, 1985). In addition, the authors suggest that it is useful for studies of acute effects of exposure in workers over short periods of time, for example, repeated administration over the workday (Letz, 1990). Because the tests are presented on computer, there is minimal interaction between the subject and the examiner, and the test environment remains the same for all subjects. Programmed into the battery are measures to flag inappropriate responses, such as holding down the reaction time button, or repeated failures to respond. Originally, 13 tests composed the NES, five of which were modifications of tests in the WHO NCTB. The 13 tests are estimated to take less than 90 minutes to administer. The battery is more heavily weighted to memory and attentional measures than the NCTB. Subsequently the NES was expanded to include four additional tests—finger tapping, serial digit learning, horizontal addition, and switching attention (Baker & Letz, 1986; see Table 1). The NES2, a second edition of the battery, has also been developed (Letz, 1991).

A pilot study using 6 of the 12 NES tests was reported for exposed and nonexposed workers (Baker, White, *et al.*, 1985). The exposed group consisted of industrial painters with a history of exposure to mixtures of organic solvents and drywall tapers with a history of exposure to asbestos. A group of nonexposed controls, bricklayers, were also evaluated. While age, education, and socioeconomic status were significantly correlated with test scores, alcohol did not relate to test performance. Among the painters, exposure to solvents was associated with reduced memory span and slower response times. A second study using nine tests from the NES was done with a larger group of painters (Baker *et al.*, 1988). Dividing the painters by lifetime exposure intensity showed that those with the most intense exposure had significantly lower scores on tests of motor speed, attention, memory, and mood. The authors concluded that the NES is sensitive to neurobehavioral changes associated with neurotoxic exposure and is useful for epidemiological settings.

Fidler also used the NES in a study of 101 construction painters (Fidler, Baker, & Letz, 1987). In this study no consistent relationship between test performance and exposure was seen. There was an association, however, between several exposure indices: Exposure in the past month and exposure in the past year were related to higher depression and increased latency, respectively, on the symbol digit test. A field study using the NES by Maizlish and co-workers examined performance in pest control workers (Maizlish, Schenker, Weisskopf, Seiber, & Samuels, 1987). Workers were divided into diazinon applicators and nonapplicators and testing was done before and after the work shift. The only significant finding was slower times on the symbol digit test in the diazinon-exposed group. A recent study using selected tests from both the NES and NCTB failed to find consistent differences between "young" and "old" painters and controls, although several exposure parameters were related to performance: In younger painters immediate memory scores were related to nonprotected painting in the past 5 years, and in older painters visuomotor performance was related to unprotected painting, and tests of learning and memory were related to more prenarcotic episodes (Hooisma, Hanninen, Emmen, & Kulig, 1993).

While the NES has been translated into several languages and thousands of subjects have been tested with the NES, many studies have used selected tests from the battery rather than the entire battery. Broadwell, Darcey, Hudnell, Otto, and Boyes (1995) selected several tests from the NES and supplemented them with additional standardized measures (e.g., grip strength) in a study of automotive workers exposed to naphtha. Several reports have also looked at performance on the NES under controlled exposure in laboratory settings. Echeverria, Fine, Langolf, Schork, and Sampaio (1989) report on 42 college students exposed to increasing levels of toluene in an inhalation chamber. Several of the NES tests, which seem to be modified somewhat for this study, showed significant declines with higher exposure levels, particularly, digit span, pattern recognition, and pattern memory. This study also included several additional tests and found mild decrements on a measure of psychomotor speed and a test of manual dexterity. Several other authors have reported adverse effects on tests from the NES with exposure to nitrous oxide (Greenberg, Moore, Letz, & Baker, 1985; Mahoney, Moore, Baker, & Letz,

1988), volatile organic compounds (Otto, 1992), as well as alcohol (Hooisma, Twisk, Platalla, Muijser, & Kulig, 1988). Finally, selected tests from the NES have recently been modified to assess children with prenatal exposure to neurotoxins (Dahl *et al.*, 1996).

Correlations between the computerized version and manual version for several tests have been reported to be in the range of 0.42 to 0.76, underscoring the need to establish normative data for the computerized tests (Letz, 1990). A reliability study of 11 NES tests and the five scales from the Profile of Mood State (POMS; McNair, Lorr, & Droppleman, 1971) demonstrated a mixed pattern of reliability (Arcia & Otto, 1992). Reliabilities above 0.80 were noted for the symbol digit substitution, continuous performance test, and switching attention, while somewhat lower coefficients were seen for finger tapping, digit span, pattern comparison, and pattern memory (0.61 to 0.82). Relatively poor reliability was found for simple reaction time and the three memory measures (0.27 to 0.67). All of the mood scales had high test–retest reliability (0.83 to 0.94). The authors point out that their population sample was young and well educated and there was a limited range of scores on several of the tests (a ceiling effect was noted on five of the tests), which will necessarily underestimate reliability. The authors make recommendations about which tests might be best suited for assessing specific cognitive domains and caution about selecting tests for various subject samples. Because the NES allows one to select certain tests and to tailor their administration (e.g., deciding how many trials of the symbol digit substitution to give), normative data will be needed for each study and it may be difficult to compare across studies if different criteria are used.

THE PITTSBURGH OCCUPATIONAL EXPOSURES TEST BATTERY

The Pittsburgh Occupational Exposures Test (POET) battery consists of 21 measures designed to evaluate neurobehavioral function in *individuals* with a history of exposure to toxic substances. It was developed so that it could be used for epidemiological studies as well as provide a set of core tests to be used for clinical evaluation of patients. Normative data from blue-collar workers with no history of neurotoxic exposure or neurological, psychiatric, or major medical disorder have been published (Ryan, Morrow, Bromet, & Parkinson, 1987). The 21 tests were factor-analyzed and five major factors were defined—learning and memory, visuospatial, psychomotor speed and manual dexterity, general intelligence, and attention (see Table 1). For ease of administration, several of the tests can be given either by card or by computer; however, the examiner is always present and there is no difference in the response given by the examinee. The entire battery can usually be completed in less than 90 minutes. The normative study found significant correlations between scores on the neuropsychological tests and age and education, but as with the NES, there was no relationship between test performance and alcohol use. Using regression techniques, coefficients for age and education were derived for each test so that predicted test scores could be calculated. In addition, cutoff values for scores below the 5th percentile are provided. Of the 182 nonexposed blue-collar workers who comprised the normative sample, 55% had a 12th-grade education and the means for several tests differed from the standard published norms. For example, the scores for the Similarities and Picture Completion subtests were somewhat lower than those published in the Wechsler Adult Intelligence Scale-Revised (WAIS-R) manual, while Block Design scores were slightly higher. The higher Block Design scores are not surprising given that the workers in the sample were employed in manufacturing positions and spatial skills were likely to be quite good. These results highlight the need to establish peer-group norms when attempting to determine impairment on specific tests.

A number of clinical, epidemiological, and case studies have been published using the POET battery. Three clinical studies from our laboratory have demonstrated the effectiveness of the POET battery in discriminating exposed from nonexposed workers as well as its usefulness in predicting long-term outcome. An initial study of 17 persons with a history of solvent exposure found that in comparison to age- and education-matched controls, exposed persons scored significantly lower on virtually all measures of cognitive function except general intelligence (Ryan, Morrow, & Hodgson, 1988). The lack of difference between the groups on the general intelligence test—comprising the WAIS-R Picture Completion, Information, and Similarities subtests—suggests that the poor performance on the other cognitive domains was not due to reduced premorbid levels of functioning in the exposed workers. It was found that several exposure-related variables, namely the presence of cacosmia and a history of peak exposure, were differentially associated with impaired performance on the cognitive measures. That is, poorer performance on measures of learning and memory were associated with the subjective experience of discomfort (dizziness) in the presence of certain odors (gasoline). Deficits on visuospatial measures were observed more frequently in workers who had sustained an exposure to a larger than normal amount of solvent—a peak exposure. A second study of 32 solvent-exposed adults replicated the earlier study and also found that elevated levels of depression and anxiety were not associated with performance on the cognitive tests (Morrow, Ryan, Hodgson, & Robin, 1990). That is, it is not the case that affective state or somatic symptomatology is responsible for the cognitive impairment or vice versa.

The POET battery has also been used in a longitudinal study of 27 solvent-exposed persons (Morrow *et al.*, 1991). Neuropsychological and psychological symptomatology were available from two separate testing sessions conducted 16 months apart. Clinical ratings were performed by providing three expert neuropsychologists with only age and the initial and follow-up test scores and asking them to rate each person as either improved, or showing no change or a decline in function. Agreement between the raters was very high and indicated that 50% of the exposed group improved over the test interval while the other half showed no improvement or actually had lower scores at the second assessment. There were two significant predictors of poor outcome. Persons in the poor outcome group were much more likely to have had a history of a peak exposure and to have reported higher levels of psychological distress at the initial testing. It was also noted that persons in the good outcome group were much more likely to have returned to work.

The POET battery has also been shown to be useful in documenting individual cases of mild to moderate solvent encephalopathy. Several case reports and series of cases have found impairments in persons with both acute and chronic exposures to solvents, pesticides, and carbon monoxide (Callender, Morrow, Subramanian, Duhon, & Ristov, 1993; Morrow, Callender, *et al.*, 1990; Hodgson, Furman, Ryan, Durrant, & Kern, 1989; Ryan, 1990).

Two epidemiological studies have also used tests from the POET battery. A large sample of randomly selected blue-collar workers with occupational exposure to low levels of inorganic lead were compared to nonexposed controls on the POET battery (Ryan, Morrow, Parkinson, & Bromet, 1987). After covarying for years of education, the only significant group differences were found for the factor measuring psychomotor speed and manual dexterity and this was limited to one test, the Grooved Pegboard. This study did not find evidence of significant cognitive impairment in persons with lead levels in the low to moderate range (average = 40 μ/dl). However, the authors point out that one cannot rule out a "healthy worker effect" since persons with lead levels above 60 μ/dl were removed from the workplace and would not have been included in the sampling. Parkinson and colleagues evaluated 567 women employed in a microelectronics plant that used a variety of solvent mixtures (Parkinson *et al.*, 1990). Workers were divided into five different categories based on exposure history. Exposure ranged from "never exposed" to "current exposure more than 50% of the time." Neurological and somatic symptoms were assessed, as well as performance on four tests from the POET battery. The study found a significant relationship between exposure and both somatic (e.g., palpitations) and CNS symptoms (e.g., memory loss, headaches), even after controlling for a number of risk factors for poor health (e.g., alcohol, age, severe obesity). There was no association, however, between performance on the cognitive tests and exposure level. Neither of the above studies finds evidence of significant neuropsychological deficits in active workers with a history of exposure to lead or organic solvent mixtures. This is not particularly surprising given that men in the lead study had fairly low lead levels, and the study of solvent-exposed women was noteworthy for levels that, according to national standards, were quite low. The latter study also used only a few of the POET battery measures, failing to administer what are probably the most sensitive measures—the paired associate learning tests. However, this study did assess symptomatology and found that increased exposure to solvents was associated with more physical and neurological symptoms. The authors suggest that symptoms such as headaches, depression, and lightheadedness may be a precursor to solvent-related cognitive changes that occur with more sustained or higher levels of exposure.

BATTERIES FOR SPECIFIC NEUROTOXINS AND FREQUENTLY USED INDIVIDUAL TESTS

As noted above, neuropsychological testing may be one of the best ways to determine subtle toxic encephalopathy. This may be especially true for CO. Patients presenting with acute CO exposure are usually assessed initially for carboxyhemoglobin (COHB) levels. This test provides a measure of the level of CO in the blood. However, symptoms of poisoning, such as headache and confusion, may not be correlated with COHB levels. That is, persons may have COHB levels that are not particularly high (e.g., 10–15%) but may still present with neurological symptoms. This can occur for a number of reasons. Persons may have moderately high levels at the time of the exposure, but treatment with oxygen during transport to the emergency room will reduce the COHB levels. Emergency room physicians must then decide if the symptoms require aggressive therapy with hyperbaric oxygen treatment. Messier and Myers (1991) have developed a battery of neuropsychological tests to assist with the evaluation of patients who present to an emergency room with symptoms of CO poisoning. The Carbon Monoxide Neuropsychological Screening Battery consists of six tests—general orientation, digit span, trail making, digit symbol, aphasia screening, and block design. Patients with CO poisoning were tested before and after treatment with hyperbaric oxygen treatment and compared to a group of matched control subjects who were tested twice on the battery. The CO-poisoned patients had significantly reduced scores compared to the controls prior to treatment and discriminant analysis found the tests to accurately identify the patient group. After treatment, the CO-poisoned patients showed significantly improved test scores and even exceeded the expected practice effect gains. The authors note that this battery can be given by trained staff in a reasonable time and may help to detect subtle symptoms that might otherwise be missed so that appropriate treatment can be initiated.

Though typically not included in a neuropsychological assessment, there are several tests that have been found to be sensitive to neurotoxic exposure. Often patients with neurotoxic exposure report changes in vestibular function—they feel off-balance and report dizziness. Kilburn and colleagues have carried out several studies investigating postural sway and found impaired balance in persons exposed to toxic chemicals (Kilburn, Warshaw, & Shields, 1989; Kilburn & Warshaw, 1995). Loss of color vision has also been associated with neurotoxic exposure, particularly organic solvent exposure. The Lanthony D-15 desaturated panel is a very easy and quick way to test for color loss, and the pattern of responses (how a patient orders 15 colored caps) can distinguish between normal vision, congenital loss, and acquired blue-yellow and red-green loss. Loss in both the blue-yellow and red-green range has been associated with solvent exposure (Mergler, 1990). Finally, chemically exposed patients often report changes in smell, particularly an increased sensitivity to rather benign odors, such as gasoline and hairspray. Several studies have found decreased olfactory function in persons working with neurotoxic chemicals (Ahlstrom, Berglund, Berglund, Lindvall, & Wennberg, 1986; Gullickson, Quinlan, Guerriero, Rosenberg, & Cone, 1988; Schartz, Doty, Monroe, Frye, & Barker, 1989). Probably the most widely used test of olfactory function is the University of Pennsylvania Smell Identification Test (Doty, Shaman, & Dann, 1984). This test is composed of 40 "scratch 'n sniff" panels that are impregnated with various odors at suprathreshold levels. The test is presented in a forced-choice format, can be self-administered, and also provides a measure of malingering.

Neurophysiological and Neuroimaging Assessment

The most recent neurofunctional and neurophysiological measurement techniques—blood flow, positron emission tomography (PET), single photon emission computed tomography (SPECT), and event-related potentials (ERPs)—have consistently shown their usefulness in documenting evidence of central nervous system (CNS) abnormalities in chemically exposed patients. Measurements of cerebral blood flow have reported decreased metabolism in frontotemporal areas in persons with a history of solvent exposure (Hagstadius, Orbaek, Risberg, & Lindgren, 1989). A PET scan, carried out on one of our patients, showed decreased glucose uptake in basal ganglia, hippocampus, amygdala, putamen, thalamus, and temporal and frontal areas following a short-term exposure to tetrabromoethane (Morrow, Callender et al., 1990). Possibly one of the most cost-effective, sensitive, and reliable indicators of CNS abnormalities following chronic or acute chemical exposure is assessment with SPECT (Heuser, 1992). Anatomical changes as measured by standard magnetic resonance imaging (MRI) or computerized tomography (CT) are not typically diagnostic of chemical exposure. It has been suggested that neurotoxic damage is not a result of anatomical changes such as demyelination and atrophy that would be picked up on structural imaging scans, but rather may be associated with a "vasculitis mechanism" that is best determined with functional imaging (Heuser, Mena, & Alamos, 1994). Several studies using SPECT have found diminished blood flow to both cortical and subcortical areas, with particular diminution of temporal and frontal cortical areas and the basal ganglia and thalamus subcortically (Callender, Morrow, & Subramanian, 1994; Callender, Morrow, Subramanian, Duhon, & Ristov, 1993; Heuser *et al.*, 1994; Heuser, Mena, Goldstein, Thomas, & Alamos, 1993; Simon *et al.*, 1994).

There are numerous studies that have found an association between organic brain dysfunction and alterations of ERP waveforms, especially the later components such as P300. Researchers have argued that assessment of ERPs may provide a better measure of subtle CNS changes than standardized neuropsychological tests (Newton, Barrett, Callanan, & Towell, 1989). While relatively little work has been conducted utilizing ERPs in toxic exposure, its usefulness has been recognized (Otto & Reiter, 1978). Several studies have investigated ERPs in exposed populations. Changes in amplitude of the P300 waveform have been reported for solvent-exposed workers and changes in latency have been demonstrated in several patients exposed to hydrogen sulfide (El Massioui, Lesevre, & Fournier, 1986; Wasch, Estrin, Yip, Bowler, & Cone, 1989). Work in our laboratory has demonstrated delays in P300 latency for solvent-exposed groups and also shown disruptions of both cardiac and pupillary reactivity. In our initial work, we reported that persons with an occupational exposure to mixtures of organic solvents had significantly increased latency of the P300 waveform and larger amplitudes of the N100, P200, and N250 components compared to normal and psychiatric control groups (Morrow, Steinhauer, Robin, & Hodgson, 1992). P300 latency was also positively correlated with length of exposure. In an ongoing study of workers with current solvent exposure—journeymen painters—we have reported longer P300 latencies in painters as compared to controls, and within the painter group, painters who were tested soon after working with solvents had longer latencies than painters who had not worked around solvents for several days (Morrow, Steinhauer, Condray, & Dougherty, 1995). Finally, changes in autonomic measures—cardiac and pupillary reactivity—have been reported in solvent-exposed adults (Morrow & Steinhauer, 1995).

PRACTICAL CONSIDERATIONS

As with any good neuropsychological assessment, one of the most important tools is the interview. This is especially true when evaluating patients with a history of chemical exposure. Most patients present with exposures to many chemicals, and obtaining an accurate exposure history, either from the job site or with biochemical markers, is very difficult. A blood draw or urine sample will be able to provide estimates of current body burden for many chemicals. Lead and heavy metals are fairly routine assays, but other unknown chemicals or solvents may be quite challenging to assess. For example, most solvents have a very short half-life, 24 to 48 hours, and if urine or blood samples are not done within this time, it is difficult to estimate the dose of exposure. In epidemiological or field studies of exposed workers, exposure may be estimated with personal monitors or with air sampling or even company records. But with patients who present for a clinical evaluation—often several weeks to months after the exposure—the interview is probably the only way to try and determine frequency and extent of exposure. There is only one published, semistructured interview that has been developed to measure exposure history, and this is only for painters. Fidler and colleagues (1987) developed a formula to estimate solvent exposure in painters based on worker information (indoor vs. outdoor painting, estimates of number of gallons per hour, use of protective equipment, etc.). Other than this questionnaire for painters, no structured interviews are available to estimate general chemical exposure.

There are several guidelines to follow if a patient with chemical exposure is referred for a clinical assessment. First, patients should also be evaluated by a physician with training and expertise in occupational medicine. An assessment by an occupational medicine specialist should help in determining what the patient was exposed to and for how long. One of the best sources of exposure data, if no actual workplace measurements are available, is the material safety data sheets (MSDS). The MSDS will provide the name and composition of the chemical as well as physical side effects, including possible neurological symptoms. Any chemical used at a workplace must have an MSDS and this must be made available to the worker. One of the first priorities in documenting exposure is to obtain the MSDS.

In the interview, the patient should be asked several key questions: What is his or her job and what is he or she required to do? What were the first symptoms, how long did they last, and what are the current symptoms? Details about the work environment itself, such as the size of the workplace, ventilation, and other types of work in the building, should also be obtained. Determining the number of hours worked is important, because many workers put in overtime. Patients should describe what chemicals they work with and how long they have worked with them during the lifetime, during the past year, and during the past month. Use of protective equipment (e.g., mask, cartridge or airline respirators) should be ascertained, as well as method of contact with chemicals (e.g., fumes, dermal). Finally, estimates of peak exposures may be particularly critical, that is, any incidents or accidents where the patient was overcome by fumes or was soaked with chemicals and had a severe, acute reaction (e.g., throwing up, prolonged headache with nausea and dizziness). Others may refer to these events as "prenarcotic episodes" (Hooisma *et al.*, 1993). In our research and that of others, peak exposures seem to play a very important and detrimental role in the development of chemical neurotoxicity.

Whether one uses a standard assessment battery that has been specifically developed for exposed populations or other assessment techniques, such as the Halstead–Reitan or a flexible process-oriented battery, several points need to be kept in mind. Batteries that have been developed specifically for exposed populations are probably best for epidemiological or field studies, especially if time is limited. The NES is probably the battery of choice if subjects are going to be tested repeatedly over the course of a workday or several weeks, and if a control group will be included. If a clinical evaluation of a chemically exposed patient is to be done, the POET battery is preferable, particularly if the patient is a blue-collar worker. If time allows, the POET battery can be supplemented with other standardized measures such as the Paced Auditory Serial Addition Test and other tests of complex information processing and attention. For many patients with a history of neurotoxic exposure, cognitive deficits are best delineated by tests that tap divided attention and complex information processing (Morrow, Steinhauer, Condray, & Hodgson, 1997). Cognitive changes may be subtle and several standardized memory tests, such as the paired associate test from the Wechsler Memory Scale, do not provide a good measure of subtle changes. Finally, careful attention needs to be paid to mood and affect. While we have consistently found that self-report measures of psychiatric symptomatology do not correlate with neuropsychological test scores in exposed populations, there is a much higher prevalence of mood-related changes in exposed populations (Morrow, Ryan, Goldstein, & Hodgson, 1989; Morrow, Kamis, & Hodgson, 1993). There are a number of short self-report psychological inventories that will provide estimates of mood and affect. Both the Profile of Mood States and the Symptom Checklist 90-R provide scores on several dimensions, including depression, anxiety, fatigue, confusion, anger, and the like. However, if time permits, more in-depth questionnaires such as the Minnesota Multiphasic Personality Inventory or the Millon Clinical Multiaxial Inventory (MCMI) may be preferable. The MCMI is very useful because it provides information on both Axis I and Axis II disorders and is much quicker than the standardized structured interviews for assessing Diagnostic Statistical Manual of Mental Disorders Fourth Edition (DSM-IV) disorders.

It is not unusual for patients with a history of toxic chemical exposure to be involved in worker's compensation or other litigation pertaining to the exposure. In these cases there is always the possibility of secondary gain or malingering. There are several tests that have been developed to determine whether a person is malingering (Binder, 1993; Niccolls & Bolter, 1991). However, all of the malingering tests have been validated by asking normal subjects to "fake" bad. The "fake bad" group is then compared to uncoached normal controls and other brain-damaged subjects. The performance of the faking group is then used as the comparison group to determine if the patient is malingering. The bias in this type of test development is that a true group of malingers has never been tested. Reitan and Wolfson (1996) recently outlined a strong case for looking at pattern or neuropsychological test scores over repeat testings as a way to possibly assess dissimulation. Tests of "malingering" may give the clinician certain information, but they cannot determine unequivocally whether someone is malingering. Functional imaging or ERPs may be one of the best ways to assess deficits in cognitive function unconfounded by motivation. That is, a person cannot consciously decrease blood flow or increase the latency of the P300 waveform. Malingering is an issue that must be carefully considered in the clinical assessment of exposed patients if there is a question of litigation, but care should be taken in interpreting tests that were not validated on the group they purport to measure.

SUMMARY

Persons exposed to toxic chemicals—both acutely and following long-term chronic exposure—can manifest neuropsychological deficits. The cognitive deficits following exposure can range from severe cognitive and behavioral alterations to relatively mild decrements in high-level information processing. Most often, the primary symptom expressed by the patient is an inability to learn and retain new information or difficulty processing simultaneous information. The few test batteries that have been published in this area consist mainly of well-known, standardized neuropsychological tests. The primary difficulty in assessing exposed persons is the lack of definitive exposure data, and in many instances, symptoms and self-report histories are the primary source of information concerning the exposure. Supplementing the neuropsychological evaluation with additional neurophysiological testing or neuroimaging may be particularly fruitful in helping to determine the nature and extent of neurobehavioral changes associated with toxic chemical exposure.

References

Agnew, J., Schwartz, B. S., Bolla, K. I., Ford, D. P., Bleecker, M. L. (1991). Comparison of computerized and examiner-administered neurobehavioral testing techniques. *Journal of Occupational Medicine, 33*(11), 1156–1162.

Ahlstrom, R., Berglund, B., Berglund, U., Lindvall, A., & Wennberg, A. (1986). Impaired odor perception in tank cleaners. *Scandinavian Journal of Work Environment Health, 12,* 574–581.

Anger, W. K., Cassitto, M. G., Liang, Y., Amador, R., Hooisma, J., Chrislip, D. W., Mergler, D., Keifer, M., Hortnagl, J., Fournier, L., Dudek, B., & Zsogon, E. (1993). Comparison of performance from three continents on the WHO-recommended neurobehavioral core test battery. *Environmental Research, 62,* 125–147.

Antti-Poika, M. (1982). Overall prognosis of patient with diagnosed chronic organic solvent intoxication. *International Archives of Occupational and Environmental Health, 51,* 127–138.

Arcia, E., & Otto, D. A. (1992). Reliability of selected tests from the neurobehavioral evaluation system. *Neurotoxicology and Teratology, 14,* 103–110.

Baker, E. L., & Letz, R. (1986). Neurobehavioral testing in monitoring hazardous workplace exposures. *Journal of Occupational Medicine, 28*(10), 987–990.

Baker, E. L., Letz, R. E., Eisen, E. A., Pothier, L. J., Plantamura, D. L., Larson, M., & Wolford, R. (1988). Neurobehavioral effects of solvents in construction painters. *Journal of Occupational Medicine, 30*(2), 116–123.

Baker, E. L., Letz, R., & Fidler, A. (1985). A computer-administered neurobehavioral evaluation system for occupational and environmental epidemiology. *Journal of Occupational Medicine, 27*(3), 206–212.

Baker, E. L., White, R. F., Pothier, L. J., Berkley, C. S., Dines, G. E., Travers, P. H., Harley, J. P., & Feldman, R. G. (1985). Occupational lead neurotoxicity: improvement in behavioural effects after reduction of exposure. *British Journal of Industrial Medicine, 42,* 507–516.

Binder, L. M. (1993). Assessment of malingering after mild head trauma with the Portland Digit Recognition Test. *Journal of Clinical and Experimental Neuropsychology, 15*(2), 170–182.

Broadwell, D. K., Darcey, D. J., Hudnell, H. K., Otto, D. A., & Boyes, W. K. (1995). Work-site clinical and neurobehavioral assessment of solvent-exposed microelectronics workers. *American Journal of Industrial Medicine, 27,* 677–698.

Callender, T. J., Morrow, L. A., & Subramanian, K. (1994). Evaluation of chronic neurological sequelae after acute pesticide exposure using spect brain scans. *Journal of Toxicology and Environmental Health, 41,* 275–284.

Callender, T. J., Morrow, L. A., Subramanian, K., Duhon, D., & Ristov, M. (1993). Three-dimensional brain metabolic imaging in patients with toxic encephalopathy. *Environmental Research, 60,* 295–319.

Camerino, D. (1991). Prevalence of abnormal neurobehavioral scores in exposed population [Abstract]. *Neurobehavioral Methods and Effects in Occupational and Environmental Health,* 50.

Cassitto, M. G., Camerino, D., Hanninen, H., & Anger, K. W. (1990). In B. Johnson (Ed.), *Advances in neurobehavioral toxicology: Applications in environmental and occupational Health* (pp. 189–202). Chelsea, MI: Lewis Publishers.

Cassitto, M. G., Gilioli, R., & Camerino, D. (1989). Experiences with the Milan Automated Neurobehavioral System (MANS) in occupational neurotoxic exposure. *Neurotoxicology and Teratology, 11,* 571–574.

Choi, I. (1983). Delayed neurologic sequelae in carbon monoxide poisoning. *Archives of Neurology, 40,* 433–435.

Dahl, R., White, R. F., Weihe, P., Sørensen, N., Letz, R., Hudnell, H. K., Otto, D. A., & Grandjean, P. (1996). Feasibility and validity of three computer-assisted neurobehavioral tests in 7-year-old children. *Neurotoxicology and Teratology, 18,* 413–419.

Doty, R. L., Shaman, P., & Dann, M. (1984). Development of the University of Pennsylvania smell identification test: a standardized microencapsulated test of olfactory function. *Physiology and Behavior, 32,* 489–502.

Echeverria, D., Fine, L., Langolf, G., Schork, A., & Sampaio, C. (1989). Acute neurobehavioral effects of toluene. *British Journal of Industrial Medicine, 46,* 483–495.

Edling, C., Ekberg, K., & Ahlborg, G. (1990). Long-term follow-up of workers exposed to solvents. *British Journal of Industrial Medicine, 47,* 75–82.

El Massioui, F., Lesevre, N., & Fournier, L. (1986). An event-related potential assessment of attention impairment after occupational exposure to organic solvents. In C. Barber & T. Blum (Eds.), *Evoked potentials III: The third international evoked-potentials symposium* (pp. 346–350). Stoneham, MA: Butterworths.

Fidler, A., Baker, E. L., & Letz, R. E. (1987). Neurobehavioral effects of occupational exposure to organic solvents among construction painters. *British Journal of Industrial Medicine, 44,* 292–308.

Greenberg, B. D., Moore, P. A., Letz, R., & Baker, E. L. (1985). Computerized assessment of human neurotoxicity: sensitivity to nitrous oxide exposure. *Clinical Pharmacology and Therapeutics, 38,* 656–660.

Gregersen, P., Klausen, H., & Uldal Elsnab, C. (1987). Chronic toxic encephalopathy in solvent-exposed painters in Denmark 1976–1980: Clinical cases and social consequences after a 5-year follow-up. *American Journal of Industrial Medicine, 11,* 399–417.

Gullickson, G., Quinlan, P., Guerriero, J., Rosenberg, J., & Crone, J. E. (1988). *Olfactory function among painters and plumbers: a pilot study* (Progress report #2). San Francisco: University of California, Hazardous Evaluation System and Information Service.

Hagstadius, S., Orbaek, P., Risberg, J., & Lindgren, M. (1989). Regional cerebral blood flow at the time of diagnosis of chronic toxic encephalopathy induced by organic-solvent exposure and after cessation of exposure. *Scandinavian Journal of Work Environmental Health, 15,* 130–135.

Hartman, D. (1995). *Neuropsychological Toxicology.* New York: Plenum Press.

Heuser, G. (1992). Diagnostic markers in clinical immunotoxicology and neurotoxicology [Editorial]. *International Journal of Occupational and Medical Toxicology, 1,* v–x.

Heuser, G., Mena, I., & Alamos, F. (1994). Neurospect findings in patients exposed to neurotoxic chemicals. *Toxicology and Industrial Health, 10,* 561–571.

Heuser, G., Mena, I., Goldstein, J., Thomas, C., & Alamos, F. (1993). NeuroSPECT findings in patients exposed to neurotoxic chemicals. *Clinical Nuclear Medicine, 18,* 923.

Hodgson, M. J., Furman, J., Ryan, C. M., Durrant, J., & Kern, E. (1989). Encephalopathy and vestibulopathy following short-term hydrocarbon exposure. *Journal of Occupational Medicine, 31*(1), 51–54.

Hooisma, J., Hanninen, H., Emmen, H. H., & Kulig, B. M. (1993). Behavioral effects of exposure to organic solvents in Dutch painters. *Neurotoxicology and Teratology, 15,* 397–406.

Hooisma, J., Twisk, D. A. M., Platalla, S., Muijser, H., & Kulig, B. M. (1988). Experimental exposure to alcohol as a model for the evaluation of neurobehavioral tests. *Toxicology, 49,* 459–467.

Juntunen, J., Antti-Poika, M., Tola, S., & Partanen, T. (1982). Clinical prognosis of patients with diagnosed chronic solvent intoxication. *Acta Neurologica Scandinavica, 65,* 488–503.

Kamrin, M. A. (1988). *Toxicology primer.* Chelsea, MI: Lewis Publishers.

Kilburn, K. H., & Warshaw, R. H. (1995). Neurotoxic effects from residential exposure to chemicals from an oil reprocessing facility and superfund site. *Neurotoxicology and Teratology, 17,* 89–102.

Kilburn, K. H., Warshaw, R. H., & Shields, M. G. (1989). Neurobehavioral dysfunction in firemen exposed to polychlorinated biphenyls (PCBs): Possible improvement after detoxification. *Archives of Environmental Health, 44,* 345–350.

Letz, R. (1990). The neurobehavioral evaluation system: An international effort. In B. Johnson (Ed.), *Advances in neurobehavioral toxicology: Applications in environmental and occupational Health* (pp. 189–202). Chelsea, MI: Lewis Publishers.

Letz, R. (1991). The neurobehavioral evaluation system 2 (NES2) [Abstract]. *Neurobehavioral Methods and Effects in Occupational and Environmental Health,* 225.

Mahoney, F. C., Moore, P. A., Baker, E. L., & Letz, R. (1988). Experimental nitrous oxide exposure as a model system for evaluating neurobehavioral tests. *Toxicology, 49,* 449–457.

Maizlish, N., Schenker, M., Weisskopf, C., Seiber, J., & Samuels, S. (1987). A behavioral evaluation of pest control workers with short-term, low-level exposure to the organophosphate diazinon. *American Journal of Industrial Medicine, 12,* 153–172.

McNair, P., Lorr, M., & Droppleman, L. (1971). *Manual for the Profile of Mood States.* San Diego, CA: Educational and Industrial Testing Service.

Mergler, D. (1990). Color vision loss: A sensitive indicator of the severity of optic neuropathy. In B. Johnson (Ed.), *Advances in neurobehavioral toxicology: Applications in environmental and occupational health* (pp. 175–182). Chelsea, MI: Lewis Publishers.

Messier, L. D., & Myers, R. A. M. (1991). A neuropsychological screening battery for emergency assessment of carbon-monoxide-poisoned patients. *Journal of Clinical Psychology, 47*(5), 675–684.

Min, S. K. (1986). A brain syndrome associated with delayed neuropsychiatric sequelae following acute carbon monoxide intoxication. *Acta Psychiatrica Scandinavica, 73,* 80–86.

Morrow, L. A., Callender, T., Lottenberg, S., Buchsbaum, M. S., Hodgson, M. J., & Robin, N. (1990).

PET and neurobehavioral evidence of tetrabromoethane encephalopathy. *Journal of Neuropsychiatry*, *2*(4), 431–435.

Morrow, L. A., Kamis, H., & Hodgson, M. J. (1993). Psychiatric symptomatology in persons with organic solvent exposure. *Journal of Consulting and Clinical Psychology*, *61*(1), 171–174.

Morrow, L. A., Ryan, C. M., Goldstein, G., & Hodgson, M. J. (1989). A distinct pattern of personality disturbance following exposure to mixtures of organic solvents. *Journal of Occupational Medicine*, *31*(9), 743–746.

Morrow, L. A., Ryan, C. M., Hodgson, M. J., & Robin, N. (1990). Alteration in cognitive and psychological functioning after organic solvent exposure. *Journal of Occupational Medicine*, *32*(5), 444–450.

Morrow, L. A., Ryan, C. M., Hodgson, M. J., & Robin, N. (1991). Risk factors associated with persistence of neuropsychological deficits in persons with organic solvent exposure. *Journal of Nervous and Mental Disease*, *179*(9), 540–545.

Morrow, L., & Steinhauer, S. R. (1995). Alterations in heart rate and pupillary response in persons with organic solvent exposure. *Biological Psychiatry*, *37*, 721–730.

Morrow, L. A., Steinhauer, S. R., Condray, R., & Dougherty, G. G. (1995). P300 latency prolongation in journeymen painters exposed to organic solvents. *Journal of the International Neuropsychological Society*, *1*, 165.

Morrow, L., Steinhauer, S. R., Condray, R., & Hodgson, M. (1997). Neuropsychological performance of journeymen painters under acute solvent exposure and exposure free conditions. *Journal of the International Neuropsychological Society*, *3*, 269–275.

Morrow, L. A., Steinhauer, S. R., Robin, N., & Hodgson, M. J. (1992). Delay in P-300 latency in patients with solvent exposure. *Archives of Neurology*, *49*, 315–320.

Needleman, H. L., Schell, A., Bellinger, D., Leviton, A., & Allred, E. (1990). The long-term effects of exposure to low doses of lead in childhood: An 11-year follow-up report. *New England Journal of Medicine*, *322*, 83–88.

Needleman, H. L., Gunnoe, C., Leviton, A., Reed, R., Peresie, H., Maher, C., & Barrett, B. S. (1979). Deficits in psychologic and classroom performance of children with elevated dentine lead levels. *New England Journal of Medicine*, *300*, 689–695.

Newton, M. P., Barrett, G., Callanan, M. M., & Towell, A. D. (1989). Cognitive event-related potentials in multiple sclerosis. *Brain*, *112*, 1637–1660.

Niccolls, R., & Bolter, J. (1991). *Multi-Digit Memory Test*. San Luis Obispo, CA: Wang Neuropsychological Laboratories.

Otto, D. A. (1992). Assessment of neurobehavioral response in humans to low-level volatile organic compound sources. *Annals of the New York Academy of Sciences*, *641*, 248–260.

Otto, D., & Reiter, L. (1978). Neurobehavioral assessment of environmental insult. In D. A. Otto (Ed.), *Multidisciplinary perspectives in event-related brain potential research* (pp. 409–416). Washington, DC: Environmental Protection Agency.

Parkinson, D. K., Bromet, E. J., Cohen, S., Dunn, L. O., Dew, M. A., Ryan, C. M., & Schwartz, J. E. (1990). Health effects of long-term solvent exposure among women in blue-collar occupations. *American Journal of Industrial Medicine*, *17*, 661–675.

Reitan, R. M., & Wolfson, D. (1996). The question of validity of neuropsychological test scores among head-injured litigants: Development of a dissimulation index. *Archives of Clinical Neuropsychology*, *11*, 573–580.

Ryan, C. M. (1990). Memory disturbances following chronic, low-level carbon monoxide exposure. (1990). *Archives of Clinical Neuropsychology*, *5*, 59–67.

Ryan, C. M., Morrow, L. A., Bromet, E. J., & Parkinson, D. K. (1987). Assessment of neuropsychological dysfunction in the workplace: Normative data from the Pittsburgh Occupational Exposures Test Battery. *Journal of Clinical and Experimental Neuropsychology*, *9*(6), 665–679.

Ryan, C. M., Morrow, L. A., & Hodgson, M. (1988). Cacosmia and neurobehavioral dysfunction associated with occupational exposure to mixtures of organic solvents. *American Journal of Psychiatry*, *145*(11), 1442–1445.

Ryan, C. M., Morrow, L. A., Parkinson, D., & Bromet, E. (1987). Low level lead exposure and neuropsychological finding in blue collar males. *International Journal of Neurosciences*, *36*, 29–39.

Schwartz, B. S., Doty, R. L., Monroe, C., Frye, R., & Barker, S. (1989). Olfactory function in chemical workers exposed to acrylate and methacrylate vapors. *American Journal of Public Health*, *79*(5), 613–618.

Simon, T. R., Hickey, D. C., Fincher, C. E., Johnson, A. R., Ross, G. H., & Rea, W. J. (1994). Single photon emission computed tomography of the brain in patients with chemical sensitivities. *Toxicology and Industrial Health*, *10*, 573–577.

Smith, J. S., & Brandon, S. (1970). Acute carbon monoxide poisoning—3 years experience in a defined population. *Postgraduate Medical Journal, 46,* 65–70.

Wasch, H. H., Estrin, W. J., Yip, P., Bowler, R., & Cone, J. E. (1989). Prolongation of the P-300 latency associated with hydrogen sulfide exposure. *Archives of Neurology, 46,* 902–904.

Weeden, R. P. (1989). Were the hatters of New Jersey mad? *American Journal of Industrial Medicine, 16,* 225–233.

11

Assessing Medically Ill Patients
Diabetes Mellitus as a Model Disease

CHRISTOPHER M. RYAN

INTRODUCTION

Medical disorders and their treatment can have a significant impact on behavior. Measurable cognitive dysfunction and psychological distress are frequently found in patients with diseases that affect organ systems throughout the body, including the heart, lungs, kidney, liver, pancreas, and the pituitary and thyroid glands. How diseases change behavior—that is, delineating the nature and extent of these behavioral abnormalities and understanding the physiological mechanisms underlying their occurrence—is the focus of *medical neuropsychology* (Tarter, Van Thiel, & Edwards, 1988), a relatively new, but increasingly influential, subspecialty area of neuropsychology.

All health care workers have come across situations where a patient has been labeled as being "uncooperative" or "obstreperous" by a physician who asks for the psychologist or social worker to provide an appropriate behavioral management intervention to make the patient more compliant. Too often, however, neuropsychological evaluation reveals evidence of disease-related or medication-induced encephalopathy. That is, these patients are "difficult" not because they do not wish to cooperate or are "acting out," but because their illness or its treatment has somehow affected normal central nervous system (CNS) function and thereby disrupted their ability to understand instructions, remember new information, and control their emotions. Frequently these CNS changes are reversible—if identified early and treated appropriately, and it is for that reason that a good neuropsychological evaluation can play a crucial role in the patient's medical management.

The goal of this chapter is to describe general assessment strategies that are applicable to individuals with a range of medical disorders. Assessing medically ill patients is complicated, however, by four major factors.

CHRISTOPHER M. RYAN   Department of Psychiatry, Western Psychiatric Institute and Clinic, Pittsburgh, Pennsylvania 15213.

Neuropsychology, edited by Goldstein *et al.* Plenum Press, New York, 1998.

First, depending on age at onset and duration of disease, the severity of physical symptoms—and the development of comorbid disorders—may vary dramatically over time and those variations may, in turn, directly or indirectly affect brain function and behavior. Similarly, the patient's psychological adjustment to a disorder may also vary over time, particularly if the disorder has a fluctuating course characterized by symptom remissions and exacerbations.

Second, most medical diseases are characterized by both acute and chronic complications. Transient metabolic derangements of the sort found in disorders like liver cirrhosis, end-stage renal disease, insulin-dependent diabetes mellitus, or hypothyroidism can confound interpretation of a typical neuropsychological assessment because they may induce dramatic alterations in mental and physical status on a day-to-day, or even an hour-to-hour basis.

Third, most patients who have a chronic illness of long duration also have other comorbid medical disorders that often arise as a complication of the initial disorder. For example, adolescents or adults who have had diabetes for more than 10 years often have disorders like hypertension, kidney disease, retinopathy, and neuropathy—all of which can independently affect performance on many neuropsychological tests.

Finally, medical management of a disorder may itself have adverse effects on cognitive functioning and mood state. Examples can be drawn from virtually any drug therapy, including use of antihypertensive or cholesterol-lowering medications, antiretroviral drugs for HIV-infected patients, antirejection drugs for patients who have had liver or kidney transplants, or cranial irradiation for brain cancer. Diabetes may provide the best illustration of this phenomenon. Although insulin therapy can prolong life by bringing blood glucose levels closer to the normal range, too much insulin, or a failure to balance food and exercise with insulin, can cause blood glucose levels to drop so low that neuroglycopenia occurs. The resulting reduction in glucose supply at the neuronal level leads to a marked decline in mental efficiency.

GENERAL ASSESSMENT STRATEGIES

Neuropsychological assessment of patients with medical diseases should not be appreciably different from the assessment of patients with other types of disorders. Ordinarily, the clinician administers a comprehensive battery of standardized, psychometrically sound tests to document cognitive strengths and weaknesses, and uses questionnaires and interview techniques to obtain information about the patient's current and past social and medical history (including psychiatric problems, alcohol and drug use, previous injuries and illnesses, and the like), current mood state, level of motivation, and any concurrent problems that could compromise test-taking performance (e.g., changes in sensory acuity, joint pain, muscle weakness, etc.). Using those data, as well as information from the medical record and from the referring health professional, the neuropsychologist then makes clinical inferences about the proximal cause of the pattern of neuropsychological test scores obtained during the examination. That is, a determination is made as to what extent these results are organogenic and reflect brain dysfunction, and to what extent they are psychogenic (or within normal limits). Finally, hypotheses are generated as to the most reasonable distal cause(s) of the patient's neuropsychological problems. These may include specific characteristics of the disease state (e.g., elevated serum ammonium levels in the person with liver cirrhosis; elevated glycosylated hemoglobin levels in the elderly patient with Type II diabetes), specific biomedical complications of the disease (e.g., development of distal symmetric polyneuropathy in the person with Type I diabetes; development of atherosclerosis in the hypertensive patient), medical treatments directed at managing the disease (e.g., dialysis for patients with end-stage kidney disease; use of calcium channel blockers as antihypertensive medications), and/or the development of a depressive or anxiety disorder that is a reaction to, or a direct physiological consequence of, the specific medical illness.

Clinicians who take this approach to the neuropsychological assessment of medical patients need to have a thorough understanding of the patient's illness—but that is not enough. They must also use a battery of psychological and neuropsychological instruments that is sufficiently "rich" to delineate a broad range of cognitive abilities. Which specific tests are used is probably of less importance than ensuring that all domains are sampled. The reason for this is quite pragmatic. Contrary to a widely held belief, there are *not* patterns of neuropsychological deficit that are uniquely diagnostic or characteristic of specific medical disorders. Research studies may indeed find that as a group, patients with a certain medical illness are more likely to show evidence of impairments in one or more cognitive domains (e.g., poor performance on visuospatial tests with nonverbal memory tests particularly affected, but with normal verbal learning abilities). Nevertheless, it is rarely the case that all or even most subjects with that particular illness will show the same pattern of deficit. In many ways this should not be surprising, because each person comes to the assessment with different competencies that are largely influenced by their genetic, developmental, social, psychological, and medical background. Indeed, more often than not, people with very different medical illnesses may show similar patterns of neuropsychological dysfunction. As an illustration, clinical researchers have reported that elderly patients with non-insulin-dependent diabetes mellitus often show evidence of impairments on memory tests, yet studies of other patient populations indicate that people with other disorders, including hypertension, liver cirrhosis, head injury, and chronic alcoholism, may also manifest those same types of deficits on formal cognitive testing. That is, neuropsychological measures may have a high level of sensitivity—they can detect moderately severe brain dysfunction in medically ill patients, but have a low level of specificity—they cannot (usually) discriminate one medical disorder from another.

If most medical disorders do not eventuate in unique patterns of cognitive impairment, why even study cognitive functioning in patients with different medical disorders? We would argue that having a working familiarity with the clinical research literature provides a map of what one might expect to find when evaluating a patient. Not only does this help ensure that the right measures will be included in the assessment, but it also permits the clinician to test more accurately hypotheses about the underlying etiology of a problem. For example, knowing that certain types of deficits are relatively rare in certain patient groups (e.g., memory disorders are usually not found in young or middle-aged adults with insulin-dependent diabetes—unless they are in very poor metabolic control), and finding evidence of such an impairment in a given patient leads us to ask certain questions in a very logical manner, as illustrated in the following case:

A 33-year-old diabetic patient manifests mild learning and memory deficits. The scientific literature indicates that memory problems are rarely found in diabetic adults—unless the person has a "brittle" form of diabetes characterized by fluctuating, or chronically high, blood glucose levels (i.e., poor metabolic control). Does this patient have a history of poor metabolic control? No? Does the patient have other medical or psychological problems that could lead to poor memory? That is, was the patient seriously depressed or anxious during the assessment? No? Does the patient have a current or extensive past history of heavy alcohol or drug use? No? Does he or she have a history of past head injury with loss of consciousness? No? Is the patient's blood pressure elevated—a not uncommon complication of diabetes? Yes? Because hypertension is frequently associated with somewhat poorer performance on measures of learning and memory, is it possible that the patient's memory problems are largely a consequence of hypertension? Will controlling the hypertension lead to improved memory function? Could this fact help motivate the patient to adhere better to his or her treatment for hypertension? All these possibilities should be addressed in the neuropsychological report.

The purpose of this illustration is to indicate that interpreting the neuropsychological test results must be undertaken in a particular context: the patient's current and past medical and psychosocial history. Collating and integrating history, medical test results, and neuropsychological data can be extraordinarily time-consuming, yet it is a critical part of the process of accurately interpreting the assessment and communicating the results, and their implications, to other health professionals.

DIABETES MELLITUS AS A MODEL MEDICAL DISORDER

For the purpose of this discussion, diabetes mellitus has been selected as the prototypical medical disorder. There are several important reasons for that choice, not the least of which is that the impact of diabetes on cognitive functioning and mood state has probably been studied more thoroughly than any other medical illness. First, this disorder affects individuals across the life span, and the medical characteristics of the disease, the neuropsychological manifestations, and the biomedical risk factors associated with neuropsychological disorder may all vary according to the age of the patient. Second, diabetes is a metabolic disorder that has both acute and chronic medical manifestations (or "complications") that are associated with acute (and largely reversible) neuropsychological dysfunction, as well chronic (and usually permanent) neuropsychological deficit. Third, patients with diabetes often have other comorbid medical illnesses which themselves can affect performance on neuropsychological tests and thereby complicate the process of accurately attributing cognitive impairment to specific disease parameters. Finally, because the medical management of diabetes, like many other chronic illnesses, is so complex, the disorder may also have a significant impact on interpersonal dynamics and hence may be associated with elevated levels of psychological distress—particularly anxiety and depression. These "psychogenic" factors may also affect performance on formal neuropsychological tests and further complicate the process of correctly identifying the etiology of the patient's cognitive impairment.

BIOMEDICAL CHARACTERISTICS OF DIABETES MELLITUS

TYPE I DIABETES. Diabetes mellitus refers to a family of metabolic disorders that together affect more than 5.5 million Americans. Of the two major forms of this disorder, Type I, or insulin-dependent, or juvenile-onset, is the best understood. This form of diabetes is one of the five most prevalent chronic diseases of childhood. Approximately 19,000 new cases are diagnosed each year in the United States; nearly all of these individuals are under the age of 30.

Type I diabetes is almost always signaled by a metabolic crisis characterized by excessive drinking, urination, and eating. If left untreated, individuals will experience severe dehydration and ketoacidosis which may result in death. It is now believed that the insulin-secreting beta cells of the pancreas are destroyed by an autoimmune process which has been triggered, in the genetically susceptible individual, by one or more viral agents. Without insulin, carbohydrates are not metabolized efficiently and blood glucose levels will rise above the normal range (hyperglycemia) to dangerously high levels several hours after a meal unless insulin is supplied exogenously. The goal of treatment is to normalize carbohydrate metabolism artificially (or maintain "good metabolic control") by balancing carbohydrate intake, energy expenditure, and insulin dose. Too much insulin, or a failure to adequately adjust insulin, food intake, and exercise may lead to dangerously low blood glucose levels (hypoglycemia). On the other hand, too little insulin, or excessive amounts of carbohydrate, will lead to hyperglycemia.

Since 1977, it has been possible to estimate the individual's level of metabolic control in the previous 2 to 3 months by measuring the glycosylated fraction of hemoglobin, also known as hemoglobin A_1 or A_{1c}. The greater the degree and duration of hyperglycemia, the higher this value, and the poorer the individual's metabolic control. The use of an objectively ascertained measure to quantify degree of metabolic control has contributed to the search for specific risk factors for complications, and numerous investigators have correlated hemoglobin A_{1c} levels with various biomedical, neuropsychological, and psychosocial outcome measures.

Diabetic individuals in poor metabolic control have a greatly increased risk of developing a wide range of biomedical complications which are mediated primarily by hyperglycemia-induced damage to both the small and large blood vessels. Microvascular damage to the retina increases the risk of impaired vision, whereas microangiopathy within the glomerular loops of the kidney increases the risk of developing end-stage renal disease. In addition, hyperglycemia-medicated neuropathy within the peripheral nervous system causes a loss of fine motor control or painful sensation in the extremities, whereas the development of autonomic neuropathy is associated with impotence and cardiac arrhythmias. Chronic hyperglycemia also leads to a predisposition to develop atherosclerosis in the large blood vessels of the heart, brain, and legs, leading to an increased risk of heart attack, stroke, and gangrene of the feet. For a variety of reasons, hypertension is also a common comorbid disorder in diabetic adults.

On the other hand, diabetic individuals in a very good or "tight" metabolic control have an increased risk of episodes of severe hypoglycemia, as does anyone who fails to balance food intake, insulin dose, and exercise regimen. A transient episode of mild to moderately severe hypoglycemia is associated with a generalized sense of discomfort characterized by sweating, weakness, motor slowing and/or incoordination, feelings of anxiety, and mental confusion. As blood glucose levels drop further, seizures or loss of consciousness may occur. Although there is no doubt that a single episode of severe hypoglycemia can produce clinically significant brain damage, a growing body of evidence suggests that repeated episodes of even mild hypoglycemia may induce structural and/or functional CNS changes that reduce cognitive efficiency significantly.

TYPE II DIABETES. This form of diabetes, also known as maturity-onset or non-insulin-dependent diabetes, is diagnosed most frequently in overweight individuals who are over the age of 40. More than 90% of all diabetics have the Type II form, and nearly 600,000 new cases are diagnosed each year. Hyperglycemia occurs in this type of diabetes for two reasons. First, an insulin resistance develops, characterized by a loss of sensitivity to insulin at peripheral cell receptor sites. In addition, there is a reduction in the amount of insulin normally secreted by the pancreatic beta cells in response to a meal. Both of these changes are thought to develop as a result of excessive insulin production that is triggered initially by chronic overeating. However, because the beta cells of the pancreas continue to secrete moderate amounts of insulin, many individuals with this type of diabetes are able to function without daily insulin injections; most can control their hyperglycemia by modifying their diet and/or taking oral hypoglycemic agents. For that reason, Type II diabetic adults are less likely to experience severe hypoglycemia. However, because of their chronically high blood glucose levels, they may develop many of the same biomedical complications associated with Type I diabetes. Unlike the insulin-requiring Type I form of the disease, however, the development of Type II diabetes is not associated with an acute medical crisis. Rather, it "sneaks up" on the individual over a number of years and is usually first diagnosed during a routine urine screen or following the development of diabetic complications. For that reason, the accurate determination of the age of onset, or the duration of disease, is usually impossible.

As described in more detail below, there is now no doubt that individuals with either form of diabetes have an elevated risk of manifesting mild cognitive deficits, the etiology of which can usually be attributed to specific disease-related biomedical complications. Nevertheless, despite the enormous body of research on this topic (for reviews see Holmes, 1990; Ryan, 1997; Strachan, Deary, Ewing, & Frier, 1977), it is important to keep in mind this statement does not mean that diabetic patients are *invariably* cognitively impaired—they are not. Indeed, the vast majority perform well within the normal range and apparently do not differ from their healthy peers on indicators of academic or vocational success. The disease of diabetes (or virtually *any* medical disease) is merely *one of many risk factors* for cognitive dysfunction.

ASSESSING CHILDREN

Children with diabetes mellitus have long been suspected of developing mild, but measurable, cognitive impairments. Stimulated by observations made by health professionals who noted that many diabetic children did not seem to be as successful as their nondiabetic peers, the first formal clinical research studies on this topic were published more than 60 years ago. Neuropsychology as a subspecialty area did not exist, and so sophisticated cognitive measures were unavailable. Typically, an IQ test was administered—usually the Stanford–Binet, and the scores from a group of diabetic patients were compared with either the published norms or, less frequently, with scores from a group of children who did not have diabetes. Early results were completely uninformative: Depending on how diabetic children were selected and with whom they were compared, researchers were able to report that they performed better than, worse than, or no different from other children without diabetes (for review see Ryan, 1990).

This early work provides several important object lessons for anyone reviewing the literature on the neuropsychology of medical illness. First, because biases in subject selection can influence study outcomes, one needs to understand who was evaluated (e.g., stage of illness), how they came to be studied (e.g., volunteers from a private clinic; drawn from a charity ward for difficult children; randomly selected from a listing of all children diagnosed with diabetes in a particular geographic region over a particular time period), and with whom they were compared (e.g., published norms; local norms of ill, but nondiabetic children recruited from the same clinic; healthy children recruited from the same community; siblings or best friends). Second, one needs to use multiple measures of cognitive functioning so that one can obtain a broad and comprehensive survey of cognitive strengths and weaknesses. A single score, like the IQ, does not provide that information, nor does a study built on a limited "screening battery." Third, one needs to conceptualize the study of cognitive functioning in the medically ill child as a search for biomedical *risk factors*. Having a systemic or metabolic disease does not *cause* cognitive impairment; rather, it is experiencing specific physiological changes that occur as a consequence of the disease that triggers CNS dysfunction. For that reason, research must be designed to identify those biomedical risk factors and explicitly examine how they influence performance. Comparisons between groups of individuals with and without a particular disease will provide few useful insights—as was the case with these early studies of diabetic children.

EARLY ONSET OF DIABETES AND COGNITIVE DYSFUNCTION. One risk factor for cognitive impairment that generalizes across diseases is age at onset. All other things being equal, children who develop certain disorders early in life are far more likely to manifest significant cognitive impairment than those who are diagnosed at a somewhat later age. Although a relationship between age at onset and cognitive dysfunction has been reported in children with diseases as diverse as epilepsy, liver disease, and end-stage renal disease, this effect has been demonstrated most convincingly in a series of studies with diabetic children and adolescents.

The first report of this effect in children with diabetes used 5 years of age as a cutoff to categorize subjects into early-onset and later-onset subgroups. Ack, Miller, and Weil (1961) hypothesized that children who developed diabetes before age 5 had a higher risk of cognitive impairment for two, somewhat different reasons. On the one hand, because the first five years of life are the period of greatest growth and change within the CNS, any serious biochemical insult occurring during that period—as could be produced by very high or very low blood sugar, in the case of diabetes—was likely to have a long-lasting adverse effect on brain function. Alternately, because very young children with a chronic illness like diabetes have fewer psychological resources available to cope with the disturbing events associated with their illness (e.g., hospitalization; injections and complex medication management), they are more likely to experience elevated levels of psychosocial distress than older children. When Ack and colleagues (1961) compared subgroups of early and later-onset diabetic children with their siblings, they found consistently lower scores on the Stanford-Binet. Moreover, these lower scores tended to be associated with an increased number of serious metabolic disturbances.

More recent studies, using far more extensive test batteries, have repeatedly confirmed the relationship between early onset of diabetes and an increased risk of neuropsychological impairment. Importantly, the pattern of cognitive deficit differs somewhat, depending on the child's age at the time of assessment. Children who developed diabetes before 4 years of age and who are 6 to 14 years old at the time of their assessment are more likely to show deficits on a wide range of visuo-spatial measures, and this is particularly pronounced in girls (Rovet, Ehrlich, & Hoppe, 1988). Standard neuropsychological tests that show the greatest sensitivity to this early-onset effect in the school-age diabetic child include the Block Design and Object Assembly subtests from the Wechsler Intelligence Scale for Children-Revised (WISC-R) and tests like the Beery–Buktenica Test of Visual Motor Integration and the Primary Mental Abilities (PMA) Spatial Relations Test.

By adolescence, cognitive deficits associated with an early disease onset may be evident on a far wider range of tests, with boys and girls equally impaired. In the largest study of this to date, our group in Pittsburgh evaluated adolescents 10 to 19 years of age who developed diabetes before age 5, and found them to be impaired on measures of attention, learning, memory, visuospatial skills, intelligence, and eye–hand coordination. Again, specific tests that best differentiated between the early- and later-onset groups included the WISC-R Block Design and Vocabulary subtests; Trail Making B, Grooved Pegboard, and Visual Reproductions from the Wechsler Memory Scale; as well as several less well known tests of learning (Symbol-Digit Learning) and visuospatial ability (Embedded Figures) that were developed initially for studies of brain-damaged adults (Ryan & Butters, 1980). Assessment of information processing strategies in adolescents with an early onset of diabetes indicates that many of them have difficulty deploying attention effectively and using appropriate learning strategies to organize and encode information efficiently (Hagen *et al.*, 1990). These conclusions are based on the use of experimental tests that are not routinely employed in clinical assessment settings.

Not all children or adolescents with an early onset of diabetes manifest cognitive impairments, but according to one estimate, 24% meet criteria for clinically significant impairment, as compared with only 6% of later-onset adolescents and 6% of nondiabetic comparison subjects (Ryan, Vega, & Drash, 1985). Certainly, age at onset cannot produce brain damage; rather, it must be a marker or surrogate for some other biomedical variable. That is, the link between early onset of disease and neuropsychological dysfunction must be mediated by some physiological process. Converging findings from a variety of sources suggest that repeated episodes of hypoglycemia disrupt the normal process of brain development. Metabolic studies have demonstrated that during the first five years of life, diabetic children have a hypersensitivity to the effects of insulin, and hence are more prone to hypoglycemia. When blood glucose levels drop, the preverbal child is unable to adequately communicate this fact to a caretaker, and hence the hypoglycemic event may go unrecognized (and untreated) for an extended time period. Because normal brain function requires consistently high levels of glucose, any prolonged reduction in glucose supply to the brain (neuroglycopenia) may induce structural brain damage. Although this could occur at any age, the very young diabetic child may be particularly vulnerable because the developing brain is unusually sensitive to any type of neurotoxic or traumatic insult. For that reason, the diabetic child with a history of repeated episodes of hypoglycemia early in life should be carefully evaluated for possible mild to moderately severe cognitive problems, especially if the child is having difficulty in school.

Children who may have a particularly elevated risk of hypoglycemia are those who were maintained early in life on a treatment regimen of three to four daily insulin injections in order to effect "tight metabolic control." This treatment strategy, which attempts to keep blood glucose levels as close to the normal range as possible, has been found in young adults to reduce the likelihood of diabetes-related microvascular and macrovascular complications like retinopathy, neuropathy, and atherosclerosis. Unfortunately, tight control is associated with a greatly increased risk of severe hypoglycemia, and for that reason, exceedingly careful blood glucose monitoring is an absolute necessity in very young patients treated with that regimen (Golden, Russell, Ingersoll, Gray, & Hummer, 1985). Given a history of tight metabolic control early in life in a child who is now manifesting mild to moderately severe cognitive problems, one needs to entertain the possibility that these neuropsychological deficits are secondary to early-onset hypoglycemia. By the same token, when evaluating a child with virtually any chronic illness, one should consider those who were first diagnosed within the first several years of life (and 5 or 6 years of age provides a reasonable upper limit) as being more likely to manifest disease-related neuropsychological impairments.

OTHER RISK FACTORS FOR COGNITIVE DYSFUNCTION IN DIABETIC CHILDREN. The modal age of diabetes onset in childhood is 10 or 11 years of age, and as a consequence, the vast majority of children with diabetes are diagnosed after 5 years of age. Are these children completely normal from a neuropsychological perspective, or do they too manifest subtle cognitive deficits that are associated with certain diabetes-related variables? Virtually all studies that have focused on this patient population have concluded that as a group these youngsters do not manifest clinically significant cognitive impairment. Nevertheless, decrements in performance frequently appear on two types of cognitive tasks: measures of psychomotor efficiency (Ryan, Vega, Longstreet, & Drash, 1984) and measures of verbal intelligence and academic achievement (Hagen *et al.*, 1990; Holmes, Dunlap, Chen, & Cornwell, 1992; Kovacs, Goldston, & Iyengar, 1992; Ryan, Longstreet, & Morrow, 1985; Ryan *et al.*, 1984).

Psychomotor slowing, as measured by performance on tasks like the Grooved Pegboard Test or the Wechsler Digit Symbol Substitution Test, is often considered to be an indicator of mild, generalized brain damage. This may not always be the case. Rather, it is possible that psychomotor slowing reflects a personality or response style that is characterized by extreme cautiousness and attention to detail. This interpretation has not been validated empirically, although it most certainly the type of response style manifested by "good" diabetic patients (and individuals with a variety of other chronic disorders) who must take responsibility for meticulous medical management of their illness. Diabetic patients, for example, must measure blood glucose levels frequently, monitor their diet carefully, readjust insulin dose in response to changes in blood glucose levels, and so on. To be reasonably successful as a patient, the diabetic child must develop a response style that borders on being "obsessive–compulsive," where accuracy, rather than speed, is paramount. We prefer this psychological interpretation, rather than a "brain damage" interpretation, to explain the almost universal finding that diabetic children tend to perform slowly, but no less accurately, on measures of psychomotor efficiency.

Although this viewpoint is not universally held, neither we nor other investigators have found any compelling evidence of brain damage (e.g., pathognomonic signs) in studies of groups of later-onset diabetic children. As pointed out above, neuropsychological tests are sensitive to brain damage, but not specific, and individuals can perform poorly on many tests for reasons that are independent of a brain lesion (e.g., elevated levels of depression or anxiety, different cultural experiences or inefficient strategies for approaching certain types of tasks). When studying medical patients who have an elevated risk of developing psychiatric disorders like depression, or who have a disorder that requires following a very strict treatment regimen—as is the case for individuals with many diseases, including diabetes and AIDS, one needs to take into account these "non-CNS" factors when interpreting neuropsychological test results.

In addition to psychomotor slowing, diabetic children tend to earn lower Verbal IQ scores—as much as 10 points lower—than their nondiabetic peers, and tend to perform more poorly on standardized measures of academic achievement, like the Wide Range Achievement Test (WRAT). Longitudinal studies (Kovacs *et al.*, 1992) have demonstrated that the decline in Verbal IQ actually reflects a deterioration in performance over time, which is evident not only on formal cognitive tests, but also in school grades. The best predictor of this decline is duration of diabetes. Because verbal intelligence tests measure a knowledge base that is acquired largely in the classroom, one might expect that those children who miss more school would perform more poorly. Support for that view has been mixed, with some researchers reporting a relationship between achievement and school absence in diabetic children. That is, diabetes, like other chronic diseases, may adversely affect academic performance because it disrupts the education process (Weitzman, Klerman, Lamb, Menary, & Alpert, 1982). In attempting to interpret unexpectedly low Verbal IQ scores or academic achievement test scores, the clinician needs as much information as possible about the child's educational experiences, including school absence. All other things being equal, children with chronic illness are likely to miss significantly more school than their healthy classmates.

In certain illnesses there is the possibility that transient metabolic fluctuations may affect brain function to such an extent that the child, although physically present in the classroom, experiences a brief "absence," which interferes with the acquisition of information and leads to poorer performance. Diabetes again provides an excellent example of this. We know that during an experimentally induced episode of hypoglycemia, many diabetic children and adolescents show a dramatic reduction in concentration and attention, as well as a decline in overall mental efficiency, which may persist for as long as an hour or more following return to euglycemia. Similar changes occurring in the classroom and unrecognized by the teacher (or the child) could have a significant impact on the child's school performance.

Transient metabolic derangements associated with other illnesses (e.g., liver disease; hypothyroidism) are also likely to disrupt cognitive functioning for a period of time. This is a particular problem for neuropsychologists who must conduct an inpatient neuropsychological evaluation of someone who is not metabolically stable. Although the resulting test data can provide a picture of how that person is functioning at that instant, little useful information is obtained about the long-term consequences of the illness on either the brain or behavior. For that reason, one needs to interpret neuropsychological results from metabolically unstable medical patients very conservatively. Whenever possible, biological markers of metabolic status should be obtained before and after the neuropsychological assessment session. For diabetic patients, this might include a fingerstick blood glucose level; for liver patients it might include blood ammonia levels. In addition, for all patients, information needs to be collected on all medications taken within the previous 24 hours.

OPTIMAL TEST BATTERY FOR MEDICALLY ILL CHILDREN. Clinicians should select those measures with which they are most familiar and comfortable and for which good norms are available. In reviewing the clinical literature, one should examine which *cognitive domain(s)* are most likely to be disrupted by a particular illness, rather than focus on one or more *specific* tests. This is a pragmatic approach because many published studies have failed to use standard clinical tests and have relied instead on tests that are experimental (and not widely available), out of date, or modifications (usually with unknown psychometric properties) of well-known tests.

Intelligence. The Wechsler Intelligence Scale for Children-Third Edition (WISC-III) can be used as a neuropsychological instrument, rather than as a means of merely estimating psychometric IQ (Lezak, 1995). That is, Information, Vocabulary, and Comprehension subtest scores can be used to estimate general intelligence; Similarities measures verbal concept formation ability; attention span is estimated from Digit Span; psychomotor efficiency is estimated from Coding; nonverbal problem-solving and visuoconstructional skills are estimated from Block Design and Object Assembly; and planning is estimated from Mazes.

Academic Achievement. The WRAT-3 is a good, quick measure of academic achievement. If there are concerns about reading comprehension, the WRAT Reading may be supplemented or replaced with the Peabody Individual Achievement Test, Revised (PIAT-R) Reading Comprehension subtest or the Gray-Oral Reading Test (GORT-3)—although both of these measures are more time-consuming. Many researchers have switched to the Wechsler Individual Achievement Test because one can readily make comparisons between academic achievement and intellectual ability (WISC-III).

Learning and Memory. The preferred measure for evaluating mnestic abilities in children is the Wide Range Assessment of Learning and Memory (WRAML). Although at this time there are few published studies of children with diabetes (or other chronic illnesses) who have been evaluated with this particular test battery, we have found it to be useful in medical clinical settings. We generally restrict our evaluation to the four screening subtests: Picture Memory, Design Memory, Verbal Learning, and Story Memory. Norms are excellent, subtests can be administered in less than 15 minutes, and the tests tap those memory skills that have been evaluated in the published literature. As an alternative, one may wish to consider the recently released Test of Memory and Learning (TOMAL).

Visuoconstructional Abilities. The Rey–Osterrieth Complex Figure test is appropriate for children 8 years and older and provides an excellent measure of visuoconstructional abilities as well as visual memory. We especially recommend mapping the patient's drawing strategy, using techniques similar to those described by Lezak (1995). Often, the presence of certain qualitative features (e.g., a very segmented or piecemeal drawing strategy in an adolescent) can provide convincing evidence of brain dysfunction or developmental delay. Excellent discussions of this approach and its utility are provided by Waber and Holmes (1985, 1986). Other useful visuoconstructional tasks for children include WISC-III Object Assembly and Block Design. Measures like the Berry and the Primary Mental Abilities Test may be particularly appropriate for younger children (Rovet *et al.*, 1988).

Attention. The Digit Vigilance Test is a "low-tech" continuous performance test that requires subjects to scan two pages of numbers and cross out all the sixes. As a measure of sustained attention and visual tracking, it places demands on both speed and accuracy, and provides independent measures of both. The myriad computerized continuous performance tasks now available are reasonable replacements. Many clinicians have also used WISC-III Digit Span, but we would argue that it is only the Digits Forward subtest (longest span) that yields a relatively "pure" measure of attention.

Trail Making. This well-known measure of attention and motor speed (Part A) and mental flexibility (Part B) should be part of any assessment. Especially informative is the patient's response style on Part B. When slowing is found, it can occur for a number of reasons: The patient can't remember what to do next, has trouble "shifting set," gets lost on the page and can't find the next target, is just slow but meticulous, or responds impulsively to an inappropriate item. This test is exquisitely sensitive to virtually any sort of biomedical or psychosocial problem; thus, evidence of any abnormally slow performance needs to be interpreted cautiously before concluding that the patient has mild brain damage. Digit Span Backwards and the Interference subtest from the Stroop Color-Word Test can also provide useful information about mental flexibility.

Psychomotor Integration. Like Trail Making, the Grooved Pegboard Test is a measure of eye–hand coordination and finger dexterity that is also sensitive to virtually any type of psychological or biomedical variable. For that reason, it is an excellent "indicator measure." Abnormally slow or clumsy performance is evidence that the patient, for whatever reason, is not functioning normally. Other excellent measures of psychomotor integration are the WISC-III Coding subtest and the Symbol Digit Modalities Test.

Deductive Reasoning. The computerized version of the Wisconsin Card Sorting Test (WCST) provides a measure of deductive reasoning that is intrinsically interesting and challenging for most older children (age 8 and over), and virtually foolproof to administer. It is important to keep in mind, however, that few studies of children with virtually any type of medical illness have found abstract reasoning skills—as measured by the WCST or tests like the Category Test—to be impaired.

Numerous other tests could be included in one's optimal neuropsychological battery, depending on assessment time availability and the nature of the referring question.

Psychopathology. Neuropsychological assessment should also include an evaluation of possible psychological distress as well as behavior problems. Good self-report measures include the Children's Depression Inventory and the Child Behavior Checklist (CBCL) or the Behavior Assessment System for Children (BASC). The latter two measures can be completed by the child, the parent, and the teacher, and we recommend administering one of these measures to all three types of informants. Finding evidence of psychopathology on screening measures like these suggests that some type of formal clinical interview (or a semistructured interview like the Schedule for Affective Disorders and Schizophrenia for school-aged children [K-SADS]) needs to be completed with the child and parent so that Diagnostic and Statistical Manual of Mental Disorders (DSM-IV) diagnoses can be made. That information should then be taken into account when interpreting neuropsychological test results.

In 1922, in one of the first neuropsychological studies of individuals with a chronic disease, diabetic adults were found to perform more poorly than a healthy control group on measures of memory, mental arithmetic, and psychomotor efficiency. This early study, conducted just prior to the discovery of insulin, suggested that these cognitive deficits may be a consequence of chronic hyperglycemia. Subsequent research with insulin-dependent (Type I) diabetic adults has continued to explore the relationship between the development of cognitive dysfunction and biomedical complications of diabetes and its treatment, and two broad candidate variables have now been identified: (1) poor metabolic control—as indexed by the development of certain diabetes-related complications (e.g., retinopathy, neuropathy) secondary to chronically high blood glucose levels, and (2) repeated episodes of severe hypoglycemia. Until recently, however, most of these studies were stymied by relatively small samples of patients, assessed with limited neuropsychological test batteries, who had diabetes-related complications that were not always well described.

POOR METABOLIC CONTROL. In spite of the limited number of studies on young and middle-aged adults with Type I diabetes, it now appears that a long history of chronic hyperglycemia is a risk factor for the occurrence of mild brain dysfunction. From a neuropsychological perspective, this is most clearly evidenced on tests that require psychomotor speed and/or spatial information processing. For example, in a study of diabetic adults drawn from a well-described community cohort, deficits were most apparent on the Object Assembly subtest, as well as on measures of time to complete the Digit Vigilance, Embedded Figures, and Grooved Pegboard tests (Ryan, Williams, Orchard, & Finegold, 1992). Importantly, the accuracy of the diabetics' performance on these same tests was comparable to that of healthy control subjects. In fact, these diabetic patients did not differ from control subjects on any other test of any other cognitive domain. Of particular interest is the observation that the best predictor of psychomotor slowing in this cohort is the presence of clinically significant neuropathy—a biomedical complication of diabetes that is associated with poor metabolic control. Patients with neuropathy were slower—even on cognitive tests that did not have a significant sensory or motor component—than patients without neuropathy.

It is reasonable to discuss the possibility of "brain dysfunction" in this instance because converging evidence from both electrophysiological and neuroimaging studies has also supported a linkage between the development of diabetic neuropathy and brain abnormalities. Auditory brain stem evoked potentials are usually abnormal in patients with diabetes of long duration, particularly when clinically significant diabetic neuropathy has been diagnosed. Moreover, patients with diabetic neuropathy have a far higher incidence of abnormalities on magnetic resonance imaging (MRI). Not only do they have a greater number of lesions, but the lesions are larger, and more diffusely distributed, than the pattern found in healthy comparison subjects (Dejgaard *et al.*, 1991).

In addition to these relatively subtle psychomotor deficits, diabetic patients in poor metabolic control may also manifest deficits on learning and memory tests, although these deficits tend to be restricted to individuals with the so-called brittle form of diabetes (Franceschi *et al.*, 1984). These are individuals whose blood glucose levels tend to fluctuate to such an extent that they are not easily controlled by standard diabetes treatment regimens. As a consequence, they have frequent hospitalizations for their very poor control. In contrast, the typical diabetic adult without a "brittle" history who is drawn not from a hospital but from the community tends to perform well within normal limits on a wide range of learning and memory tests (Ryan & Williams, 1993).

REPEATED EPISODES OF HYPOGLYCEMIA. A single episode of severe hypoglycemia (blood glucose level less than 40 mg/dl) may induce coma or seizure, and lead to permanent brain damage (Chalmers *et al.*, 1991; Gold *et al.*, 1994). Although most diabetic patients and their physicians are well aware of (and appropriately apprehensive about) this association, only a few cases of hypoglycemia-induced brain damage have actually been reported in the clinical literature. That is, the incidence and prevalence of such catastrophic disorders is unknown, but probably lower than one might expect. Of greater importance from a public health perspective is whether repeated episodes of mild to moderately severe hypoglycemia (blood glucose levels of 65 to 40 mg/dl) may result in measurable neuropsychological dysfunction. Unfortunately, this question remains unsettled: Smaller cross-sectional studies have suggested that five or more episodes of hypoglycemia may lead to a decline in mental efficiency whereas large-scale longitudinal studies have found no association between hypoglycemia and cognitive impairment.

In what may be the best cross-sectional study on this topic to date, small but statistically significant correlations were reported between number of moderately severe hypoglycemic episodes and measures of choice reaction time and Wechsler Adult Intelligence Scale-Revised (WAIS-R) Performance IQ, leading to the conclusion that recurrent hypoglycemia produces a decline in "fluid intelligence" or adaptive problem-solving abilities (Deary *et al.*, 1993). Other investigators, using very different measures, have also reported declines in mental efficiency with speed of responding, rather than accuracy, most disrupted by repeated episodes of hypoglycemia (Wredling, Levander, Adamson, & Lins, 1990).

Longitudinal studies have not succeeded in replicating these results. Both the Diabetes Control and Complications Trial (DCCT) and the Stockholm Diabetic Intervention Study (SDIS) examined the effects of intensive metabolic therapy on diabetic adults followed for 5 to 9 years (Diabetes Control and Complications Trial, 1996; Reichard, Berglund, Britz, Levander, & Rosenqvist, 1991). Although intensive therapy was associated with a greatly increased risk of moderate to severe hypoglycemia, there was no corresponding increase in neuropsychological dysfunction. One can always argue that null results of this sort may be a consequence of biases in subject selection (for example, the DCCT patients were selected to be healthy and highly motivated); the nature of the experimental intervention (participating research subjects were more carefully monitored than would ordinarily be the case in patients drawn from a nonstudy population); or the sensitivity of the neuropsychological assessment. Given these disparate findings and potential methodological problems, it may be best to conclude that the link between severe hypoglycemia and cognitive dysfunction is not yet proven (Deary & Frier, 1996). Nevertheless, when evaluating a patient with a past history of recurrent hypoglycemia, the clinician needs to be especially alert for possible mild neuropsychological dysfunction, especially on WAIS-R Performance IQ subtests.

NON-INSULIN-DEPENDENT DIABETES MELLITUS. Our understanding of the effects of non-insulin-dependent diabetes mellitus (NIDDM) on cognitive functioning is based almost exclusively on studies that have been conducted primarily on elderly adults over the age of 60. This is somewhat surprising, since this disorder is quite common in individuals who are 40 to 60 years of age—especially those who are overweight and/or have a family history of NIDDM. Early research demonstrated that in elderly individuals, NIDDM was associated with significant impairments on measures of learning and memory. In general, diabetic adults learned fewer items across trials on experimental serial learning tests (Perlmuter *et al.*, 1984) as well as on free-recall learning tests like the Rey Auditory Verbal Learning Test (Mooradian, Perryman, Fitten, Kavonian, & Morley, 1988). Memory, assessed by measures like the Benton Visual Retention Test, has also been found to be impaired (Mooradian *et al.*, 1988). The best predictor of poor mnestic performance has consistently been found to be degree of chronic hyperglycemia: Diabetics with poorer metabolic control (i.e., higher glycosylated hemoglobin values) learned less information despite repeated study/test trials. This basic pattern of results has been replicated by a number of investigators, using very different measures (for review see Strachan *et al.*, 1977), and at least one study has demonstrated that medical regimens that improve older NIDDM patients' metabolic control may be associated with corresponding improvements in learning and memory ability (Gradman, Laws, Thompson, & Reaven, 1993). In general, psychomotor and visuospatial skills appear to be intact in NIDDM patients; this is quite the opposite of what has been noted in younger patients with insulin-dependent diabetes.

OPTIMAL TEST BATTERY FOR MEDICALLY ILL ADULTS. When evaluating the neuropsychological status of an adult with a medical disorder, measures should sample a broad array of cognitive processes, the clinician should be familiar and comfortable with the actual tests, and the available norms should be appropriate for the particular patient population. Comparing a young adult medically ill outpatient with norms obtained from middle-aged psychiatric inpatients is completely inappropriate. Because of a need to evaluate cognitive processes as comprehensively as possible, any "basic" test battery should include measures of attention, learning, memory, problem-solving, visuospatial skills, psychomotor efficiency, and "mental flexibility." In addition, premorbid intelligence needs to be estimated.

Because performance on certain subtests from the WAIS-R—particularly those comprising Performance IQ—can be influenced by disease-related parameters (hypoglycemia; elevated ammonia levels in cirrhotic patients), most if not all of the WAIS-R or WAIS-III should be incorporated into any assessment. When used to assess neuropsychological functioning (rather than serve merely to calculate IQ), subtests from the Wechsler intelligence scales can provide important insights into the patient's problem-solving and abstract reasoning skills (Similarities; Block Design), their visuospatial abilities (Block Design, Object Assembly), attention (Digit Span), psychomotor efficiency (Digit Symbol Substitution), and premorbid intelligence (Information, Vocabulary, Comprehension) (Lezak, 1995). An increasing number of clinicians have begun to use the National Adult Reading Test (NART) to estimate premorbid intelligence, and there is much converging evidence to suggest that this may be a better measure for that purpose than the WAIS-R Vocabulary or Information subtests (Crawford, 1992).

A thorough assessment of learning and memory functioning must always be part of the adult neuropsychological evaluation. List learning tests like the Rey Auditory Verbal Learning Test or the California Verbal Learning Test are reasonable choices, as are story recall tests like the Logical Memory subtest from the

Wechsler Memory Scale-Revised (WMS-R). Immediate and delayed visual memory skills can best be assessed with measures like the Rey Complex Figure Test or the WMS-R Visual Reproductions. We have not routinely used the entire WMS-R because we have found certain subtests (e.g., Associate Learning) to be relatively insensitive to the mild cognitive impairment typically found in most young or middle-aged medically ill patients. The Memory and Location subtests from the Tactual Performance Test (TPT) also provide an excellent measure of "incidental memory" in medically ill adults.

Adults with a variety of chronic illnesses, including liver disease and diabetes mellitus, frequently manifest performance decrements on tasks requiring sustained attention, mental flexibility, and psychomotor efficiency. For that reason we also administer the Digit Vigilance (or a similar computer-based continuous performance test), Trail Making, Grooved Pegboard, and WAIS-R Digit Symbol Substitution (or Symbol Digit Modalities) tests. To assess deductive reasoning, we generally administer the Category Test (on the computer) rather than the Wisconsin Card Sorting Test, as we have found it to be more sensitive to mild neuropsychological impairment. Including this measure is particularly important when assessing patients with certain disorders, like cirrhosis, that often have a major impact on abstract reasoning skills. The TPT may also be useful in assessing cognitive integrity in high-functioning adults (including those with diabetes). Not only has the standard quantitative score (e.g., time taken to complete the TPT) been associated with certain biomedical variables, but observation of the individual's behavior during the test (e.g., type of search strategies used) can yield important qualitative information on how systematically patients behave while performing an often frustrating task. Despite our enthusiasm, some controversy remains over the use of this test, and Lezak (1995) has made a strong argument for not routinely administering the TPT (e.g., excessively long time requirement; patient discomfort induced by blindfold). Based on the specific referral question or a clinical observation, other tests may be added as well (e.g., language processing), although in response to recent efforts within the United States to reduce medical costs, we try to restrict the battery to a relatively focused assessment that can be completed with the majority of patients in less than 3 hours.

Rates of concomitant psychopathology are quite high in medically ill patients. Because clinically significant depression or anxiety can have a major impact on neuropsychological test performance, it is critical that current level of psychological distress be ascertained. For the medically ill adult patient, the Symptom Checklist 90 (SCL-90-R) serves as an excellent screening measure. If patients manifest clinically significant elevations on the Global Severity Index (T score > 67) or show elevations on either the Anxiety or Depression subscales, the neuropsychologist may request a psychiatric consult or administer a semistructured interview like the Structured Clinical Interview for DSM-IV (SCID-IV).

The primary aims of neuropsychological assessment are not only to document cognitive strengths and weaknesses, but to make attributions about causality. Discovering that a patient has comorbid neuropsychological deficits and psychiatric symptomatology often leads clinicians to conclude that the poorer neuropsychological test performance must necessarily be secondary to depression. That may be true, but only under certain, relatively restricted conditions. When patients, following a face-to-face interview, meet formal clinical criteria for a psychiatric disorder like Major Depression or Anxiety Disorder, it is indeed likely that performance on some if not all of the neuropsychological tests *may* have been affected adversely by their psychopathology. This may be most evident on tests of learning and memory, or tests requiring rapid responding. In contrast, there is actually little empirical support for a link between poor performance on neuropsychological tests and depressive *symptomatology* when depression is measured *solely* by self-report questionnaires like the "D" scale from the Minnesota Multiphasic Personality Inventory (MMPI), the Depression subscale, or the Global Severity Index from the SCL-90, or the Beck Depression Inventory (Gass & Russell, 1986). It is for this reason that when elevations are found on self-report measures of psychopathology, a formal clinical interview should be conducted to determine whether the patient meets DSM-IV criteria for a diagnostic disorder. As numerous investigators have pointed out, patients with medical disorders often manifest both cognitive deficits and significant psychiatric distress that can be completely independent of one another (Hinkin *et al.*, 1992).

Concluding Remarks

Assessing the patient with a medical illness is not a trivial endeavor. Clinicians need to evaluate their patients as thoroughly as possible with measures of both cognitive functioning and psychological distress. However, more important than the actual tests used is the clinician's ability (and willingness) to integrate results from this comprehensive neuropsychological evaluation into the body of information that has been obtained on the patient's current and past medical disorders, as well as his or her developmental, educational, vocational, and social history. Studies of medically ill patients have alerted us to the fact that these individuals have a higher risk of neuropsychological (and psychosocial) problems than their healthy peers. Unfortunately, in the absence of neuropsychological profiles that are sensitive and specific to individual diseases, it is impossible to ever conclude that a particular patient's performance is what one would expect from a diabetic adult—or a cirrhotic patient, or someone with end stage renal disease, or a person with AIDS. The best we can do is to document cognitive strengths and weaknesses in any given patient and consider the possibility that certain "weaknesses" may be attributable to one or more medical disorders, as opposed to some idiopathic brain disorder.

Indeed, when asked to evaluate a patient with a neurologic disorder, or someone who is referred for a neuropsychological evaluation but who does not have any obvious reason for cognitive impairment, we have an obligation to review that person's medical history in the same way that we ordinarily ask about the person's history of head injury or past and current alcohol and drug use. Does the person have a history of hypertension? Are all hormone levels, particularly thyroid levels, normal? Was there a past history of chronic obstructive pulmonary disease? Did the person have open-heart surgery? By using open-ended questions about current and past medical disorders, and by examining the current literature on the extent to which those disorders *may* affect neuropsychological test performance, the clinician can complete the most critical, and difficult part of the formal neuropsychological assessment: making a causal attribution when impairments are found.

References

Ack, M., Miller, I., Weil, W. (1961). Intelligence of children with diabetes mellitus. *Pediatrics, 28*, 764–770.

Chalmers, J., Risk, M. T. A., Kean, D. M., Grant, R., Ashworth, B., & Campbell, I. W. (1991). Severe amnesia after hypoglycemia. *Diabetes Care, 10*, 922–925.

Crawford, J. R. (1992). Current and premorbid intelligence measures in neuropsychological assessment. In J. R. Crawford, D. M. Parker, & W. W. McKinlay (Eds.), *A handbook of neuropsychological assessment* (pp. 21–49). Hove, UK: Lawrence Erlbaum Associates.

Deary, I., Crawford, J., Hepburn, D. A., Langan, S. J., Blackmore, L. M., & Frier, B. M. (1993). Severe hypoglycemia and intelligence in adult patients with insulin-treated diabetes. *Diabetes, 42,* 341–344.

Deary, I. J., & Frier, B. M. (1996). Severe hypoglycaemia and cognitive impairment in diabetes: link not proven. *British Medical Journal, 313,* 767–768.

Dejgaard, A., Gade, A., Larsson, H., Balle, V., Parving, A., & Parving, H. (1991). Evidence for diabetic encephalopathy. *Diabetic Medicine, 8,* 162–167.

Diabetes Control and Complications Trial. (1996). Effects of intensive diabetes therapy on neuropsychological function in adults in the Diabetes Control and Complications Trial. *Annals of Internal Medicine, 124,* 379–388.

Franceschi, M., Cecchetto, R., Minicucci, F., Smizne, S., Baio, G., & Canal, N. (1984). Cognitive processes in insulin-dependent diabetes. *Diabetes Care, 7,* 228–231.

Gass, C. S., & Russell, E. W. (1986). Differential impact of brain damage and depression on memory test performance. *Journal of Consulting and Clinical Psychology, 54,* 261–263.

Gold, A. E., Deary, I. J., Jones, R. W., O'Hare, J. P., Reckless, J. P. D., & Frier, B. M. (1994). Severe deterioration in cognitive function and personality in five patients with long-standing diabetes: a complication of diabetes or a consequence of treatment? *Diabetic Medicine, 11,* 499–505.

Golden, M., Russell, B., Ingersoll, G., Gray, D., & Hummer, K. (1985). Management of diabetes mellitus in children younger than 5 years of age. *American Journal of Diseases of Children, 139,* 448–452.

Gradman, T. J., Laws, A., Thompson, L. W., & Reaven, G. M. (1993). Verbal learning and/or memory improves with glycemic control in older subjects with non-insulin-dependent diabetes mellitus. *Journal of the American Geriatric Society, 41,* 1305–1312.

Hagen, J. W., Barclay, C. R., Anderson, B. J., Feeman, D. J., Segal, S. S., Bacon, G., & Goldstein, G. W. (1990). Intellective functioning and strategy use in children with insulin-dependent diabetes mellitus. *Child Development, 61,* 1714–1727.

Hinken, C. H., Van Gorp, W. G., Satz, P., Weisman, J. D., Thommes, J., & Buckingham, S. (1992). Depressed mood and its relationship to neuropsychological test performance in HIV-1 seropositive individuals. *Journal of Clinical and Experimental Neuropsychology, 14,* 289–297.

Holmes, C. S. (1990). Neuropsychological sequelae of acute and chronic blood glucose disruption in adults with insulin-dependent diabetes. In C. S. Holmes (Ed.), Neuropsychological and behavioral aspects of diabetes (pp. 122–154). New York: Springer-Verlag.

Holmes, C. S., Dunlap, W. P., Chen, R. S., & Cornwell, J. M. (1992). Gender differences in the learning status of diabetic children. *Journal of Consulting and Clinical Psychology, 60,* 698–704.

Kovacs, M., Goldston, D., & Iyengar, S. (1992). Intellectual development and academic performance of children with insulin-dependent diabetes mellitus: A longitudinal study. *Developmental Psychology, 28,* 676–684.

Lezak, M. D. (1995). Neuropsychological assessment (3d ed.). New York: Oxford University Press.

Mooradian, A. D., Perryman, K., Fitten, J., Kavonian, G. D., & Morley, J. E. (1988). Cortical function in elderly non-insulin dependent diabetic patients. Behavioral and electrophysiologic studies. *Archives of Internal Medicine, 148,* 2369–2372.

Perlmuter, L. C., Hakami, M. K., Hodgson-Harrington, C., Ginsberg, J., Katz, J., Singer, D. E., & Nathan, D. M. (1984). Decreased cognitive functioning in aging non-insulin-dependent diabetic patients. *American Journal of Medicine, 77,* 1043–1048.

Reichard, P., Berglund, B., Britz, A., Levander, S., & Rosenqvist, U. (1991). Hypoglycemic episodes during intensified insulin treatment: increased frequency but no effect on cognitive function. *Journal of Internal Medicine, 229,* 9–16.

Rovet, J., Ehrlich, R., & Hoppe, M. (1988). Specific intellectual deficits associated with the early onset of insulin-dependent diabetes mellitus in children. *Child Development, 59,* 226–234.

Ryan, C. M. (1990). Neuropsychological consequences and correlates of diabetes in childhood. In C. S. Holmes (Ed.), *Neuropsychological and behavioral aspects of diabetes* (pp. 58–84). New York: Springer-Verlag.

Ryan, C. M. (1997). Effects of diabetes mellitus on neuropsychological functioning: A lifespan perspective. *Seminars in Clinical Neuropsychiatry, 2,* 4–14.

Ryan, C., & Butters, N. (1980). Further evidence for a continuum-of-impairment encompassing male alcoholic Korsakoff patients and chronic alcoholic men. *Alcoholism: Clinical and Experimental Research, 4,* 190–198.

Ryan, C., Longstreet, C., & Morrow, L. A. (1985). The effects of diabetes mellitus on the school attendance and school achievement of adolescents. *Child: Care, Health, and Development. 11,* 229–240.

Ryan, C., Vega, A., & Drash, A. (1985). Cognitive deficits in adolescents who developed diabetes early in life. *Pediatrics, 75,* 921–927.

Ryan, C., Vega, A., Longstreet, C., & Drash, L. (1984). Neuropsychological changes in adolescents with insulin-dependent diabetes mellitus. *Journal of Consulting and Clinical Psychology, 52,* 335–342.

Ryan, C. M., & Williams, T. M. (1993). Effects of insulin-dependent diabetes on learning and memory efficiency in adults. *Journal of Clinical and Experimental Neuropsychology, 15,* 685–700.

Ryan, C. M., Williams, T. M., Orchard, T. J., & Finegold, D. N. (1992). Psychomotor slowing is associated with distal symmetrical polyneuropathy in adults with diabetes mellitus. *Diabetes, 41,* 107–113.

Strachan, M. W. J., Deary, I. J., Ewing, F. M. E., & Frier, B. M. (1997). Is Type 2 (non-insulin-dependent) diabetes mellitus associated with an increased risk of cognitive dysfunction? *Diabetes Care, 40,* 438–445.

Tarter, R. E., Van Thiel, D. H., & Edwards, K. L. (Eds.). (1988). Medical neuropsychology: The impact of disease on behavior. New York: Plenum Press.

Waber, D. P., & Holmes, J. M. (1985). Assessing children's copy productions of the Rey–Osterrieth Complex Figure. *Journal of Clinical and Experimental Neuropsychology, 7,* 264–280.

Waber, D. P., & Holmes, J. M. (1986). Assessing children's memory productions of the Rey–Osterrieth Complex Figure. *Journal of Clinical and Experimental Neuropsychology, 8,* 565–580.

Weitzman, M., Klerman, L., Lamb, G., Menary, J., & Alpert, J. (1982). School absence: A problem for the pediatrician. *Pediatrics, 69,* 739–746.

Wredling, R., Levander, S., Adamson, U., & Lins, P. E. (1990). Permanent neuropsychological impairment after recurrent episodes of severe hypoglycaemia in man. *Diabetologia, 33,* 152–157.

12

Evaluation of Neoplastic Processes

Richard A. Berg

Cerebral tumors are common causes of severe neuropsychologic deficit in humans. In many cases, these individuals are not seen by neuropsychologists because they may initially present with strong signs of increases in intracranial pressure, strong focal neurologic signs, or seizure activity which can rapidly lead to medical care and treatment. Other neoplasms, however, may present very subtle signs which can be missed or ignored when the patient is evaluated. The symptoms may, in fact, superficially suggest an apparently clear-cut psychiatric disorder. Although this occurs rarely in behaviorally disturbed populations, the presence of psychiatric or cognitive symptoms with such tumors is frequent; the exact symptoms depend upon such factors as the location of the tumor, the rate of growth, the stage of development, and the effects of brain organization and metabolism as a whole (Lishman, 1978).

Over the past several decades, the treatment for various neoplastic disorders has become more effective resulting in longer survival times and higher survival rates for those with such disorders. Comprehensive clinical management now must address not only the medical status of the patient, but also the overall psychological condition and social adjustment of the individual who has (had) a neoplasm. Emotional reactions, subjective feelings, functional behavioral capacities, interpersonal relationships, and psychosocial adjustment are encompassed by the *quality of life* construct. One of the important factors influencing quality of life is the cognitive status of the individual. The treatment of neoplasms can result in a variety of cognitive impairments ranging from mild to severe, as can the neoplasm itself (Oxman & Silverfarb, 1980). Determining the cognitive capabilities of these patients can be especially important because cognitive dysfunction in patients with cancer is often indicative of poor medical prognosis (Folstein, Fetting, Lobo, Niaz,

Richard A. Berg   Coastal Rehabilitation Hospital and Wilmington Health Associates, Wilmington, North Carolina 28401.

Neuropsychology, edited by Goldstein *et al.* Plenum Press, New York, 1998.

& Capozolli, 1984). Cognitive impairment also can profoundly affect the individual in numerous facets of daily life, compromising the patient's capacity for independent living. With increasing cognitive impairment, the ability to provide information and comply with a treatment regimen may decline. Valuable information can thus be obtained from a neuropsychologic examination which can be used to maximize the efficacy of medical management.

HISTORY

The investigation of behavioral sequelae of neoplasms can be traced back over a century. In 1880, the associations between disease involving the territory of the precentral gyrus and motor disability and between language disorders and lesions of the foot of the left third frontal gyrus and its surrounding area were universally recognized (Benton, 1991). The first paper to deal with clinical aspects of cerebral localization discussed frontal lobe symptomatology in terms of a specific form of dementia characterized by an oddly cheerful agitation in patients with tumors of the frontal lobe (Jastrowitz, 1888). A comprehensive paper on the clinical manifestations of brain tumors dealt with among other things childish and inappropriate joking in patients with brain tumors. Clearly, the early investigations of changes that occurred in the presence of cerebral neoplasms relied primarily on observation of overt behavioral changes.

Bricker (1934, 1936) presented one of the earlier reports assessing more than simply overt behavior. In this case report of a patient with what amounted to a total excision of the prefrontal region in the course of a surgical resection of a large meningioma, Bricker presented a detailed account of the patient's social behavior, intellectual functioning, and personality characteristics. These studies established that distinctive changes in personality, behavior, and cognitive ability could be related to disease of the brain.

As knowledge of brain–behavior relationships and testing techniques have expanded, so too has the ability to determine the effects of brain tumors in increasingly more specific terms. In a recent review of the progress of neuropsychology, Boll (1985) points out the rapidity with which tests have become more sophisticated. Cancer has been a major health concern for several years. Unfortunately, the neuropsychological sequelae of this broad class of diseases remain relatively unstudied. This is due in part to the diffuse nature of the disease and its many different forms. Neoplasms outside the nervous system may affect any organ system and thus may be highly idiosyncratic in their involvement of CNS activities. One way in which tumors may act on the CNS suggests that autoimmune antibodies that attack the tumor may pass the blood–brain barrier and compromise CNS integrity (Lorig, 1992). Another mechanism is alteration of endocrine systems by tumors that secrete endocrine-like substances which mimic naturally occurring material and thus throw the entire system into dysregulation. Both autoimmune and endocrine-related problems may produce symptoms of cortical dementia in some patients (Davis *et al.*, 1987). Seventy-one percent of a group of treated cancer victims referred for neuropsychiatric evaluations in this study met criteria for organic mental disease. Other investigators have reported smaller proportions on otherwise normal groups of individuals treated for various forms of cancer (Derogatis, Morrow, & Fettig, 1983).

There are a variety of dysfunctions that should come to mind when working with an individual who has been diagnosed with a neoplasm. These include iatrogenic treatment effects and metastases. Metastatic tumors arising from tumors in the lungs, breasts, and gastrointestinal tract may invade the nervous system and produce focal, space-occupying lesions. Metastatic tumors of the lung seem to have a particular affinity for nervous system tissue. One study reports on two patients with metastatic tumors of the limbic system arising from small-cell tumors of the lung (den Hollander, Van Hulst, Meerwaldt, and Haasjes, 1989). This particular form of tumor is comparatively rare; however, this study points to the possible need for using psychiatric symptoms as a warning sign of metastatic tumors (Lorig, 1992).

Currently, there are three primary treatment approaches for neoplasms: resection, chemotherapy, and radiation. Surgical resection is the most localized treatment and may produce the most limited CNS involvement, particularly if the original tumor is located outside of the nervous system. The neuropsychological sequelae of chemotherapy has been studied only recently. A variety of commonly used antineoplastic agents have been reported to produce memory, motor, and other CNS deficits (e.g., Goff, Anderson, & Cooper, 1980; Weiss, Walker, & Wienik, 1974). Similar effects have been found for the drug Ara-C which is frequently used in cancer treatment (Hwang, Yung, Estey, & Fields, 1985). The long-held notion that chemotherapeutic agents do not cross the blood–brain barrier needs to be reassessed (Davis *et al.*, 1987).

Radiation as a treatment for cancer has been studied for a number of years as a result of general health concerns regarding the effects of radioactive fallout or whole body irradiation (Lorig, 1992). The use of current radiotherapy technology in cancer treatment is, however, somewhat different. In these cases, X or gamma rays are focused on specific body areas with little incidental irradiation of other body parts. Such incident irradiation may have minor effects on neuropsychological functioning. In recent investigations with children, little direct effects of radiotherapy were found (Bordeaux *et al.*, 1988). Even limited cranial irradiation may have few measurable neuropsychologic consequences (e.g., Berg *et al.*, 1983). One problem in the neuropsychological literature is the inconsistency of results. In a subsequent investigation, it was found that cranial irradiation interacted with chemotherapy to produce deficits in neuropsychological performance which ultimately were attributed to peripheral neuropathy induced by chemotherapy (Copeland *et al.*, 1988).

Finally, accurate diagnosis of cognitive functioning in individuals who have neoplasms is confounded by the typical flattening of affect associated with a diagnosis of cancer. Cavanaugh and Wettstein (1989) reported that cancer victims often experience profound depression.

TYPES OF NEOPLASMS

The most prevalent form of cancer in adults is carcinoma. It represents approximately 88.4% of all cancers reported in adults (Young *et al.*, 1992). Carcinoma is a neoplasm of epithelial tissue (i.e., tissue lining various organs). The most frequent organs involved are the lungs, liver, stomach, and breast. Leukemias are the second most prevalent variety of neoplasm (2.7%), followed by malignancies in the central nervous system (1.4%). In children, leukemia is the most prevalent form of neoplasm followed by brain tumors and lymphomas. Table 1 presents the epidemiological data for the various types of cancer.

Mortality rates vary with the form of tumor. For adults with carcinoma, such a lung and breast cancers, the prognosis is poor despite advances in prolonging

TABLE 1. NEOPLASM INCIDENCE FOR ADULTS
AND CHILDREN[a]

Type	Adult	Child
Carcinoma	88.4	5.7
Brain	1.4	19.8
Leukemia	2.7	31.0
Other sites (combined)	7.5	43.5

[a]Adapted from Young *et al.* (1982).

survival that have been made since the mid-1960s. The same is generally true for some types of CNS tumors. The prognosis for survival of childhood cancer has dramatically improved over the past two or three decades. Certain tumors in children have a 5-year survival rate of over 70% while the rate for the most prevalent form of childhood cancer, leukemia, is about 50% (Mauer, 1980).

ANATOMY AND PHYSIOLOGY

CARCINOMAS

The etiology of carcinomas is still uncertain to a large extent. It is likely that there are multiple factors involved. Current theories suggest that the presence of a tumor leads to production of autoimmune antibodies in a number of tissues including the CNS (Wilkinson, 1964). It is hypothesized that these antibodies are able to penetrate the blood–brain barrier or attach themselves to peripheral nerve or muscle tissue. If this is the case, pathologic changes may arise in the affected tissues that would not be incompatible with some of the cognitive changes/dysfunction which will be described. This theory has received some support from the description of antibrain antibodies in "carcinomatous neuromyopathy" and antibodies against sensory carcinomatous neuropathy (Wilkinson & Zeromski, 1965). Changes in structure may also be the result of a chronic viral infection (Gilroy & Meyer, 1975). Finally, some of the abnormalities associated with malignancies may be the result of concomitant endocrine or metabolic disturbances.

CEREBRAL TUMORS

Neoplasia can affect the CNS in three primary ways: Primary tumors may develop in the brain, spinal cord, or surrounding structures; metastatic tumors may spread to the CNS from primary cancer elsewhere; or the brain and spinal cord may be damaged indirectly by the presence of a tumor elsewhere in the body. Tumors arising within the CNS or surrounding structures, including the meninges, blood vessels, embryonic cell rests, and bone, constitute roughly 10% of all tumors (Gilroy, 1990). Gliomas are the most common type followed by meningiomas and pituitary adenomas. Tumors arising from embryonic tissue such as the medulloblastoma tend to occur early in life while gliomas can occur at any age with increasing frequency up to about 65 years of age. Metastatic tumors are usually seen among older patients with a steady increase in incidence after the age of 50.

ETIOLOGY. Brain tumors have a number of etiologies and may arise following head injury, infection, metabolic and other systemic diseases, and exposure to toxins and radiation (Joseph, 1990). Some can be caused by tumors in other parts of the body that metastasize to the brain. Others are believed to have begun in embryonic cells left in the brain during development. In some cases tumors are a consequence of embryonic timing and migration errors; for example, if germ layers differentiate too rapidly or if cells migrate to the wrong location, they can exert neoplastic influences within their unnatural environment (Joseph, 1990). Among children, many tumors are congenital, developing from displaced embryonic cells, dysplasia of developing structures, and the altered development of primitive cells that are normally precursors to neurons and glia. Most of these tumors tend to occur within the brain stem, cerebellum, and midline structures, including the third ventricle. Among older persons, most tumors are due to dedifferentiation of adult elements (Shapiro, 1986).

Symptoms Associated with Tumor Formation

Initially, tumor development is associated with localized cerebral displacement, compression, edema, and regional swelling, particularly in the white matter. Because of this, the symptoms produced are more dependent on the tumors' rate of growth and location rather than the tumor type per se. Symptoms tend to be site-specific such that initial localized effects are the most prominent (Joseph, 1990). Physical and some cognitive symptoms associated with the presence of neuroplastic processes are presented in Table 2. Hence the most common modes of onset include progressive focal symptoms such as focal seizure activity, monoplegia, hemiplegia, language disruption, cerebellar deficiency, and symptoms of increased intracranial pressure (ICP) (Gilroy, 1990; Joseph, 1990). Most individuals also experience periodic bifrontal and bioccipital headaches that can awaken the individual at night or that are present upon waking. Vomiting, mental lethargy, unsteady gait, sphincter incontinence, and papilledema also are frequent sequelae.

As ICP increases, disturbances of consciousness, cognition, and neurological and neuropsychological functioning become progressively widespread. Generalized symptoms, however, tend to occur late in the process or not at all. Even when this occurs, functional disruptions associated with tumor formation are usually slow to appear and develop, taking months or years. The sudden appearance of symptoms is generally indicative of vascular disease. However, in some cases, a "silent," slow-growing tumor may hemorrhage, in which case the symptoms are of sudden onset.

Generalized Neoplastic Syndromes and Symptoms

The space within the cranial vault is limited and generally fixed. As tumors develop, ICP is likely to increase. As ICP increases, disturbances of consciousness, cognition, and widespread neurological functioning become increasingly diffuse. Development and enlargement of any space-occupying lesion occurs at the expense of brain tissue which becomes increasingly compressed. Tumor growth can not only displace underlying cerebral tissue, but also compress blood vessels as well as decrease the amount of cerebrospinal fluid in the ventricles and subarachnoid space. With increased displacement and edema, CSF pressure can eventually increase. This tend tends to compress midline structures as well as contralateral brain tissue.

As the tumor enlarges and ICP rises, cognitive abilities tend to decline. Comprehension slows and there can be a loss of capacity to sustain continuous mental

TABLE 2. SIGNS ASSOCIATED WITH LOCALIZED LESIONS[a]

Location of lesion	Associated signs
Prefrontal area	Loss of judgment, failure of memory, inappropriate behavior, apathy, poor attention span, increased distractiblity, release phenomena
Frontal eye fields	Failure to sustain gaze to the opposite side, saccadic eye movements, impersistence, seizures with forced deviation of the eyes to the opposite side
Precentral gyrus	Partial motor seizures, Jacksonian seizures, generalized seizures, hemiparesis
Superficial parietal lobe	Partial sensory loss, loss of cortical sensation including two-point discrimination, tactile localization, stereognosis, and graphism
Angular gyrus	Agraphia, acalculia, finger agnosia, right–left confusion (Gerstmann syndrome)
Broca's area	Expressive aphasive
Superior temporal gyrus	Receptive aphasia
Midbrain	Early hydrocephalus, loss of upward gaze, pupil abnormalities, third cranial nerve involvement (ptosis, exteral strabismus, diploplia), ipsilateral cerebellar signs, contralateral hemiparesis, parkinsonism, akinetic mutism
Cerebellar hemisphere	Ipsilateral cerebellar ataxia with hypotonia, dysmetria, intention tremor, nystagmus to side of lesion
Pons	Contralateral hemiparesis, contralateral hemisensory loss, ipsilateral cerebellar ataxia, locked-in syndrome, sixth and seventh cranial nerve involvement
Medial surface of the frontal lobe	Apraxia of gait, urinary incontinence
Corpus callosum	Left-hand apraxia and agraphia, generalized tonic-clonic seizures
Thalamus	Contralateral hemisensory loss
Temporal lobe	Partial complex seizures, contralateral homonymous upper quadrantanopsia
Deep parietal lobe	Autotopagnosia, anosognosia, contralateral homonymous lower quadrantanopsia
Third ventricle	Paroxysmal headache, hydrocephalus
Fourth ventricle	Hydrocephalus, progressive cerebellar ataxia, progressive spastic hemiparesis or quadraparesis
Optic chiasm	Incongruous bitemporal field defects, bitemporal hemianopsia
Orbital frontal surface	Partial complex seizures
Internal capsule	Contralateral hemiplegia, hemisensory loss, homonymous hemianopsia
Occipital lobe	Partial seizures with elementary visual phenomena, homonymous hemianopsia with macular sparing

[a]Adapted from Gilroy (1990).

activity, although specific localized signs of disease may not be apparent. Adams and Victor (1981) noted a slowing of various symptoms associated with tumor growth, including thought processes and reaction time, increased and undue irritability, emotional lability, inertia, faulty insight, forgetfulness, reduced range of mental activity, indifference to common social practices, as well as inordinate drowsiness, apathy, and stoicism.

The most common psychological change that occurs in the presence of a neoplasm is a disturbed level of consciousness (Golden, Moses, Coffman, Miller, & Strider, 1983). When minimal, it can appear as diminished attention and concentration, faulty memory, decreased responsiveness, and easy mental fatigue. In early stages of tumor development, the degree of impairment is variable, and lucid intervals are common. Subtle changes of this type may be the first, and for long periods, the only indicator of a lesion. As the condition progresses, drowsiness may appear, followed by further decline and coma if the condition remains untreated.

More general intellectual disorganization can commonly be seen with cerebral neoplasms even in the absence of impaired consciousness (Joseph, 1990). Such difficulties can be diffuse as in a mild dementia with faulty abstractions, slowed thinking, decreased judgment, impaired recent memory, and impoverished associations.

INTRINSIC VERSUS EXTRINSIC NEOPLASMS

As has been noted, the effect of neoplasms on neuropsychological processes tends to be a function of tumor size, location, and rate of growth. As a general rule, large tumors produce more widespread deficits, both due to the direct effects of the tumor and the more widespread effects that result from increases in ICP. As tumors increase in size, there is a gradual progression in the impairment, from confusion to dementia in more advanced cases. Focal effects of neoplastic growth are determined by the specific location (Farr, Greene, & Fisher-White, 1986). Thus, virtually any neuropsychological deficit can result from a tumor, depending on the tumor's location within the cerebral hemispheres. A fast-growing tumor is likely to lead to greater disruption of neuropsychological abilities, as slower-growing neoplasms allow compensatory processes to occur.

Tumors tend to be progressive. Intrinsic tumors may grow rapidly; their onset, however, is less rapid than that of a head injury or a stroke. Extrinsic neoplasms grow very slowly, and their sequelae are frequent not seen until later in left despite the involvement of a rather large mass. This suggests that the slow increase in size allows for some adjustment in functioning.

Certain apparent differences between the performance of individuals with extrinsic tumors and that of patients with intrinsic tumors would be expected on both the Wechsler Adult Intelligence Scale (WAIS) and the Halstead–Reitan Neuropsychological Test Battery (HRND) (Farr *et al.*, 1986). Extrinsic tumors would be expected to produce less severe or less focused deficits, they should involve a greater area of the brain, and therefore, they should involve more cognitive abilities. Intrinsic tumors, conversely, should result in more dramatic performance decrements in specific neuropsychological abilities while other abilities are left virtually unaffected. Hom and Reitan (1982, 1984), however, do note that patients with intrinsic tumors sometimes have severe motor impairments corresponding with other indications of lateralized cerebral damage, but only a mild degree of cognitive deficit.

A summary of much of the quantitative data available on the WAIS and the HRNB is presented in Tables 3, 4, and 5. Interpretation of the available data requires a few general comments, however. There is little published quantitative data on the neuropsychological effects of neoplastic growths. Second, although it seems reasonable to separate tumors into subgroups depending on the hemisphere involved, in the recent past, small sample sizes and apparently limited differences by cerebral section (i.e., frontal, temporal, parietal, occipital) suggest that it may be more efficacious to combine much of the available data to try to delineate the more global aspects of impairment. Syndromes associated with more focal tumors are discussed in the following section as are the extremely limited data. Third, it also would seem that whether the neuropsychological examination was performed pre- or postsurgically should have direct implications for results found. The limited data available in the literature suggest no such differences. Tables 3, 4, and 5 in effect, offer what amounts to a generic neoplasm group.

TABLE 3. Summary of Relevant Variables in Research Studies of Patients with Tumors[a]

Study	Patient type	N		Age		Education		Patient status	Time post	Assessment device
		M	F	M	SD	M	SD			
Diffuse										
Boll (1974)	Mixed	40		42.7	11.5	10.5	3.5	—	—	WAIS, HRNB
Farr & Greene (1983)	Postsurgery (mixed)	8		28.6	7.62	—	—	Outpatient university hospital	44 months	WAIS, HRNB
Right hemisphere										
Haaland & Delaney (1981)	Presurgery	15		47	13	11	3	—	—	WAIS, HRNB
Hochberg & Slotnick (1980)	Postsurgery	6		43.2	—	—	—	Outpatients	12–30 months	HRNB
McGlone (1977)		9	0	44.8	—	11.6	—	Inpatients	3.0 months	WAIS
McGlone (1977)		0	3	46.0	—	10.9	—	Inpatients	2.5 months	WAIS
Left hemisphere										
Haaland & Delaney (1981)	Presurgery	15		47	13	11	3	—	—	WAIS, HRNB
Hochberg & Slotnick (1980)	Postsurgery	6		43.2	—	—	—	Outpatients	12–30 months	HRNB
McGlone (1977)	Presurgery	5	0	58.3	—	10.1	—	Inpatients	1.0 month	WAIS
McGlone (1977)	Postsurgery	0	6	42.5	—	11.6	—	Inpatients	1.4 month	WAIS

[a]WAIS, Wechsler Adult Intelligence Scale; HRNB, Halstead–Reitan Neuropsychological Test Battery.

TABLE 4. PERFORMANCE OF PATIENTS WITH TUMORS ON THE WECHSLER ADULT INTELLIGENCE SCALE (WAIS)[a,b]

Study	VIQ	PIQ	FSIQ	Info	Comp	Arith	Sim	DSp	Voc	DSy	PC	BD	PA	OA
Diffuse														
Boll (1974)	92.0	92.4	90.5	—	—	—	—	—	—	—	—	—	—	—
Farr & Greene (1983)	94.3	83.8	89.4	9.0	9.8	7.8	11.0	6.8	10.3	6.1	9.0	7.3	8.2	6.7
Right Hemisphere														
Haaland & Delaney (1981)	94.0	94.0	94.0	—	—	—	—	—	—	—	—	—	—	—
McGlone (1977)	106.7	92.0	—	—	—	—	—	—	—	—	—	—	—	—
McGlone (1977)	96.3	92.0	—	—	—	—	—	—	—	—	—	—	—	—
Left Hemisphere														
Haaland & Delaney (1981)	94.0	94.0	94.0	—	—	—	—	—	—	—	—	—	—	—
McGlone (1977)	84.2	92.0	—	—	—	—	—	—	—	—	—	—	—	—
McGlone (1977)	106.0	109.0	—	—	—	—	—	—	—	—	—	—	—	—
Weighted trials	95.6	93.6	92.7	9.0	9.8	7.8	11.0	6.8	10.3	6.1	9.0	7.3	9.3	6.7
N	100	100	77	8	8	8	8	8	8	8	8	8	8	8

[a]Adapted from Farr, Greene, & Fisher-White (1986).
[b]VIQ, Verbal IQ; PIQ, Performance IQ; FSIQ, Full Scale IQ; Info, Information; Comp, comprehension; Arith, Arithmetic; Sim, Similarities; DSp, Digit Span; Voc, Vocabulary; DSy, Digit Symbol; PC, Picture Completion; BD, Block Design; PA, Picture Arrangement; OA, Object Assembly.

TABLE 5. PERFORMANCE OF PATIENTS WITH TUMORS ON THE HALSTEAD–REITAN NEUROPSYCHOLOGICAL TEST BATTERY[a]

	Categories	TPT[b] total time	TPT Memory	TPT Localization	TPT time right	TPT time left	TPT time both	Rhythm (errors)	Speech perception	Tapping right	Tapping left	Trails A	Trails B	Impairment Index (Halstead)
Diffuse														
Boll (1974)	—	—	—	—	—	—	—	—	—	—	—	122.2	207.7	—
Farr & Greene (1983)	85.0	26.2	5.0	0.3	9.9	8.3	8.1	7.9	6.1	45.3	35.0	83.6	229.6	—
Right hemisphere														
Haaland & Delaney (1981)	—	—	—	—	—	—	—	—	—	46.0	43.0	—	—	0.6
Hochberg & Slotnick (1980)	85.3	21.2	7.0	1.8	8.6	7.8	5.0	—	—	47.2	39.7	—	106.8	—
Left hemisphere														
Haaland & Delaney (1981)	—	—	—	—	—	—	—	—	—	42.0	48.0	—	—	0.7
Hochberg & Slotnick (1980)	89.0	22.0	8.0	3.0	9.2	7.6	5.3	—	—	26.5	33.5	—	96.5	—
Weighted totals	85.6	23.8	6.1	1.2	9.3	8.0	6.6	7.9	6.1	43.9	42.3	115.8	196.1	0.6
Rating equivalents	3	3	1	3	2	3	4	2	1	2	2	5	4	
N	31	16	16	16	16	16	16	8	8	45	45	48	56	29

[a]Adapted from Farr, Greene, & Fisher-White (1986).
[b]TPT, Tactual Performance Test.

The limited nature of the data available makes any conclusions tentative. As can be seen in Table 4, there is very little variation among the three IQ scores, despite marked variation among the subtests. Verbal IQ is just slightly higher than performance IQ, and Verbal subtest scores tend to be higher than those on Performance subtests. Digit Span and Arithmetic are the most adversely affected Verbal subtest suggesting potential deficits in attention and/or concentration. These subtests comprise a significant portion of the WAIS freedom from distractibility factor. Performance subtest scores are much more variable and, consequently, harder to summarize. Digit Symbol, Object Assembly, and Block Design appear to be the most adversely affected subtests.

The HRNB data tend to be more variable and apparently more susceptible to the effects of the neoplasm and/or its treatment (see Table 5). Russell, Neuringer, and Goldstein (1970) ratings of the various tests in the HRNB place performance in the mild to moderate range of impairment. An exception to this is Trails A which was in the very severely impaired range with Trails B performance in the severe range. Categories, the Tactual Performance Test Total Time score, and the TPT Localization score were in the moderate range.

As can be seen, it is difficult to detect a reliable performance pattern of neuropsychological deficit in individuals with various neoplasms; however, more distinct deficits do appear to be present with respect to complex problem solving and new learning. This pattern is not unlike that which is commonly seen in those individuals who have sustained a closed head injury. Motor speed and verbal abilities seem to be intact in these study groups (Farr *et al.*, 1986).

FOCAL SYNDROMES AND NEOPLASMS

Data on the effects of tumors in specific brain regions is limited to an even greater degree than that of the more generalized effects of neoplasms on neuropsychological integrity. The studies conducted have been limited in scope and population. The information presented in this section represents a compilation of data from a variety of sources. It is important to remember, however, with very limited exceptions, that the information presented is based on very limited research. As more data are gathered, a clearer picture of cognitive functioning will likely emerge.

FRONTAL LOBE TUMORS

Depending on which portion of the frontal lobes is involved, symptoms associated with neoplastic growth can be quite variable (Levin, Eisenberg, & Benton, 1991) or strikingly absent with regard to neurological signs (Golden *et al.*, 1983). During the early stages of tumor formation, patients tend to complain of headache. During later stages, papilledema and frequent vomiting are common, usually as a result of increased ICP.

If the tumor is within the midline region, the more prominent features noted include apathy, mutism, and depressive-like features with posturing and grasp reflexes. A number of individuals with tumors in this location are misdiagnosed as being depressed (Joseph, 1990). Downward pressure on the olfactory nerve may lead to anosmia on the side of the lesion. Corticospinal fiber involvement leads to contralateral weakness that is most marked in the face and tongue (Gilroy, 1990).

As the tumor extends to the medial motor regions, the individual's lower extremity may become paralyzed.

Left frontal lobe tumors often give rise to speech disruption, some degree of motor impairment, and depressive-like symptomatology, while right frontal lobe tumors may initially result in manic and delusional symptoms (Levin *et al.*, 1991). With increasing tumor size, a progressive dementia becomes increasingly apparent (Neary & Snowden, 1991). About 50% of patients with frontal lobe tumors can be expected to develop seizures (Joseph, 1990). Lesions of the dominant hemisphere are most likely to disrupt the capacity to plan and organize goal-directed behavior. This impairment may represent a decrement in the brain's capacity to exercise what Luria (1966) refers to as the "executive functions."

Dementia may be the most frequent diagnosis in patients with frontal lobe neoplasms (Golden *et al.*, 1983). Memory disturbances also have been noted. In an early report, Hecaen and Ajuriaguerra (1956) found that "amnésie de fixation" was prominent in 10 of 80 cases. Memory failure, however, often occurs in conjunction with profound apathy making it difficult to determine if the noted difficulties represent "true" impairment or a lack of motivation.

Other notable symptoms include a loss of spontaneity, inertia, and a generalized slowing of mental efficiency (Luria, 1966). Speech may be slow and labored even in the absence of dysphasia or dominant frontal lobe involvement. Certain specific cognitive disturbances are associated with frontal lobe tumors but only in the absence of a massive or acute lesion. In diffuse lesions, the ability to organize behavior is reduced or even lost, especially in cases where the dominant frontal lobe is involved (Luria, 1966). Deficits are seen on complex tasks in which the patient must use internal language to mediate behavior and flexibly respond to the task demands (Luria, 1966). Consequently, response perseveration is commonly seen on problem-solving tests such as the Category Test (Reitan & Wolfson, 1993a) and the Wisconsin Card Sorting Test (Milner, 1963). Moreover, deficits can readily be elicited on such tests as the Stroop Color-Word Test (Perrett, 1974). This tasks requires that the patient be fluent verbally as well as suppress the effects of a distracting stimulus during performance.

Right frontal lobe deficits are much less well defined than those associated with the left frontal lobe. As a result, they are less likely to be noted during examination. Impairments, when manifest, may be seen in speech-related processes. Subtle deficits may also be present on other aspects of cognition such as musical expression (Botez & Wertheim, 1959), behavior sequencing ability, and performing complex spatial tasks (Corkin, 1965; Joseph, 1990; Milner, 1971).

Verbal fluency appears to be affected by frontal lobe tumors. Butler, Rorsman, and Hill (1993) reported that complex verbal fluency was a comparatively sensitive indicator of frontal lobe dysfunctioning; however, simple verbal fluency offered a more accurate assessment of the extent of frontal lobe involvement. More traditional assessment devices did not effectively discriminate between patients with brain disease and normal controls (Butler, Rorsman, Hill, & Tuma, 1993).

Overall, many of the traditional neuropsychological measures are insensitive to frontal lobe involvement (Bigler, 1988). In studying the results of testing with four patients with clearly documented frontal lobe lesions as a result of neoplasms, no consistent pattern of test results was noted. Tests traditionally considered to assess frontal lobe functioning were often misleading in their usual interpretation (Bigler, 1988). Damage to the frontal regions of the brain as a result of neoplastic processes can alter behavior in a multifaceted fashion, but, in terms of current knowledge and testing technology, no one clinical syndrome can encompass all the potential signs and symptoms ascribed to frontal damage. Similarly, because of this complexity, no current neuropsychological battery is going to be uniformly sensitive to frontal lobe impairment. This necessitates that the clinician use a broad spectrum of measures. The clinician must focus on behavioral and cognitive changes that accompany frontal damage. Careful observation and family/spouse interviewing become indispensable in qualifying changes because traditional psychological measures appear to be generally insensitive to alterations in these areas and individuals with frontal lobe injury are poor judges of the changes that may have occurred. Since many neuropsychological tests and tests thought to assess frontal lobe function are not uniformly affected by frontal damage, there is a lack of appropriate methods to assess the full constellation of deficits that accompany neoplastic changes in the frontal regions (Bigler, 1988).

Temporal Lobe Tumors

Tumors of the temporal lobe can cause receptive and expressive language deficits, receptive amusia and agnosia for environmental and emotional sounds, and visual field defects. Of all the cerebral tumors, those that occur in the anterior temporal lobes may result in the highest frequency of mental disorder (Golden *et al.*, 1983). This is due, in large part, to symptoms associated with temporal lobe epilepsy, although numerous cognitive deficits are present even when there is no seizure disturbance (Keschner, Bender, & Strauss, 1936). Approximately 50% of individuals with temporal lobe neoplasms experience seizures (Strobos, 1953). Apart from symptoms associated with temporal lobe epilepsy, no specific mental disturbance has been reported when the tumor is localized in the left frontotemporal region. Some investigators have reported an early onset and rapidly progressive dementia to be characteristic of frontotemporal tumors (Kerschner *et al.*, 1936); however, this may be the result of a dysphasia (Hecaen & Ajuriaguerra, 1956). Neoplastic processes that occur in the nondominant frontotemporal region may remain clinically "silent" until they are very advanced and large.

In one of the few large-scale studies available, Bingley (1958) concluded that tumors in the dominant hemisphere result in substantially greater intellectual impairment, involving both verbal and nonverbal abilities, than nondominant hemisphere tumors. Verbal slowing and decreased speech spontaneity have been reported in individuals with temporal lobe tumors. Reitan and Wolfson (1993b) report that deficits associated with temporal region malignancies can include dysnomia, spelling dyspraxia, dyslexia, dyscalculia, dysgraphia, and right–left confusion. Deficits were found to be particularly evident on the verbal sections of the Wechsler scales as well as on a number of the tests of the HRNB. In one of the few reports on right temporal lobe tumors, Bondi and Kazniak (1990) noted aphasic symptoms in a left-handed individual with a right temporal lobe astrocytoma. Memory as assessed by the Wechsler Memory Scale was normal as was performance on the Category Test, Judgement of Line of Orientation Test, and a test of facial recognition. Language impairment was noted on the Boston Diagnostic Aphasia Examination, Token Test, and Porch Index of Communicative Ability.

Tumors of the middle to posterior dominant temporal lobe are most notably featured by disturbances of language comprehension whereas nondominant temporal lobe tumors usually result in more subtle symptoms. The most common disturbances are in visual and auditory pattern analysis (Joseph, 1990; Kimura, 1967; Meier & French, 1965).

Parietal Lobe Tumors

Complex cognitive deficits are commonly associated with parietal lobe tumors (Joseph, 1990). A dysphasia or ideational dyspraxia may occur in individuals if the lesion is in the dominant hemisphere (Brown, 1974). The Gerstmann syndrome, consisting of finger agnosia, dyscalculia, dysgraphia, and right–left confusion, also may be seen as a circumscribed disorder or as part of a more global cognitive disruption (Berg, 1988). Nondominant parietal lobe neoplasms are more likely to be associated with visual–spatial disorientation, dressing dyspraxia, and topographic disturbance (Brown, 1974; Joseph, 1990). Nondominant tumors have also been associated with disorders of body image and awareness. These disturbances can include denial of the existence of a limb paralyzed secondary to a neurological lesion as well as attribution of the affected limb to another individual.

Parietal lobe neoplasms are invariably associated with evidence of tactile, kinesthetic, and proprioceptive disorders. These disturbances can include the loss or inability to analyze sensation, locate body parts, recognize objects by touch, and recognize the location of sensation of the body (Chusid, 1973; Golden, 1981; Reitan & Wolfson, 1993). A large lesion in either parietal lobe may result in a bilateral disorder, particularly in cases of a rapidly growing tumor and where the dominant hemisphere is involved (Corkin, Milner, & Taylor, 1973). Parietal postcentral tumors are also associated with hypotonia and wasting of various parts of the body, a phenomenon referred to as "parietal wasting."

Occipital Tumors

The dominant symptom associated with occipital lobe tumors is a visual halffield impairment. Small tumors may result in a scotoma. Tumors located in the dominant occipital lobe commonly produce a loss of verbal–visual skills and a visual agnosia (Brown, 1974). Occipital neoplasms tend not to lead to a disproportionate amount of emotional disorders compared to tumors in other brain locations (Lishman, 1978).

Cerebral Tumors in Children

Several important factors have relevance for the study of neurobehavioral effects in childhood neoplasms, including age at diagnosis and treatment, and the location, extent, and permanence of the lesion (Boll & Barth, 1981). The presence of hydrocephalus or undergoing surgery, irradiation, and chemotherapy also may have a substantial impact on functional cognitive abilities. The assessment of children surviving brain tumors has traditionally been restricted to a global clinical evaluation of neurological status at various intervals following treatment, without reference to cognitive factors.

It has been estimated that between 70% to 80% of surviving children function at or near their premorbid level (Abramson, Raben, & Cavanaugh, 1974; Bloom, Wallace, & Henk, 1969; Gjerris, 1976). In-depth evaluations have, on the other hand, found a greater prevalence of functional impairment after treatment (Fennell *et al.,* 1993; Fletcher & Copeland, 1988; Kun, Mulhern, & Crisco, 1983).

AGE AT DIAGNOSIS. The importance of the developmental status of the CNS at the time of injury is still a cause for debate in neuropsychology. Evidence in support of, as well as opposition to, the contention that the immature brain is more "plastic" with respect to cognitive recovery and future development of psychological functioning has been extensively documented (Chelune & Edwards, 1981; Satz & Fletcher, 1981; St. James-Roberts, 1979). A review of the literature on childhood brain lesions suggests that age at the time of injury and type of lesion (focal vs. diffuse) may interact to determine the level of cognitive impairment. Young children with focal injuries tend to demonstrate better recovery of functioning than older children (Chelune & Edwards, 1981).

The research on the outcome of brain neoplastic processes in childhood has not systematically addressed the issue of age at the time of diagnosis (Mulhern, Crisco, & Kun, 1983; Stehbens *et al.,* 1991). In one of the few studies investigating age effects, Kun and colleagues (1983) were unable to find a relationship between age at diagnosis and subsequent neuropsychological function; however, clinically apparent deficiencies in attention and concentration were more frequently observed in younger children. A more recent review by Stehbens and colleagues (1991) suggests that diagnosis and treatment prior to 5 years of age may result in a greater prevalence of various learning disorders.

TUMOR LOCATION. The relationship between the location of cerebral neoplasms and neurobehavioral functioning in children has not been extensively investigated. Among those studies that classified children by tumor location, the investigations were confined primarily to patients with infratentorial tumors (e.g., Berry, Jenkin, Keen, Nair, & Simpson, 1981; Tamaroff, Salwen, Allen, Murphy, & Nir, 1982). The prevalence of neurological deficit in children with these types of lesions is quite variable. This is due, in part, to the diversity of treatment strategies such as the differential use of cranial rather than local or no irradiation and the use of chemotherapy for advanced or especially malignant neoplasms. Infratentorial tumors treated by surgery result in a 27%–59% incidence of neuropsychological deficit and behavioral disturbance (Hirsch, Reneir, Czernichow, Benveniste, & Pierre-Kahn, 1979; Ron, 1988).

Supratentorial tumors tend to result in a greater prevalence of neurological and neuropsychological deficit than is found in infratentorial neoplasms (Gjerris, 1976). Kun and co-workers (1983) reported twice the incidence of dysfunction in supratentorial tumors as compared to infratentorial brain tumors. Deficits observed in children in these studies are highly varied and include visuospatial, attentional, and auditory processing impairment (Mulhern *et al.,* 1983). A large discrepancy between Verbal and Performance IQ scales also has been noted; however, this finding is not consistent across studies (Mulhern *et al.,* 1983).

A retrospective study of the neurobehavioral sequelae of childhood craniopharyngioma (Kerr, Smith, DaSilva, Hoffman, & Humphries, 1991) found normal functioning on measures of intelligence. Moderate to severe impairments in delayed recall of verbal or nonverbal information were seen in about 45% of patients studied. Frontal lobe functioning as assessed by the Wisconsin Card Scoring Test was noted to be intact in 67% of those children tested. Kerr and colleagues concluded that neuropsychological deficits are not inevitable consequences associated with treatment of the tumor.

CHILDHOOD LEUKEMIA

Although not a neoplasm in the strictest sense, childhood leukemia deserves a brief mention in this discussion. Childhood acute lymphocytic leukemia (ALL) is the most widely studied of all cancers (adult and child) with respect to neuropsychological functioning. A number of early reports have indicated that CNS prophylaxis produces long-term cognitive deficits, particularly when radiation is combined with chemotherapy (Eiser, 1980; Eiser & Lansdown, 1977). Goff and colleagues (1980) noted a distinct pattern of attention and concentration deficits in long-term leukemia survivors, and language-related abilities have also been reported to decline in children with leukemia (Moehle, Berg, Ch'ien, & Lancaster, 1983).

Fletcher and Copeland (1988) reviewed 41 studies of the effect of prophylactic cranial irradiation and concluded that CNS irradiation impairs cognitive development. In a comprehensive review of the cognitive effects of irradiation in children with ALL, Cousens, Waters, and Stevens (1988) surveyed 31 studies and performed a meta-analysis of pre- and post-IQ comparisons. The results demonstrated a substantial average decrement (more than two thirds of one standard deviation) in intellectual functioning in irradiated subjects compared to a variety of controls not receiving CNS prophylactic irradiation. The specific dysfunctions suggested on IQ measures included short-term memory impairments, distractibility, specific processing deficits, and disturbances in motor speed and perception. Conversely, Williams and Davis (1986), in a large review paper, reported that adverse effects of cranial irradiation on the developing child's intellectual abilities have not been conclusively demonstrated. Investigations also have indicated that children who are receiving systemic and intrathecal chemotherapy without CNS irradiation may be at risk for subsequent learning disorders (Brown *et al.*, 1990) particularly those involving right hemisphere simultaneous processing (Brown *et al.*, 1992).

NEUROPSYCHOLOGICAL MEASURES

As can be inferred from the paucity of comprehensive studies investigating the neuropsychological effects of neoplasms and the treatment modalities used in the care of neoplastic disorders, there is likely some debate as to what constitutes the appropriate measure to use in the investigation of tumor-related neurobehavioral dysfunction. As is the case in the study of HIV and its related deficits, some research advocates a comparatively brief screening battery of 25 to 30 minutes duration (e.g., Miller *et al.*, 1990), while other investigators employ batteries that require several hours (e.g., McCaffrey *et al.*, 1991; Saykin *et al.*, 1988). Not surprisingly, the likelihood of identifying a problem is related to the time spent in examination of that problem. Conversely, studies employing large test batteries must protect against the spurious findings associated with the administration of a large number of tests (Bornstein, 1994). It is almost always necessary to strike a balance between the breadth and depth of an examination, and the amount of time before the patience and cooperation of research subjects are exhausted. Studies in which neurocognitive issues are not the primary focus usually face significant constraints in the time allotted for a neuropsychological examination. In addition, individuals who have been diagnosed with, or are receiving treatment for, a neoplasm are often physically ill and unable to complete a lengthy test battery.

Perhaps the most comprehensive neuropsychological battery for this purpose has been offered by Butters and colleagues (1990) in the study of HIV-related cognitive deficits. Recognizing the impracticality of such a battery, a 1- to 2-hour screening battery was proposed by Butters and colleagues that was felt to be more feasible in terms of patient tolerance and requirements for the investigation of other areas of inquiry such as psychosocial and psychiatric sequelae of the disease. This approach can be generalized to patients with neoplastic processes. A comprehensive screening battery must, at a minimum, tap all areas of neuropsychological functioning—premorbid intelligence, attention, speed of processing, memory, abstraction, language, visual–perceptual processing, constructional abilities, motor abilities, and psychiatric status. The battery proposed by Butters and co-workers (1990) incorporates traditional measures and emphasizes divided and sustained attention, speed of processing, and retrieval from working and long-term memory, however, certain aspects of cognitive functioning have not been addressed. For this reason, a proposed screening battery of tests is presented in Table 6. All tests were drawn from the original comprehensive battery proposed by Butters et al. (1990).

The pertinence of neuropsychological test findings to daily life also requires consideration. In some studies reporting mean differences among groups of patients with tumors, the means of the test scores are still within normal limits (e.g., Berg et al., 1983; Kerr et al., 1991). The interpretation of mean differences may depend to some extent on demonstrating that such differences actually reflect compromise of daily skills and functions. Bornstein (1994) recommends that studies of neurobehavioral dysfunction include ratings by patients and/or others of daily functioning. One such measure, the Sickness Impact Profile (Bergner, Bobbit, Pollard, Martin, & Gibson, 1976), provides subjective ratings of the impact of symptoms on various aspects of daily life. Demonstration of a relationship between these ratings and neuropsychological dysfunction would lend credence to the notion that deficits identified on neuropsychological tests do affect an individual's daily functioning. (An extended discussion of the ecological validity of neuropsychological techniques can be found in Tupper and Cicerone [1990]).

In the relatively few studies conducted to evaluate the question of neuropsychological dysfunction in patients with tumors, most methods have involved con-

TABLE 6. PROPOSED SCREENING BATTERY FOR ASSESSMENT OF NEOPLASTIC PROCESSES[a]

Cognitive domain	Test
Indication of premorbid intelligence	Vocabulary (Wechsler Adult Intelligence Scale-Revised)
Memory	California Verbal Learning Test
Speed of Processing	Paced Auditory Serial Addition Test (PASAT)
Attention	Visual Memory Span (Wechsler Memory Scale-Revised)
Abstraction	Trail Making Test, Parts A & B
Language	Boston Naming Test
Visual–spatial processing	Digit Symbol Substitution
Construction abilities	Block Design (Wechsler Adult Intelligence Scale-Revised)
Motor abilities	Grooved Pegboard
Psychiatric status	Hamilton Depression Scale

[a]Adapted from Butters et al. (1990).

ventional parametric analyses of group means. It is likely that there is considerable variability in the performance assessed. It becomes necessary, therefore, to examine the homogeneity of variance before using parametric procedures. In those cases where parametric techniques are inappropriate, nonparametric statistics are indicated and may demonstrate differences not readily apparent by the misapplication of parametric procedures (Franzen, 1989).

In addition to the analysis of measures of central tendency, it is becoming increasingly important to determine the extent to which those means are reflective of the individuals who compose the groups. Because of the likely heterogeneity in the nature and pattern of neuropsychological dysfunction in those with cerebral neoplastic processes, analysis of group means should be accompanied by analysis of the proportion of groups who exceed some defined criterion level for impairment (Bornstein, 1994). Such analyses can help to better define the frequency of dysfunction in these various areas and provide a more accurate characterization of the pattern of cognitive deficits associated with cerebral tumors.

The criteria used to define impaired performance are a central issue in the interpretation of neuropsychological performance and have been the topic of extensive debate. A wide range of criteria have been used that differ markedly in their threshold for identification of impairment over the past several years. Bornstein (1994) notes that most studies have defined impairment in terms of deviations from some standard mean score while others have been based on clinical ratings by experienced clinicians. To a large extent, the criteria used in any given setting is a function of the conceptual model of neuropsychological dysfunction underlying the investigation. If, for example, a study is investigating the problem at hand as a true dementia, more severe levels of impairment will be required to satisfy a criterion. Conversely, when investigating more subtle dysfunction, less severe levels of impairment will be needed to reach the criterion. There are three principal factors inherent in the definition of criteria for impairment: the reference base against which the subjects are to be compared, the extent of deviation from the reference base, and the proportion of measures with the appropriate levels of deviation required to satisfy the criteria.

In most cases it is preferable to base definitions of "normal" performance on a group of control subjects that is large enough and as well-matched as possible to the study group of interest. Some studies are based either implicitly or explicitly on normal samples whose concordance with subjects with cerebral neoplasms on important characteristics is typically not completely known. Also, the normative data on various tests comprising of test battery are usually based on different subject groups which raises the problem of cohort effects (Franzen, 1989). The use of well-matched controls to define normal performance on the broad range of measures avoids these difficulties.

The extent of deviation from normative standards required to meet the criteria for impairment is linked to the conceptual model underlying the investigation. Models using a subtle degree of impairment tend to use deviations of one standard deviation from the control mean whereas models hypothesizing a more severe impairment have used a standard deviation of 1.5 or 2 from the mean as the criterion for impairment (Bornstein, 1994). With increasingly stringent criteria for impairment, the proportion of individuals identified as impaired will decrease. This becomes an important issue when attempting to identify impairments in the individual.

Almost all investigations of the effects of cancer on cognitive functioning have required evidence of impairment on more than one measure for an individual's neuropsychological performance to be considered abnormal. Some studies have grouped the measures according to domains of function and require impairment in two or more areas to meet the criteria for overall impairment. Other studies have focused on individual test measures and require a certain number of tests to be abnormal by various criteria in order for an individual to be identified as impaired. It is also possible to base criteria for impairment on a summary neuropsychological rating measure. Problems associated with impairment criteria that focus on individual measures are the considerable overlap among various tests and the number of variables that may be derived from an individual test (Franzen, 1989; Bornstein, 1994). It may be necessary, therefore, to combine measures from particular tests or groups of tests to avoid undue influence of any particular test or area of dysfunction on estimates of impairment.

THE FUTURE

Given the research available to date, however limited in its scope, there is little doubt that neuropsychological dysfunction can be associated with cerebral neoplastic processes. While little consistent evidence exists as yet for distinct performance deficit patterns with specific types and locations of tumors, there is sufficient evidence from a number of studies to conclude that at the very least, subtle forms of neurobehavioral dysfunction can be observed in a number of individuals with neoplasms. To some extent, the current lack of consistent findings can be attributed to different conceptualizations of the problem and to the resulting methodologies derived from those models. Many of the studies appear to be addressing different questions, for example, the prevalence of severe deficits (e.g., dementia) versus subtle impairment. This entails the use of differing measures; therefore it becomes somewhat less surprising that different findings are obtained.

Continued arguments about which battery or tests to use or what proportion of patients are impaired would appear to be counterproductive. Rather, focusing on the nature and pattern of impairment may be more productive as would attempts to identify the factors that influence the rate of neuropsychological deterioration. There have been no reported investigations of different patterns of initial deficit or the potential prognostic significance of those different patterns. The majority of studies have been conducted retrospectively. Additional prospective, longitudinal examinations would help to identify the course of impairment.

There are a number of conceptual and methodological factors that have defined and contributed to the debate regarding the incidence and prevalence of neuropsychological dysfunction accompanying cerebral neoplasms. There is sufficient evidence to indicate that individuals who have neoplasms have cognitive impairment; however, it remains unclear as to whether the deficits to be expected are subtle or represent significant dementia. It is becoming less productive to continue to debate whether a certain proportion of patients with cerebral cancer will develop deficits. Further investigations addressing the more essential questions about the nature of the cognitive deterioration will increase the understanding of a series of disease processes that have had and will continue to have a significant impact on public health. As the number of survivors of brain tumors increases, issues regarding quality of life and functional capabilities will become paramount.

REFERENCES

Abramson, N., Raben, M., & Cavanaugh, P. J. (1974). Brain tumors in children: Analysis of 136 cases. *Radiology, 112*, 669–672.

Adams, R. D., & Victor, M. (1981). *Principles of neurology.* New York: McGraw-Hill.

Benton, A. L. (1991). The prefrontal region: Its early history. In H. S. Levin, H. M. Eisenberg, & A. L. Benton (Eds.), *Frontal lobe function and dysfunction.* New York: Oxford University Press.

Berg, R. A. (1988). Cancer. In R. E. Tarter, D. H. Van Thiel, & K. L. Edwards (Eds.), *Medical neuropsychology: The impact of disease on behavior.* New York: Plenum Publishing Corporation.

Berg, R. A., Ch'ien, L. T., Bowman, W. P., Ochs, J., Lancaster, W., Goff, J. R., & Anderson, H. R. Jr. (1983). The neuropsychological effects of acute lymphocytic leukemia and its treatment—A three year report: Intellectual functioning and academic achievement. *Clinical Neuropsychology, 5*, 9–13.

Bergner, M., Bobbit, R. A., Pollard, W. E., Martin, D. P., & Gibson, B. S. (1976). The sickness impact profile: Validation of a health status measure. *Medical Care, 14*, 57–67.

Berry, M. P., Jenkin, D. T., Keen, G. W., Nair, B. D., & Simpson, W. J. (1981). Radiation for medulloblastoma. *Journal of Neurosurgery, 55*, 43–51.

Bigler, E. D. (1988). Frontal lobe damage and neuropsychological assessment. *Archives of Clinical Neuropsychology, 3*, 279–298.

Bingley, T. (1958). Mental symptoms in temporal lobe epilepsy and temporal lobe glioma. *Acta Psychiatrica et Neurologica Scandinavica, Supplement 120*, 1–151.

Bloom, H. J. G., Wallace, E. N. K., & Henk, J. M. (1969). The treatment and prognosis of medulloblastoma in children. *American Journal of Roentgenology, 105*, 43–62.

Boll, T. J. (1974). Psychological differentiation of patients with schizophrenia versus lateralized cerebrovascular, neoplastic, or traumatic brain damage. *Journal of Abnormal Psychology, 83*, 456–458.

Boll, T. J. (1985). Developing issues in clinical neuropsychology. *Journal of Clinical and Experimental Neuropsychology, 7*, 475–485.

Boll, T. J., & Barth, J. T. (1981). Neuropsychology of brain damage in children. In S. B. Filskov & T. J. Boll (Eds.), *Handbook of clinical neuropsychology.* New York: Wiley.

Bondi, M. W., & Kaszniak, A. W. (1990). Language functioning following excision of a right hemisphere astrocytoma. National Academy of Neuropsychology Abstracts of the Annual Meeting, Washington, DC. *Archives of Clinical Neuropsychology, 5*, 157.

Bordeaux, J. D., Dowell, R. E. Jr., Copeland, D. R., Fletcher, J. M., Francis, D. J., & van Eys, J. (1988). A prospective study of neuropsychological sequelae in children with brain tumors. *Journal of Child Neurology, 3*, 63–68.

Bornstein, R. A. (1994). Methodological and conceptual issues in the study of cognitive change in HIV infection. In I. Grant and A. Martin (Eds.), *Neuropsychology of HIV infection.* New York: Oxford University Press.

Botez, M. I., & Wertheim, N. (1959). Expressive aphasia and amnesia following right frontal lesion in a right-handed man. *Brain, 82*, 186–202.

Brickner, R. M. (1934). An interpretation of frontal lobe function based upon the study of a case of bilateral frontal lobectomy. In S. T. Orton, J. F. Fulton, & T. K. Davis (Eds.), *Localized function in the cerebral cortex.* Baltimore, MD: Williams & Wilkins.

Brickner, R. M. (1936). *The intellectual functions of the frontal lobes.* New York: Macmillan.

Brown, J. W. (1974). *Aphasia, apraxia, and agnosia: Clinical and theoretical aspects.* Springfield, IL: Charles C. Thomas.

Brown, R. T., Madan-Swain, A., Paris, R., Lambert, R. G., Baldwin, K., Casey, R., Frank, N., Sexson, S. B., Ragab, A., & Kamphaus, R. W. (1992). Cognitive status of children treated with central nervous system prophylactic chemotherapy for acute lymphocytic leukemia. *Archives of Clinical Neuropsychology, 7*, 481–497.

Brown, R. T., Madan-Swain, A., Ragab, A., Good, L., Wilson, L., & Pais, R. (1990). Sequential and simultaneous processing in pediatric oncology patients treated with chemotherapy. National Academy of Neuropsychology Ninth Annual Meeting Abstracts. *Archives of Clinical Neuropsychology, 5*, 158.

Butler, R. W., Rorsman, I., & Hill, J. M. (1993). The effects of frontal brain tumors on simple and complex fluency. Division of Clinical Neuropsychology of the American Psychological Association Abstracts of the Annual Meeting, Toronto, Ontario, August, 1993. *The Clinical Neuropsychologist, 7*, 336.

Butler, R. W., Rorsman, I., Hill, J. M., & Tuma, R. (1993). The effects of frontal brain impairment on fluency; Simple and complex paradigms. *Neuropsychology, 7*, 519–529.

Butters, N., Grant, I., Haxby, J., Judd, L. L., Martin, A., McClelland, J., Pequegnat, W., Schacter, D., &

Stover, E. (1990). Assessment of AIDS-related cognitive changes: Recommendations of the NIMH workshop on neuropsychological assessment approaches. *Journal of Clinical and Experimental Neuropsychology, 12*, 963–978.

Cavanaugh, S. V., & Wettstein, R. M. (1989). Emotional and cognitive dysfunction associated with medical disorders. *Journal of Psychosomatic Research, 33*, 505–514.

Chelune, G. J., & Edwards, P. (1981). Early brain lesions: Ontogenetic-environmental considerations. *Journal of Consulting and Clinical Psychology, 49*, 777–790.

Chusid, J. G. (1973). *Correlative neuroanatomy and functional neurology* (15th ed.). Los Altos, CA: Lange Medical Publications.

Copeland, D. R., Dowell, R. E. Jr., Fletcher, J. M., Bordeaux, J. D., Sullivan, M. P., Jaffee, N., Frankel, L. S., Reid, H. L., & Cangir, A. (1988). Neuropsychological effects of childhood cancer treatment. *Journal of Child Neurology, 3*, 53–62.

Corkin, S. (1965). Tactually guided maze-learning in man. Effects of unilateral cortical excisions and bilateral hippocampal lesions. *Neuropsychologia, 3*, 339–351.

Corkin, S., Milner, B., & Taylor, L. (1973). Bilateral sensory loss after unilateral cerebral lesion in man. *Transactions of the American Neurological Association, 98*, 118.

Cousens, P., Waters, J. S., & Stevens, M. (1988). Cognitive effects of cranial irradiation in leukemia: A survey and meta-analysis. *Journal of Child Psychology and Psychiatry, 29*, 839–852.

Davis, B. D., Fernandez, F., Adams, F., Holmes, V., Levy, J. K., Lewis, D., & Neidhart, J. (1987). Diagnosis of dementia in cancer patients. *Psychosomatics, 28*, 175–179.

den Hollander, A., Van Hulst, A., Meerwaldt, J., & Haasjes, J. (1989). Limbic encephalitis: A rare presentation of the small-cell lung carcinoma. *General Hospital Psychiatry, 11*, 388–392.

Derogatis, L. R., Morrow, S. R., & Fettig, J. (1983). The prevalence of psychiatric disorders among cancer patients. *Journal of the American Medical Association, 249*, 751–757.

Eiser, C. (1980). Effects of chronic illness on intellectual development. *Archives of Disease in Children, 55*, 766–770.

Eiser, C., & Lansdown, R. (1977). Retrospective study of intellectual development in children treated for acute lymphoblastic leukemia. *Archives of Disease in Children, 52*, 525–529.

Farr, S. P., & Greene, R. L. (1983). *Disease type, onset, and process and its relationship to neuropsychological performance.* Unpublished manuscript.

Farr, S. P., Greene, R. L., & Fisher-White, S. P. (1986). Disease process, onset, and course and their relationship to neuropsychological performance. In S. B. Filskov & T. J. Boll (Eds.), *Handbook of clinical neuropsychology* (Vol. 2). New York: Wiley.

Fennell, E. B., Mann, L. W., Maria, B., Fiano, K., Booth, M., Mickle, J. P., & Quisling, R. (1993). Neuropsychology of posteriorfossa versus other focal brain tumors in children. International Neuropsychological Society Annual Meeting Abstracts. *Journal of Clinical and Experimental Neuropsychology, 15*, 58.

Fletcher, J. M., & Copeland, D. R. (1988). Neurobehavioral effects of central nervous system prophylactic treatment of cancer in children. *Journal of Clinical and Experimental Neuropsychology, 10*, 495–538.

Folstein, M. F., Fetting, J. H., Lobo, A., Niaz, U., & Capozzoli, K. D. (1984). Cognitive assessment of cancer patients. *Cancer, 31*, 2250–2257.

Franzen, M. D. (1989). *Reliability and validity in neuropsychological assessment.* New York: Plenum Publishing Corporation.

Gilroy, J. (1990). *Basic neurology* (2nd ed.). New York: Pergamon Press.

Gilroy, J., & Meyer, J. S. (1975). *Medical neurology* (2nd ed.). New York: Macmillan.

Gjerris, F. (1976). Clinical aspects and long-term prognosis of intracranial tumors in infancy and childhood. *Developmental Medicine and Child Neurology, 18*, 145–159.

Goff, J. R., Anderson, H. R., & Cooper, P. F. (1980). Distractibility and memory deficit in long-term survivors of acute lymphoblastic leukemia. *Journal of Developmental and Behavioral Pediatrics, 1*, 158–161.

Golden, C. J. (1981). *Diagnosis and rehabilitation in clinical neuropsychology* (2nd ed.). Springfield, IL: Charles C. Thomas.

Golden, C. J., Moses, J. A. Jr., Coffman, J. A., Miller, W. R., & Strider, F. D. (1983). *Clinical neuropsychology: Interface with neurologic and psychiatric disorders.* New York: Grune & Stratton.

Haaland, K. Y., & Delaney, H. D. (1981). Motor deficits after left or right hemisphere damage due to stroke or tumor. *Neuropsychologia, 19*, 17–27.

Hecaen, H., & Ajuriaguerra, J. (1956). *De troubles mentaux au cours des tumeurs intracraniennes.* Paris: Masson.

Hirsch, J. F., Reneir, D., Czernichow, R., Benveniste, L., & Pierre-Kahn, A. (1979). Medulloblastoma in childhood. Survival and functional results. Survival and functional results. *Acta Neurochirurgica, 48,* 1–15.

Hochberg, F. H., & Slotnick, B. (1980). Neuropsychologic impairment in astrocytoma survivors. *Neurology, 30,* 172–177.

Hom, J., & Reitan, R. M. (1982). Effect of lateralized cerebral damage upon contralateral and ipsilateral sensorimotor performances. *Journal of Clinical Neuropsychology, 4,* 249–268.

Hom, J., & Reitan, R. M. (1984). Neuropsychological correlates of rapidly vs. slowly growing intrinsic neoplasms. *Journal of Clinical Neuropsychology, 6,* 309–324.

Hwang, T., Yung, A. W. K., Estey, E. H., & Fields, W. S. (1985). Central nervous system toxicity with high-dose Ara-C. *Neurology, 35,* 1475–1479.

Jastrowitz, M. (1888). Beiträge zur localization im grosshirn und über deren praktische verwerthung. *Deutsche Medizinische Wochenschrift, 14,* 81–83.

Joseph, R. (1990). *Neuropsychology, neuropsychiatry, and behavioral neurology.* New York: Plenum Press.

Kerr, E. N., Smith, M. L., DaSilva, M., Hoffman, H. J., & Humphries, R. P. (1991). Neuropsychological effects of craniopharyngioma treated by microsurgery. International Neuropsychological Society Annual Meeting Abstract. *Journal of Clinical and Experimental Neuropsychology, 13,* 57.

Keschner, M., Bender, M. B., & Strauss, I. (1936). Mental symptoms in cases of the tumor of the temporal lobe. *Archives of Neurology and Psychiatry, 110,* 572–596.

Kimura, D. (1967). Right temporal lobe damage: Perception of unfamiliar stimuli after damage. *Archives of Neurology, 8,* 264–271.

Kun, L. E., Mulhern, R. K., & Crisco, J. J. (1983). Quality of life in children treated for brain tumors: Intellectual, emotional, and academic function. *Journal of Neurosurgery, 58,* 1–6.

Levin, H. S., Eisenberg, H. M., & Benton, A. L. (1991). *Frontal lobe function and dysfunction.* New York: Oxford University Press.

Lishman, W. (1978). *Organic psychiatry: The psychological consequences of cerebral disorder.* Oxford: Blackwell.

Lorig, T. S. (1992). Cardiovascular and somatic disorders. In A. E. Puente & R. J. McCaffrey (Eds.), *Handbook of neuropsychological assessment: A biopsychosocial perspective.* New York: Plenum Press.

Luria, A. R. (1966). *Higher cortical functions in man.* New York: Basic Books.

Matarazzo, J. (1972). *Wechsler's measurement and appraisal of adult intelligence.* Baltimore, MD: Williams & Wilkins.

Mauer, A. E. (1980). Therapy of acute lymphocytic leukemia in childhood. *Blood, 56,* 1–10.

McCaffrey, R. J., Orsillo, S. M., Lefkowicz, D. P., Ortega, A., Haase, R. F., Wagner, H., & Ruckdeschel, J. C. (1991). Neuropsychological sequelae of chemotherapy and prophylactic cranial irradiation: An extension of earlier findings. National Academy of Neuropsychology Abstracts of the Annual Meeting, Washington, DC. *Archives of Clinical Neuropsychology, 6,* 205.

McGlone, J. (1977). Sex differences in cerebral organization of verbal functions in patients with unilateral brain lesions. *Brain, 100,* 775–793.

Meier, M. J., & French, L. A. (1965). Lateralized deficits in complex visual discrimination and bilateral transfer of reminiscence following unilateral temporal lobotomy. *Neuropsychologia, 3,* 261–272.

Miller, E. N., Selnes, O. A., McArthur, J. C., Satz, P., Becker, J. T., Cohen, B. A., Sheridan, K., Machado, A. M., Van Gorp, W. G., & Visscher, B. (1990). Neuropsychological performance in HIV-1 infected homosexual men: The Multicenter AIDS Cohort Study (MACS). *Neurology, 40,* 197–203.

Milner, B. (1963). Effects of different brain lesions on card scoring. *Archives of Neurology, 9,* 90.

Milner, B. (1971). Interhemispheric differences in the localization of psychological processes in man. *British Medical Bulletin, 27,* 272.

Moehle, K. A., Berg, R. A., Ch'ien, L. T., & Lancaster, W. (1983). Language-related skills in children with acute lymphocytic leukemia. *Journal of Developmental and Behavioral Pediatrics, 4,* 257–261.

Mulhern, R. K., Crisco, J. J., & Kun, L. E. (1983). Neuropsychological sequelae of childhood brain tumors: A review. *Journal of Clinical Child Psychology, 12,* 66–73.

Neary, D., & Snowden, J. S. (1991). Demential of the frontal lobe type. In H. S. Levin, H. M. Eisenberg, & A. L. Benton (Eds.), *Frontal lobe function and dysfunction.* New York: Oxford University Press.

Oxman, T. E., & Silverfarb, P. M. (1980). Serial cognitive testing in cancer patients receiving chemotherapy. *American Journal of Psychiatry, 137,* 1263–1265.

Perret, E. (1974). The left frontal lobe of man and the suppression of habitual responses in verbal categorical behavior. *Neuropsychologia, 12,* 323–330.

Reitan, R. M., & Wolfson, D. (1993). *The Halstead–Reitan Neuropsychological Test Battery: Theory and clinical interpretation* (2nd ed.). Tucson, AZ: Neuropsychology Press.

Reitan, R. M., & Wolfson, D. (1993b). *Neuropsychological evaluation of older children.* Tucson, AZ: Neuropsychology Press.

Ron, E. (1988). Tumors of the brain and nervous systems after radiotherapy in childhood. *New England Journal of Medicine, 319,* 1033–1039.

Russell, E. W., Neuringer, C., & Goldstein, G. (1970). *Assessment of brain damage: A neuropsychological key approach.* New York: John Wiley & Sons.

Satz, P., & Fletcher, J. M. (1981). Emergent trends in neuropsychology: An overview. *Journal of Consulting and Clinical Psychology, 49,* 851–865.

Saykin, A. J., Janssen, R. A., Sprehn, G. C., Kaplan, J. E., Spira, T. J., & Weller, P. (1988). Neuropsychological dysfunction in HIV infection: Characterization in a lymphadenopathy cohort. *International Journal of Clinical Neuropsychology, 10,* 81–95.

Shapiro, J. R. (1986). Biology of gliomas: Heterogeneity, oncogenes, growth factors. *Seminars in Oncology, 13,* 4–15.

Stehbens, J. A., Kaleita, T. A., Noll, R. B., MacLean, W. E. Jr., O'Brien, R. T., Waskerwitz, M. J., & Hammond, G. D. (1991). CNS prophylaxis of childhood leukemia: What are the long-term neurological, neuropsychological, and behavioral effects? *Neuropsychology Review, 2,* 147–177.

St. James-Roberts, I. (1979). Neurological plasticity, recovery from brain insult and child development. In H. W. Reese (Ed.), *Advances in child development and behavior* (Vol. 14). New York: Academic Press.

Strobos, R. R. J. (1953). Tumors of the temporal lobe. *Neurology, 3,* 752–760.

Tamaroff, M., Salwen, R., Allen, J., Murphy, M., & Nir, Y. (1982). Neuropsychological test performance in children treated for malignant tumors of the posterior fossa. *American Society of Clinical Oncology Abstracts, 23,* 50.

Tupper, D. E., & Cicerone, K. D. (1990). *The neuropsychology of everyday life: Assessment and basic competencies.* Norwell, MA: Kluwer Academic Publishers.

Weiss, H. D., Walker, M. D., & Wienik, P. H. (1974). Neurotoxicity of commonly used antineoplastic agents. *New England Journal of Medicine, 291,* 75–127.

Wilkinson, P. C. (1964). Serological findings in carcinomatous neuromyopathy. *Lancet, 1,* 1301–1303.

Williams, J. M., & Davis, K. S. (1986). Central nervous system prophylactic treatment to childhood leukemia: Neuropsychological outcome studies. *Cancer Treatment Reviews, 13,* 1–15.

Wilkinson, P. C., & Zeromski, J. (1965). Immunofluorescent detection of antibodies against neurons in sensory carcinomatous neuropathy. *Brain, 88,* 529–538.

Young, J. L., Percy, C. L., Asire, A. J., Berg, J. W., Cusano, M. M., Gloecker, L. A., Horm, J. W., Lourie, W. I., Pollack, E. S., & Shambaughm, E. M. (1982). Cancer incidence and mortality in the United States, 1973–1977 (Monograph No. 57). Washington, DC: National Cancer Institute.

13

Evaluation of Patients with Epilepsy

Michelle C. Dolske, Gordon J. Chelune, and Richard I. Naugle

Introduction

The study of epilepsy may have contributed more to the understanding of human brain–behavior relationships than the study of any other central nervous system disorder (Novelly, 1992). J. Hughlings Jackson was among the first scientists to relate clinical or behavioral observations to brain morphology (Reynolds, 1988). Jackson maintained it was possible to ascertain the neuroanatomical location of a focal seizure focus by observing the physical progression of the seizure (Haynes & Bennett, 1992). He subsequently developed the hypothesis that epilepsy was caused by abnormal discharges from lesions in the brain, establishing a theory of epilepsy that has led to our contemporary understanding of epilepsy (Reynolds, 1988). Further advances in understanding localization of brain functioning were made through cortical stimulation performed on conscious patients undergoing neurosurgery for removal of epileptogenic lesions. This procedure allowed for the development of somatotopic maps of the peri-rolandic motor–sensory cortex and the concept of the homunculus (Penfield & Rasmussen, 1950). The study of children undergoing hemispherectomies to treat intractable epilepsy has contributed to our understanding of the plasticity of speech and language. Specifically, critical periods for developing speech and language and transfer of speech and language following hemispherectomy have been identified (Goodman & Whitaker, 1985). Models of memory have been studied extensively among epilepsy patients undergoing temporal lobectomy and hippocampus removal. Studies from as early as the 1950s

Michelle C. Dolske Florida Hospital, Medical Psychology Section, Orlando, Florida 32804. Gordon J. Chelune and Richard I. Naugle Department of Psychiatry and Psychology, Cleveland Clinic Foundation (P-57), Cleveland, Ohio 44195.

Neuropsychology, edited by Goldstein *et al.* Plenum Press, New York, 1998.

demonstrated a link between bilateral mesial temporal resection and amnesia (Scoville & Milner, 1957). Severe memory disturbance following unilateral temporal resection was reported in patients with suspected lesions in the contralateral hippocampal zone (Penfield & Milner, 1958).

This chapter offers an overview of many different aspects of evaluating patients with epilepsy. First, an overview of the medical aspects of epilepsy is presented in order to understand how medical factors may relate to a patient's neuropsychological status. Basic information related to incidence, etiology, medical diagnosis, classification, and treatment is reviewed. Next, a general overview of the neuropsychological evaluation of patients with epilepsy is presented. Commonly encountered referral questions are then reviewed in greater detail.

MEDICAL ASPECTS OF EPILEPSY

Epilepsy, in itself, is not a disorder but a symptom of abnormal brain activity. It can be operationally defined as the occurrence of two or more unprovoked nonfebrile seizures during one's life, with 45 per 100,000 newly diagnosed cases of epilepsy per year (Annegers, 1994; Sander, 1993).

MEDICAL DIAGNOSIS OF EPILEPSY

The general medical evaluation for diagnosis of epilepsy consists of three major components: (1) history and, if possible, observation of a seizure, (2) physical examination, and (3) laboratory evaluation (Engel, 1989; Gilroy, 1990). The seizure history should include information about any auras or prodromal symptoms, the seizure semiology or symptomatology, and postictal symptoms as well as information about precipitating factors and frequency. In addition to obtaining a seizure history, a general medical history may reveal information regarding etiology of the seizure disorder. Onset of seizures in the neonatal period are commonly due to intracranial infection, intracranial hemorrhage, perinatal asphyxia, drug withdrawal, primary metabolic defects, or developmental defects (Engel, 1989; Gilroy, 1990). Seizures having an onset later in infancy and childhood may be related to trauma, intracranial infection, developmental defects, migrational disorders (e.g., dysplasia), or inherited central nervous system disorders (Engel, 1989; Gilroy, 1990). Febrile seizures may be a predisposing factor for mesial temporal sclerosis (Engel, 1989; Gilroy, 1990). Trauma and intracranial infections are common etiologies of seizure onset in adolescence (Engel, 1989). In adults, seizures may be related to trauma, intracranial infection, neoplasm, or vascular occlusion (Engel, 1989; Gilroy, 1990). Cerebrovascular disease and metastatic carcinoma may be related to seizure onset in the elderly (Engel, 1989; Gilroy, 1990).

The physical examination is conducted to identify possible disease processes or physical malformations that may cause seizures (Engel, 1989; Gilroy, 1990). For example, certain dermatological findings may be significant for specific diseases. Port wine stains may indicate Sturge–Weber syndrome, café-au-lait spots are often associated with neurofibromatoses, and adenoma sebaceum occurs with tuberous sclerosis. Although often normal, the clinical neurological examination may provide useful information (Engel, 1989). For example, focal signs of a sensory or motor deficit may reflect a partial epilepsy and may help distinguish it from a generalized epilepsy.

A number of different laboratory tests may be performed to confirm the diagnosis of epilepsy and to detect underlying disease processes that may be causing seizures (Engel, 1989; Gilroy, 1990). Electroencephalography (EEG) is the primary test for the diagnosis of epilepsy and is performed to detect electrical alterations in the brain associated with epilepsy (Engel, 1989; Lesser, 1994). Hyperventilation, photic stimulation, sleep, and sleep deprivation may be used to elicit EEG abnormalities diagnostic of epilepsy (Engel, 1989; McIntosh, 1992). An EEG recording of a seizure may provide confirming evidence of whether the seizure is focal (partial) or generalized, and may have implications for antiepileptic drug (AED) treatment (Lesser, 1994). Video/EEG monitoring allows for correlation of the EEG record with observed clinical phenomena. The use of video/EEG monitoring is likely to allow for increased diagnostic accuracy and subsequently decrease misclassification of seizure type, including pseudoseizures (Lesser, 1994).

Because seizure control may be optimized when seizures are correctly classified, additional studies may be performed to determine possible etiology (Engel, 1989). More specifically, neuroimaging techniques, such as magnetic resonance imaging (MRI), are used to investigate structural abnormalities that may be related to the etiology of the seizures (Jack, 1994). Other medical diagnostic studies may be conducted to rule out syncopal disorders, toxic/metabolic disorders, or migraine headaches as the etiology for altered behavior that is suspected to be a seizure (McIntosh, 1992).

CLASSIFICATION

In the past, seizures were often labeled "grand mal" or "petit mal." However, this dichotomy has been supplanted by a more complex classification system. In 1969, the Commission on Classification and Terminology of the International League Against Epilepsy (ILAE) introduced a unified system for classifying seizure types, which was revised in 1981 (ILAE, 1981). The current classification system relies on data obtained from clinical observation and ictal EEG expression, and does not take etiology into account. The first step in determining seizure type involves identifying whether the initial abnormal brain activity detected by EEG is limited to one or both hemispheres. If its onset is localized to one hemisphere, then the seizure is classified as a *partial* seizure. In contrast, the seizure with onset in both hemispheres is classified as *generalized*. With respect to partial seizures, the term *simple* is applied if the patient remains fully conscious during the seizure, whereas the term *complex* is applied if the patient experiences a loss or alteration of consciousness during the seizure. Thus, simple partial seizures are those that start in a localized area of the brain and no impairment of consciousness is observed. Complex partial seizures also begin in a localized part of the brain, but result in an impairment of consciousness at the onset of the seizure or progress to an impairment of consciousness. Some partial seizures will progress to involve both hemispheres and, consequently, are classified as secondarily generalized seizures.

In the generalized seizure category, six different types of seizures have been identified: absence seizures, myoclonic seizures, clonic seizures, tonic seizures, tonic-clonic seizures, and atonic seizures. Another category exists for seizures that are unclassified. Seizures that defy classification, such as those that often occur in infancy, or seizures that cannot be classified due to inadequate or incomplete data, constitute a separate category. A number of seizure syndromes have been described that take into account characteristic clusters of symptoms associated with

TABLE 1. SEIZURE CLASSIFICATION

I. Partial (focal, local) seizures
   A. Simple partial seizures
      • with motor signs
      • with somatosensory or special sensory symptoms
      • with autonomic symptoms or signs
      • with psychic symptoms
   B. Complex partial seizures (automatism, or aberrations of behavior may occur)
      • simple partial onset followed by impairment of consciousness
      • with impairment of consciousness at onset
   C. Partial seizures evolving to secondarily generalized seizures
      • simple partial seizures evolving to generalized seizures
      • complex partial seizures evolving to generalized seizures
      • simple partial seizures evolving to complex partial seizures evolving to generalized seizures
II. Generalized seizures (convulsive or nonconvulsive)
   A. Absence seizures
      • sudden cessation of ongoing activity; brief duration typically lasting a few seconds; unresponsive during seizure; no or minimal postictal confusion
      • absence seizures, with impairment of consciousness only, or with mild clonic components, or with atonic components
      • atypical absence seizures
   B. Myoclonic seizures
      • brief, shock-like muscle contractions or jerks occurring singly or repetitively, predominantly prior to sleep or after sleep
   C. Clonic seizures
      • repetitive clonic contractions occurring without a tonic phase, with a shorter postictal period
   D. Tonic seizures
      • bilateral myoclonic jerk followed by residual tonic contraction of brief duration, with impaired consciousness during seizure
   E. Tonic-Clonic seizures
      • First stage is a tonic contraction of muscles (often leading to a fall), second stage is clonic convulsive movements that are followed by relaxation of muscles, consciousness is lost, patients often sleep following the seizure
   F. Atonic seizures
      • Significant decrease or loss of postural and muscle tone occurs resulting in dropping of the head, limb, or entire body with a brief, if any, loss of consciousness
III. Unclassified epileptic seizures

epilepsy such as West's syndrome and Lennox-Gastaut syndrome (ILAE, 1989). Table 1 provides an overview of this classification system and more detailed descriptions of the different seizure types.

PHARMACOLOGICAL TREATMENT

After a patient is diagnosed as having epilepsy, AED therapy is typically initiated, and is the primary method to achieve seizure control (McIntosh, 1992; Pellock, 1994; Sander, 1993). Follow-up studies of patients with newly diagnosed epilepsy report that between 65% and 80% of the patients have complete seizure control with AED treatment at 1 year. After a substantial period (e.g., 2–4 years) of being seizure-free, AEDs may be discontinued. The risks of discontinuing AED treatment were evaluated in a large study of AED discontinuation (Medical Research Council [MRC] AED Withdrawal Study Group, 1991). Patients who had been seizure-free for 2 years were randomly assigned to one of two groups: AED withdrawal or continued AED treatment. Two years later, 41% of the patients who were withdrawn from AEDs experienced a recurrence of seizures, whereas only 22% of the patients who continued to take medication had a recurrence (MRC AED Withdrawal Study Group, 1991).

Many people have excellent seizure control while being maintained on AEDs. In general, of the patients who do not have adequate control with monotherapy, about half achieve improved seizure control with polytherapy, although the likelihood of side effects is greater (Mattson, 1994). The remaining patients have seizures that are difficult to control with the currently available AEDs (Mattson, 1994).

The selection of an AED is often based on the patient's seizure type or epilepsy syndrome, as certain AEDs are more effective with certain types of seizures (Pellock, 1994). The use of one AED (i.e., monotherapy) often provides better seizure control with fewer adverse effects than the use of multiple AEDs in the majority of patients (McIntosh, 1992; Pellock, 1994; Willmore & Wheless, 1993). Therapeutic serum blood levels must be assessed and patients must be closely monitored for adverse side effects in order to achieve maximal seizure control. Neurotoxicity associated with AED therapy is the most frequently described side effect, and may vary with regard to specific complaints depending on the specific AED. The symptoms most commonly associated with neurotoxicity are nystagmus, diplopia, blurred vision, ataxia, tremor, sedation, or lethargy (Pellock, 1994).

In 1912 the first AED, phenobarbital, was introduced into use. At the present time, there are over 20 different AEDs, but only six of these medications are used as primary AEDs: phenobarbital, phenytoin, carbamazepine, valproate, ethosuximide, and primidone (Pellock, 1994). Other medications are used as adjunctive treatments. Table 2 summarizes common AED usages (Homan & Rosenberg, 1993; McIntosh, 1992; Pellock, 1994).

Surgical Treatment

While the majority of patients diagnosed with epilepsy are successfully treated with AEDs, approximately 20% to 30% of patients are refractory to pharmacological treatment. Patients with disabling seizures that are intractable to treatment with AEDs may be considered for surgical intervention. Selection of surgical candidates

TABLE 2. COMMON ANTIEPILEPTIC DRUGS

Generic name	Trade name	Common use
Phenobarbital	Luminal	GTC, GT, GC, partial seizures, neonatal seizures
Phenytoin	Dilantin	GTC, partial seizures
Carbamazepine	Tegretol	GTC, partial seizures
Valproate	Depakene, Depakote	GTC, GM, GA, partial seizures with secondary generalization
Ethosuximide	Zarotin	GA
Primidone	Mysoline	All seizure types except GA
Diazepam	Valium	SE
Clonazepam	Klonopin	SE
Felbamate	Felbatol	Partial and secondarily generalized seizures, Lennox–Gastaut syndrome
Gabapentin	Neurontin	Partial and secondarily generalized seizures
Lamotrigine	Lamictal	Partial and secondarily generalized seizures

Note. GTC, generalized tonic-clonic seizures; GT, generalized tonic seizures; GC, generalized clonic seizures; GM, generalized myoclonic seizures; GA, generalized absence seizures; SE, status epilepticus.

is a process that involves a number of specific steps (Duchowny, 1993). First, patients must have failed all major AED trails in which doses were "pushed to their therapeutic limits" (Duchowny, 1993, p. 999).

Since the success of the surgery is largely dependent on removing the epileptogenic region, patients next undergo rigorous evaluation to carefully determine and define the epileptogenic region to be resected. The evaluation typically consists of several days of video/EEG monitoring, neuroimaging, and neuropsychological evaluation for a "multimodal analysis of physiologic, structural, and functional data" (Duchowny, 1993). Patients may require additional invasive evaluation with depth or subdural electrodes if the data from the scalp EEG evaluation were non-localizing or if the seizure focus is thought to lie close to eloquent brain areas, such as language or motor areas (Duchowny, 1993). Further studies are commonly conducted to provide additional data regarding the epileptogenic region. Since hippocampal atrophy has been shown to correlate with seizure focus and outcome following epilepsy surgery, hippocampal volumetric assessment, based on MRI images, is often conducted (Garcia, Laxer, Barbaro, & Dillon, 1994). Preliminary investigations of the use of functional MRI techniques to map partial motor seizures and subclinical events have yielded promising results (Jackson, Connelly, Cross, Gordon, & Gadian, 1994). Because paroxysmal changes in electrical activity alter the metabolic state of the tissue in which they occur, single photon emission tomography (SPECT) imaging may be useful in lateralizing and localizing epileptogenic regions (Spencer, 1994). Positron emission tomography (PET) imaging also may be helpful in identifying areas of altered metabolic activity (Spencer, 1994). Epileptogenic areas are likely to appear hypometabolic interictally and hyper-metabolic during ictal events.

If the epileptogenic region is well defined and the removal of that area is not likely to cause disruption of important neuropsychological or physical functioning, then the patient may proceed to surgery with a good likelihood of reduction in seizure frequency or seizure remission. Success rates for improvement following anterior temporal lobectomy, amygdalohippocampectomy, and lesionectomies have improved with the availability of new surgical techniques and more clearly established selection criteria. Approximately two thirds of the patients may become seizure-free and many of the remaining third have a significant reduction in seizure frequency (Lesser, 1994). Rates of improvement following surgery are often related to the location of the epiloptogenic region that is resected. Temporal lobectomy is generally associated with better success rates (e.g., approximately 80% seizure-free outcome) compared to extratemporal procedures (Lesser, 1994; Olivier, 1988).

NEUROPSYCHOLOGICAL EVALUATION

Since epilepsy is a manifestation of central nervous system dysfunction, it is likely to have an effect on a person's cognitive status. A number of factors can influence the neuropsychological status of persons with epilepsy (Dodrill & Matthews, 1992). Age at onset of epilepsy has been associated with neuropsychological performance in that late onset is associated with better intellectual abilities compared to early onset of epilepsy (Dikmen, 1980; O'Leary, Seidenberg, Berent, & Boll, 1981). Duration of the seizure disorder, or number of years with seizures, is inversely correlated with neuropsychological performance. A higher number of tonic-clonic seizures that have occurred over the course of the seizure disorder has been associated with diminished mental abilities (Dodrill, 1986a). The observation that neuropsychological abilities appear more diminished in patients with major generalized seizure disorders has been documented in both adult and pediatric populations (Matthews & Kløve, 1967; Zimmerman, Burgemeister, & Putnam, 1951). Antiepileptic drugs and their adverse effects on cognitive performance have been studied for many years and are discussed in a subsequent section of this chapter.

Patients with epilepsy are referred for neuropsychological evaluation for a variety of reasons. Neuropsychological evaluation may be requested to describe cognitive functioning, differentiate pseudoepilepsy from epilepsy, monitor the effects of seizures over time, monitor the effects of AEDs, establish a baseline prior to seizure surgery, determine the degree of postoperative neuropsychological change, or address school and/or work issues. As with any comprehensive neuropsychological evaluation, the evaluation of a person with epilepsy is typically broad-based to evaluate global cognitive functioning as well as specific deficits in cognitive functioning. An evaluation of adjustment and psychosocial factors is often desirable as well.

A comprehensive approach to testing multiple neuropsychological functions is accomplished in a number of different ways. The use of a general neuropsychological battery, such as the Halstead–Reitan Neuropsychological Test Battery (HRB), has been used across a variety of patient groups, including epilepsy patient groups (Reitan & Wolfson, 1993). The HRB assesses general intelligence, simple sensory–perceptual functions, motor and psychomotor functions, language and communication, visual–spatial, problem-solving, concept formation, and emotional status (Boll, 1981; Dikmen, 1980; Reitan & Wolfson, 1993).

In a survey of epilepsy centers, the HRB was the only general neuropsychological battery reported to be used (Jones-Gotman, Smith, & Zatorre, 1993). Many centers use components of the HRB (e.g., 66% use the Trail Making Test, 57% use Finger Oscillation Test, 14% use the Category Test), but less than 10% of the centers use the full Halstead–Reitan battery (Jones-Gotman et al., 1993). As is often the case, component measures of the neuropsychological test battery evolve over time as ongoing evaluation of the efficacy of different measures leads to the addition and deletion of tests to meet a specific evaluation need of any given test battery.

An example of a neuropsychological battery designed specifically for epilepsy populations is the Neuropsychological Battery for Epilepsy (NBE) developed by Dodrill (1978). Dodrill evaluated over 100 test variables, many taken from the HRB, and reduced them to 16 variables that discriminated between epilepsy and control groups without excessive test overlap. He then cross-validated these variables on a second sample of patients with epilepsy and control subjects. In addition to the 16 tests found to be sensitive to the effects of epilepsy, Dodrill (1978) also uses the Wechsler Adult Intelligence Scale (Wechsler, 1955), the Lateral Dominance Exam (Reitan, 1966), and the Minnesota Multiphasic Personality Inventory (MMPI; Hathaway & McKinley, 1943). Table 3 outlines the tests comprising the NBE. The NBE has demonstrated its utility in a variety of research areas including evaluating the cognitive effects of seizures (Dodrill, 1986a), EEG epileptiform discharges (Wilkus & Dodrill, 1976), non-epileptiform EEG abnormalities (Dodrill & Wilkus, 1978), AEDs (Dodrill & Temkin, 1989), seizure surgery (Ojemann & Dodrill, 1985), and seizure history variables (Dodrill, 1986a).

Another approach to evaluating a person with epilepsy would be to use a

TABLE 3. DODRILL'S NEUROPSYCHOLOGICAL BATTERY
FOR EPILEPSY

General measures
    Wechsler Adult Intelligence Scale
    Lateral Dominance Exam
    Minnesota Multiphasic Personality Inventory
Discriminative measures
    Category Test
    Tactual Performance Test, Total Time
    Tactual Performance Test, Memory
    Tactual Performance Test, Localization
    Tapping, Total
    Trail Making Test, Part B
    Aphasia Screening Test, Errors
    Constructional Dyspraxia
    Perceptual Examination, Errors
    Seashore Tonal Memory Test
    Stroop, Part I
    Stroop, Part II − Part I
    Name Writing, Total
    Wechsler Memory Scale, Logical Memory
    Wechsler Memory Scale, Visual Reproduction

flexible battery (Lezak, 1995). This approach involves administering a core neuro-psychological test battery, sampling multiple domains. The core neuropsychologi-cal test battery is subject to change, with the addition or deletion of tests, so that the resulting battery is suitable, practicable, and useful for answering the referral ques-tion for any given patient. Because patients with epilepsy are referred for neuro-psychological evaluation for a number of different reasons, selection of the test battery and examination procedures is frequently geared toward the specific refer-ral question. Seven common reasons for referral will be discussed separately in the following sections, highlighting issues that are germane to each.

NEUROPSYCHOLOGICAL EVALUATION IN THE INITIAL NEUROLOGICAL DIAGNOSTIC WORK-UP

When a person with suspected epilepsy is undergoing the initial medical eval-uation, the physician must consider several diagnostic issues. The neuropsycholo-gist may be consulted to assist with identifying the neuropsychological and psycho-social sequelae of epilepsy for the patient. In addition, the neuropsychologist may be asked to determine whether the patient's neuropsychological functioning is consistent with any epilepsy syndrome. In these ways, the neuropsychological eval-uation may complement other diagnostic procedures (Dikmen, 1980). As an illus-tration of this point, one can consider neuroimaging studies. Magnetic resonance imaging may be the procedure of choice in detecting the parameters of a brain lesion, but MRI cannot detect the *effect* of that lesion on a person's cognition. The neuropsychological examination provides unique information regarding the cogni-tive sequelae of epileptogenic lesions or epilepsy itself.

When devising a neuropsychological test battery, a common standard of prac-tice is to include tests that assess the major domains of neuropsychological function-ing: intellectual, higher-level problem-solving, executive functions, attention, learning and memory, language, visual–spatial, psychomotor, and emotional/

personality. Following the administration of the test battery, the neuropsychological test results can be examined to determine whether test results are within the expected range given the patient's age, education, and other background history. If the test results are not within normal limits, one may examine the profile to determine whether generalized dysfunction or focal dysfunction of a specific area or system (e.g., language or memory) is suggested. Doing so enables the neuropsychologist to delineate the cognitive sequelae of epilepsy and determine if the profile is consistent with the neuropsychological characteristics of an epilepsy syndrome. Some epilepsy syndromes, which are described below, have characteristic cognitive features that can be addressed by neuropsychological evaluation. While differential diagnosis is obviously not likely to be made solely on the basis of neuropsychological data, in some cases, the neuropsychological profile may confirm or disconfirm the presence of certain cognitive aspects that would be expected with a given epilepsy syndrome.

Patients with complex partial seizures (temporal lobe epilepsy) may have greater deficits in memory functioning, with material-specific deficits corresponding to the side of the lesion (Bornstein, Pakalnis, & Drake, 1988). In other words, patients with epilepsy of left temporal lobe origin may have verbal memory deficits; patients with right temporal lobe epilepsy may have nonverbal memory deficits, although this finding is not nearly as robust as the association between verbal memory deficits and left temporal lobe epilepsy (Bornstein *et al.*, 1988; Saykin, Gur, Sussman, O'Connor, & Gur, 1989). Patients with generalized epilepsy tend to show greater neuropsychological impairment than patients with complex partial seizures, and they are more likely to have attentional problems evident on neuropsychological evaluation (Trimble, 1987).

Landau–Kleffner syndrome (LKS) is characterized by two features: acquired aphasia and temporal or temporoparieto-occipital paroxysmal EEG changes occurring during sleep (Landau & Kleffner, 1957; Roger, Genton, Bureau, & Dravet, 1993). The LKS is rare, and in the 200 cases reported since its identification in 1957, an auditory verbal agnosia appears to begin before the age of 6 in 70% of the patients; expressive difficulties are also apparent early in the course of LKS. Intellectual abilities are reportedly unaffected in the majority of patients, but language deficits may occur progressively or in a stepwise course.

A number of childhood epilepsy syndromes are characterized by developmental delay and/or mental retardation. West syndrome, which is defined by infantile spasms, mental deterioration, and hypsarrhythmia, is associated with a high incidence (71%–81%) of mental retardation, with severe mental retardation in over 50% of patients (Dulac & Plouin, 1993; Favata, Leuzzi, & Curatolo, 1987). Speech and language disorders are also frequent as well as psychiatric disorders such as autism in children with West syndrome or infantile spasms (Riikonen & Amnell, 1981). Mental retardation is commonly associated with severe myoclonic epilepsy of infancy. An intellectual deficit is usually acquired between the ages of 2 and 4, and mental retardation is reported to be severe in half of the patients over 10 years of age (Roger *et al.*, 1993). Lennox–Gastaut syndrome (LGS) is characterized by tonic, atonic, or atypical absence generalized seizures, occurring with a high seizure frequency and frequent status epilepticus (Farrell, 1993). It is associated with diffuse cognitive dysfunction and mental retardation, with more severe cognitive deficits related to earlier age of onset and episodes of status epilepticus (Chevrie & Aicardi,

1972; Farrell, 1993; Gastaut, 1982). Cognitive functioning in patients with progressive myoclonus epilepsy (PME) is likely to deteriorate over time and is thought to be associated with the spread of cortical neuronal loss (Genton & Roger, 1993).

In contrast to the syndromes described above, in the benign focal epilepsies of childhood (BFEC) and juvenile myoclonic epilepsy, no long-term cognitive impairment is expected in the majority of patients (Loiseau, 1993; Roger *et al.*, 1993; Serratosa & Delgado-Escueta, 1993). Epilepsy with continuous spikes and waves during sleep (CSWS) is often characterized by premorbid normal development with a concomitant onset of neuropsychological problems at the time of CSWS onset (Roger *et al.*, 1993). In children with CSWS, Verbal IQ tasks are more severely impaired than Performance IQ tasks, and half of the children studied demonstrate memory deficits and disorientation of time and space (Roger *et al.*, 1993). Epilepsy with CSWS is similar to Landau–Kleffner syndrome in that speech disturbance occurs in approximately 50% of the cases, but differs from Landau–Kleffner in that other neuropsychological impairments are also seen (Hirsch *et al.*, 1990).

DIFFERENTIAL DIAGNOSIS OF PSEUDOEPILEPSY

Not all patients with a presenting complaint of suspected seizures actually have epilepsy. A certain portion of patients presenting for evaluation of epilepsy actually have pseudoseizures (also known as psychogenic seizures, nonepileptic seizures, or pseudoepilepsy), sometimes in combination with actual seizures. Pseudoseizures by definition are nonepileptic events or "spells" that are thought to have a psychiatric rather than organic etiology (Özkara & Dreifuss, 1993; Porter, 1993). Patients who are suspected of having pseudoseizures may be referred for neuropsychological evaluation to assist with differential diagnosis. However, the results of several studies reviewed below suggest that neuropsychological evaluation, including assessment of personality variables, may not be sensitive enough to differentiate pseudoepilepsy from epilepsy on a case-by-case basis.

In a multicenter study, Sackellares and colleagues (1985) compared the neuropsychological functioning of three groups: pseudoseizure patients (n = 19), epilepsy patients (n = 20), and patients with both pseudoseizures and epilepsy (n = 18). Their results indicated that pseudoseizure patients tended to have higher IQ scores and were less impaired on the HRB than the other groups. Nonetheless, the pseudoseizure group Impairment Index was 0.45, which falls in the borderline range and indicates that this group was not functioning in the expected range. In contrast to Sackellares *et al.*, Wilkus, Dodrill, and Thompson (1984) found no differences in intellectual or neuropsychological functions between a group of patients with pseudoseizures (n = 25) and patients with epilepsy (n = 25) who were comparable in age, gender, and years of education on Dodrill's (1978) NBE. In a later study, Wilkus and Dodrill (1989) replicated the results from their previous study and the results of the Sackellares *et al.* study (1985). In other words, when patient groups were matched on age, gender, and education, no neuropsychological differences emerged between the groups; when the groups were not matched on these variables, the pseudoseizure group appeared less impaired. Hermann (1993) reported on a series of 12 patients who had been previously diagnosed with pseudoseizures. After extensive evaluation with invasive EEG monitoring (e.g., with subdural strip electrodes), half of the patients were found to have epilepsy and the other half continued to carry the diagnosis of pseudoseizures. Both groups were similar in terms of intellectual functioning and neuropsychological functioning, with abnormal neuropsychological results observed in all six epilepsy patients and five out of six pseudoseizure patients (Hermann, 1993). To summarize, neuropsychological discrimination between patients with pseudoseizures and actual seizures can be difficult as it appears that compromised neuropsychological function is common in both groups (Hermann & Connell, 1992).

In terms of psychological functioning, Wilkus and Dodrill (1989) found that pseudoseizure patients appeared more emotionally disturbed than epilepsy patients on the basis of their MMPI profiles. However, the MMPI could not reliably distinguish between pseudoseizure and epilepsy patient groups (Wilkus & Dodrill, 1989). Review of the MMPI literature by Hermann and Connell (1992) yields a similar conclusion in that the "hit rate" for correct diagnostic classification of pseudoepilepsy is not high enough to warrant its use in classifying individual patients. Nonetheless, psychosocial factors, including history of childhood loss, somatization disorder, personality disorder, and to a lesser extent, presence of a model for seizure symptoms, distinguished a group of pseudoseizure patients from epileptic patients (Eisendrath & Valan, 1994). Interestingly, two variables that have been reported to have a high frequency in pseudoseizure populations, the presence of secondary gain and a history of sexual abuse, did not differentiate between the two groups (Eisendrath & Valan, 1994).

MONITORING THE EFFECTS OF REPEATED SEIZURES ON BRAIN FUNCTIONING

Monitoring the effects of cognitive functioning over time may be useful in clinical decision-making. If a patient has documented cognitive decline over time, then more aggressive forms of treatment, including seizure surgery, may need to be explored. Cognitive deterioration is often associated with specific epilepsy syndromes. For example, deteriorating cognitive functioning is expected in PME (Serratosa & Delgado-Escueta, 1993). Neuropsychological assessment during the initial evaluation is critical in order to establish a baseline against which to compare future test results. Repeated neuropsychological evaluation can "prove or disprove deteriorating intelligence" with deteriorating cognitive functioning suggesting PME and stability of cognitive functioning suggesting a juvenile myoclonic epilepsy (Serratosa & Delgado-Escueta, 1993, p. 562). In addition, different forms of PME have differential rates of cognitive deterioration (Genton & Roger, 1993). Changes in cognition are also thought to be associated with certain seizure variables.

Cognitive deterioration in some patients with epilepsy has been documented since the 1920s (Hermann, 1991; Lesser, Lüders, Wyllie, Dinner, & Morris, 1986). In a review of the literature, Lesser *et al.* (1986) concluded that cognitive deterioration occurs in a small minority of patients, and is most often related to measures of severity of the seizure disorder. A prospective study conducted by Bourgeois, Prensky, Palkes, Talent, and Busch (1983) examined intellectual functioning in 72 children with epilepsy. Children were administered IQ tests within 2 weeks of the initial diagnosis of epilepsy, and follow-up evaluations were conducted yearly for an average of 4 years. Siblings were tested as control subjects in 45 cases. At the beginning of the study, the mean IQ of the patient group was 99.7 and 104.4 for the sibling group. At the time of the last follow-up evaluation, the mean IQ of the group remained stable, but this was not true on a case-by-case basis, with 11% of the patients demonstrating a decrease in IQ over time. Follow-up data on siblings was not reported. Variables related to cognitive deterioration were higher incidence of drug toxicity, polytherapy, difficulty with seizure control, and earlier age at onset of the seizure disorder.

Another risk factor for cognitive deterioration is status epilepticus (SE), which is an episode of prolonged, uninterrupted seizure activity. In a review of the literature on intellectual impairment as an outcome of SE, Dodrill and Wilensky (1990) found only 13 studies that actually attempted to quantify mental status in patients with a history of SE. The majority of these studies involved children, and only a few of the studies used psychological or neuropsychological testing to quantify cognitive abilities. These studies were retrospective and relied on review of hospital records, neurological examination, and clinical mental status examination. Nonetheless, the majority of these studies concluded that SE was related to negative cognitive effects.

In a five-year prospective study of 143 patients with epilepsy, nine patients with medical record documentation of SE involving at least 30 minutes of altered consciousness were identified (Dodrill & Wilensky, 1990). These nine patients were matched for age and education with patients who clearly did not have any SE episodes in the intervening 5 years. Seizure control was equivalent across the two groups. Although not statistically significant, the SE group had a slightly younger average age at seizure onset and took a larger number of medications. At baseline, the mean IQ score for the SE group (M = 90.44, SD = 22.16) was less than that of the control group (M = 100.00, SD = 10.69). In addition, at baseline the SE group showed mild to moderate impairment on an extensive neuropsychological battery, whereas the control group demonstrated only mild impairment. At 5 years follow-up, a significant interaction effect was found for Full Scale IQ with the control group showing a slight increase in IQ scores (M = 103.30, SD = 19.63), likely due to practice effect, whereas the SE group demonstrated a deterioration in IQ scores (M = 85.89, SD = 23.90).

Thus, it appears that SE may be a contributing factor to intellectual decline in adults or an alteration of normal developmental rates in children with epilepsy. Regarding the extent of the effect of SE on cognitive functioning, it has been suggested that SE produces neurological damage especially in certain neuronal regions such as CA1 and CA3 of the hippocampus, neocortex, and substantia nigra (DeGiorgio, Tomiyasu, Gott, & Treiman, 1992; Hoch, Hill, & Oas, 1994). It is one of many risk factors for negative cognitive sequelae associated with epilepsy, assuming an otherwise static neurologic course.

MONITORING THE EFFECTS OF MEDICATION

Dodrill (1993) reports that prior to the 1970s, the adverse cognitive effects of AEDs were not given much attention. Since that time, the issue of negative neurocognitive effects of AEDs has become increasingly important in selecting pharmacological treatments for patients, especially as the emphasis on quality of life issues has emerged. Antiepileptic medications are known to potentially impair cognitive functioning, with greater impairment occurring with polytherapy and at higher dosages or increased blood levels (Meador & Loring, 1991; Trimble, 1987). Neuropsychological evaluation has proven helpful in understanding the cognitive sequelae of AEDs (Bennett, 1992).

Phenobarbital, a barbiturate, has been shown to impair nonepileptic volunteers' performance on tests of psychomotor speed, verbal learning, vigilance, and sustained attention (Hutt, Jackson, Belsham, & Higgins, 1968). Hyperactivity has also been reported as a paradoxical side effect of phenobarbital (McGowan, Neville, & Reynolds, 1983). Primidone (Mysoline), a barbiturate analogue, has not been studied as extensively as phenobarbital, although it is believed to have similar side effects (Bennett, 1992).

In healthy adults, phenytoin (Dilantin) and carbamazepine (Tegretol) have been shown to produce memory impairment (Meador, Loring, Huh, Gallagler, & King, 1990; Meador et al., 1993). Phenytoin has also been shown to have negative effects on psychomotor functioning, concentration, memory, and speeded problem-solving, whereas sodium valproate (Depakote) and carbamazepine have been associated with minimal adverse cognitive effects (Trimble, 1987). Significant improvements in concentration and psychomotor function were observed in a group of 17 patients after withdrawal from phenytoin in comparison to a group of 12 control patients (May, Bulmahn, Wohluter, & Rambeck, 1992). It has been argued that some of the neuropsychological deficits seen with the use of phenytoin may be a function of psychomotor slowing rather than a true cognitive deficit (Dodrill & Temkin, 1989). In a comparison of phenobarbital, phenytoin, and carbamazepine utilizing a randomized double-blind, triple crossover design, Meador and colleagues found that subjects had comparable performances on most neuropsychological measures (Meador et al, 1990). The only difference that emerged was on the Digit Symbol subtest of the Wechsler Adult Intelligence Scale-Revised (Wechsler, 1981), with lower performances noted when subjects were on phenobarbital. When individual performances were examined, no differences emerged between phenytoin and carbamazepine. However, when phenobarbital was compared to phenytoin and carbamazepine, more instances of poor performance were observed when subjects were on phenobarbital.

In summary, as AEDs are known to affect neurocognitive functioning, neuropsychological evaluation can prove helpful in delineating specific cognitive sequelae of AED use. In addition, neuropsychological evaluation may be used to detect subtle effects of neurotoxicity of AEDs. A brief, repeatable battery may be preferred when monitoring for neurotoxicity. When selecting tests for such a battery, particular emphasis is typically given to tests sensitive to speeded information processing and psychomotor functioning, as these neuropsychological domains may be the most susceptible to the effects of AEDs.

Preoperative Evaluation

Among patients with intractable seizure disorders, neuropsychological evaluations are commonly obtained prior to surgical interventions to (1) characterize baseline functioning; (2) identify baseline deficits that may decrease understanding of or compliance to a complex medical regimen; (3) assist with lateralization or localization of the seizure focus; and (4) predict postoperative seizure control (Chelune, 1994). First, a preoperative evaluation can be used to characterize a patient's current level of neurocognitive functioning as a baseline against which future comparisons can be made. Obtaining a preoperative neuropsychological baseline is essential for identifying potential areas of neuropsychological morbidity resulting from the planned neurosurgery. The results of the neuropsychological evaluation can be used to make predictions of postoperative outcome in terms of seizure control and possible decrements in cognitive functioning (Chelune, Naugle, Lüders, & Awad, 1991; Dodrill, Wilkus, & Ojemann, 1992). With this data-based information, patients can make informed decisions regarding benefits and risks related to seizure surgery. For example, many investigations have demonstrated differences in neuropsychological outcome between patients undergoing left and right temporal lobe surgery (Chelune *et al.*, 1991; Ivnik, Sharbrough, & Laws, 1987; Novelly *et al.*, 1984; Ojemann & Dodrill, 1985; Rausch & Crandall, 1982). That is, these two patient groups typically have different patterns of postsurgical neuropsychological decrements. In general, at the time of preoperative evaluation, right temporal lobectomy patients tend to perform at higher levels than left temporal lobectomy patients on neuropsychological examinations. Postsurgically, right temporal lobectomy patients may appear unchanged or slightly improved. In contrast, left temporal lobectomy patients tend to show greater decrements on verbal memory and reasoning tasks after surgery. In addition, a differential effect appears postsurgically between higher-functioning and lower-functioning dominant lobectomy patients. Higher-functioning patients are more likely to have greater postsurgical decrements, which cannot be entirely attributed to statistical regression to the mean (Bauer *et al.*, 1994; Chelune *et al.*, 1991; Kneebone, Chelune, Dinner, Awad, & Naugle, 1995). In summary, from a preoperative perspective, patients who appear most at risk for neuropsychological morbidity related to seizure surgery appear to be those with a dominant temporal lobe seizure focus and higher rather than lower baseline functioning.

Baseline neuropsychological evaluation might also be used to identify deficits that may decrease the patient's ability to fully comply with his or her medical regimen or presurgical evaluation. If such behavioral deficits are identified, then the neuropsychologist can recommend environmental manipulations designed to enhance the patient's ability to understand what is expected from him or her in order to participate more fully in his or her own medical care. For example, if severe memory deficits are identified in a patient, then several modifications to the medical regimen can be recommended, including but not limited to having the patient accompanied by another adult to all doctor's appointments, instructing the patient in the use of external memory aids, and monitoring medication administration.

At some epilepsy centers, another purpose of the preoperative evaluation is to assist with localization and lateralization of the epileptogenic focus (Chelune, 1994; Jones-Gotman, 1991). If the results of the neuropsychological evaluation suggest circumscribed functional deficits, then inferences may be made regarding lateralization and localization of the epileptogenic focus. When results from the neuropsychological evaluation are concordant with results of other studies (i.e., EEG/video monitoring, MRI, PET, etc.) then the confidence that the correct localization of the seizure focus has been identified increases. Nonetheless, discordance between the neuropsychological test results and other neurodiagnostic tests also occurs on a fairly frequent basis. This type of discrepancy may occur due to the inherent differences between data sources. For example, EEG detects neural electrical abnormalities, MRI detects structural abnormalities, and PET detects metabolic abnormalities, whereas neuropsychological tests reflect the behavioral output of the brain, which may or may not be specific to the epileptogenic focus or the condition producing it (Chelune, 1994). Thus, discordant neuropsychological findings may reflect difficulty with cognitive functioning related to areas other than the seizure focus.

Systematic attempts have been made to use preoperative neuropsychological data to predict postoperative seizure control. Bengzon, Rasmussen, Gloor, Dussault, and Stephens (1968) studied 104 patients who had undergone surgery at least 5 years earlier. Of these patients, 50 had significant improvements in seizure control and were essentially seizure-free. The remaining 54 patients demonstrated little or no improvement following surgery. Those patients with good seizure outcome had preoperative neuropsychological data suggesting restricted temporal lobe dysfunction, whereas the poor outcome group was more likely to demonstrate evidence of bilateral cerebral dysfunction on the preoperative neuropsychological examination. These findings also paralleled the EEG data in that bilateral or widespread EEG abnormalities were also associated with a less favorable surgical outcome (Bengzon *et al.*, 1968). Wannamaker and Matthews (1976) also found that patients with greater preoperative generalized neuropsychological impairment had a poorer prognosis for seizure control following surgery compared to patients with less preoperative neuropsychological impairment. In summary, patients with generalized or diffuse neuropsychological dysfunction at baseline may be less likely to achieve good seizure control following surgery than patients with more focal patterns of neuropsychological deficit.

POSTOPERATIVE FOLLOW-UP

The National Institutes of Health Consensus Development Conference (1990) on Surgery for Epilepsy recommends multiple outcome measures, including neuropsychological assessment, when evaluating the surgical treatment of epilepsy. However, examination of postsurgical neuropsychological status cannot be used to infer the effects of surgery without having established a presurgical baseline in order to differentiate preexisting or presurgical deficits from postsurgical changes. Only in the context of a presurgical baseline can postsurgical comparisons be made to identify changes in cognition (Chelune, 1994). However, caution must be applied when interpreting postoperative neuropsychological change. Generally, in a test–retest paradigm, a positive change score is interpreted as an improvement and a negative change score is interpreted as a decrement. However, in order to accurately interpret change scores, it is important to consider the reliability of the measures and the base rate of expected practice effects (Chelune, Naugle, Lüders, Sedlak, & Awad, 1993). A positive change score might reflect a practice effect alone or, alternatively, a practice effect in addition to real change. Conversely, a negative change may actually reflect a decrement *and the absence of a practice effect*. A score that remains unchanged may actually reflect a cancellation phenomenon in which a mild decrement is offset by expected practice effects.

Several areas of neuropsychological functioning have shown consistent patterns of change following seizure surgery. In addition to postsurgical memory change discussed in the previous section, alterations in intellectual abilities and language have been studied. In a review of the literature, Naugle (1991) surmised that surgical intervention does not result in significant deterioration of intellectual ability in most cases. A number of studies have noted hemisphere-specific results in that patients who undergo left hemisphere resections demonstrate greater gains in Performance IQ, whereas patients who undergo right hemisphere resections show increases in Verbal IQ (Olivier, 1988; Rausch & Crandall, 1982).

Evaluation of language functioning following dominant temporal lobectomy has not been as extensively studied as intellectual ability or memory. In a well-controlled prospective study, Hermann and Wyler (1988) evaluated language function in a group of 15 dominant temporal lobectomy patients and 14 nondominant temporal lobectomy patients. Prior to surgery, language functioning was noted to be lower in patients with dominant temporal lobe seizure foci. Following surgery, neither group demonstrated losses in language function, but the dominant lobectomy group appeared to improve in receptive language comprehension and associative verbal fluency. These positive changes were attributed to the elimination of the epileptogenic focus that may have adversely affected language functions via association pathways prior to surgery. In his review, Naugle (1991) also reported that subtle deficits in verbal fluency and other aspects of language have been noted following dominant temporal lobectomy.

As the technology of seizure surgery evolves, more patients are having positive outcomes so that approximately 80% of temporal lobe epilepsy patients are seizure-free at follow-up (Olivier, 1988). Many patients who were not able to work or fully carry out activities of daily living given the severity of their seizure disorder prior to surgery are able to do so following successful surgical treatment (Augustine et al., 1984). Many of these patients may need vocational habilation or rehabilitation following surgery in order to bridge the gap between being disabled and being more functionally independent, including but not limited to gaining full-time employment. In a review of the available research, Fraser and Dodrill (1994) found that postsurgical employment was highest among patients who were employed prior to surgery and among those patients who achieve a seizure-free outcome following surgery. In addition, Fraser and Dodrill (1994) report that employment outcomes may be mediated by impairments in cognitive functioning, especially in motor and/or visual–spatial functioning, as well as psychological state.

SCHOOL AND WORK ISSUES

The possibility that epilepsy, the treatment for epilepsy, and the psychosocial consequences of epilepsy can affect school and work functioning is obvious. Early onset in children can affect the acquisition of school-based skills, and therefore, may have negative implications for school and eventual occupational achievement. Onset of epilepsy in adulthood may affect those abilities and skills needed to perform a job or obtain employment.

SCHOOL OUTCOME. Compared to children in the general population, children with epilepsy have higher rates of emotional and behavioral problems, and are at greater risk for developing learning problems that can potentially have an adverse impact on future vocational functioning (Bolter, 1986). After reviewing studies of intellectual functioning of children with epilepsy, Bolter (1986) reported that the distribution of IQ scores is positively skewed, with more scores falling in the lower end of the normal distribution. In terms of academic achievement, arithmetic skills were most affected, followed by spelling skills in a group of school-age children (Seidenberg et al., 1986). Reading comprehension typically falls at least 1 year below grade level by the age of 10 to 11 (Bolter, 1986). With respect to psychosocial functioning, school-age children with epilepsy tend to show more emotional instability, school problems, social adaptation, and poorer self-concepts (Matthews, Barabas, & Ferrari, 1982; Ziegler, 1981). Given that epilepsy has a number of different etiologies and manifestations, it is not surprising that no unique pattern of cognitive dysfunction has been found in school-age children (Bolter, 1986). Although these children as a group demonstrate mild and nonspecific neuropsychological impairments, the diagnosis of epilepsy does not necessarily imply the presence of functional deficits. Intellectual evaluation alone is not sufficient to determine the specific nature of cognitive deficits or educational needs of the child (Bolter, 1986). Rather, a comprehensive and broad-based assessment of neuropsychological and psychoeducational functioning is required for the school-age child with epilepsy. In addition to neuropsychological and achievement testing, information is commonly obtained from parents and teachers regarding general psychosocial adjustment, behavioral difficulties, and academic functioning. The results of the neuropsychological evaluation can be used to formulate specific recommendations to enhance learning at school and to address any psychological or behavioral problems the child may have as a result of epilepsy, and as such may be useful in contributing practical suggestions to an individualized educational plan (IEP) or for management of children in regular classroom placements. For example, if a child has difficulty attending to tasks, it would be important to assist the teacher in identifying ways to decrease distractions in the classroom and to make sure adequate breaks are provided for the child. Complex or multifaceted tasks may need to be divided into smaller units to enhance comprehension and completion of the task. Identification and utilization of the child's primary mode of learning (e.g., verbal vs. visual) may lead to greater academic success (Hooper & Willis, 1989). The degree of repetition needed to optimize new learning can be ascertained from the neuropsychological examination and integrated into the classroom. In addition to neuropsychological issues, attention to the child's emotional functioning and how it interferes with or complements classroom learning should be taken into account.

VOCATIONAL OUTCOME. A person's successes or failures in school may have an impact on future vocational functioning. Hauser and Hesdorffer (1990) reported that individuals with epilepsy are more likely to experience difficulty with employment in comparison to their peers. This was reflected in a study by Clemmons and Dodrill (1983) who examined vocational functioning in 42 people within the first 5 years following high school graduation. They found that 31% of the subjects in their sample were receiving supplemental social security income, with only 43% being economically independent. In another study, Dodrill and Clemmons (1984) found that neuropsychological tests predicted later difficulty in vocational adjustment and daily life functioning. This suggests that early identification and intervention may make it possible to positively influence vocational outcome. Accordingly, recent legislation requires public schools to provide more focused attention to the process of transition from school to work in the development of IEPs (20 U.S.C. § 1401, 1990). School-based interventions targeted for students with epilepsy have demonstrated their effectiveness in decreasing school dropout rates by one third to one half and decreasing the unemployment rate by 50% 1 to 2 years after graduation compared to students in a regular curriculum (Freeman, Jacobs, Vining, & Rabin, 1984). However, this school-based vocational intervention does not address the needs of individuals past school age or those in need of vocational rehabilitation services.

The neuropsychological evaluation can provide information for identifying the cognitive strengths and weaknesses of an individual that may affect work functioning (Fraser, 1980). In addition, strengths may be used to compensate for areas of relative weakness. When exploring potential vocational options, it may be important to consider the cognitive demands of a specific job and how those demands will interact with a patient's cognitive and behavioral strengths and weaknesses in order to increase the likelihood of success in a job (Chelune & Moehle, 1986). Determining realistic vocational goals may be aided when guided by the results of the neuropsychological examination. Thus, it may be optimal to "match" the abilities of the patient to the demands of the job (Dikmen, 1980). If a patient is having difficulty on the job or in real-life situations, a neuropsychological evaluation (including evaluation of emotional factors) may yield results that identify factors contributing to the patient's difficulty (Dikmen, 1980). Documenting neuropsychological deficits and making interventions to minimize their effects may allow the patient to maximize his or her vocational activities.

Vocational settings can vary widely, from a supervised, sheltered workshop setting to supported employment (e.g., job coaching) to fully independent and gainful employment. Assistance with securing employment can be obtained from a number of agencies including state vocational rehabilitation programs, Epilepsy Foundation of America Training and Placement Services Program (TAPS), programs developed by local epilepsy organizations, rehabilitation programs, or Epilepsy Center programs (Hauser & Hesdorffer, 1990).

Despite the availability of services, adults with epilepsy may still have difficulty securing and maintaining employment. Dodrill (1986b) has speculated that this may be due to an overemphasis on skills training and an underemphasis on job development and placement strategies. Typically, vocational programs work with the client in obtaining a position, and follow-up assistance with maintaining the job is often overlooked. Nonetheless, supported employment may be optimal for individuals with any combination of social, emotional, or neuropsychological impairments. Ongoing support by a job coach or vocational counselor is necessary when the individual has difficulty learning job tasks without special assistance or difficulty in adjusting to changes on the job. A job coach can be instrumental in providing the extra training needed beyond what would normally be expected from the employer. In addition, the job coach can provide support as needed for problem-solving or adjusting to new demands of the job. One of the most important roles of the job coach is to interface with the employer and the employee, providing education regarding epilepsy and the sequelae of epilepsy in order to prevent problems in the workplace (Fraser, 1980).

Conclusion

Epilepsy is a heterogeneous entity and varies in its etiology, classification, clinical course, and sequelae. Therefore, it is not surprising that the neuropsychological evaluation of epilepsy is often complex and depends on the background of the patient and the purpose of the evaluation. Neuropsychologists may be asked to evaluate patients as part of their initial medical evaluation for epilepsy; to monitor the effects of repeated seizures on brain functioning, to monitor the effects of AEDs; as part of the preoperative or postoperative evaluation for surgical management of epilepsy; to address school and/or work issues; or to assist with the differential diagnosis of pseudoseizures, although the sensitivity of the neuropsychological evaluation in differential diagnosis has yet to be proven. Typically, the neuropsychological evaluation yields important information in answering these referral questions. It is common practice to administer a test battery that measures the major areas of neuropsychological functioning, as well as personality or emotional adjustment. A broad-based assessment is important in most cases as the neuropsychological evaluation is often the basis for a number of interventions, whether it be counseling a patient regarding the cognitive risks attendant to cortical resection of a suspected seizure focus or assisting with vocational planning. Thus, the neuropsychological evaluation can be used for a number of purposes and should therefore be carefully constructed to maximize its utility.

REFERENCES

Annegers, J. F. (1994). The natural course of epilepsy: An epidemiologic perspective. In A. R. Wyler & B. P. Hermann (Eds.), *The surgical management of epilepsy*. Boston: Butterworth-Heinemann.

Augustine, E. A., Novelly, R. A., Mattson, R. H., Glaser, G. H., Williamson, P. D., Spencer, D. D., & Spencer, S. S. (1984). Occupational adjustment following neurosurgical treatment of epilepsy. *Annals of Neurology, 15*, 68–72.

Bauer, R. M., Breier, J., Crosson, B., Gilmore, R., Fennell, E. B., & Roper, S. (1994). Neuropsychological functioning before and after unilateral temporal lobectomy for intractable epilepsy. *Program and Abstracts of the International Neuropsychological Society, 60*.

Bengzon, A. R. A., Rasmussen, T., Gloor, P., Dussault, J., & Stephens, M. (1968). Prognostic factors in the surgical treatment of temporal lobe epileptics. *Neurology, 18*, 717–731.

Bennett, T. L. (1992). Cognitive effects of epilepsy and anticonvulsant medications. In T. L. Bennett (Ed.), *The neuropsychology of epilepsy*. New York: Plenum Press.

Boll, T. J. (1981). The Halstead–Reitan neuropsychological battery. In S. B. Filskov & T. J. Boll (Eds.), *Handbook of clinical neuropsychology*. New York: Wiley-Intersciences.

Bolter, J. (1986). Epilepsy in children: neuropsychological effects. In J. E. Obrzut & G. W. Hynd (Eds.), *Child neuropsychology*. San Diego: Academic Press, Inc.

Bornstein, R. S., Pakalnis, A., & Drake, M. E. (1988). Verbal and nonverbal memory and learning in patients with complex partial seizures of temporal lobe origin. *Journal of Epilepsy, 1*, 203–208.

Bourgeois, B. F. D., Prensky, A. L., Palkes, H. S., Talent, B. K., & Busch, S. G. (1983). Intelligence in epilepsy: a prospective study in children. *Annals of Neurology, 14*, 438–444.

Chelune, G. J. (1994). The role of neuropsychological assessment in the presurgical evaluation of the epilepsy surgery candidate. In A. R. Wyler & B. P. Hermann (Eds.), *The surgical management of epilepsy*. Boston: Butterworth-Heinemann.

Chelune, G. J., & Moehle, K. A. (1986). Neuropsychological assessment and everyday functioning. In D. Wedding, A. M. Horton, & J. Webster (Eds.), *The neuropsychology handbook: Behavioral and clinical perspectives*. New York: Springer Publishing Company.

Chelune, G. J., Naugle, R. I., Lüders, H., & Awad, I. A. (1991). Prediction of cognitive change as a function of ability status among temporal lobectomy patients seen at 6-month follow-up. *Neurology, 41*, 399–404.

Chelune, G. J., Naugle, R. I., Lüders, H., Sedlak, J., & Awad, I. A. (1993). Individual change after epilepsy surgery: Practice effects and base-rate information. *Neuropsychology, 7*, 41–52.

Chevrie, J. J., & Aicardi, J. (1972). Childhood epileptic encephalopathy with slow spike-wave: A statistical study of 80 cases. *Epilepsia, 13*, 259–271.

Clemmons, D. C., & Dodrill, C. B. (1983). Vocational outcomes of high school students with epilepsy. *Journal of Applied Rehabilitation Counseling, 14*, 49–53.

DeGiorgio, C. M., Tomiyasu, U., Gott, P. S., & Treiman, D. M. (1992). Hippocampal pyramidal cell loss in human status epilepticus. *Epilepsia, 33*, 23–27.

Dikmen, S. (1980). Neuropsychological aspects of epilepsy. In B. P. Hermann (Ed.), *A multidisciplinary handbook of epilepsy*. Springfield, IL: Thomas Books.

Dodrill, C. B. (1978). A neuropsychological battery for epilepsy. *Epilepsia, 19*, 611–623.

Dodrill, C. B. (1986a). Correlates of generalized tonic-clonic seizures with intellectual, neuropsychological, emotional and social function in patients with epilepsy. *Epilepsia, 27*, 399–411.

Dodrill, C. B. (1986b). Psychological consequences of epilepsy. In S. B. Filskov & T. J. Boll (Eds.), *Handbook of clinical neuropsychology* (Vol. 2). New York: John Wiley & Sons.

Dodrill, C. B. (1993). Cognitive and psychosocial effects of epilepsy on adults. In E. Wyllie (Ed.), *The treatment of epilepsy: Principles and practices*. Philadelphia: Lea & Febiger.

Dodrill, C. B., & Clemmons, D. (1984). Use of neuropsychological tests to identify high school students with epilepsy who later demonstrate inadequate performances in life. *Journal of Consulting and Clinical Psychology, 52*, 520–527.

Dodrill, C. B., & Matthews, C. G. (1992). The role of neuropsychology in the assessment and treatment of persons with epilepsy. *American Psychologist, 47*, 1139–1142.

Dodrill, C. B., & Temkin, N. R. (1989). Motor speed is a contaminating factor in evaluating the "cognitive" effects of phenytoin. *Epilepsia, 30,* 453–457.

Dodrill, C. B., & Wilensky, A. J. (1990). Intellectual impairment as an outcome of status epilepticus. *Neurology, 40*(Suppl. 2), 23–27.

Dodrill, C. B., & Wilkus, R. J. (1978). Neuropsychological correlates of the electroencephalogram in epileptics: III. Generalized nonepileptiform abnormalities. *Epilepsia, 19,* 453–462.

Dodrill, C. B., Wilkus, R. J., & Ojemann, L. M. (1992). Use of psychological and neuropsychological variables in selection of patients for epilepsy surgery. *Epilepsy Research, 5* (Suppl.), 71–75.

Duchowny, M. (1993). Identification of surgical candidates and timing of operation: An overview. In E. Wyllie (Ed.), *The treatment of epilepsy: Principles and practices.* Philadelphia: Lea & Febiger.

Dulac, O., & Plouin, P. (1993). Infantile spasms and West syndrome. In E. Wyllie (Ed.), *The treatment of epilepsy: Principles and practice.* Philadelphia: Lea & Febiger.

Eisendrath, S. J., & Valan, M. N. (1994). Psychiatric predictors of pseudoepileptic seizures in patients with refractory seizures. *Journal of Neuropsychiatry and Clinical Neurosciences, 6,* 257–260.

Engel, J. (1989). *Seizures and epilepsy.* Philadelphia: F. A. Davis Company.

Farrell, K. (1993). Secondary generalized epilepsy and Lennox–Gastaut syndrome. In E. Wyllie (Ed.), *The treatment of epilepsy: Principles and practice.* Philadelphia: Lea & Febiger.

Favata, I., Leuzzi, V., & Curatolo, P. (1987). Mental outcome in West syndrome: Prognostic value of some clinical factors. *Journal of Mental Deficiency Research, 31,* 9–15.

Fraser, R. T. (1980). Vocational aspects of epilepsy. In B. P. Hermann (Ed.), *A multidisciplinary handbook of epilepsy.* Springfield, IL: Thomas Books.

Fraser, R. T., & Dodrill, C. B. (1994). Vocational outcome. In A. R. Wyler & B. P. Hermann (Eds.), *The surgical management of epilepsy.* Boston: Butterworth-Heinemann.

Freeman, J. M., Jacobs, H., Vining, E., & Rabin, C. E. (1984). Epilepsy and the inner city schools: A school-based program that makes a difference. *Epilepsia, 25,* 438–442.

Garcia, P. A., Laxer, K. D., Barbaro, N. M., & Dillon, W. P. (1994). Prognostic value of qualitative magnetic resonance imaging hippocampal abnormalities in patients undergoing temporal lobectomy for medically refractory seizures. *Epilepsia, 35,* 520–524.

Gastaut, H. (1982). The Lennox–Gastaut syndrome: Comments on the syndrome's terminology and nosological position among the secondary generalized epilepsies of childhood. *Electroencephalography and Clinical Neurophysiology, 35*(Suppl.), S71–S84.

Genton, P., & Roger, J. (1993). The progressive myoclonus epilepsies. In E. Wyllie (Ed.), *The treatment of epilepsy: Principles and practice.* Philadelphia: Lea & Febiger.

Gilroy, J. (1990). Epilepsy. In *Basic neurology* (2nd ed., pp. 67–81). Pergamon Press: New York.

Goodman, R. A., & Whitaker, H. A. (1985). Hemispherectomy: A review (1928–1981) with special reference to the linguist abilities and disabilities of the residual right hemisphere. In C. Best (Ed.), *Hemispheric function and collaboration in the child.* Orlando, FL: Academic Press.

Hathaway, S. R., & McKinley, J. C. (1943). *Manual for the Minnesota Multiphasic Personality Inventory.* New York: Psychological Corporation.

Hauser, W. A., & Hesdorffer, D. C. (1990). *Epilepsy: Frequency, causes, and consequences.* New York: Demos Publishing.

Haynes, S. D., & Bennett, T. L. (1992). Historical perspective and overview. In T. L. Bennett (Ed.), *The neuropsychology of epilepsy.* New York: Plenum Press.

Hermann, B. P. (1991). Contributions of traditional assessment procedures to an understanding of the neuropsychology of epilepsy. In W. E. Dodson, M. Kinsbourne, & B. Hiltbrunner (Eds.), *The assessment of cognitive function in epilepsy.* New York: Demos Publications.

Hermann, B. P. (1993). Neuropsychological assessment in the diagnosis of non-epileptic seizures. In A. J. Rowan & J. R. Gates (Eds.), *Non-epileptic seizures.* Boston: Butterworth-Heinemann.

Hermann, B. P., & Connell, B. E. (1992). Neuropsychological assessment in the diagnosis of nonepileptic seizures. In T. L. Bennett (Ed.), *The neuropsychology of epilepsy.* New York: Plenum Press.

Hermann, B. P., & Wyler, A. R. (1988). Effects of anterior temporal lobectomy on language function: A controlled study. *Annals of Neurology, 23,* 585–588.

Hirsch, E., Marescaux, C., Maquet, P., Metz-Lutz, M. N., Kiesmann, M., Salmon, E., Franck, G., & Kurtz, D. (1990). Landau–Kleffner syndrome: A clinical and EEG study of five cases. *Epilepsia, 31,* 756–767.

Hoch, D. B., Hill, R. A., & Oas, K. H. (1994). Epilepsy and mental decline. *Neurologic Clinics, 12,* 101–113.

Homan, R. W., & Rosenberg, H. C. (1993). Benzodiazepines. In E. Wyllie (Ed.), *The treatment of epilepsy: Principles and practice.* Philadelphia: Lea & Febiger.

Hooper, S. R., & Willis, S. R. (1989). Treatment. In *Learning disability subtyping* (pp. 164–192). New York: Springer-Verlag.

Hutt, S. J., Jackson, P. M., Belsham, A., & Higgins, G. (1968). Perceptual-motor behaviour in relation to blood phenobarbitone level: A preliminary report. *Developmental Medicine and Child Neurology, 10,* 626–632.

International League Against Epilepsy. Commission on Classification and Terminology (1981). Proposal for revised clinical and electroencephalographic classification of epileptic seizures. *Epilepsia, 22,* 489–501.

International League Against Epilepsy. Commission on Classification and Terminology (1989). Proposal for revised classifications of epilepsies and epileptic syndromes. *Epilepsia, 30,* 389–399.

Ivnik, R. J., Sharbrough, F. W., & Laws, E. R. (1987). Effects of anterior temporal lobectomy on cognitive function. *Journal of Clinical Psychology, 43,* 128–137.

Jack, C. R. (1994). MRI-based hippocampal volume measurements in epilepsy. *Epilepsia, 35* (Suppl. 6), S21–S29.

Jackson, G. D., Connelly, A., Cross, J. H., Gordon, I., & Gadian, D. G. (1994). Functional magnetic resonance imaging of focal seizures. *Neurology, 44,* 850–856.

Jones-Gotman, M. (1991). Presurgical neuropsychological evaluation for localization and lateralization of seizure focus. In H. Lüders (Ed.), *Epilepsy surgery.* New York: Raven Press.

Jones-Gotman, M., Smith, M. L., & Zatorre, R. J. (1993). Neuropsychological testing for localizing and lateralizing the epileptogenic region. In E. Wyllie (Ed.), *The treatment of epilepsy: Principles and practice.* Philadelphia: Lea & Febiger.

Kneebone, A. C., Chelune, G. J., Dinner, D., Awad, I. A., & Naugle, R. I. (1995). Use of the intracarotid amobarbital procedure to predict material specific memory change following anterior temporal lobectomy. *Epilepsia, 36,* 857–865.

Landau, W. M., & Kleffner, F. R. (1957). Syndrome of acquired aphasia with convulsive disorder in children. *Neurology, 7,* 523–530.

Lesser, R. P. (1994). The role of epilepsy centers in delivering care to patients with intractable epilepsy. *Neurology, 44,* 1347–1352.

Lesser, R. P., Lüders, H., Wyllie, W., Dinner, D. S., & Morris, H. H. (1986). Mental deterioration in epilepsy. *Epilepsia, 27*(Suppl. 2), S105–S123.

Lezak, M. D. (1995). *Neuropsychological assessment* (3rd ed.). New York: Oxford University Press.

Loiseau, P. (1993). Benign focal epilepsies of childhood. In E. Wyllie (Ed.), *The treatment of epilepsy: Principles and practice.* Philadelphia: Lea & Febiger.

Matthews, W. S., Barabas, G., & Ferrari, M. (1982). Emotional concomitants of childhood epilepsy. *Epilepsia, 23,* 671–681.

Matthews, C. G., & Kløve, H. (1967). Differential psychological performances in major motor, psychomotor and mixed seizure classification of known and unknown etiology. *Epilepsia, 8,* 117–128.

Mattson, R. H. (1994). Current challenges in the treatment of epilepsy. *Neurology, 44*(Suppl. 5), S4–S9.

May, T. W., Bulmahn, A., Wohluter, M., & Rambeck, B. (1992). Effects of withdrawal of phenytoin on cognitive and psychomotor functions in hospitalized epileptic patients on polytherapy. *Acta Neurologica Scandinavica, 86,* 165–170.

McGowan, M. E. L., Neville, B. G. R., & Reynolds, E. H. (1983). Comparative monotherapy trial in children with epilepsy. *British Journal of Clinical Practice, Symposium Supplement, 27,* 115–118.

McIntosh, G. C. (1992). Neurological conceptualizations of epilepsy. In T. L. Bennetts (Ed.), *The neuropsychology of epilepsy.* New York: Plenum Press.

Meador, K. J., & Loring, D. W. (1991). Cognitive effects of antiepileptic drugs. In O. Devinsky & W. Theodore (Eds.), *Epilepsy and behavior* (pp. 151–170). New York: Wiley-Liss.

Meador, K. J., Loring, D. W., Abney, O. L., Allen, M. E., Moore, E. E., Zamrini, E. Y., & King, D. W. (1993). Effects of carbamazepine and phenytoin on EEG and memory in healthy adults. *Epilepsia, 34,* 153–157.

Meador, K. J., Loring, D. W., Huh, K., Gallagher, B. B., & King, D. W. (1990). Comparative cognitive effects of anticonvulsants. *Neurology, 40,* 391–394.

Medical Research Council (MRC) Antiepileptic Drug Withdrawal Study Group. (1991). Randomised study of antiepileptic drug withdrawal in patients in remission. *Lancet, 337,* 1175–1180.

National Institutes of Health Consensus Development Conference. (1990). Surgery for Epilepsy. *Consensus Statement, 8(2).*

Naugle, R. I. (1991). Neuropsychological effects of surgery of epilepsy. In H. Lüders (Ed.), *Epilepsy surgery.* New York: Raven Press.

Novelly, R. A. (1992). The debt of neuropsychology to the epilepsies. *American Psychologist, 47,* 1126–1129.

Novelly, R. A., Augustine, E. A., Mattson, R. H., Glaser, G. H., Williamson, P. D., Spencer, D. D., & Spencer, S. S. (1984). Selective memory improvement and impairment in temporal lobectomy for epilepsy. *Annals of Neurology, 15,* 64–67.

Ojemann, G., & Dodrill, C. B. (1985). Verbal memory deficits after temporal lobectomy for epilepsy. *Journal of Neurosurgery, 62,* 101–107.

O'Leary, D. S., Seidenberg, M., Berent, S., & Boll, T. J. (1981). Effects of age of onset of tonic-clonic seizures on neuropsychological performance in children. *Epilepsia, 22,* 197–204.

Olivier, A. (1988). Risk and benefit in the surgery of epilepsy: Complications and positive results on seizures tendency and intellectual function. *Acta Neurologica Scandinavica, 78,* 114–121.

Özkara, C., & Dreifuss, F. E. (1993). Differential diagnosis in pseudoepileptic seizures. *Epilepsia, 34,* 294–298.

Pellock, J. M. (1994). Standard approach to antiepileptic drug treatment in the United States. *Epilepsia, 35*(Suppl. 4), S11–S18.

Penfield, W., & Milner, B. (1958). Memory deficit produced by bilateral lesions in the hippocampal zone. *Archives of Neurology and Psychiatry, 79,* 475–497.

Penfield, W., & Rasmussen, T. (1950). *The cerebral cortex of man.* New York: Macmillan.

Porter, R. J. (1993). Epileptic and non-epileptic seizures. In A. J. Rowan & J. R. Gates (Eds.), *Non-epileptic seizures.* Boston: Butterworth-Heinemann.

Rausch, R., & Crandall, P. H. (1982). Psychological status related to surgical control of temporal lobe seizures. *Epilepsia, 23,* 191–202.

Reitan, R. M. (1966). A research program on the psychological effects of brain lesions in human beings. In N. R. Ellis (Ed.), *International review of research in mental retardation* (Vol. 1). New York: Academic Press.

Reitan, R. M., & Wolfson, D. (1993). *The Halstead–Reitan Neuropsychological Test Battery: theory and clinical interpretation* (2nd ed.). Tucson, AZ: Neuropsychology Press.

Reynolds, E. H. (1988). Hughlings Jackson: A Yorkshireman's contribution to epilepsy. *Archives of Neurology, 45,* 675–678.

Riikonen, R., & Amnell, G. (1981). Psychiatric disorders in children with earlier infantile spasm. *Developmental Medicine and Child Neurology, 23,* 747–760.

Roger, J., Genton, P., Bureau, M., & Dravet, C. (1993). Less common epileptic syndromes. In E. Wyllie (Ed.), *The treatment of epilepsy: Principles and practice.* Philadelphia: Lea & Febiger.

Sackellares, J. C., Giordani, B., Berent, S., Seidenberg, M., Dreifuss, F., Vanderzant, C. W., & Boll, T. J. (1985). Patients with pseudoseizures: Intellectual and cognitive performance. *Neurology, 35,* 116–119.

Sander, J. W. A. S. (1993). Some aspects of prognosis in the epilepsies: a review. *Epilepsia, 34,* 1007–1016.

Saykin, A. J., Gur, R. C., Sussman, N. M., O'Connor, M. J., & Gur, R. E. (1989). Memory deficits before and after temporal lobectomy: Effect of laterality and age of onset. *Brain and Cognition, 9,* 191–200.

Scoville, W. B., & Milner, B. (1957). Loss or recent memory after bilateral hippocampal lesions. *Journal of Neurology, Neurosurgery and Psychiatry, 20,* 11–21.

Seidenberg, M., Beck, N., Geisser, M., Giordani, B., Sackellares, J. C., Berent, S., Dreifuss, F. E., & Boll, T. J. (1986). Academic achievement of children with epilepsy. *Epilepsia, 27,* 753–759.

Serratosa, J. M., & Delgado-Escueta, A. V. (1993). Juvenile myoclonic epilepsy. In E. Wyllie (Ed.), *The treatment of epilepsy: Principles and practice.* Philadelphia: Lea & Febiger.

Spencer, S. S. (1994). The relative contributions of MRI, SPECT, and PET imaging in epilepsy. *Epilepsia, 35*(Suppl. 6), S72–S89.

Trimble, M. R. (1987). Anticonvulsant drugs and cognitive function: A review of the literature. *Epilepsia, 28*(Suppl. 3), S37–S45.

Wannamaker, B. B., & Matthews, C. G. (1976). Prognostic implications of neuropsychological test performance for surgical treatment of epilepsy. *The Journal of Nervous and Mental Disease, 163,* 29–34.

Wechsler, D. (1955). *Manual for the Wechsler Adult Intelligence Scale.* New York: The Psychological Corporation.

Wechsler, D. (1981). *WAIS-R manual.* New York: The Psychological Corporation.

Wilkus, R. J., & Dodrill, C. G. (1976). Neuropsychological correlates of the electroencephalogram in epileptics: I. Topographical distribution and average rate of epileptiform activity. *Epilepsia, 17,* 89–100.

Wilkus, R. J., & Dodrill, C. G. (1989). Factors affecting the outcome of MMPI and neuropsychological assessments of psychogenic and epileptic seizure patients. *Epilepsia, 30,* 339–347.

Wilkus, R. J., Dodrill, C. G., & Thompson, P. M. (1984). Intensive EEG monitoring and psychological studies of patients with pseudoepileptic seizures. *Epilepsia, 25,* 100–107.

Willmore, L. J., & Wheless, J. W. (1993). Adverse effects of antiepileptic drugs. In E. Wyllie (Ed.), *The treatment of epilepsy: Principles and practice.* Philadelphia: Lea & Febiger.

Ziegler, R. G. (1981). Impairments of control and competence in epileptic children and their families. *Epilepsia, 22,* 339–346.

Zimmerman, F. T., Burgemeister, B. B., & Putnam, T. J. (1951). Intellectual and emotional makeup of the epileptic. *Archives of Neurology and Psychiatry, 65,* 545–556.

14

Evaluation of Neuropsychiatric Disorders

Doug Johnson-Greene and Kenneth M. Adams

Questions relating to cognitive functioning in persons with psychiatric illness comprise one of the most common referrals to neuropsychologists, second currently only to referrals for evaluation of closed head injury (Grant & Adams, 1996). The well-informed reader may ask, particularly in this age of managed health care, what advantages are present in conducting neuropsychological evaluations with psychiatric patients given that the incidence of overt neurological dysfunction is relatively small in these populations, aside from the iatrogenic effects of neuroleptic medication. Neuropsychological evaluations can be construed as relatively time-consuming as well as costly; and occasionally they yield more questions than answers. These criticisms not withstanding, there are several important rationales for conducting neuropsychological assessments in neuropsychiatric populations.

Rationale for Evaluating Neuropsychiatric Patients

First of all, many psychiatric disorders are associated with cognitive deficits and, not surprisingly, these deficits have been found to correlate with structural changes in the brain. A given psychiatric condition, as we will see later in this chapter, may have a unique constellation of cognitive deficits, frequently in the absence of significant mental status or neurological findings. Mental status and neurological examinations may lack the sensitivity and specificity necessary to elucidate fully the cognitive concerns in psychiatric patients. As such, cursory screening approaches tend to act more as a litmus test for organicity, indicating in a less than reliable manner whether dysfunction is present without specific elucidation of the

Doug Johnson-Greene Department of Physical Medicine and Rehabilitation, Johns Hopkins University School of Medicine, Baltimore, Maryland 21239. Kenneth M. Adams Psychology Service, VA Medical Center, and Department of Psychiatry–Division of Psychology, University of Michigan Medical Center, Ann Arbor, Michigan 48109-0840.

Neuropsychology, edited by Goldstein *et al.* Plenum Press, New York, 1998.

296

DOUG
JOHNSON-
GREENE AND
KENNETH M.
ADAMS

nature and extent of the dysfunction. Simply stated, neuropsychological evaluation is a more effective and sensitive method than medical and mental status examinations for assessing cognitive deficits among psychiatric populations. Once characterized, cognitive deficits can then be taken into account in clinical practice, such as when making recommendations for treatment or discharge planning.

In addition, despite the unique constellation of cognitive concerns that frequently accompanies a psychiatric disturbance, individuals with the same general psychiatric diagnosis may differ with regard to the relative pattern of deficits. The heterogeneity of deficits within a given symptom constellation, thought to be reflective of the heterogeneity of diagnostic entities, suggests that a "cookie cutter approach" to providing assessment and treatment in psychiatric patients is ill advised and is more suited to the lowest common denominator within a patient group. Methods of research on psychiatric disorders involving "screens," rating scales, and self-report measures have enjoyed superficial advantages involving little professional time investment as well as a face-valid appeal concerning cognition. Unfortunately, the fundamental knowledge yield of such methods in scientific terms and improved clinical decision-making has been minimal.

Quantification of a patient's cognitive strengths and deficits may suggest appropriate individualized clinical interventions and will presumably provide the basis for well-informed recommendations tailored to the patient's specific needs and strengths. For example, a common referral question that is received by neuropsychologists pertaining to this population is: "Is this patient capable of managing his or her medication independently?" For some psychiatric patients who have significant deficits in higher cortical functioning or memory, the answer is no. Yet, an alarmingly large number of treatment facilities tend to presume that a patient is competent to manage his or her own medication, often relying more on gut instinct than data-based judgments. One can only speculate as to whether this is a contributing factor to the "revolving door" phenomenon associated with the chronically mentally ill. Neuropsychological data can also become particularly important when the specific needs of a patient change and fluctuate over time in response to the changing characteristics of their psychiatric illness, as many psychiatric illnesses are episodic or seldom remain static over time. In this regard, a neuropsychological evaluation can serve as a baseline or benchmark whose purpose is to document progression of the illness through comparison of serial neuropsychological assessments.

A third role of neuropsychological evaluation is its use as a tool in making differential diagnoses. Some psychiatric conditions can mimic neurological conditions, or vice versa. In fact, psychiatric symptoms are often the first signs of some underlying neurological or medical condition. General medical conditions such as diabetes (Ryan & Williams, 1993) and more pervasive neurological conditions such as HIV infection (Bornstein, Nasrallah, Para, & Whitacre, 1993) are known to impact cognitive function. Subtle cognitive deficits may represent the first clinical signs of an underlying neurological illness. For example, the first signs of dementia are often found within the cognitive domain (Bondi, Monsch, Galasko, & Butters, 1994), sometimes long before there are other symptoms associated with the condition. A related issue is that interpretation of neuropsychological performance is predicated on garnering maximal motivation from the patient. Psychiatric conditions, such as depression and anxiety, can and do compromise test performance. One function of neuropsychological assessment then is to discern organicity from a patient's complex symptom constellation. Neuropsychological assessment can provide additional data to evaluate these conative influences, which in turn can be taken into consideration in evaluating differential diagnoses.

Finally, from a heuristic standpoint neuropsychological assessment may facilitate our understanding of the etiology, nosology, characteristics, prognosis, and treatment of psychiatric illness. Despite the vast resources that have been provided for research, most psychiatric illnesses remain poorly understood. A host of extraneous variables are endemic in research involving psychiatric patients, such as the effects of medication, comorbid substance abuse, difficulty characterizing patient symptoms and diagnostic entities, and developmental abnormalities and risk factors such as learning disabilities, to name a few. Neuropsychological investigations afford the possibility of understanding brain–behavior relationships in these complex patients, which in turn may yield greater insight as to the mechanisms associated with psychiatric illness.

From the aforementioned arguments the reader should be able to discern the basis for conducting neuropsychological assessment with psychiatric patients. The goal of the remainder of this chapter is to describe cognitive deficits associated with some of the most common psychiatric illnesses. We also attempt to provide information concerning conative influences of psychiatric illness that complicate assessment and interpretation of neuropsychological data throughout the chapter. As substance abuse is described elsewhere in this book we focus specifically on anxiety disorders, affective disturbances, and schizophrenia. Similarly, we refer the reader to other chapters throughout this series for information pertaining to psychiatric manifestations associated with neurological conditions. Obviously there are many different psychiatric conditions encountered by clinicians, but the three we have listed are among the most prevalent. Also, neuropsychological research with other neuropsychiatric disorders has been notably absent. Further rationale for curtailing our discussion to these patient groups comes from the knowledge that among psychiatric conditions, these diagnostic entities are most reliably associated with neurological deficits, structural changes in the brain, and changes in cognition.

SPECIAL CONSIDERATIONS FOR EVALUATING NEUROPSYCHIATRIC DISORDERS

There is no one test that can be used to assess, with any degree of certainty, the presence of "organicity." Questioning the very value of this term in the scientific literature has been ubiquitous over the past 25 years; yet the term has near miraculous staying power in psychiatric parlance. As we have pointed out there are problems of sensitivity and specificity associated with screening instruments, such as tests of mental status. A more valuable goal in evaluating the patient with a neuropsychiatric illness is understanding the nature and extent of a patient's cognitive deficits rather than attempting to determine where the "lesion" resides. Any battery of tests used with a patient with psychiatric illness should assess an appropriately broad spectrum of ability domains, with particular emphasis on assessment instruments that can assist in answering the referral question or questions and that have the potential for suggesting meaningful recommendations.

Unfortunately, many patients with significant psychiatric illnesses are unable to sustain maximum effort for more than a few hours at a time, and sometimes for much less. In our own practice we have found that smaller blocks of time spread

298

DOUG
JOHNSON-
GREENE AND
KENNETH M.
ADAMS

over a day or several days can be more effective than administering a lengthy neuropsychological battery. This also decreases the likelihood of being in the awkward position of trying to discern whether decreased performance on neuropsychological tests represents genuine deficits or is secondary to the effects of fatigue and/or decreased motivation to perform. The overriding emphasis in any evaluation, particularly with neuropsychiatric patients, must be the procurement of optimal performance and motivation from the patient. For this reason, comprehensive testing must sometimes give way to measures that maintain the patient's maximal performance.

It is also recommended that patients hospitalized with psychotic symptoms be assessed near the end of their inpatient stay, presumably once they have been stabilized on medication. This will ensure that they are at or near baseline functioning and that their psychiatric illness does not interfere with the actual testing process. Anyone who has evaluated a patient who has florid paranoia, for example, can attest to the difficulty in acquiring responses, not to mention motivated performance. Patients who are seen on an outpatient basis should also be at or near baseline functioning, though this is often more difficult to determine. Obtaining a good clinical history, with particular emphasis on medication compliance, along with reports from significant others or care providers should provide the most direct indication of the patient's current mental status and the extent to which this current status is similar to baseline functioning. It is perhaps more important to obtain collateral reports of psychological functioning and well-being for patients with psychiatric illnesses than for any other patient group. As any seasoned neuropsychologist can attest, reports from patients are frequently discrepant from those of their relatives and significant others along a wide range of dimensions.

NEUROPSYCHOLOGICAL FINDINGS

MOOD DISORDERS

Affective disorders represent the most common general psychiatric condition for which mental health treatment is sought. It has been estimated that nearly 70% of all persons will experience a major depressive episode sometime during their lifetime. For the individual who is referred for evaluation of a suspected organic condition, affective states can become an important variable to consider for several reasons. Depression can mimic neurological conditions, especially dementia in the elderly (McAllister, 1983; Spar, 1982). While the concept of "pseudodementia" may have been overgeneralized and applied in questionable cases in the 1980s, it remains an interesting phenomenon observed particularly in inpatient settings. It will be of interest to see how mental health services provided in an emerging, near exclusively outpatient/ambulatory environment will cope with such patients.

It should also be noted that depression is often the first reported symptom for a number of medical conditions (Hall, 1980) as well as several neurological conditions. For example, right-handed patients (who typically have left hemisphere dominant language function) who suffer left frontal hemispheric strokes quite often display symptoms of depression. Obviously, it is important to know what proportion of the patient's cognitive loss or excess is related to affective disturbance and how much is related to organic pathology.

Research regarding the cognitive effects of affective conditions has had to contend with a number of methodological hurdles. Subject groups can be quite heterogeneous, despite having the same diagnosis. The effect of medications (e.g., antidepressants, lithium) and electroconvulsive therapy (ECT) is not well known in these patients and many studies have failed to control for these factors, although enduring neuropsychological effects of unilateral ECT or medications for affective disorders have not been clearly identified. Nonetheless, a number of researchers retain the hypothesis that these factors alter neuropsychological status.

Finally, elderly patients, a group of individuals who are more prone to develop depression, frequently have concurrent medical conditions which are known to affect cognitive functioning. Also, research has employed a variety of psychometric instruments making comparison research results across studies almost impossible. In general, well-controlled studies have not been available until recently and more recent studies have produced more questions than they have answered.

UNIPOLAR DEPRESSION. There have been a number of studies that have investigated neuropsychological performance among neurologically intact patients with depression. It is somewhat unclear if depression affects global functioning uniformly or if depression produces decreased cognitive performance in specific ability realms, although, at present there is more support for the latter. Fisher, Sweet, and Pfaelzer-Smith (1986) studied 15 depressed inpatients both before and after treatment, and 15 normal controls. They found that relative to normal control subjects, depressed patients showed impaired performance on 10 of 14 measures, suggesting general rather than specific impairment. The authors further suggest that clinicians may need to adjust their impairment criteria when working with depressed patients in order to prevent diagnostic misclassifications. Similar results have been found by other researchers in support of global cognitive deterioration for depressed patients (Siegal & Gershon, 1986; Siegal, Gurevich, & Oxenkrug, 1989; Siegfried, 1985). In contrast, some researchers have been unable to document cognitive declines in depressed patients who exhibit symptoms and characteristics of depression, but who are not clinically depressed according to current diagnostic classifications (Bieliauskas, 1993; Niederehe & Camp, 1985; Williams, Little, Scates, & Blockman, 1987). Therefore, there is some suggestion that isolated depressive symptoms do not affect cognition to the same extent as severe clinical depressive syndromes. Interestingly, one well-controlled study of 73 elderly depressed patients and 110 normal controls by Boone and colleagues (1995) found that more severely depressed patients had mildly decreased executive and information-processing speed deficits compared to mildly depressed patients, but they did not exhibit learning and memory, visuospatial, naming, or attentional deficits as did the mild depression group.

These studies would seem to suggest that the extent of cognitive impairment is directly related to the intensity of the depressive episode (Weingartner & Silberman, 1982). Presumably, then, one method for differentiating organic from non-organic depressive states is to conduct repeat testing once the patient's depressive condition improves. Several studies support the notion that persisting cognitive deficits, particularly following alleviation of depressive symptoms, more strongly suggest genuine neurological impairment (Boyar, 1981; Fisher et al., 1986; Sweet, 1983).

Still others suggest that patients with depression exhibit deficits in specific ability realms. Neuropsychological investigations of hospitalized patients with severe depression have consistently found deficits in specific domains such as learning and

300

DOUG
JOHNSON-
GREENE AND
KENNETH M.
ADAMS

memory (Brand, Jolles, & Gispen-de Wied, 1992; King, Caine, Conwell, & Cox, 1991; King, Caine, & Cox, 1993; Speedie, Rabins, & Pearlson, 1990), psychomotor speed (Cassens, Wolfe, & Zola, 1990; Miller, 1975; Raskin, 1986; Weingartner & Silberman, 1982), and visuospatial skills (Cassens *et al.*, 1990; Richards & Ruff, 1989), and evidence is mixed for naming deficits (Emery & Breslau, 1989; Hill *et al.*, 1992; King *et al.*, 1991). Some have suggested that the cognitive effects evidenced by these individuals are the result of concurrent medical conditions, advancing age, or both. At least one study suggests that age may influence tasks of psychomotor function, copying, and perceptual integration, but presence of medical illness apparently was found to have minimal influence (King, Cox, Lyness, & Caine, 1995).

BIPOLAR AFFECTIVE DISORDER. In comparison to unipolar depression, the cognitive effects of bipolar affective disorder have been much less extensively researched. One reason for this may be simply that patients who are hospitalized in the midst of a manic episode are not as complacent and cooperative with testing as depressed patients. Since cooperation and task orientation are required components of a valid examination, patients who are manic, or for that matter bipolar affective patients who are hospitalized during a pervasive depressive cycle of their illness may not be amenable to testing. Furthermore, medications can interfere with cognition making interpretation of test results difficult. Lithium and carbamazepine, the most commonly prescribed medications for the disorder, may have an untoward effect on cognitive functioning and neuropsychological testing, particularly at toxic blood levels.

While most researchers agree that cognitive impairments are present in patients with bipolar illness, the nature and extent of those impairments remain quite unclear. As is the case with many other diagnostic entities, the diagnosis of bipolar disorder comprises a heterogeneous group of illnesses. At present the diagnostic criteria for bipolar disorder are all encompassing and the course of the disorder can vary from a single manic episode years before, with the presence of continuous depression since, to persistent unremitting mania without incidence of significant depression.

Several studies have attempted to ascertain the nature and extent of cognitive difficulties within this group of patients. One study by Coffman, Bornstein, Olson, Schwarzkopf, and Nasrallah (1990) found significant and diffuse cognitive impairments among a group of patients with bipolar affective disorder. However, this group may have represented an extreme subset of bipolar patients in that the authors noted that the subjects had all experienced psychotic symptoms. Another study by Jones, Duncan, Mirsky, Post, and Theodore (1994) investigating the neuropsychological profiles of 26 neurologically intact patients with bipolar affective disorder found that compared to normal controls, these patients had more difficulty on tests from the "focusing-execution domain" (i.e., Trail Making Test, Stroop Color-Word Interference Test, and Talland Letter Cancellation Task) and on a task of psychomotor speed and fine motor movement (i.e., Purdue Pegboard). Surprisingly, and in contrast to studies of patients with unipolar depression, the patients in this study did not differ from normal controls in terms of memory functioning. The impaired neuropsychological test performance in the Jones *et al.* (1994) study could also be considered global since the Trails test and Stroop Color-Word test are thought to reflect diffuse rather than specific impairment.

Several studies have looked at the relationship between cortical abnormalities detected with computed tomography (CT) and magnetic resonance imaging (MRI) with cognitive impairments exhibited by these patients. Imaging studies of bipolar patients have yielded abnormalities that are highly similar to those found in patients with schizophrenia (Nasrallah, Coffman, & Olson, 1989). Patients with bipolar disorder have shown increased cerebrospinal fluid (CSF) filled spaces; ventricular enlargement (especially enlargement of the third ventricle); and hyperdensity of anterior frontal regions, subcortical structures such as the caudate and thalamus, and right temporal lobe. In the best controlled study to date, Dupont and colleagues (1990) studied 19 patients with bipolar affective disorder and 10 normal controls with MRI. The results of this study suggest that there is an unusually high incidence of subcortical T2-weighted signal hyperintensities in the brains of bipolar patients compared to normal controls (i.e., 9 of 19 bipolar patients had subcortical signal hyperintensities but none of the normal control subjects matched for age and education did). The significance of these findings are unknown since repeat scanning of bipolar patients failed to reveal any changes from initial MRI scans. However, comparison of bipolar patients with and without MRI abnormalities revealed significantly worse performance on verbal fluency and Digit Symbol tests, findings which could be consistent with a subcortical degenerative process.

SUMMARY OF AFFECTIVE DISORDERS. In summary, there is minimal evidence that suggests that patients with bipolar affective disorder exhibit global deficits, unlike unipolar depression. The degree of cognitive impairment may be related to the severity of both illnesses. Unlike patients with unipolar depression, memory may be well preserved in patients with bipolar affective disorder. Given the paucity of studies with these populations, well-controlled investigations correcting for extraneous variables such as medication and severity of illness would greatly enhance our extant knowledge of mood disorders.

ANXIETY DISORDERS

The only psychiatric condition more prevalent than anxiety disorders is mood disturbance. Consequently, in addition to their cognitive complaints, patients undergoing neuropsychological evaluation may also present with anxiety-based concerns. There is a large body of laboratory literature concerning the effect of subclinical anxiety on neuropsychological performance. Most of these studies have found slightly decreased performance on tasks requiring focused attention and concentration, as well as motor performance (Hodges & Spielberger, 1969).

There are some indications that anxiety may differentially affect neuropsychological performance by males for some tasks. For example, Martin and Franzen (1989) found that among 56 college students exposed to either an anxiety stimulus or neutral condition, males performed more poorly on the Stroop test and several measures of the Randt Memory Battery than did females. However, anxious males seem to perform better than females on motor tasks, such as the Finger Tapping test (King, Hannay, Masek, & Burns, 1978; Martin & Franzen, 1989). Interestingly, a study by Dodrill (1979) found significant anxiety effects related to gender for normal control subjects, but not for patients with neurological impairment, which may suggest that the effects of anxiety related to a person's gender in a normal population become less pronounced in neurologically impaired patients. This raises interesting questions concerning the applicability of sex-based norms for neurologically impaired and nonimpaired patients in clinical practice.

Clinically significant anxiety, on the other hand, appears to have a more pro-

302

DOUG
JOHNSON-
GREENE AND
KENNETH M.
ADAMS

nounced effect on cognitive functioning. A review of the literature pertaining to the effect of anxiety on cognitive function vis-à-vis research with patients with posttraumatic stress disorder and obsessive–compulsive disorder follows.

POSTTRAUMATIC STRESS DISORDER. Posttraumatic stress disorder (PTSD) is a disorder characterized by a mixture of heightened anxiety-related symptoms following exposure to a traumatic event. While virtually any severe traumatic event can lead to development of PTSD symptoms, most research has focused upon combat-related PTSD among veterans. The disorder is quite often intractable as evidenced by its high incidence of recidivism. Increasingly, patients have been treated in intensive impatient PTSD programs. One possible reason for the tenacity of this condition is that it is associated with a number of other comorbidities including substance abuse and increased risk of traumatic head injuries. At least one study found that patients with alcoholism were two to four times more likely to have a previous history of head injury than the general population (Hillbom & Holm, 1986). Assessment of cognitive functioning of PTSD patients can be confounded by these factors, and even the astute neuropsychologist may have difficulty assessing the relative contributions of anxiety and genuine CNS impairment in these patients.

In general, research suggests that neurologically intact patients with PTSD perform similarly on neuropsychological tests to individuals with other anxiety disorders. That is, patients with PTSD perform less well on tasks that have a strong attention or psychomotor speed component. One study by Dalton, Pederson, and Ryan (1989) administered a comprehensive neuropsychological test battery to 100 veterans being treated for PTSD in an inpatient setting. Their study found that patients had slightly decreased performance on tests sensitive to anxiety and deficits in attention/concentration, such as the Digit Symbol and Digit Span subtests of the Wechsler Adult Intelligence Scale-Revised (WAIS-R) and the Stroop color and word test.

OBSESSIVE–COMPULSIVE DISORDER. One of the most debilitating anxiety disorders, and one that is highly resistant to treatment, is obsessive–compulsive disorder (OCD). The disorder is characterized by ritualistic recurring behaviors (i.e., compulsions) and persistent intrusive thoughts (i.e., obsessions). From a cognitive perspective, these individuals appear to have difficulty with the brain mechanism responsible for assigning relative importance to environmental stimuli (Gray, 1982; Pitman, 1987). Specifically, they have difficulty filtering out irrelevant stimuli and place excessive meaning and importance to other stimuli such that emotional arousal and behavior are maintained over prolonged periods of time regardless of feedback the organism receives, resulting in a perseverative behavior pattern.

The structures most involved in emotional regulation and modulation of behavior and environmental stimuli are the frontal lobes and basal ganglia. Not surprisingly, neuroimaging studies utilizing positron emission tomography (PET) have found that local rates for glucose metabolism are increased in the orbitofrontal gyri of the frontal lobe and the caudate nucleus relative to control samples (Baxter et al., 1987, 1988; Nordahl, Benkelfat, Semple, & Gross, 1989). In addition, lesions to the frontal lobes such as those seen following head injury can result in behavior patterns which mimic obsessive–compulsive disorder (Jenicke, 1984).

Several studies have found frontal lobe dysfunction among OCD patients in the absence of deficits in other ability domains (Behar, Rappaport, Berg, & Fairbanks, 1984; Head, Bolton, & Hymas, 1989). These studies suggest that when compared to normal controls, OCD patients show greater impairment on tests sensitive to executive functioning, such as the Wisconsin Card Sorting Test. In contrast, several studies have found intact performance on neuropsychological tests of frontal lobe function, but impaired performance on tests of visuospatial skill (Boone, Ananth, Philpott, & Kaur, 1991; Zielinski, Taylor, & Juzwin, 1991).

Otto (1992) completed an extensive review of OCD and cognitive dysfunction. He highlights the possibility that cognitive dysfunction, when present, may be a result rather than a cause of OCD. A potential confounding factor in evaluating cognitive deficits in patients with OCD is that there is a high incidence of comorbid disorders, particularly depression. These comorbid disorders may, for some individuals, account for some of the variance in neuropsychological performance exhibited by these patients. Secondly, OCD is a heterogeneous disorder comprising several different symptom clusters. At least one researcher has found that patients with different experimental subtypes of OCD perform differently on neuropsychological tasks of executive functioning (Malloy, 1987).

SUMMARY OF ANXIETY DISORDERS. In laboratory studies, anxiety has been shown to both hinder and facilitate neuropsychological performance within several domains depending upon the subject's gender, though these differences may be absent in clinical populations. Both laboratory-induced anxiety and clinically diagnosed anxiety disorders appear to predominantly affect the attention/concentration domain, though other cognitive domains have also been implicated. For example, patients with OCD appear to display pronounced deficits in executive function. Patients with PTSD perform more poorly on tests of attention/concentration and motor function, though comorbid conditions such as substance abuse and psychiatric illnesses represent a considerable obstacle to the conduct of research with this group.

SCHIZOPHRENIA

Schizophrenia is the most chronic of the psychiatric illnesses from the standpoint that it affects more domains than any other illness and is more prone to affect long-term social functioning. The disorder usually first appears between the ages of 18 and 25 and there are no apparent gender differences in terms of overall prevalence rates (Reiger *et al.*, 1988). The etiology of schizophrenia remains unclear. A number of theories have been postulated including genetic predisposition (Kendler & Diehl, 1993), immunological and viral abnormalities (Kirch, 1993) and obstetric and perinatal risk factors (Jacobsen & Kinney, 1980; Lewis & Murray, 1987; Mednick, 1970), to name a few. The current conceptualization of the etiology of the disorder could be best described as a neurodevelopmental genetic risk model reflecting the interplay between genetic and nongenetic contributions (Goldstein, 1994).

A recent nosological system for classification of schizophrenia incorporates the presence of "negative" and "positive" symptoms. This distinction has also been incorporated into the nomenclature for the most recent diagnostic criteria for the disorder (*Diagnostic and Statistical Manual of Mental Disorders,* Fourth Edition; American Psychiatric Association, 1994). Patients with schizophrenia display not only behavioral deficits (i.e., negative symptoms) such as flattened affect and social withdrawal, but also behavioral excesses (i.e., positive symptoms) such as delusions

304

DOUG
JOHNSON-
GREENE AND
KENNETH M.
ADAMS

and hallucinations. The fact that not all patients with schizophrenia display the same symptoms underscores the heterogeneous nature of this illness and the probable existence of multiple subtypes of the disorder. While positive symptoms have been found to be quite responsive to neuroleptic treatment, negative symptoms remain relatively unresponsive to medication. In addition, negative symptoms are more reliably associated with both structural brain abnormalities, decreased performance on neuropsychological tests of cognitive function (Addington, Addington, & Maticka-Tyndale, 1991; Breier, Schreiber, Dyer, & Pickar, 1991; Williamson, Pelz, Merskey, Morrison, & Conlon, 1991), poorer premorbid adjustment and neurological impairment (Buchanan, Kirkpatrick, Heinrich, & Carpenter, 1990), and poorer social outcome and course of illness (Walker & Lewine, 1988).

Schizophrenia is associated with elevated rates of neurological abnormalities (Nasrallah, 1990; Weinberger, 1987; Woods, Kinney, & Yurgelun-Todd, 1991), though it is unclear how these abnormalities specifically relate to the etiology and pathogenesis of the disorder. Elevated rates of neurological impairment have also been found in nonaffected relatives of schizophrenia patients, particularly on tests of attention and executive skills (Cannon *et al.*, 1994; Kinney, Yurgelun-Todd, & Woods, 1991; Kremen *et al.*, 1994), which suggests that these abnormalities may represent a behavioral marker for familial risk. One study by Levin, Yurgelun-Todd, and Craft (1989) found that offspring of schizophrenic parents raised by foster parents also show decreased performance on tests of cognitive function compared to normal control groups with similar age and sex distributions. The implication of these findings is that cognitive deficits in these subjects are probably independent of extrinsic factors, such as socioeconomic status.

COGNITIVE DEFICITS. One primary manifestation of schizophrenia is cognitive impairment. Kraepelin's (1919/1971) early account of schizophrenia, which he termed "dementia praecox," underscores the observation that cognitive abnormalities are a central feature of the disorder. One recent investigation examining cognitive function in discordant monozygotic twin pairs found greater impairment in affected twins compared to nonaffected twins (Goldberg, *et al.*, 1990) which suggests that in genetically identical persons, the presence of schizophrenia is more strongly associated with cognitive deficits. The severity of cognitive deficits does not appear to be gender-related (Goldberg, Gold, Torrey, & Weinberger, 1995), despite several studies which suggest that the course of schizophrenia in women is considerably less chronic compared to that in men, in terms of symptom severity and onset (Childers & Harding, 1990; Goldstein, 1988; Lewine, 1981) and premorbid functioning (Addington & Addington, 1993). Another potential confound in evaluating cognitive deficits in schizophrenia, and in neuropsychology in general, concerns the method chosen for assessment of premorbid functioning. While there have been few studies examining this issue among schizophrenics, at least one investigation suggests that performance on an achievement test of sight reading is a better estimate of premorbid intelligence than the patient's educational history (Kremen *et al.*, 1996).

There has been considerable debate over the nature and extent of cognitive deficits in patients with schizophrenia, and even over whether these impairments are secondary to psychotic symptoms or represent an independent abnormality (Strauss, 1993). The search for a neuropsychological signature of schizophrenia has for the most part remained elusive. Early investigations seemed to suggest that deficits were global and that virtually all ability realms were affected (Chelune,

Heaton, Lehman, & Robinson, 1979; Heaton & Crowley, 1981; Seidman, 1984).
However, these conclusions may have been inaccurate in that the majority of studies at that time failed to control for multiple confounding variables, although several well-controlled studies have yielded similar results (Blanchard & Neale, 1994; Braff *et al.*, 1991). Blanchard and Neale (1994) found in their study of 28 schizophrenic patients, after controlling for potential confounding variables, a generalized dysfunction, though the authors make a valid point in suggesting that differential cognitive deficits may exist that are undetected by currently available neuropsychological instruments. Recently, in an attempt to find a neuropsychological signature of schizophrenia, research has focused on disruption of differential cognitive processes and efforts to define diffuse nonlocalizing impairments have lost favor.

Deficits in executive skills are widely thought to reflect frontal lobe dysfunction (Milner, 1963) and are a common finding in patients with schizophrenia (Bornstein *et al.*, 1990; Braff *et al.*, 1991). These deficits appear to be correlated with the degree of dorsolateral prefrontal cortex (DLPFC) atrophy measured with MRI (Seidman *et al.*, 1994) and decreased DLPFC cerebral blood flow (Berman, Zec, & Weinberger, 1986; Weinberger, Berman, & Illowsky, 1988). Executive skills comprise a number of cognitive abilities including concept and rule formation, modulation of affective states, planning and organization, and utilization of feedback. Studies incorporating factor analytic approaches suggest that executive deficits are independent of attentional deficits and represent two distinct factors in schizophrenic patients (Kremen, Seidman, Faraone, Pepple, & Tsuang, 1992; Mirsky, 1988; Mirsky, Anthony, Duncan, Ahearn, & Kellam, 1991). One investigation by Heinrichs and Awad (1993) using neuropsychological evaluation with 104 schizophrenic patients analyzed with hierarchical and disjoint clustering procedures found five clusters with cluster 1 comprising 24 patients with executive prefrontal dysfunction and another 20 patients comprising a cluster related to executive motor deficits. Though the researchers used relatively few neuropsychological measures in the study, these results highlight the strong presence of executive deficits in factor analytic studies of patients with schizophrenia. Similar cluster profiles have been found by Goldstein (1994) who has made an attempt to explain symptom heterogeneity in terms of neurobehavioral subtypes. The most extensively researched instrument used to assess executive deficits in patients with schizophrenia is the Wisconsin Card Sorting Test (WCST). Beatty, Jocic, Monson, and Katzung (1994) found that schizophrenic patients executed fewer concepts, had inaccurate and incomplete verbal explanations for concepts, and made more perseverative responses compared to normal control subjects on the Wisconsin and California Card Sorting tests.

Several investigations suggest possible explanations for the etiology of executive deficits in schizophrenics. One area of investigation that has shown promise is the presence of executive deficits in relatives of affected schizophrenics, suggesting that the presence of cognitive deficits are not necessarily associated with overt clinical symptoms of schizophrenia. For example, Yurgelin-Todd and Kinney (1993) found that performance on the WCST and Trails test discriminated schizophrenic patients from both siblings and nonschizophrenic psychiatric normal controls. Keefe and colleagues (1994) failed to find deficits on the WCST, but did find deficits on other tasks related to executive function in relatives of 23 schizophrenic probands. In a study by Sullivan, Shear, Zipursky, Sagar, and Pfefferbaum (1994) researchers found that executive function deficits could be uniquely accounted for

306

DOUG
JOHNSON-
GREENE AND
KENNETH M.
ADAMS

by disease duration, whereas memory deficits were related to symptom severity in a group of 34 schizophrenic patients, after controlling for age-related differences. Still other studies have identified a relationship between negative symptoms and decreased performance on tests of executive function (Hammer, Katsanis, & Iacono, 1995; Perlick, Mattis, Stastny, & Silverstein, 1992; Silverstein, Harrow, & Bryson, 1994; Stolar, Berenbaum, Banich, & Barch, 1994), though this relationship could conceivably be related to a third as yet unidentified variable relating to illness duration or intensity.

Cognitive deficits have also been found in several other domains. A second area of impairment common among schizophrenics is attention (Nuechterlein & Dawson, 1984). One aspect of attention, vigilance, has been found to be impaired in both acute (Asarnow & MacCrimmon, 1978) and in post-acute stages and persistent stages of the illness (Neuchterlein, Edell, Norris, & Dawson, 1986). Other processes of attention, such as automation of visual detection responses as measured by patients' reaction time in dual task paradigms, are similarly affected (Granholm, Asarnow, & Marder, 1996). Memory is another cognitive domain that has been consistently impaired in schizophrenics, which is consistent with neuropathology studies indicating reduced tissue volume in temporal-limbic structures (Crow, 1990), most notably within the left hemisphere. Multiple studies have found disproportionate deficits in most aspects of memory function (Calev, Korin, Kugelmass, & Lerer, 1987; Gold, Randolph, Carpenter, Goldberg, & Weinberger, 1992a, 1992b; Saykin *et al.*, 1991; Tamlyn *et al.*, 1992), but have failed to identify more pronounced deficits in any specific aspects of memory functioning, such as increased deficits in verbal versus spatial memory or explicit versus implicit memory.

STABILITY OF NEUROPSYCHOLOGICAL DEFICITS. It is still not clear which aspects of schizophrenia are static and which aspects change over time (Waddington, 1993). One important question in schizophrenia research has been whether neuropsychological deficits are stable or progressive in patients with schizophrenia and whether these changes relate to symptom intensity or extrinsic factors, such as the untoward effects of medication (Spohn & Strauss, 1989). The original diagnostic entity described by Kraepelin (1919/1971) suggests a disease of progressive deterioration (i.e., dementia). Several factors have contributed to the notion that cognitive deficits in schizophrenics remain stable. Data from neuroimaging studies suggests that structural changes, such as ventricular enlargement, remain static over time (Illowsky, Juliano, Bigelow, & Weinberger, 1988), but are sometimes discordant with the severity of neuropsychological deficits (Adams, Jacisin, Brown, Boulos, & Silk, 1984). There is also anecdotal evidence that first-episode and chronic schizophrenics have comparable levels of cognitive impairment (Cantor-Graae, Warkentin, & Nilsson, 1995; Hoff, Riordan, O'Donnell, Morris, & Delisi, 1992), which is contrary to the concept that schizophrenia has a chronic course. A third factor challenging the notion of progressive deterioration comes from longitudinal studies.

Despite the importance of these questions there has been surprisingly very little longitudinal research conducted on the neuropsychological function of patients with schizophrenia. Mixed findings have been found in studies examining the stability of cognitive deficits in these patients, which have typically utilized serial neuropsychological evaluation paradigms. Sweeney, Haas, Keilp, and Long (1991) examined neuropsychological function in 39 stable inpatients and again 1 year later during a nonacute period. They found improvement in neuropsychological measures, suggesting that improvement in cognitive function can occur following an acute episode, even after initial clinical improvement has occurred. There is at least some support for this observation from another study examining clinical improvement and cognitive status following treatment of acute episodes (Cantor-Graae *et al.*, 1995). In this study, separate analyses were performed using first-episode and chronic patients, which revealed no significant differences between the groups. Other studies involving longitudinal neuropsychological function in first-episode schizophrenics have been conducted with similar results (Addington & Addington, 1993; Bilder *et al.*, 1991; McGlashan & Fenton, 1993). Addington and Addington (1993) found in their evaluation of 39 patients at the time of hospitalization and again 6 months later that premorbid history of positive and negative symptom severity was the single best predictor of current cognitive status, with cognitive deficits being more strongly correlated with severity of negative symptoms. Longitudinal symptom heterogeneity has also been observed in other studies examining the influence of clinical subtype (Carpenter & Kirkpatrick, 1988; McGlashan & Fenton, 1993).

SUMMARY OF RESEARCH ON SCHIZOPHRENIA. Schizophrenia has a debilitating effect on the cognitive processes of those who are afflicted, and probably the relatives of schizophrenics as well. Some early investigations seemed to suggest that schizophrenics have diffuse cognitive deficits, though many studies failed to control for confounding variables. More recent studies have identified deficits in executive skills, attention, memory, and motor skills. These ability domains appear to be differentially affected, though considerable heterogeneity exists among schizophrenics of similar age and demographic characteristics. Heterogeneity of cognitive symptoms may reflect neurobehavioral subtypes of the disorder, or as yet undefined intrinsic and extrinsic factors. The stability of these deficits has been debated, with the majority of studies challenging traditional notions of a chronic progressive decline of cognitive function in patients with schizophrenia.

CONCLUSIONS AND RECOMMENDATIONS FOR FUTURE RESEARCH

The purpose of this chapter was to elucidate clinical issues surrounding assessment of neuropsychiatric disorders. We focused predominantly on the cognitive sequelae of specific psychiatric conditions as a means of providing the reader with a basic understanding of cognitive deficits commonly associated with these illnesses. We also examined the cognitive effects of transient subclinical psychiatric entities, such as depressed mood and intermittent anxiousness, as a way of differentiating the cognitive effects of transient mood states from bonafide psychiatric illness.

One prevailing conclusion from the majority of investigations reported in this chapter is the tenuous nature of their findings. As was our forecast in our opening statements, research involving neuropsychological assessments sometimes yields more questions than answers. Research with these populations, perhaps more than any other group of patients, has been plagued by unintended effects of confounding variables. Among some of the extraneous variables researchers contend with are side effects of medication, heterogeneity of patients' illnesses, comorbid substance abuse or psychiatric illness, genetic variability, and premorbid neurological and psychological history. While some researchers have been successful in eliminating the influence of the majority of these factors, ultimately all research pertaining to these patient populations is affected, usually without the explicit knowledge of

308

DOUG
JOHNSON-
GREENE AND
KENNETH M.
ADAMS

the researcher. Conflicting research and the appearance of heterogeneous characteristics among patient groups, when they exist, may be due at least in part to these issues.

Another observation may help the reader to understand the relative paucity of strong studies on the neuropsychology of these neuropsychiatric disorders, beyond the complexities just noted. A perusal of available references in this realm shows the majority to be published in the psychiatric literature. While this makes sense, a closer look shows neuropsychological issues, methods, and outcomes to be secondary or side issues in psychiatric context. This is sometimes reflected in very limited neuropsychological assessment paradigms attached to larger psychiatric investigations aimed at what psychiatric researchers usually see as the main issues (most often neuropsychopharmacological or currently popular diagnostic ones). These studies have in many instances favored cognitive screening measures or neuropsychological instruments which may have strong face validity but which may have little or no relevance to the specific hypotheses of the research. The wide assortment of neuropsychological instruments used by researchers, some of which may lack acceptable psychometric properties and appropriate standardization, make comparison between studies a problematic endeavor.

There remains a need to develop primary neuropsychological research in these disorders which will relate more practically to case management issues. The expectation that these problems will be revealed as "brain diseases" remains unfulfilled while millions of patients suffering from these disorders attempt to cope not only with symptoms, but with disordered cognitive, perceptual, motor, and personality problems best addressed through neuropsychological assessment.

Acknowledgments

This work was supported in part by NIH grants AA07378 and AG08671. We thank the staff of the Ann Arbor VAMC Psychology Service for their assistance and Dr. Lisa Johnson-Greene for her editorial suggestions and proofreading.

REFERENCES

Adams, K. M., Jacisin, J. J., Brown, G. G., Boulos, R. S., & Silk, S. (1984). Neurobehavioral deficit and computed tomographic abnormalities in three samples of schizophrenic patients. *Perceptual and Motor Skills, 59,* 115–119.

Addington, J., & Addington, D. (1993). Premorbid functioning, cognitive functioning, symptoms and outcome in schizophrenia. *Journal of Psychiatric Neurosciences, 18,* 18–23.

Addington, J., Addington, D., & Maticka-Tyndale, E. (1991). Cognitive functioning and positive and negative symptoms in schizophrenia. *Schizophrenia Research, 5,* 123–134.

American Psychiatric Association. (1994). *Diagnostic and statistical manual of mental disorders* (4th Ed.). Washington, DC: Author.

Asarnow, R. F., & MacCrimmon, D. J. (1978). Residual performance deficit in clinically remitted schizophrenics: A marker of schizophrenia? *Journal of Abnormal Psychology, 87,* 597–608.

Baxter, L. R., Phelps, M. E., Mazziotta, J. C., Guze, B. H., Schwartz, J. M., & Selin, C. E. (1987). Local cerebral metabolic rates in obsessive–compulsive disorder. *Archives of General Psychiatry, 44,* 211–218.

Baxter, L. R., Phelps, M. E., Mazziotta, J. C., Guze, B. H., Schwartz, J. M., & Selin, C. E. (1988). Cerebral glucose metabolic rates in nondepressed patients with obsessive–compulsive disorder. *American Journal of Psychiatry, 145,* 1560–1563.

Beatty, W. W., Jocic, Z., Monson, N., & Katzung, V. (1994). Problem solving by schizophrenic and schizoaffective patients on the Wisconsin and California Card Sorting Tests. *Neuropsychology, 8*(1), 49–54.

Behar, D., Rappaport, J. L., Berg, C. J., & Fairbanks, L. (1984). Computerized tomography and neuropsychological test measures in adolescents with obsessive–compulsive disorder. *American Journal of Psychiatry, 141,* 363–369.

Berman, K. F., Zec, R. F., & Weinberger, D. R. (1986). Physiologic dysfunction of dorsolateral prefrontal cortex in schizophrenia: II. Role of neuroleptic treatment, attention and mental effort. *Archives of General Psychiatry, 43,* 126–135.

Bieliauskas, L. A. (1993). Depressed or not depressed? That is the question. *Journal of Clinical and Experimental Neuropsychology, 15,* 119–134.

Bilder, R. M., Kipschultz-Broch, L., Reiter, G., Mayerfoff, D., Loebel, A., Degreef, D., Ashtari, M., & Lieberman, J. A. (1991). Neuropsychological studies of first episode schizophrenia. *Schizophrenia Research, 4,* 381–382.

Blanchard, J. J., & Neale, J. M. (1994). The neuropsychological signature of schizophrenia: Generalized or differential deficit? *American Journal of Psychiatry, 151,* 40–48.

Bondi, M. W., Monsch, A. U., Galasko, D., & Butters, N. (1994). Preclinical cognitive markers of dementia of the Alzheimer's type. *Neuropsychology, 8* (3), 374–384.

Boone, K. B., Ananth, J., Philpott, L., & Kaur, A. (1991). Neuropsychological characteristics of non-depressed adults with obsessive–compulsive disorder. *Neuropsychiatry, Neuropsychology, and Behavioral Neurology, 4* (2), 96–109.

Boone, K. B., Lesser, I. M., Miller, B. L., Wohl, M., Berman, N., Lee, A., Palmer, B., & Back, C. (1995). Cognitive functioning in older depressed outpatients: Relationship of presence and severity of depression to neuropsychological test scores. *Neuropsychology, 9*(3), 390–398.

Bornstein, R. A., Nasrallah, H. A., Olson, S. C., Coffman, J. A., Torello, M., & Schwartzkopf, S. B. (1990). Neuropsychological deficit in schizophrenic subtypes: Paranoid, nonparanoid and schizoaffective subgroups. *Psychiatry Research, 31,* 15–24.

Bornstein, R. A., Nasrallah, H. A., Para, M. F., & Whitacre, C. C. (1993). Neuropsychological performance in symptomatic and asymptomatic HIV infection. *AIDS, 7*(4), 519–524.

Boyar, J. D. (1981). Coping with non-organic factors on neuropsychological examination: Utility of repeated neuropsychological measures. *Clinical Neuropsychology, 3,* 15–17.

Braff, D. L., Heaton, R., Kuck, J., Cullum, M., Moranville, J., Grant, I., & Zisook, S. (1991). The generalized pattern of neuropsychological deficits in outpatients with chronic schizophrenia with heterogeneous Wisconsin Card Sorting Test results. *Archives of General Psychiatry, 48,* 891–898.

Brand, A. N., Jolles, J., & Gispen-de Wied, C. (1992). Recall and recognition memory deficits in depression. *Journal of Affective Disorders, 25,* 77–86.

Breier, A., Schreiber, J. L., Dyer, J., & Pickar, D. (1991). National Institute of Mental Health longitudinal study of chronic schizophrenia. *Archives of General Psychiatry, 48,* 239–246.

Buchanan, R. W., Kirkpatrick, B., Heinrich, D. W., & Carpenter, W. T. (1990). Clinical correlates of the deficit syndrome of schizophrenia. *American Journal of Psychiatry, 147,* 290–294.

Calev, A., Korin, Y., Kugelmass, S., & Lerer, B. (1987). Performance of chronic schizophrenics on matched word and design recall tasks. *Biological Psychiatry, 22,* 699–709.

Cannon, T. D., Zorrilla, E., Shtasel, D., Gur, R. E., Gur, R. C., Marco, E. J., Moberg, P., & Price, R. A. (1994). Neuropsychological functioning in siblings discordant for schizophrenia and healthy volunteers. *Archives of General Psychiatry, 51,* 651–661.

Cantor-Graae, E., Warkentin, S., & Nilsson, A. (1995). Neuropsychological assessment of schizophrenic patients during a psychotic episode: Persistent cognitive deficit? *Acta Psychiatrica Scandinavica, 91,* 283–288.

Carpenter, W. T., & Kirkpatrick, B. (1988). The heterogeneity of the long-term course of schizophrenia. *Schizophrenia Bulletin, 14,* 645–652.

Cassens, G., Wolfe, L., & Zola, M. (1990). The neuropsychology of depressions. *Journal of Neuropsychiatry and Clinical Neurosciences, 2,* 202–213.

Chelune, G. J., Heaton, R. K., Lehman, R. A. W., & Robinson, A. (1979). Level versus pattern of neuropsychological performance among schizophrenic and diffusely brain-damaged patients. *Journal of Consulting and Clinical Psychology, 47,* 155–163.

Childers, S. E., & Harding, C. M. (1990). Gender, premorbid functioning, and long-term outcome in DSM-III schizophrenia. *Schizophrenia Bulletin, 16,* 309–318.

Coffman, J. A., Bornstein, R. A., Olson, S. C., Schwarzkopf, S. B., & Nasrallah, H. A. (1990). Cognitive impairment and cerebral structure by MRI in bipolar disorder. *Biological Psychiatry, 27,* 1188–1196.

Crow, T. J. (1990). Temporal lobe asymmetries as the key to etiology of schizophrenia. *Schizophrenia Bulletin, 16,* 433–443.

310

DOUG
JOHNSON-
GREENE AND
KENNETH M.
ADAMS

Dalton, J. E., Pederson, L. S., & Ryan, J. J. (1989). Effects of post-traumatic stress disorder on neuropsychological test performance. *International Journal of Clinical Neuropsychology, 11*(3), 121–124.

Dodrill, C. B. (1979). Sex differences on the Halstead–Reitan Neuropsychological Battery and other neuropsychological measures. *Journal of Clinical Psychology, 35*(2), 236–241.

Dupont, R. M., Jernigan, T. L., Butters, N., Delis, D., Hesselink, J. R., Heindel, W., & Gillin, J. C. (1990). Subcortical abnormalities detected in bipolar affective disorder using magnetic resonance imaging. *Archives of General Psychiatry, 47*, 55–59.

Emery, O. B., & Breslau, L. D. (1989). Language deficits in depression: Comparisons with SDAT and normal aging. *Journal of Gerontology: Medical Sciences, 44*, 85–92.

Fischer, D. G., Sweet, J. J., & Pfaelzer-Smith, E. A. (1986). Influence of depression on repeated neuropsychological testing. *The International Journal of Clinical Neuropsychology, 8*(1), 14–18.

Gold, J. M., Randolph, C., Carpenter, C. J., Goldberg, T. E., & Weinberger, D. R. (1992a). Forms of memory failure in schizophrenia. *Journal of Abnormal Psychology, 101*, 487–494.

Gold, J. M., Randolph, C., Carpenter, C. J., Goldberg, T. E., & Weinberger, D. R. (1992b). The performance of patients with schizophrenia on the Wechsler Memory Scale-Revised. *The Clinical Neuropsychologist, 6*, 367–373.

Goldberg, T. E., Gold, J. M., Torrey, E. F., & Weinberger, D. R. (1995). Lack of sex differences in the neuropsychological performance of patients with schizophrenia. *American Journal of Psychiatry, 152*, 883–888.

Goldberg, T. E., Ragland, J. D., Torrey, E. F., Gold, J. M., Bigelow, L. B., & Weinberger, D. R. (1990). Neuropsychological assessment of monozygotic twins discordant for schizophrenia. *Archives of General Psychiatry, 47*, 1066–1072.

Goldstein, G. (1994). Neurobehavioral heterogeneity in schizophrenia. *Archives of Clinical Neuropsychology, 9*, 265–276.

Goldstein, J. M. (1988). Gender differences in the course of schizophrenia. *American Journal of Psychiatry, 145*, 684–689.

Granholm, E., Asarnow, R. F., & Marder, S. R. (1996). Dual-task performance operating characteristics, resource limitations, and automatic processing in schizophrenia. *Neuropsychology, 10*(1), 11–21.

Grant, I., & Adams, K. A. (1996). *Neuropsychological assessment of neuropsychiatric disorders* (2nd ed.). New York: Oxford University Press.

Gray, J. A. (1982). Précis of the neuropsychology of anxiety: An inquiry into the functions of the septo-hippocampal system. *Behavior and Brain Science, 5*, 469–534.

Hall, R. C. (1980). Depression. In R. C. Hall (Ed.), *Psychiatric presentations of medical illness: Somatopsychia disorders* (pp. 37–63). New York: SP Medical and Scientific Books.

Hammer, M. A., Katsanis, J., & Iacono, W. G. (1995). The relationship between negative symptoms and neuropsychological performance. *Biological Psychiatry, 37*, 828–830.

Head, D., Bolton, D., & Hymas, N. (1989). Deficit in cognitive shifting ability in patients with obsessive–compulsive disorder. *Biological Psychiatry, 25*, 929–937.

Heaton, R. K., & Crowley, T. J. (1981). Effects of psychiatric disorders and somatic treatments on neuropsychological test results. In S. B. Filskov & T. J. Boll (Eds.), *Handbook of clinical neuropsychology* (pp. 481–525). New York: John Wiley.

Heinrichs, R. W., & Awad, A. G. Z. (1993). Neurocognitive subtypes of chronic schizophrenia. *Schizophrenia Research, 9*, 49–58.

Hill, C. D., Stoudemire, A., Morris, R., Martino-Saltzman, D., Markwalter, H. R., & Lewison, B. J. (1992). Dysnomia in the differential diagnosis of major depression, depression-related cognitive dysfunction, and dementia. *Journal of Neuropsychiatry and Clinical Neurosciences, 4*, 64–69.

Hillbom, M., & Holm, L. (1986). Contribution of traumatic head injury to neuropsychological deficits in alcoholics. *Journal of Neurology, Neurosurgery, and Psychiatry, 49*, 1348–1353.

Hodges, W. F., & Spielberger, C. D. (1969). Digit Span: An indicant of trait or state anxiety? *Journal of Clinical and Consulting Psychology, 33*, 430–434.

Hoff, A. L., Riordan, H., O'Donnell, D. W., Morris, L., & Delisi, L. E. (1992). Neuropsychological functioning of first-episode schizophreniform patients. *American Journal of Psychiatry, 149*, 898–903.

Illowsky, B. P., Juliano, D. M., Bigelow, L. B., & Weinberger, D. R. (1988). Stability of CT scan findings in schizophrenia: Results of an 8-year follow up study. *Journal of Neurology, Neurosurgery and Psychiatry, 51*, 209–213.

Jacobsen, B., & Kinney, D. K. (1980). Perinatal complications in adopted and non-adopted samples of schizophrenics and controls. *Acta Psychiatric Scandinavia, 62* (Suppl. 285), 3337–3346.

Jenicke, M. A. (1984). Obsessive–compulsive disorder: A question of neurologic lesion. *Comprehensive Psychiatry, 24*, 99–115.

Jones, B. P., Duncan, C. C., Mirsky, A. F., Post, R. M., & Theodore, W. H. (1994). Neuropsychology profiles in bipolar affective disorder and complex partial seizure disorder. *Neuropsychology, 8*(1), 55–64.

Keefe, R. S. E., Silverman, J. M., Roitman, S. E. L., Harvey, P. D., Duncan, M. A., Alroy, D., Siever, L. J., Davis, K. L., & Mohs, R. C. (1994). Performance of nonpsychotic relatives of schizophrenic patients on cognitive tests. *Psychiatry Research, 53*, 1–12.

Kendler, K. S., & Diehl, S. R. (1993). The genetics of schizophrenia: A current, genetic epidemiological perspective. *Schizophrenia Bulletin, 19*, 261–286.

King, D. A., Cox, C., Lyness, J. M., & Caine, E. D. (1995). Neuropsychological effects of depression and age in an elderly sample: A confirmatory study. *Neuropsychology, 9*(3), 399–408.

King, G. D., Caine, E. D., Conwell, Y., & Cox, C. (1991). The neuropsychology of depression in the elderly: A comparative study with normal aging and Alzheimer's disease. *Journal of Neuropsychiatry and Clinical Neurosciences, 3*, 163–168.

King, G. D., Caine, E. D., & Cox, C. (1993). Influence of depression and age on selected cognitive functions. *Clinical Neuropsychologist, 7*, 443–453.

King, G. D., Hannay, H. J., Masek, B. J., & Burns, J. W. (1978). Effects of anxiety and sex on neuropsychological tests. *Journal of Consulting and Clinical Psychology, 46*(2), 375–376.

Kinney, D. K., Yurgelun-Todd, D. A., & Woods, B. T. (1991). Hard neurological signs and psychopathology in relatives of schizophrenia patients. *Psychiatry Research, 39*, 45–53.

Kirch, D. G. (1993). Infection and autoimmunity at etiological factors in schizophrenia: A review and reappraisal. *Schizophrenia Bulletin, 19*, 355–370.

Kraepelin, E. (1971). *Dementia praecox and paraphenia* (R. M. Barclay, Trans.) (G. M. Robertson, Ed.) (pp. 282–329). Huntington, NY: R. E. Krieger Publishing Company, Inc. (Original work published 1919).

Kremen, W., Seidman, L., Faraone, S., Pepple, J. R., & Tsuang, M. T. (1992). Attention/information processing factors in psychotic disorders: Replication and extension of recent neuropsychological findings. *Journal of Nervous and Mental Disease, 180*, 89–93.

Kremen, W. S., Seidman, J., Pepple, J. R., Lyons, M. J., Tsuang, M. T., & Faraone, S. V. (1994). Neuropsychological risk indicators for schizophrenia: A review of family studies. *Schizophrenia Bulletin, 20*(1), 103–118.

Kremen, W. S., Seidman, L. J., Faraone, S. V., Pepple, J. R., Lyons, M. J., & Tsuang, M. T. (1996). The "3 Rs" and neuropsychological function in schizophrenia: An empirical test of the matching fallacy. *Neuropsychology, 10*(1), 23–31.

Levin, S., Yurgelun-Todd, D. A., & Craft, S. (1989). Contributions of clinical neuropsychology to the study of schizophrenia. *Journal of Abnormal Psychology, 98*, 341–356.

Lewine, R. J. (1981). Sex differences in schizophrenia: Timing or subtypes? *Psychological Bulletin, 90*, 432–444.

Lewis, S. W., & Murray, R. M. (1987). Obstetric complications, neurodevelopmental deviance and risk of schizophrenia. *Journal of Psychiatric Research, 21*, 413–421.

Malloy, P. (1987). Frontal lobe dysfunction in obsessive–compulsive disorder. In E. Perecman (Ed.), *The frontal lobes revisited*. New York: IRBN Press.

Martin, N. J., & Franzen, M. D. (1989). The effect of anxiety of neuropsychological function. *International Journal of Clinical Neuropsychology, 11*(1), 1–8.

McAllister, T. W. (1983). Overview: Pseudodementia. *American Journal of Psychiatry, 140*, 528–532.

Mednick, S. A. (1970). Breakdown in individuals at high risk for schizophrenia: Possible predispositional perinatal factors. *Mental Hygiene, 54*, 50–63.

Miller, W. R. (1975). Psychological effects of depression. *Psychological Bulletin, 82*, 238–260.

Milner, B. (1963). Effects of different brain lesions on card sorting. *Archives of Neurology, 9*, 90–100.

Mirsky, A. F. (1988). Research on schizophrenia in the NIMH Laboratory of Psychology and Psychopathology, 1954–1987. *Schizophrenia Bulletin, 14*, 151–156.

Mirsky, A. F., Anthony, B. J., Duncan, C. C., Ahearn, A. B., & Kellam, S. G. (1991). Analysis of the elements of attention: A neuropsychological approach. *Neuropsychology Review, 2*, 109–145.

Nasrallah, H. A. (1990). Brain structure and functions in schizophrenia: Evidence for fetal neurodevelopmental impairment. *Current Opinions in Psychology, 3*, 75–78.

Nasrallah, H. A., Coffman, J. A., & Olson, S. C. (1989). Structural brain imaging findings in affective disorders: An overview. *Journal of Neuropsychiatry and Neurosciences, 1*, 21–26.

Niederehe, G., & Camp, C. J. (1985). Signal detection analysis of recognition memory in depressed elderly. *Experimental Aging Research, 11*, 207–213.

312

DOUG
JOHNSON-
GREENE AND
KENNETH M.
ADAMS

Nordahl, T. E., Benkelfat, C., Semple, W. E., & Gross, M. (1989). Cerebral metabolic glucose rates in obsessive–compulsive disorder. *Neuropsychopharmacology, 2*, 23–28.

Nuechterlein, K. H., & Dawson, M. E. (1984). Information processing and attentional functioning in the developmental course of schizophrenic disorders. *Schizophrenia Bulletin, 10*, 160–203.

Nuechterlein, K. H., Edell, W. S., Norris, M., & Dawson, M. E. (1986). Attentional vulnerability indicators, thought disorder, and negative symptoms. *Schizophrenia Bulletin, 12*, 408–426.

Otto, M. W. (1992). Normal and abnormal information processing—A neuropsychological perspective on obsessive compulsive disorder. *Psychiatric Clinics of North America, 15*(4), 825–848.

Perlick, D., Mattis, S., Stastny, P., & Silverstein, B. (1992). Negative symptoms are related to both frontal and nonfrontal neuropsychological measures in chronic schizophrenia. *Archives of General Psychiatry, 49*, 245.

Pitman, R. K. (1987). A cybernetic model of obsessive–compulsive psychopathology. *Comprehensive Psychiatry, 28*, 334–343.

Raskin, A. (1986). Partialling out the effects of depression and age on cognitive functions: Experimental data and methodologic issues. In L. W. Poon (Ed.), *Clinical memory assessment of older adults* (pp. 244–256). Washington, DC: American Psychological Association.

Regier, D. A., Boyd, J. H., Burke Jr., J. D., Rae, D. S., Myers, J. K., Kramer, M., Robins, L. N., George, L. K., Karno, M., & Locke, B. Z. (1988). One-month prevalence of mental disorders in the United States: Based on five epidemiological catchment area sites. *Archives of General Psychiatry, 45*, 977–986.

Richards, P. M., & Ruff, R. M. (1989). Motivational effects of neuropsychological functioning: Comparison of depressed verses nondepressed individuals. *Journal of Consulting and Clinical Psychology, 57*, 396–402.

Ryan, C. M., & Williams, T. M. (1993). Effects of insulin-dependent diabetes on learning and memory efficiency in adults. *Journal of Clinical and Experimental Neuropsychology, 15*(5), 685–700.

Saykin, A. J., Gur, R. C., Gur, R. E., Mozley, P. D., Mozley, L. H., Resnick, S. M., Kester, D. B., & Stafiniak, P. (1991). Neuropsychological function in schizophrenia: Selective impairment in memory and learning. *Archives of General Psychiatry, 48*, 618–624.

Seidman, L. J. (1984). Schizophrenia and brain dysfunction: An integration of recent neurodiagnostic findings. *Psychological Bulletin, 94*, 195–238.

Seidman, L. J., Yurgelun-Todd, D., Kremen, W. S., Woods, B. T., Goldstein, J. M., Faraone, S. V., & Tsuang, M. T. (1994). Relationship of prefrontal and temporal lobe MRI measures to neuropsychological performance in chronic schizophrenia. *Biological Psychiatry, 35*, 235–246.

Siegal, B., & Gershon, S. (1986). Dementia, depression, and pseudodementia. In H. J. Altman (Ed.) *Alzheimer's disease* (pp. 29–44). New York: Plenum Press.

Siegal, B., Gurevich, D., & Oxenkrug, G. F. (1989). Cognitive impairment and cortisol resistance to dexamethasone suppression in elderly depression. *Biological Psychiatry, 25*, 229–234.

Siegfried, K. (1985). Cognitive symptoms in late-life depression and their treatment. *Journal of Affective Disorders, 9* (Suppl. 1), S33–S40.

Silverstein, M. L., Harrow, M., & Bryson, G. J. (1994). Neuropsychological prognosis and clinical recovery. *Psychiatry Research, 52*, 265–272.

Spar, J. E. (1982). Dementia in the aged. *Psychiatric Clinics of North America, 5*, 67–86.

Speedie, L. J., Rabins, P. V., & Pearlson, G. D. (1990). Confrontation naming deficit in dementia of depression. *Journal of Neuropsychiatry and Clinical Neurosciences, 2*, 59–63.

Spohn, H. E., & Strauss, M. E. (1989). Relation of neuroleptic and anticholinergic medication to cognitive functions in schizophrenia. *Journal of Abnormal Psychology, 98*, 367–380.

Stolar, N., Berenbaum, H., Banich, M. T., & Barch, D. (1994). Neuropsychological correlates of alogia and affective flattening in schizophrenia. *Biological Psychiatry, 35*, 164–172.

Strauss, M. E. (1993). Relation of symptoms to cognitive deficits in schizophrenia. *Schizophrenia Bulletin, 19*, 215–231.

Sullivan, E. V., Shear, P. K., Zipursky, R. B., Sagar, H. J., & Pfefferbaum, A. (1994). A deficit profile of executive, memory and motor functions in schizophrenia. *Biological Psychiatry, 36*, 641–653.

Sweeney, J. A., Haas, G. L., Keilp, J. G., & Long, M. (1991). Evaluation of the stability of neuropsychological functioning after acute episodes of schizophrenia: One-year follow-up study. *Psychiatry Research, 38*, 63–76.

Sweet, J. J. (1983). Confounding effects of depression on neuropsychological testing: Five illustrative cases. *Clinical Neuropsychology, 4*(4), 103–109.

Tamlyn, D., McKenna, P. J., Mortimer, A. M., Lund, C. E., Hammond, S., & Baddeley, A. D. (1992).

Memory impairment in schizophrenia: Its extent, affiliations and neuropsychological character. *Psychological Medicine, 22,* 101–115.

Waddington, J. L. (1993). Neurodynamics of abnormalities in cerebral metabolism and structure in schizophrenia. *Schizophrenia Bulletin, 14,* 645–652.

Walker, E., & Lewine, R. L. (1988). Negative symptom distinction in schizophrenia: Validity and etiological relevance. *Schizophrenia Research, 1,* 315–328.

Weinberger, D. R. (1987). Implications of normal brain development for the pathogenesis of schizophrenia. *Archives of General Psychiatry, 4,* 660–669.

Weinberger, D. R., Berman, K. F., & Illowsky, B. P. (1988). Physiological dysfunction of dorsolateral prefrontal cortex in schizophrenia: III. A new cohort and evidence for a monoaminergic mechanism. *Archives of General Psychiatry, 45,* 609–615.

Weingartner, H., & Silberman, E. (1982). Models of cognitive impairment: Cognitive changes in depression. *Psychopharmacology, 18,* 27–42.

Williams, J. M., Little, M. M., Scates, S., & Blockman, N. (1987). Memory complaints and abilities among depressed older adults. *Journal of Consulting and Clinical Psychology, 55,* 595–598.

Williamson, P., Pelz, D., Merskey, H., Morrison, S., & Conlon, P. (1991). Correlation of negative symptoms in schizophrenia with frontal lobe parameters on magnetic resonance imaging. *British Journal of Psychiatry, 159,* 130–134.

Woods, B. T., Kinney, D. K., & Yurgelun-Todd, D. A. (1991). Neurological "hard" signs and family history of psychosis in schizophrenia. *Biological Psychiatry, 30,* 806–816.

Yurgelin-Todd, D. A., & Kinney, D. K. (1993). Patterns of neuropsychological deficits that discriminate schizophrenic individuals from siblings and control subjects. *The Journal of Neuropsychiatry and Clinical Neurosciences, 5,* 294–300.

Zielinski, C. M., Taylor, M. A., & Juzwin, K. R. (1991). Neuropsychological deficits in obsessive-compulsive disorder. *Neuropsychiatry, Neuropsychology, and Behavioral Neurology, 4,* 110–126.

III

Specialized Assessment

15

Neuropsychological Assessment of Abstract Reasoning

GERALD GOLDSTEIN

INTRODUCTION AND HISTORICAL BACKGROUND

Abstract reasoning is no doubt the most advanced of the cognitive abilities. While animals may be capable of problem-solving, only humans can abstract. Thus, abstraction and problem-solving are not synonymous, and problems can be solved without abstraction. However, formation of an abstract concept is often the most elegant way of solving a problem. The word abstraction connotes abstracting some unifying idea or principle on the basis of observation of diverse material. It is therefore an activity that is removed from direct sensory experience, and constitutes a representation of such experience. The term abstraction is often contrasted to concreteness, the latter term indicating cognitive activity associated with direct experience, and without such representation. Concreteness is direct interaction with the "real world" without additional processing.

The relationship between brain function and abstract reasoning was probably first discussed during the late 19th century by the neurologists Henry Head and Hughlings Jackson. However, this relationship had its first full theoretical development in the work of Kurt Goldstein and Martin Scheerer, and is best articulated in their 1941 monograph on abstract and concrete behavior (Goldstein & Scheerer, 1941). As an orientation for this chapter, we will use their eight points describing the abstract attitude.

1. To detach our ego from the outerworld or from inner experiences
2. To assume a mental set
3. To account for acts to oneself; to verbalize the account
4. To shift reflectively from one aspect of the situation to another

GERALD GOLDSTEIN   VA Pittsburgh Healthcare System, Highland Drive Division (151R), Pittsburgh, Pennsylvania 15206-1297.

Neuropsychology, edited by Goldstein *et al.* Plenum Press, New York, 1998.

5. To hold in mind simultaneously various aspects
6. To grasp the essential of a given whole; to break up a given whole into parts, to isolate and synthesize them
7. To abstract common properties reflectively; to form hierarchic concepts
8. To plan ahead ideationally; to assume an attitude towards the "mere possible" and to think or perform symbolically (p. 4)

With respect to neuropsychology, patients with various forms of brain damage or disease lose all or some of these characteristics, as do some patients with psychiatric disorders, notably schizophrenia. Some of Goldstein and Scheerer's (1941) points are made in existential terms, while others are phrased in terms of abilities or attitudes. Goldstein and Scheerer use the term attitude to characterize abstract and concrete behavior, but they use it in an unconventional sense. To them, it is a phenomenological concept having to do with whether one views the outer world or inner experience in an abstract or concrete way. Thus, to Goldstein and Scheerer, the abstract attitude is a core aspect of the personality, and unlike contemporary neuropsychology, they do not treat it merely as one cognitive domain among others. Thus, one can remember, use language, exercise spatial abilities, solve problems, deploy attention, or perceive inner states or the outer world abstractly or concretely. The abstract attitude, or its absence, pervades all of the cognitive domains and the modalities. Thus, tests of abstract ability are not viewed as part of a profile of numerous abilities, but are meant to document or illustrate the points made above about the nature of the abstract attitude, and about behavior when the abstract attitude is impaired or absent.

METHODOLOGICAL CONSIDERATIONS

Perhaps the key methodological problem involving tests of abstraction is construct validity. The issue becomes particularly important when a procedure has the appearance of being a measure of abstraction when, in fact, that need not be the case. The theoretical issue is that many tasks may be approached, and many problems may be solved, either abstractly, or concretely. Tests of abstraction with good construct validity are those that contain problems that can only be solved by abstract reasoning. Alternatively, tests and observational procedures that permit determination of the way in which the problem is being solved can also be useful tests of abstraction. The following examples illustrate this:

Scheerer, Rothman, and K. Goldstein (1945) described a boy who was an idiot savant. One remarkable thing the boy could do, despite his substantial mental retardation, was solve verbal analogy problems. However, further analysis of his performance suggested that he may not have been using actual analogical reasoning (i.e., abstraction) to solve these problems, but simple word association, clearly a lower functional level of behavior. Thus, observation of test performance permitted an evaluation of whether or not the abstract attitude was functioning. Subsequently, G. Goldstein (1962) and Willner (1971), in collaboration with Martin Scheerer, developed a new analogy test procedure that was resistant to solution by word association. They took verbal analogy problems, removed the first two words, and asked subjects to solve the problem with the third word only. Thus, in free choice format, the item "hot is to cold as up is to _____" was changed to "up _____," and the subject was asked to respond with the word that she or he thought was the right answer without having seen, no less solved, the analogy. In this research, we discovered that many analogy items could be solved in this manner at statistically significant levels. Willner went on to develop a new analogies test, called the Conceptual Level Analogy Test (CLAT) (Willner, 1971) that was developed on the basis of excluding items that could be solved by word association at beyond chance levels. A children's version of this test was developed by Goldstein (1962). The point of this research is that correct answers to analogies problems can be obtained by two different processes at very different functional levels.

A second example is a study of the Halstead Category Test (Halstead, 1947) conducted by Simmel and Counts (1957). The Category Test is widely accepted as a challenging measure of abstraction ability, but these investigators make the following comment about that belief: "Our own data have consistently refused to bear out the above assumptions. We have found many correct responses given by subjects who had clearly not grasped the principle to be applied; we have also observed numerous incorrect answers of subjects who had apparently demonstrated their knowledge of the relevant principle by long errorless runs of immediately preceding items" (pp. 7–8). In a detailed and meticulous analysis of this test, Simmel and Counts showed that sometimes correct and sometimes incorrect responses to the test may be determined by perceptual characteristics of the stimuli, application to a new set of items of a previously learned principle, mental sets, and response tendencies learned during the course of the individual's lifetime. Based on their analyses, they concluded that a good score on the Category Test does not necessarily mean the identification of principles, and conversely, errors may be made through misapplication of the correct principles. We would only note that while one may not have to identify the correct principles to do well on the Category Test, it nevertheless is probably the most effective way of doing well. These two examples are intended to illustrate that the construct validity of tests purported to assess abstraction ability may be weak, and that performance levels on these tests may be associated with other, usually lower, functional level abilities that do not require abstraction.

There are two major methodological approaches to determine whether or not abstract reasoning is being used to solve a problem. One of them, and perhaps the most desirable one, is to construct tests such that they are resistant to solution by other processes. This was the method used by Willner and G. Goldstein in the case of their analogies tests. Simmel and Counts give some suggestions as to how such tests should be constructed. First, the coincidence of an initially preferred and the correct response should be avoided. Essentially that is what was done in the case of the CLAT, since correct responses were never common associations to the third word of the analogy. The correct response should involve weaker response tendencies. Simmel and Counts also suggest that the test should be long enough to allow for eventual rejection of initially preferred responses in favor of responding in a way that is consistent with the correct principle. Another effective method of assessing genuine abstract reasoning is the procedure used in the Wisconsin Card Sorting Test (Grant & Berg, 1948), in which the principle is changed without the knowledge of the test-taker. Thus, the first category may be readily achieved on the basis of a preferred response tendency, but it is difficult to see how such tendencies would lead to solution after the first shift of the principle. Still another way of determining whether abstract reasoning is being used is to use principles that are not direct stimulus properties, such as color or shape. Higher-order abstractions involving combinations of stimulus properties or relationships among stimuli may be used for this purpose. For example, G. Goldstein (1962) used the concept of causality in his verbal analogy item "Wind is to Sailboat as Gasoline is to (1) Car (2) Fuel (3) Tank (4) Oil (5) Station." It is also useful in multiple choice analogies tests to include at least one associative foil, or incorrect answer that is a common association with the third word. For example, in the item "Slow is to Fast as Bicycle is to (1) Wheels (2) Airplane (3) Wagon (4) Ride (5) Play." "Wheels" and "Ride" are probably more common associates of bicycle than is the correct answer "Airplane."

The second major method of assessing abstraction ability is performance analysis, or systematic observation of how the individual takes the test (Scheerer, 1946). This systematic observation is guided by theoretically and clinically derived methods of analysis. While performance analysis, sometimes described as analysis of process (Werner, 1937), typically involves observation of test behavior, it can also be applied to behavior in everyday life. Goldstein and Scheerer (1941) provide numerous anecdotal examples of concreteness in everyday life, such as a patient who was requested to bring someone a comb, but could not do so without combing her own hair. This example would illustrate forced responsiveness, or being stimulus-bound, an aspect of concrete behavior.

Performance analysis is a method of determining whether some given problem-solving task is approached abstractly or concretely. A good example is the analysis provided by Goldstein and Scheerer (1941) of the Goldstein–Scheerer Cube Test (Block Design). Essentially, this analysis begins with the assumption that the models printed in the booklets for the Cube Test are representations of the cube pattern that is two- rather than three-dimensional, reduced in size, and without dividing lines. One can approach the task abstractly by sharing that assumption, or concretely by a matching or copying procedure in which one tries to match the impression of the design through manipulating the blocks until a match is experienced. Thus, a correct solution may be "happened upon," but is not reached by conceptual processes. Goldstein and Scheerer provide numerous examples of errors made by patients when they approach the task concretely. There might be concrete dependence upon size, seen when the subject bases the construction on the actual size of the model. The patient may have a preference for a triangle in its usual orientation with the base at the bottom, and thus may be unable to reproduce a model that contains an inverted triangle. These errors and numerous others reflect an inability to break the pattern up imaginatively into squares. For each square, the corresponding block side must be found and the design becomes constructed by single block sides. A crucial point involves the imposing of an imaginal network upon the design, and holding in mind the square units contained in this network while matching them with the turned-up block sides. The extent of abnormal concreteness is evaluated by reducing the abstract challenge of the task. Models are presented that are the same size as the construction to be completed, or in reduced size but with the lines drawn in. Finally, the examiner reproduces the design with blocks, and the subject is asked to use that reproduction as a model.

The trend through the entire series of Goldstein–Scheerer tests is consistent with the above example. The point of the tests is to provide a number of problem-solving situations where one can observe whether the task is being approached abstractly or concretely. For example, the Stick Test is a simple procedure in which subjects are asked to reproduce stick patterns from memory immediately after they are presented. Normally, individuals learn to comprehend geometric directions in space, even when the object has no particular meaningful content. Thus, we can imagine a line tilted at a 45 degree angle. When such a line is displayed and removed, we readily reproduce it at the correct angle. However, an individual who has lost this abstract ability to appreciate directions in space may fail to make a correct reproduction. The interesting point is that when such an individual is shown a more complex figure with substantive content, performance may be normal. Thus, a patient may successfully reproduce a nine-stick stimulus assembled into a "house," but may fail to reproduce a single stick at the correct angle. This example illustrates not only a point about concreteness, but also suggests that abstract content and complexity are not always synonymous. The house was perceptually more complex than the single stick, but was associated with a normal response, while the single stick was not. Thus, increasingly complexity of a task does not necessarily increase its difficulty from the standpoint of abstraction.

Levels of Abstraction

The Qualitative versus Quantitative Debate

Two statements, one made by Kurt Goldstein and the other by Martin Scheerer, reflect a philosophical view of the abstract attitude that is of great theoretical significance, but not without controversy. K. Goldstein (1951) said:

> Even in its simplest form, however, abstraction is separate in principle from concrete behavior. There is no gradual transition from the one to the other. (p. 60)

Scheerer (1962) said:

> The average adult can, and often does, perform at either the level of successive functions or the level of simultaneous functions. However, the young and brain injured are capable of performing only at the first level. (p. X2)

Thus, the theory is that abstraction does not lie on a continuum with concreteness, but exists at a level of behavior that is qualitatively distinct. Thus, if one loses the abstract attitude as a result of brain damage, one does not simply become impaired in a still present ability; the ability is lost entirely. Put more philosophically, the abstract attitude is an emergent property, and not the sum of more basic, lower-level cognitive abilities. As Boring (1950) summarized nicely, emergence is rooted in gestalt psychology, and particularly the concept that many properties of wholes are emergent, or inhere in no single part. The gestalt psychologists repeatedly asserted that the whole is more than, or different from, the sum of its parts. With regard to abstraction, the process is viewed as an emergent entity, and not the sum of more elementary cognitive processes, such as association.

Reitan (1958, 1959) took issue with this view, showing that brain-damaged patients improved their performance on Series VI of the Category Test relative to Series V. Both series involve the same principle. Furthermore, correlation matrices based on neuropsychological tests taken by brain-damaged and control subjects revealed the same organization of abilities in both groups. Therefore, Reitan concluded that abstraction existed on a continuum of deviation from normal. The issue was of particular importance, because measurement of abstraction ability in the clinic or laboratory would be compromised if it were an all-or-none phenomenon, and it would be difficult to justify use of quantitative tests of abstract reasoning such as the Category Test or Wisconsin Card Sorting Test. Subsequent consideration and study led us to conclude that both arguments were flawed. As Simmel and Counts (1957) have shown, there are ways of obtaining many correct answers on the Category Test without learning the principle, and it might not be correct to say that the brain-damaged subjects in the Reitan (1959) study improved in abstraction. Furthermore, some of the subjects in the Reitan study did not improve on Series VI.

In an attempt to help resolve this matter, G. Goldstein, Neuringer, and Olson (1968) administered a concept-identification task that was resistant to solution by means other than learning the principle. They found that many brain-damaged subjects showed no evidence of learning even a simple task, but that some brain-damaged patients were able to learn both a simple and a more complex task. They therefore concluded that some brain-damaged patients did have apparently complete absence of abstraction ability, while others demonstrated clear learning ability, reflecting a quantitative deficit. Therefore, the Goldstein–Scheerer theory appeared to be overgeneralized, but on the other hand, there appear to be some brain-damaged patients who have a qualitative loss, and who cannot assume the abstract attitude even in simple situations. Clinical or diagnostic distinctions between these groups were not identified in the Goldstein *et al.* (1968) study, but in a follow-up, Neuringer, Goldstein, and Jannes (1973) found that qualitative deficit appeared with greater frequency among older than younger subjects. Age itself was not viewed as the explanation for the difference, but may have been associated with a number of factors including type of neurological disorder and age-associated changes in cognitive function. Many years later, this same distinction between qualitative and quantitative impairment emerged in the case of schizophrenic patients. It was found in several studies that some schizophrenic patients performed quite normally on the Wisconsin Card Sorting Test, while others demonstrated severe impairment (Braff *et al.*, 1990; Goldstein & Shemansky, 1995; Goldstein, Beers, & Shemansky, 1996). Correspondingly, several studies reported success in teaching schizophrenic patients to improve on the Wisconsin Card Sorting Test (Bellack, Mueser, Morrison, Tierney, & Podell, 1990; Goldman, Axelrod, & Tompkins, 1992; Green, Satz, Ganzell, & Vaclav, 1992; Summerfelt *et al.*, 1991), while another study could find no such improvement, despite substantial effort to obtain it (Goldberg, Weinberger, Berman, Pliskin, & Podd, 1987). Again, we appear to have variability with regard to existence of qualitative or quantitative impairment.

The data would indicate that it is acceptable practice to measure abstraction ability with psychometric procedures, as most clinicians now do, but substantial, perhaps crucial, clinical information can be obtained through characterization of concreteness. On Block Design, does the patient use only one-color sides or become fixed on sensory cohesion between similarly colored half sides and stick to that fixation, though it may not lead to progress in reproducing the design? On the Stick Test, does the patient successfully reproduce items that can be named, but fail items that cannot be understood as concrete "things"? Efforts have been made to "score" these observations, but the common practice remains that of obtaining only the conventional scores involving number of errors and performance time.

DIFFERENT KINDS OF ABSTRACT REASONING

As we have seen, within neuropsychological assessment, we have the conventional tests of abstract reasoning as well as tests generally classified as assessing other abilities, such as the visual–spatial measures of Block Design or the Stick Test, that can be interpreted from the standpoint of abstract and concrete behavior through qualitative observation. While the abstract attitude may be involved in all of these procedures, the conventional tests provide a direct assessment of the individual's ability to learn or form an abstract concept. Some of these procedures are paper and pencil tests that use language directly as the test medium. The most commonly used tests of this type are analogies and proverbs tests. The Raven Progressive Matrices Test (1982) contains analogy items that use pictorial material, but factor analytic studies have shown that the test has a strong verbal component, apparently because many of the pictures of objects are nameable (Lezak, 1995). However, the tests used most commonly in neuropsychological assessment are the performance tests, which should not be characterized as nonverbal tests for various reasons, but which use nonverbal media, such as colored blocks or geometric forms. The major reason for not characterizing these tests as nonverbal is that while the media used are generally not linguistic symbols, the test solution process may place heavy reliance on language. We will refer to them as performance tests, for want of a better term.

The most commonly used of these performance tests are sorting tasks. Many years ago Egon Weigl (1927) invented the prototype of these tasks, which are still referred to as Weigl-type sorting tests. The first of these tests was of the free-sorting type in which a variety of objects are placed on a table and the subject is asked to group the objects through such instructions as "Sort those figures which you think belong together" or "Put those together which you think can be grouped together." After the first sorting, the subject is asked to put the figures together in another way. In the Goldstein–Scheerer series there is one relatively simple task, The Weigl–Goldstein–Scheerer Color Form Sorting Test, and two more complex tasks, the Gelb–Goldstein Color Sorting Test and the Gelb, Goldstein, Weigl and Scheerer Object Sorting Test. The test method, however, is the same in all cases; the materials are set out, the subject is asked to sort them, and then to resort them. These tests assess the general capacities to form an abstraction or concept as the basis for the initial sorting, and also evaluate cognitive flexibility, or the capacity to shift concepts. The administration of the Color Sorting Test is somewhat different from the other sorting tests. The test material consists of a large number of skeins of wool (Holmgren Wools) that vary in hue and brightness. The subject is asked to select a skein of her or his preference, and to pick out the other skeins that can be grouped with it (e.g., different shades of green). When this procedure is completed, the examiner picks out a skein of a different hue, and the subject is asked to pick out the other skeins that go with it. This procedure is followed by triple matching. Three skeins at a time are placed before the subject varying in hue and brightness. The left and center skeins have the same hue but different brightnesses, and the right skein has the same brightness as the center skein, but differs in hue. The examiner points to the left and right skein and asks about where the center one belongs. The shift relates to whether the subject can sort according to both hue and brightness. Shifting from hue to brightness is difficult for some normal people, and prompting about the idea of brightness is permissible, the point being whether or not the subject accepts the shift, and the idea of common brightness.

These free categorization tests provide abundant opportunity for qualitative assessment and variations of the procedure to elicit various features of concreteness. However, they differ from the concept identification procedures to be described below in the sense that they are true measures of concept *formation*. That is, the subject is provided with an array of diverse material out of which the abstraction has to be formed. The concept has to be self-initiated, and the subject makes up the rule that provides the basis for grouping. The rule may be very simple (e.g., color or shape) or quite complex as in the brightness or hue concept involved in the Color Sorting Test. Nevertheless, the subject is required to initiate his or her own categorization, or may fail to do so.

In the tests that followed the Weigl-type sorting tests, notably the Halstead Category Test and the Wisconsin Card Sorting Test, there was an important change. The concepts in these tests are not formed by the subject, but are inherent in the test materials themselves. The subject's task has changed from forming concepts to identifying concepts formed by the test-maker. Investigators in this area have therefore made a distinction between concept *formation,* which can be assessed with free-sorting tests, and concept *identification,* which is involved in the Category Test and card sorting procedures. In a series of studies by Bourne and collaborators, the process of concept identification was studied, mainly in normal individuals, in order to provide a detailed understanding of its relevant parameters, such as complexity and the role of informative feedback (Bourne, 1966). In a sense the difference between a concept formation and a concept identification procedure is analogous to the difference between a projective and an objective test. In the former, the subject can exercise free self-expression, while in the latter, there is a requirement for adhering to a particular structure.

Perrine (1993) has made the important distinction in concept identification tests between attribute identification and rule learning. The Wisconsin Card Sorting Test stresses attribute identification. The correct answer is the stimulus attribute of form, color, or number. In the case of the Category Test, the correct principle is a rule, regardless of the attributes of the stimuli. For example, the correct answer is the odd object in an array. Interestingly, Perrine reported only 30% shared variance between the two procedures.

The Bourne procedure involves primarily rule-learning tasks in which test stimuli may vary in color, size, shape, orientation, or location. Tasks are constructed with varying levels of complexity, with complexity defined as number of dimensions. Dimensions may be relevant or irrelevant, but complexity is generally increased by adding to the number of irrelevant dimensions. For example, geometric figures may vary in only color or shape. Color may be the relevant and shape the irrelevant dimension. If size also varies, color may remain as the relevant dimension, but both size and shape become irrelevant dimensions, thereby making the task more complex. Other stimulus parameters such as delay of informative feedback or interstimulus interval can also be varied.

There are some tests that incorporate aspects of both concept formation and concept identification. The Hanfmann–Kasanin Concept Formation Test (Hanfmann & Kasanin, 1937) is a challenging procedure in which the subject is asked to do a number of sorts, much like the Color Form Sorting Test. However, there is a correct answer that the subject must learn through making sorts and obtaining information from the examiner concerning the correctness of the solution. The task is challenging because the concept is not a directly perceivable attribute, but is a second-order principle that has to be derived from the characteristics of the attributes. The Twenty Questions Task (Minshew, Siegel, Goldstein, & Weldy, 1994) also has a correct answer, but the subject has to self-initiate sorting strategies to arrive at that answer. The procedure is much like the Twenty Questions parlor game in which a target object must be named based on questions that can only be answered yes or no. The strategy for narrowing down the possibilities and arriving at the right answer has to be formed by the player.

Another way of studying abstraction is through the examination of generalization. When the same response is made to a continuum of stimuli, the phenomenon is referred to as stimulus generalization. At a conceptual level, stimulus generalization allows for classification, such that all objects with the same invariant characteristics may be classed into specific categories. Thus, a table is still a table regardless of wide variations in size, color, shape, and other characteristics. When tasks are of a conceptual nature, stimulus generalization has been referred to as equivalence range (Gardner & Schoen, 1962). Equivalence range problems assess an individual's tolerance for variability in stimulus characteristics within some category. In the case of the Color Sorting Test, for example, the equivalence range would be the amount of variation in brightness accepted to categorize a skein as being of a common hue. In a study by Olson, Goldstein, Neuringer, and Shelly (1969), the task involved presenting geometric figures, half of which were permutations of a circle and the other half of which were permutations of a diamond. The permutations reduced the figures in width in the direction of a common shape. Subjects were shown the figures one at a time, and were asked to indicate if it was a circle, a diamond, or neither. The measure of equivalence range was correctly classified figures. A modified version of the Color Sorting Test was also administered. The literature suggested that brain-damaged individuals have narrow equivalence ranges, and that was what was found for both the color sorting and visual forms tasks. Thus, it would appear that abstraction of common properties by brain-damaged individuals has a narrow focus, probably limited to specific, concrete stimulus properties.

These tests of abstract reasoning are all pertinent to Goldstein and Scheerer's eight points characterizing the abstract attitude. They require to a greater or lesser degree the ability to maintain a mental set, to shift reflectively, to hold in mind simultaneously various aspects of a task, to abstract common properties, and to grasp essentials. Most scholars in the field would agree that these tests may be treated quantitatively, and would not agree with the relatively extreme view taken by Goldstein and Scheerer regarding numerical scoring. However, contemporary neuropsychology does not eschew the use of qualitative observation, and efforts are being made to make such observations objective, reliable, and perhaps quantifiable.

We have emphasized the distinction within abstract reasoning between those tasks in which the test-taker has to generate concepts and those in which an established concept has to be identified through experiencing a series of positive and negative instances. While self-initiated concept formation, attribute identification, and rule learning may all require the abstract attitude, they nevertheless appear to be separable cognitive abilities that may have differing clinical and adaptive implications. Absence of the abstract attitude, and consequent concreteness, may prevent solution of even the simplest conceptual tasks, but the capability of abstract reasoning can exist at numerous levels. As we will see in our review of some pertinent research, ability to identify relevant and irrelevant perceptual attributes and the ability to learn rules does not guarantee intact ability to generate conceptual strategies in "open-field" novel problem-solving situations.

The importance of flexibility was also stressed, since attainment of a perfectly correct concept may not be adaptive when environmental circumstances necessitate a change. The Wisconsin Card Sorting Test stresses this latter consideration. The symptom of fixed perseverative rigidity is perhaps the endpoint of this failure to reconceptualize under changing circumstances.

GERALD
GOLDSTEIN

Impairment of abstraction ability, or features of abnormal concreteness, can occur in a broad variety of brain disorders, and in a variety of types of psychopathology, notably schizophrenia, alcoholism, and several of the developmental disorders. Earlier in the history of the field of neuropsychology there appears to have been a consensus that abstraction ability was localized in the frontal lobes, and consequently, that neuropsychological tests of abstract reasoning are particularly sensitive to frontal lobe dysfunction. While it is not being suggested that the frontal lobes have nothing to do with abstraction, substantial evidence does not support specific localization. Impairment of abstract reasoning is found among patients with generalized brain damage, and is often found when there is focal brain damage outside of the frontal lobes. There is impressive evidence from lesion (Milner, 1963) and imaging studies (Goldberg *et al.*, 1987) that there is an association between performance on the Wisconsin Card Sorting Test and dysfunction of the dorsolateral surface of the frontal lobe. However, that finding may pertain to the specific symptom of perseverative rigidity, or incapacity to shift sets, and not to abstraction per se. Patients with a variety of disorders that do not involve the dorsolateral surface of the frontal lobes also do poorly on the Wisconsin Card Sorting Test, but abnormal concreteness may be manifested in different error patterns than those found in patients with focal frontal lobe lesions. While the specific pathways have not been identified, abstraction is probably best thought of as being mediated by a widespread network involving numerous cortical, and perhaps subcortical, structures. This view of the frontal lobes and abstraction is supported in a review paper by Reitan and Wolfson (1994) in which there is a detailed analysis of the research associated with the frontal lobes and tests of abstraction. Reitan concludes his review by stating in reference to the field of clinical neuropsychology: "It appears that such an approach would also lead investigators to recognize that other cerebral areas share essentially all of the higher level cognitive functions of the frontal lobes" (p. 192). If this is the case, clinicians should exercise caution in characterizing tests involving abstraction, such as the Wisconsin Card Sorting Test or the Trail Making Test, as specific indicators of frontal lobe dysfunction. Perhaps a more productive line of inquiry would involve study of particular disorders in which there are interesting patterns of impaired abstract reasoning that could help in clarifying the nature of the underlying neurobiological substratum. We therefore turn our attention to two such disorders, schizophrenia and autism.

ABSTRACT REASONING IN SCHIZOPHRENIA

There have been several generations of research concerning thought disorder in schizophrenia, much of it related to the schizophrenic patient's incapacity to reason conceptually. At least at a quantitative level, schizophrenic patients were characterized as performing like brain-damaged patients on sorting tests and related measures of abstract reasoning (Goldstein, 1978). However, in many ways, their concreteness was reported as being different from what was found in patients with structural brain damage. It has been described as more physiognomic, personalized, paleological, or bizarre than the more primitive and simplified thinking of the brain-damaged patient. The equivalence range of schizophrenics may be overly broad rather than overly narrow. Thus, when asked how a dog and a lion are alike, while the brain-damaged patient may say "They both have tails," the schizophrenic may say "They both exist in the universe." Thus, schizophrenics have been reported to overgeneralize or overinclude rather than undergeneralize (Payne, 1961). In recent years, great emphasis has been placed on perseverative rigidity in the thinking of schizophrenics, based largely on their tendency to produce a large number of perseverative errors and responses on the Wisconsin Card Sorting Test. Studies utilizing cerebral blood flow and position emission tomography (PET) demonstrated that unlike normals, schizophrenic patients failed to activate the dorsolateral surface of the frontal lobes when they took that test (Weinberger, Berman, & Zec, 1986). However, it has been suggested that schizophrenics do not entirely lose the abstract attitude as does the seriously brain-damaged patient, but fail to use it under conditions of high anxiety (K. Goldstein, 1959). Normal individuals under conditions of severe stress may also behave concretely. This behavior may be sustained in the schizophrenic because of the chronic high anxiety level associated with the disorder. Recent findings of structural brain abnormalities in schizophrenics tended to discourage the idea that concreteness in schizophrenia is entirely associated with anxiety, but the question remains as to the availability of abstract behavior. One way of evaluating this matter is to determine whether or not there is an improvement in abstract reasoning as a result of training. As indicated above, there have been several reports on attempts to teach schizophrenic patients to improve their performance on the Wisconsin Card Sorting Test, with mixed results. Shemansky and Goldstein (1996) have recently applied Reitan's method of comparing performance on Series V and VI of the Category Test to schizophrenic, brain-damaged, and control subjects, finding that the improvement by the schizophrenic subjects was comparable to that of the controls, with both groups improving significantly more than the brain-damaged group.

Based on the more recent literature, it would appear that impairment of abstraction ability in schizophrenia is a heterogeneous phenomenon. Some schizophrenic patients show no clear impairment of abstraction on the Wisconsin Card Sorting Test and Category Test, while others show severe impairment. The conflicting literature on the effects of training also suggest heterogeneity, since some studies showed that training was effective while others did not. The basis for this heterogeneity is not well understood, but is not a simple matter of variation in age, education, medication status, or chronicity with regard to length of illness or of hospitalization (Goldstein, 1990). Most of the evidence suggests that the impairment is quantitative, and does not reflect a complete loss of the abstract attitude. Nevertheless, the consensus of opinion appears to be that the deficit is primarily produced by impairment of brain function, and is not directly associated with anxiety. One widely held view is that the nature of the concreteness is fixed, perseverative rigidity because of the high number of perseverative errors and responses often made by schizophrenics on the Wisconsin Card Sorting Test. It has been suggested that this impairment is associated with a more fundamental deficit in working memory (Kimberg & Farah, 1993). The denseness of concrete thinking in schizophrenia may actually be severity-dependent. In a cluster analytic study, we identified a subgroup of patients that, on the average, completed only one out of six categories on the Wisconsin Card Sorting Test, and made an average of 110 errors on the Category Test. Other subgroups achieved greater numbers of card sorting categories, and made substantially fewer errors on the Category Test. We speculated that the former subgroup consisted largely of individuals with very poor outcome or "Kraepelinian" schizophrenia (Keefe *et al.*, 1987). Their inability to make even one card sorting conceptual shift and their performance at essentially a chance level on the Category Test suggest that they have a qualitative impairment of the abstract attitude. The other patients either did not have measurable impairment of abstract reasoning, or differed only quantitatively from normal. In summary, the clinician should expect to find diverse results when assessing abstraction ability in schizophrenic patients, reflecting the heterogeneity of the disorder. The basis for that heterogeneity, and whether it reflects actual subtypes, a continuum of severity, or a combination of the two remains a matter for future research.

ABSTRACT REASONING IN AUTISM

Autism is a developmental disorder marked by inability to make affective social contact, a restricted range of interest, and delay in language development. In the more severe cases, there may be total lack of language ability and mental retardation. It is not possible to assess abstract reasoning abilities in this group, and so our discussion will be limited to so-called high-functioning autistics, generally defined in terms of meeting all diagnostic criteria, and having a Verbal and Full Scale IQ score of 70 or greater (Schopler & Mesibov, 1992). In these cases, communicative language is present, and the individual is able to participate in testing. There is an extensive literature on cognitive ability in high-functioning autism, but here we will restrict our discussion to abstraction ability.

Studies using the Wisconsin Card Sorting Test have produced mixed results, with some reports of normal (Minshew, Goldstein, Muenz, & Payton, 1992) and some of abnormal findings (Rumsey & Hamburger, 1988). The most likely explanation for the discrepancies are differences in general intelligence (IQ) levels among the samples used in the various studies. Autism, while a single disorder, may vary extensively in severity. Rutter and Schopler (1987) initially proposed that IQ is a reasonable index of severity of autism, and that view has been widely accepted. The major point we wish to make, however, relates to those individuals with autism who perform normally on the Wisconsin Card Sorting Test. Minshew et al. (1992) reported that their individually age- and IQ-matched, carefully diagnosed autistic older children and young adults and controls did not differ from each other on the perseverative errors score from the Wisconsin Card Sorting Test, the total errors score from the Category Test, or time to complete Part B of the Trail Making Test. Thus, relative to controls, they did not differ from matched controls on the commonly used neuropsychological tests of abstraction ability. However, they were also administered the Goldstein–Scheerer Object Sorting Test. On this test, they were able to do an initial sorting comparable to the controls, but unlike the controls, when asked to find another basis for sorting the objects, they were unable to do so. Thus, autistic subjects who performed normally on the Wisconsin Card Sorting Test, the Category Test, and Part B of the Trail Making Test were unable to shift concepts on a sorting procedure, producing a highly significant difference from controls on this measure.

In a subsequent study, this group of investigators used the twenty questions procedure described above, and found that the autistic group was impaired relative to controls on this task (Minshew et al., 1994). Rather than logically approaching the task by asking questions that reduced the number of possible correct answers, they tended to guess, or to ask questions that did not lead to a solution. In other words, they did not self-initiate a productive strategy. Taking these results together, and recalling our distinction among concept formation, attribute identification, and rule-learning aspects of abstraction, it appears that individuals with high-functioning autism are not impaired with regard to identifying relevant attributes, or learning preestablished rules, but are substantially impaired in concept formation ability. Thus, they do well on the more recently developed concept identification tests, but poorly on the classic Weigl-type sorting tests. Since autism is now thought to be produced by an as yet not fully understood abnormality of brain development, the implication is that this distinction among types of abstraction may have neurobiological significance, since we have a disorder in which they can be dissociated. More specifically, the cognitive processes involved in self-initiating a schema for problem-solving may be quite different from those concerned with learning to separate relevant from irrelevant aspects of the environment, or from learning preexisting rules on the basis of repeated experience. Since this is a chapter about abstraction and not about autism we will not pursue this matter further, except to indicate that there is evidence that the difficulty autistic individuals have with organizing schemata or strategies is found in a number of areas of cognitive function including memory and language comprehension (Ameli, Courchesne, Lincoln, Kaufman, & Grillon, 1988; Minshew & Goldstein, 1993; Minshew, Goldstein, Siegel, 1995).

SUMMARY

Neuropsychological assessment of abstract reasoning may be accomplished with specialized tests of conceptual ability, or with a variety of tasks that may be accomplished abstractly or concretely. The former tasks are generally quantitative procedures, while the latter method typically involves qualitative observation. In assessment of abstraction ability, it is important to ascertain that the patient passes or fails the task because of a deficit in abstract reasoning ability. There are tasks that appear to measure that ability, but that may actually be accomplished through a variety of other methods. Impairment of abstraction ability in brain-damaged patients may be a qualitative loss of the abstract attitude or a quantitative impairment of level of ability. Furthermore, there are varying levels of abstraction, characterized as concept formation and concept identification. Concept formation refers to those tasks in which the subject must self-initiate concepts, while concept identification describes the situation in which the subject must learn an established concept. Within concept identification, learning may involve identifying relevant attributes or learning rules that organize diverse stimuli. Abstraction ability is not represented in any particular locus in the brain, and may be impaired by generalized brain damage or by focal brain lesions throughout the cerebral hemispheres. We have suggested that analysis of the differing patterns of impairment of abstract reasoning in schizophrenia and autism can contribute to the further delineation of the neurobiology of this uniquely human ability.

REFERENCES

Ameli, R., Courchesne, E., Lincoln, A., Kaufman, A. S., & Grillon, C. (1988). Visual memory processes in high-functioning individuals with autism. *Journal of Autism and Developmental Disorders, 18,* 601–615.

Bellack, A. S., Mueser, K. T., Morrison, R. L., Tierney, A., & Podell, K. (1990). Remediation of cognitive deficits in schizophrenia. *American Journal of Psychiatry, 147,* 1650–1655.

Boring, E. G. (1950). *A history of experimental psychology* (2d ed.). New York: Appleton-Century-Crofts.

Bourne, L. E., Jr. (1966). *Human conceptual behavior*. Boston: Allyn & Bacon.

Braff, D. L., Heaton, R. K., Kuck, J., Cullum, M., Moranville, J., Grant, I., & Zisook, S. (1990). The generalized pattern of neuropsychological deficits in outpatients with chronic schizophrenia with heterogeneous Wisconsin Card Sorting Test results. *Archives of General Psychiatry, 48*, 891–898.

Gardner, R. W., & Schoen, R. A. (1962). Differentiation of abstraction in concept formation. *Psychological Monographs, 76* (41, Whole No. 560).

Goldberg, T. E., Weinberger, D. R., Berman, K. F., Pliskin, N. H., & Podd, M. H. (1987). Further evidence for dementia of the prefrontal type in schizophrenia? *Archives of General Psychiatry, 44*, 1008–1014.

Goldman, R. S., Axelrod, B. N., & Tompkins, L. M. (1992). Effect of instructional cues on schizophrenic patients' performance on the Wisconsin Card Sorting Test. *American Journal of Psychiatry, 149*, 1718–1722.

Goldstein, G. (1962). *Developmental studies in analogical reasoning*. Unpublished doctoral dissertation, University of Kansas, Lawrence.

Goldstein, G. (1978). Cognitive and perceptual differences between schizophrenics and organics. *Schizophrenia Bulletin, 4*, 160–185.

Goldstein, G. (1990). Neuropsychological heterogeneity in schizophrenia: A consideration of abstraction and problem solving abilities. *Archives of Clinical Neuropsychology, 5*, 251–264.

Goldstein, G., Beers, S. R., & Shemansky, W. J. (1996). Neuropsychological differences between schizophrenic patients with heterogeneous Wisconsin Card Sorting Test performance. *Schizophrenia Research, 21*, 13–18.

Goldstein, G., Neuringer, C., & Olson, J. L. (1968). Impairment of abstract reasoning in the brain damaged: Qualitative or quantitative? *Cortex, 4*, 372–388.

Goldstein, G., & Shemansky, W. J. (1995). Influences on cognitive heterogeneity in schizophrenia. *Schizophrenia Research, 18*, 59–69.

Goldstein, K. (1951). *Human nature in the light of psychopathology*. Cambridge, MA: Harvard University Press.

Goldstein, K. (1959). The organismic approach. In S. Arieti (Ed.), *American handbook of psychiatry* (Vol. 2, pp. 1333–1347). New York: Basic Books.

Goldstein, K., & Scheerer, M. (1941). Abstract and concrete behavior: An experimental study with special tests. *Psychological Monographs, 53* (2, Whole No. 239).

Grant, D. A., & Berg, E. A. (1948). A behavioral analysis of the degree of reinforcement and ease of shifting to new responses in a Weigl-type card sorting problem. *Journal of Experimental Psychology, 38*, 404–411.

Green, M. F., Satz, P., Ganzell, S., & Vaclav, J. F. (1992). Wisconsin Card Sorting Test performance in schizophrenia: Remediation of a stubborn deficit. *American Journal of Psychiatry, 149*, 62–67.

Halstead, W. C. (1947). *Brain and intelligence*. Chicago: University of Chicago Press.

Hanfmann, E., & Kasanin, J. (1937). A method for the study of concept formation. *Journal of Psychology, 3*, 521–540.

Keefe, R. S. E., Mohs, R. C., Losonczy, M. F., Davidson, M., Silberman, J. M., Kendler, K. S., Horvath, T. B., Nora, N., & Davis, K. L. (1987). Characteristics of very poor outcome schizophrenia. *American Journal of Psychiatry, 144*, 889–895.

Kimberg, D. Y., & Farah, M. J. (1993). A unified account of cognitive impairments following frontal lobe damage: The role of working memory in complex, organized behavior. *Journal of Experimental Psychology: General, 122*, 411–428.

Lezak, M. D. (1995). *Neuropsychological assessment* (3d ed.). New York: Oxford University Press.

Milner, B. (1963). Effects of different brain lesions on card sorting. *Archives of Neurology, 9*, 90–100.

Minshew, N. J., & Goldstein, G. (1993). Is autism an amnesic disorder? Evidence from the California Verbal Learning Test. *Neuropsychology, 7*, 209–216.

Minshew, N. J., Goldstein, G., Muenz, L. R., & Payton, J. B. (1992). Neuropsychological functioning in non-mentally retarded autistic individuals. *Journal of Clinical and Experimental Neuropsychology, 14*, 740–761.

Minshew, N. J., Goldstein, G., & Siegel, D. J. (1995). Speech and language in high functioning autistic individuals. *Neuropsychology, 9*, 255–261.

Minshew, N. J., Siegel, D. J., Goldstein, G., & Weldy, S. (1994). Verbal problem solving in high functioning autistic individuals. *Archives of Clinical Neuropsychology, 9*, 31–40.

Neuringer, C., Goldstein, G., & Jannes, D. T. (1973). The relationship between age and qualitative or quantitative impairment of abstract reasoning in the brain damaged. *Journal of Genetic Psychology, 123*, 195–200.

Olson, J. L., Goldstein, G., Neuringer, C., & Shelly, C. H. (1969). Relation between equivalence range and concept formation ability in brain-damaged patients. *Perceptual and Motor Skills, 28,* 743–749.

Payne, R. W. (1961). Cognitive abnormalities. In H. J. Eysenck (Ed.), *Handbook of abnormal psychology: An experimental approach* (pp. 193–261). New York: Basic Books.

Perrine, K. (1993). Differential aspects of conceptual processing in the Category Test and Wisconsin Card Sorting Test. *Journal of Clinical and Experimental Psychology, 15,* 461–473.

Raven, J. C. (1982). *Revised manual for Raven's Progressive Matrices and Vocabulary Scale.* Windsor, UK: NFER Nelson.

Reitan, R. M. (1958). Qualitative versus quantitative mental changes following brain damage. *Journal of Psychology, 46,* 339–346.

Reitan, R. M. (1959). Impairment of abstraction ability in brain damage: Quantitative versus qualitative changes. *Journal of Psychology, 48,* 97–102.

Reitan, R. M., & Wolfson, D. (1994). A selective and critical review of neuropsychological deficits and the frontal lobes. *Neuropsychology Review, 4,* 161–198.

Rumsey, J. M., & Hamburger, S. D. (1988). Neuropsychological findings in high-functioning men with infantile autism, residual state. *Journal of Clinical and Experimental Neuropsychology, 10,* 201–221.

Rutter, M., & Schopler, E. (1987). Autism and pervasive developmental disorders: Concepts and diagnostic issues. *Journal of Autism and Developmental Disorders, 17,* 159–186.

Scheerer, M. (1946). Problems of performance analysis in the study of personality. *Annals of the New York Academy of Science, 46,* 653–678.

Scheerer, M. (1962). *Seminar in abnormal psychology.* Unpublished manuscript.

Scheerer, M., Rothman, E., & Goldstein, K. (1945). A case of "idiot savant": An experimental study of personality organization. *Psychological Monographs, 58* (Whole No. 269).

Schopler, E., & Mesibov, G. B. (Eds.). (1992). *High-functioning individuals with autism.* New York: Plenum Press.

Shemansky, W. J., & Goldstein, G. (1996, November). *Evidence for conceptual learning in schizophrenic patients.* Poster presented at the annual meeting of the Reitan Society, New Orleans, LA.

Simmel, M. L., & Counts, S. (1957). Some stale response determinants of perception, thinking, and learning: A study based on the analysis of a single test. *Genetic Psychology Monographs, 56,* 3–157.

Summerfelt, A. T., Alphs, L. D., Wagman, A. M. I., Funderburk, F. R., Hierholzer, R. M., & Strauss, M. E. (1991). Reduction of perseverative errors in patients with schizophrenia using monetary feedback. *Journal of Abnormal Psychology, 100,* 613–616.

Weigl, E. (1927). Zur psychologie sogenannter abstraktionsprozess. *Zeitschrift für psychologie, 103,* 1–45.

Weinberger, D. R., Berman, K., & Zec, R. (1986). Physiological dysfunction of dorsolateral prefrontal cortex in schizophrenia: I. Regional cerebral blood flow evidence. *Archives of General Psychiatry, 43,* 114–125.

Werner, H. (1937). Process and achievement. *Harvard Educational Review, 7,* 353–368.

Willner, A. E. (1971). *Conceptual Level Analogy Test.* New York: Author.

16

Neuropsychological Assessment of Memory

Joel H. Kramer and Dean C. Delis

The ability to encode, store, and later retrieve information is a highly complex cognitive function requiring a host of memory-related skills that can be disrupted by any number of neurological and psychiatric disorders. Memory deficits are often the first neuropsychological symptom in progressive dementing disorders and can be the only finding in disorders such as mild head trauma and multiple sclerosis. In addition, memory problems are among the most frequent complaints of the elderly (Craik, 1984) and of patients with major psychiatric disorders such as depression and schizophrenia (Caine, 1981; Weingartner & Silberman, 1984). The frequency of these problems underscores the importance of learning and memory assessment in any intellectual or neuropsychological evaluation.

Memory assessment has long been a part of mental status examinations. This is usually accomplished by testing the patient's immediate and short-delay recall of brief word lists (Cullum, Thompson, & Smernoff, 1993). Screening of memory functions was also included in some of the earlier IQ measures (Binet & Henri, 1895). Memory assessment in clinical psychology received a significant boost in 1945 with the publication of the Wechsler Memory Scale (WMS; Wechsler, 1945). Although its emphasis on the "memory quotient" invited the misleading supposition that a patient's memory functioning could be described by a single composite score, the WMS did survey a range of memory abilities. The 1940s also saw the introduction of the Rey Auditory Verbal Learning Test (RAVLT; Rey, 1964) and the Rey–Osterrieth Complex Fixture (Osterrieth, 1944), two tests that continue to enjoy widespread use by neuropsychologists.

In the past decade, the introduction of several new memory tests has paralleled an increasing appreciation of the complexity and importance of learning and memory. Three important trends in neuropsychology have contributed to this increased

Joel H. Kramer   Department of Psychiatry, University of California Medical Center, San Francisco, California 94143.   Dean C. Delis   Psychology Service (116B), VA Medical Center, San Diego, California 92161.

Neuropsychology, edited by Goldstein *et al.* Plenum Press, New York, 1998.

appreciation. First, neuropsychology has moved away from localizing lesions and more toward delineating the patient's spared and impaired abilities. Second, the past decade has generated the integration of principles of cognitive science with psychometric assessment techniques. In this approach, the goal of test construction has shifted from global achievement scores toward identifying important constructs developed in cognitive psychology and neuroscience and empirically establishing methods for quantifying them. Finally, the work of Butters (Butters, Cairns, Salmon, Cermak, & Moss, 1988; Heindel, Salmon, & Butters, 1989) and others (Squire, 1986, 1987) has shown how different neuropsychological syndromes produce dramatically different types of memory disorders. Collectively, these trends have served as a mandate for neuropsychology to acknowledge the complexity of memory in clinical evaluations and to develop memory assessment procedures that capture the most pertinent memory constructs.

Models and Constructs

This section reviews several useful models that have been developed to describe the various components and processes involved in learning and memory. The study of memory has inspired several different theories, each of which divides memory dichotomously (Squire, 1987). These divisions are not mutually exclusive, of course, and often the differences between them reflect differing areas of emphasis rather than opposing views about the underlying nature of memory. Nevertheless, an understanding of how these dissociations may be manifested in patients' test performance can enrich the clinical neuropsychologist's assessment of memory disorders.

The following section describes several proposed dichotomous models of memory. The emphasis here will be on constructs developed by cognitive psychologists and neuroscientists, although their clinical relevance will be highlighted. A clinical model for assessing memory is offered in a later section of this chapter.

Cognitive Models of Memory

Declarative versus Nondeclarative Memory. One of the most important advances in memory research in recent years has been the distinction between declarative and nondeclarative memory systems (Squire, 1993). Declarative memory refers to memories that are directly accessible to conscious recollection. It deals with facts, data, and experiences that are acquired through learning; the retrieval of this information is usually intentional and within the awareness of the individual. Because the act of remembering within the declarative system is typically explicit, the term *explicit memory* is often used (Schacter, 1992). Neuroanatomical structures mediating the declarative memory system are presumed to include neocortex, mesial temporal regions (hippocampus and surrounding cortex), and diencephalon (Squire & Zola-Morgan, 1991).

Nondeclarative memory refers to several different memory systems that are distinct from the declarative memory system. The most studied types of nondeclarative memory are priming and skills learning. Priming is the phenomenon by which prior exposure of material facilitates an individual's later performance with that same material. For example, subjects may be shown a list of words, and later be given a word fragment and asked to identify what the word might be. Even amnesic patients who display no conscious recollection of the previous list of words are faster at completing word fragments when the word was on the previous list. Skills learning or procedural memory is another type of nondeclarative memory. Procedural memory is demonstrated when patients exhibit learning of a skill (Squire, 1987). For example, patients with amnesia and dementia are often capable of learning new perceptual and motor skills, and can even become more proficient at complex problem-solving. Importantly, this increased proficiency can take place in the absence of the patient's conscious recollection of any learning trials. Priming and procedural learning are sometimes referred to as implicit memory because memory for the to-be-learned material is implicit in the patient's normal performance (Schacter, 1992).

The distinction between declarative and nondeclarative memory is most dramatic in patients with amnesic disorders. Typically, amnesic patients have declarative memory impairment, while nondeclarative memory systems are quite intact. Studies of the memory impairment associated with dementia, however, have revealed a more complex picture. While all dementia patients have declarative memory deficits, they differ in whether or not they maintain abilities in the area of nondeclarative memory. For example, Alzheimer's patients show impaired word-completion priming (Shimamura, Salmon, Squire, & Butters, 1987) but may show exaggerated repetition priming and preserved procedural learning (Heindel, Salmon, & Butters, 1989). Huntington's patients, on the other hand, typically perform normally on priming tasks but exhibit impairments on measures of procedural memory (Heindel *et al.*, 1989).

At present, almost all clinically available memory tests evaluate declarative memory. Even when incidental learning is being assessed (e.g., immediate recall of the Rey–Osterrieth figure), the patient's task is to consciously recollect the to-be-learned information. The recent development of standardized, normed, and validated measures of priming (e.g., word stem completion) and procedural learning (e.g., Tower of Hanoi) offers exciting new assessment possibilities with significant clinical and research applications (Davis, Bajszar, & Squire, 1994).

EPISODIC VERSUS SEMANTIC MEMORY. Some investigators have further subdivided declarative memory into episodic and semantic memory (Tulving, 1972). Episodic memory refers to memory for specific events, that is, episodes in one's life that can be assigned to a particular point in time. Semantic memory, on the other hand, refers to our storehouse of general information. Unlike episodic memory, semantic memory is not temporally coded; it holds information that does not depend on a particular time or place. Semantic memory also holds information such as words, symbols, and grammatical rules that are necessary in the use of language. Thus, our concept of what a sandwich is forms part of semantic memory; our recollection of the sandwich we ate for lunch yesterday is stored within episodic memory.

In clinical neuropsychology, the tests we typically use to evaluate memory are assessing episodic memory. Even when the to-be-remembered information is a word list, the meaning of the words (i.e., semantic memory) is presumed to be intact; what is intended to be measured is the patient's memory for the contents of that particular list (i.e., episodic memory). There are no clinical neuropsychological measures that were explicitly designed to assess semantic memory. Tasks such as the Vocabulary and Information subtests of the Wechsler scales, however, although not usually thought of as memory tests, do reflect the development and integrity of a patient's semantic memory.

JOEL H.
KRAMER AND
DEAN C. DELIS

SHORT-TERM VERSUS LONG-TERM MEMORY. In 1890, William James proposed one of the first and most enduring dichotomies to characterize human memory: short-term and long-term memory (or primary and secondary memory; Atkinson & Schiffrin, 1971; James, 1890; Waugh & Norman, 1965). Short- and long-term memory are generally thought of as being part of the declarative memory system (Squire, 1987). Short-term memory (STM) generally refers to recall of material *immediately* after it is presented or during uninterrupted rehearsal of the material; it is thought to be limited in its capacity (Miller, 1956) and, in the absence of rehearsal, undergoes rapid decay (Peterson & Peterson, 1959). Long-term memory (LTM) refers to recall of information after a delay interval during which the examinee's attention is focused away from the target items. It is thought to have an extraordinarily large capacity and to be fairly durable over time. The clinical significance of this distinction is exemplified in patients with amnesic syndromes. These patients tend to perform within the normal range on immediate recall of non-supraspan material but are unable to remember any information once a delay interval is imposed between presentation and recall.

Despite the efforts within psychology to clarify the STM–LTM distinction, the terms STM And LTM are generally used differently by other disciplines and the lay public. Typically, when patients refer to problems with their "short-term memory," they are describing difficulties remembering recently acquired information after delays (i.e., what cognitive psychologists call LTM). When patients cannot remember events from a long time ago (what neuropsychologists generally refer to as remote memory), they typically label it a problem with "long-term memory." Consequently, we recommend that the terms STM and LTM not be used in clinical practice. Much less confusion is likely if neuropsychologists are more descriptive in their characterizations of patient's memory performance, and use terms like "remote" versus "recent" versus "immediate" memory, or task-specific terms such as "immediate recall" and "30-minute delayed recall."

The major emphasis in a clinical neuropsychological evaluation is the patient's ability to learn new information, that is, his or her LTM (for episodic information) because LTM is the memory component that is most vulnerable to neuropsychological disorders. Neuropsychologists should be cognizant, however, of how well a particular clinical procedure permits a clear differentiation between the various theoretical constructs. For example, during most immediate recall tasks, examinees are typically relying on a combination of STM And LTM, particularly when the amount of information to be learned exceeds the individual's immediate memory span. Inferences about the integrity of a patient's LTM, therefore, cannot be based solely on the results of an immediate memory task.

RETROGRADE VERSUS ANTEROGRADE AMNESIA. Acute insult to neural systems responsible for memory can disrupt recall of information acquired prior to the injury (retrograde amnesia), as well as information presented for learning after the insult (anterograde amnesia). Retrograde memory can be assessed by asking the patient to recall autobiographical information that the examiner has obtained from relatives and other sources (Barbizet, 1970). Assessing retrograde memory overlaps with *remote memory* because both are concerned with information that has been encoded and stored in LTM prior to central nervous system injury or disease. Both of these constructs further overlap with semantic memory to the extent that semantic memory is composed largely of information stored prior to the onset of most memory disorders. Anterograde memory is evaluated whenever a patient is asked to learn new material and thus overlaps significantly with the constructs of LTM and episodic, declarative memory.

Patients vary greatly in terms of their relative levels of retrograde and anterograde amnesia (Butters & Miliotis, 1985; Squire, 1987). Some patients show greater anterograde than retrograde amnesia (Corkin, 1984; Zola-Morgan, Squire, & Amaral, 1986), some show greater retrograde than anterograde amnesia (Goldberg, Antin, Bilder, Hughes, & Mattis, 1981), and others show equally severe anterograde and retrograde amnesia (Damasio, Eslinger, Damasio, Van Hoesen, & Cornell, 1985). Given the heterogeneity in the patterns of anterograde and retrograde memory loss among neurologic patients, it behooves the clinician to assess both aspects of memory.

CLINICAL MODELS OF MEMORY

The models of memory offered by cognitive neuroscientists provide a broad framework for understanding the various forms memory takes and the ways in which learning and memory can be disrupted by neurological and psychiatric disorders. This section is devoted to a discussion of clinical assessment of memory.

Clinical neuropsychologists should be aware of two important points when evaluating a patient's memory. First, the spared and impaired components of a patient's learning and memory must be inferred from test performance. An examinee's performance on any task, however, depends on a range of attributes and skills, many of which are unrelated to the construct the clinician wants to measure. Test scores are end points that summarize the contributions of a variety of cognitive and other factors. Therefore, assessment of learning and memory cannot take place in isolation but should be part of a more comprehensive neuropsychological evaluation that includes not only assessment of other cognitive abilities, but also an identification of noncognitive factors (e.g., motivational, sensory, emotional) that influence memory tests performance.

The second important point is that the act of learning and memory is a series of events that take place over time. Thus, it is probably not reasonable or accurate to talk about a patient's "memory"; there are multiple component processes that each make a contribution to a patient's overall level of memory functioning. The clinician's goal is to tease apart the various contributions to patient's performance in order to most effectively diagnose and make treatment recommendations.

In the following section, we propose a clinical model of learning and memory to assist the clinician in weighing the several factors that influence an examinee's performance on memory tests. The model's five components, attention, processing, encoding, storage, and retrieval, are discussed (see Figure 1).

ATTENTION. An examinee's ability to learn and remember will be limited by how well he or she is paying attention to the to-be-learned information. An inattentive patient is likely to do poorly on most memory measures, regardless of how

Figure 1. A proposed model for the clinical evaluation of memory.

intact or impaired that patient's memory might be. This is particularly true for visual memory tasks that afford only a single, time-limited exposure (e.g., 10 seconds) of the stimuli, and verbal memory tasks involving a single oral presentation of supraspan information (e.g., WMS-R Logical Memory).

Attentional problems can be primary or secondary. Primary attentional problems include those biological conditions that disrupt a patient's capacity to attend globally or within a restricted portion of their spatial or sensory world (e.g., left neglect). In the most extreme cases, it would be unrealistic to expect patients with left neglect to recall information presented to their left hemispace. Children and adults with attention deficit–hyperactivity disorder (ADHD) are also frequently reported to perform less well than normals on memory tests (Loge, Staton, & Beatty, 1990) although their performance may improve when their attention is better focused.

Secondary attentional problems refer to a host of other factors such as anxiety, depression, and low motivation that can affect how attentive a patient is during a memory evaluation. It is particularly important to consider the role of depression, because patients with mood disorders often do less well than normals, and their memory test performance can be misinterpreted as reflecting underlying memory deficits. Anxiety is another important factor to rule out, because even very transient anxiety can severely disrupt performance on tasks that present information quickly (Cannon, 1997).

PROCESSING. Just as attention is a necessary precursor to learning and remembering, so is the capacity to process the to-be-learned information. For example, patients with aphasia are unlikely to do well on verbal memory tasks, and patients with visuospatial problems will do poorly on measures of visual memory. Inferences about an aphasic patient's memory should not be made on the basis of performance on a verbal memory task, however, since the capacity to learn and retain the information will be limited by the patient's inability to adequately process the information.

Even a cursory neuropsychological evaluation is usually sufficient to identify the more severe information processing impairments that will adversely affect memory test performance. A more pressing challenge for clinicians is appreciating how patients without obvious deficits are processing the to-be-learned material, since patients are likely to vary in terms of how long and how extensively they process incoming information. The degree and depth of information processing clearly affects how well it is learned and remembered. Craik and Lockhart (1972) have shown that information that is processed more superficially (e.g., orthographically) is generally not remembered as well as information processed more deeply (e.g., semantically). The implications of this for clinical assessment of memory can be quite profound. Children with dyslexia, for example, generally perform less well than controls on verbal learning tasks, but this may be attributable more to their subtle language or auditory processing deficiencies than to a memory deficit per se (Knee, Wittenburg, Burns, DeSantes, & Keenan, 1990). The extent to which an examinee processes the to-be-learned material varies along several parameters, including the amount of time spent, the degree and way in which the material is organized, and the number of associations made. The way in which information is processed (and subsequently encoded) is also influenced by the examinee's preexisting knowledge. When familiarity with the to-be-learned material is high, there is already a rich network of associations that can reduce processing demands and facilitate encoding (Chi, 1981).

Attention and processing can be thought of as precursors to learning and memory. Because of the large role they play in memory test performance, attention and processing must be considered by clinicians. There are at least two ways in which the clinician can accomplish this. First, as noted above, the memory evaluation should take place in the context of a comprehensive neuropsychological evaluation. Independent assessment of the patient's attention and processing abilities will enable the clinician to know when these factors can or cannot be ruled out as factors in the patient's memory test performance. Clinicians also can use memory tests that help control for attentional and processing abilities. For example, when testing visual memory with the Rey–Osterrieth Complex Figure, the examinee copies the design prior to being asked to draw it from memory. One important advantage of this procedure is that it is possible from an analysis of the examinee's copy to determine whether perceptual and constructional abilities are sufficiently intact to permit a meaningful assessment of memory. In addition, the degree to which the examinee perceptually organizes the stimulus can be evaluated, and the relationship between the examinee's copying strategy and subsequent recall can be assessed. Tests like the California Verbal Learning Test (Delis, Kramer, Kaplan, & Ober, 1987) also quantify several parameters associated with the patient's learning and organizational strategies. Tests that yield only a single achievement score can often be adapted to test hypotheses about which factors may be influencing patient performance (Kaplan, Fein, Morris, & Delis, 1991).

ENCODING. Encoding refers to the process by which physical information is transformed into a stored, mental representation. Encoding is central to the process of learning new information, that is, for the transfer of information into LTM.

Assessment of encoding requires that memory be assessed after a delay of at least 30 seconds during which the examinee has been prevented from rehearsing. If recall or recognition is assessed immediately following presentation of the stimulus, or if the patient has been allowed to rehearse during the delay, the information may still be available in STM. Patients with even severe encoding deficits can do deceptively well in their immediate recall provided their attention and STM are preserved.

Encoding is the presumed locus of impairment in most amnesic disorders that involve the medial-temporal structures or dorsal-medial nucleus of the thalamus. Patients with Korsakoff's disease and Alzheimer's disease, and famous patients HM and NA have been characterized as having profound deficits in the ability to encode new information into LTM.

Encoding is inferred rather than measured directly. If a patient displays impaired recall *and* recognition of the information, and attention and processing are intact, then an encoding deficit is a viable inference. If recall is impaired but the patient is able to adequately recognize the information, then some encoding can be presumed to have taken place.

STORAGE. Once information is learned (i.e., attended to, processed, and encoded), it becomes a part of LTM. Ideally, this information is accessible when needed; however, personal and clinical experience amply demonstrate that this is not always the case. There are several aspects of storage that neuropsychologists should be aware of when evaluating a patient's learning and memory. First, storage does not occur instantaneously; rather, it is a process that occurs gradually. There is much experimental and clinical data supporting this notion. For example, Squire (1984) showed that electroconvulsive shock therapy (ECT) can affect the storage of memories that were acquired as long as 1 to 3 years prior to the ECT. It is also common for closed head injury patients to have no recollection of events that occurred minutes or even hours prior to the moment of trauma. These data imply that the process of storing information takes place over an undetermined time interval, and that this process, referred to as consolidation, is vulnerable to disruption.

Stored information is also subject to distortion. Bartlett's classic studies (1932) demonstrated that the need to reduce incompatibilities within LTM can produce alterations in the to-be-remembered information. Loftus (1979) also showed that information stored in LTM can be changed by new information that the examinee learns after the to-be-remembered information has been encoded, and even by response sets established by the examiner. Alterations in LTM can also be observed during the clinical evaluation. For example, on the California Verbal Learning Test (Delis *et al.*, 1987), a commonly used list-learning task, a patient may tentatively offer an intrusion response on the short-delay cued recall trial. Twenty minutes later during the long-delay free recall trial, this intrusion response will be given again, but with a high degree of confidence. It appears that by virtue of having given the response earlier, the patient's "memory" for the list has been altered.

There is no question that even neurologically intact individuals forget information that has been encoded and stored in LTM. Ebbinghaus (1885) and Squire and Slater (1975) demonstrated that after an initial rapid decline, forgetting can occur gradually over a period of years. There has been considerable debate about why this forgetting occurs, however. There are two different but not mutually exclusive explanations. First, information in LTM is subject to decay, particularly if the information is not used for extended periods of time. A second mechanism that explains forgetting is interference, that is, the information is still in LTM but competing information is making the target information inaccessible.

Two types of interference have been proposed. Proactive interference is the decremental effect of prior learning on the retention of subsequently learned material; retroactive interference is the decremental effect of subsequent learning on the retention of previously learned material (Postman, 1971). Interference tends to be greatest when the interfering material is similar to the target information. Evaluation of both types of interference effects can play an important role in differential diagnosis. Patients with Korsakoff's disease, for example, are vulnerable to proactive interference (and also fail to release from proactive interference) whereas chronic alcoholics show normal release from proactive interference but display vulnerability to retroactive interference (Blusewicz, Kramer, & Delmonico, 1996).

RETRIEVAL. As we have all experienced, having information that is stored does not equal ready access. The information must be retrieved, that is, the examinee must engage in an active search through memory structures using whatever cues are available (Klatzky, 1980). Brain-damaged patients vary greatly in terms of their capacity for retrieval. If information has been encoded but cannot be adequately retrieved, then performance, relative to normal controls, should be poor on free recall (which places maximal demands on retrieval) but significantly better on recognition testing (which maximally aids retrieval). Evidence for this pattern of memory performance has been reported in patients with Parkinson's disease (Breen, 1993) and Huntington's disease (Butters, Wolfe, Granholm, & Martone, 1986; but see Brandt, Corwin, & Krafft, 1992, and Kramer, Delis, Blusewicz *et al.,* 1988, for a contrasting view). If the primary locus of impairment is in encoding, or if the stored information has decayed, then aiding retrieval through recognition testing will not enhance performance. This pattern of performance has been reported in Alzheimer and Korsakoff patients (Delis, Massman *et al.,* 1991).

Ideally, both recall and recognition will be assessed for the same material. Tests that evaluate only free recall will not discriminate between patients with retrieval deficits but who have the information stored from patients who do not have the information in storage. Free recall tasks can underestimate the amount of information encoded and stored because they fail to document the level of recognition memory. On the other hand, tests that evaluate only recognition run the risk of overestimating an examinee's memory abilities because they will be insensitive to deficits in recall.

SUMMARY

A patient's performance on a memory test can be impaired for several different reasons. The neuropsychologists's goal is to identify the spared and impaired cognitive components that influenced the patient's performance and to use this information for more refined and precise diagnoses and treatment plans.

It should be clear that the five component processes described here, attention, processing, encoding, storage, and retrieval, are highly interrelated. For example, a patient with a very passive learning style will not devote a great deal of mental energy thinking about and organizing the target information. This patient is less likely, therefore, to form associations during encoding that can assist retrieval later on. The paucity of associations and lack of subsequent processing further make this information vulnerable to decay during storage and to interference during encoding, storage, and retrieval. Thus, poor processing is associated with problems with encoding, storage, and retrieval. With patients like this, however, it cannot be assumed that the patient's impaired memory test performance is revealing very much about his or her capacity for new learning or about the integrity of the neural structures underlying learning and memory. If poor information processing is identified as a primary problem, the neuropsychologist can attempt to increase the patient's processing time and efficiency in order to assess more reliably new learning ability.

CLINICAL TESTS OF LEARNING AND MEMORY

Because memory is such as intricate function, it follows that a learning and memory test should acknowledge this complexity. The ideal assessment procedure would ensure attention, allow inferences about processing, assess retention of the information over delays, and utilize both recall and recognition formats. The degree to which extant memory tests approximate this ideal varies. Different procedures also emphasize different components of the information processing chain.

In this section, some of the most commonly used memory tests will be described. Due to space limitations, only a handful of instruments can be discussed;

the reader is referred to Lezak (1995) and Spreen and Strauss (1997) for reviews of additional tests.

Most memory tests measure the patient's ability to encode, store, and later retrieve new information. Because assessing immediate and remote memory are also important, however, these domains will be discussed briefly first.

IMMEDIATE MEMORY

The most commonly used test of immediate memory span is the Digit Span subtest of the Wechsler intelligence scales. On this and other digit span tasks, the patient is asked to repeat progressively longer sequences of numbers. There also are nonverbal immediate memory span tests (De Renzi, Faglioni, & Previdi, 1977; Kaplan *et al.*, 1991; Lezak, 1995; Milner, 1971; Wechsler, 1987). In these tests, small blocks are placed randomly on a board or are contained on a card. The examiner touches increasingly long sequences of the blocks or squares, and the examinee is asked to repeat each sequence in the same order.

Measures of digit span have alternately been referred to as tests of immediate or short-term memory, attention, freedom from distractibility, and auditory processing. How can a single procedure measure so many different constructs? In fact, it may be more accurate to state that digit and spatial span memory tests are sensitive to impairment in a variety of domains but do not necessarily serve as a metric for any one of them. Clinicians should therefore not reflexively assume that a poor digit span "reveals" any single deficit. Rather, they need to take a hypothesis testing approach to decipher which of the many potential (dis)abilities contributed to a particular examinee's performance. By administering a spatial span test, for example, the clinician can differentiate between a specific auditory processing problem (poor digit span; intact spatial span) and a more global impairment in immediate memory or attention (poor digit and spatial span).

REMOTE MEMORY

One of the major goals of a remote memory assessment is to characterize the nature of the retrograde memory deficit. Retrograde amnesia often adheres to a temporal gradient in which memory for recent events is more impaired than memory for remote events. Temporal gradients in retrograde amnesia have been reported in patients with Korsakoff's syndrome and those with mild Alzheimer's disease (Albert, Butters, & Levin, 1979; Squire, Haist, & Shimamura, 1989). In contrast, patients with Huntington's disease (Albert, Butters, & Brandt, 1981a) and with moderate to severe Alzheimer's disease (Wilson, Kaszniak, & Fox, 1981) have been found to display a "flat" retrograde amnesia (i.e., they are equally impaired in recalling events from different past decades).

Assessment of patients' memory for events that occurred prior to the onset of their brain dysfunction is often done informally by eliciting autobiographical recall. Because confabulations and inaccuracies in memory-impaired patients are common, verification of the information with relatives or by asking for the information a second time after a delay interval is recommended. Kopelman, Wilson, and Baddeley (1989) have also published guidelines for a semistructured interview that assesses recall of facts and specific events and incidents from the individual's past life.

Memory tests of public information have also been developed to quantify a patient's retrograde or remote memory. Because of large individual differences in exposure to public information, however, these tests are most appropriate for making only tentative inferences about retrograde memory.

WAIS-R INFORMATION SUBTEST. This subtest of the Wechsler Adult Intelligence Scale-Revised (Wechsler, 1981) assesses recall of cultural, biographical, scientific, and geographic facts likely to have been learned in school or in early life experiences. As with all WAIS-R subtests, one of its strengths is the fact that it has been normed on 1,800 adults representative of the U.S. population. For middle-aged and older individuals, this subtest can serve as a gross measure of remote memory. Because the items were neither selected systematically from different decades nor equated for ease of recall, important qualitative features of a patient's remote memory impairment such as the presence or absence of a temporal gradient are difficult to evaluate. Kaplan *et al.* (1991), however, have developed a recognition test for this subtest in order to help differentiate retrieval and naming problems from a remote memory deficit.

BOSTON RETROGRADE AMNESIA BATTERY (BRAB). The BRAB (Albert *et al.,* 1979) assesses memory for public events and famous actors, politicians, celebrities, and other individuals who were in the public spotlight from the 1930s to the 1970s. There are three components to the battery: a famous faces test, a verbal recall questionnaire, and a multiple-choice recognition questionnaire. An examinee's performance level in remembering information from each decade is graphed, which reveals whether a patient's retrograde amnesia is equally severe across all decades (flat retrograde amnesia) or less severe for more remote decades (temporal gradient). A creative feature of the test is that some of the pictures of famous people were taken during different decades (e.g., Jimmy Stewart during the 1930s and 1970s); patients whose retrograde amnesia adhere to a temporal gradient will often recognize the earlier but not the later picture.

Comparison data from normal control subjects are available. The BRAB has been used in numerous investigations which have documented the nature of retrograde memory loss in various patient populations such as alcoholic Korsakoff's disease, Huntington's disease, and Alzheimer's disease (Albert *et al.,* 1979; Albert, Butters, & Brandt, 1980, 1981b).

ASSESSMENT OF NEW LEARNING

VERBAL TESTS

MEMORY FOR WORD LISTS. Several verbal tests of memory use the word list format in which the subject is asked to recall words from a list.

Rey Auditory Verbal Learning Test (RAVLT). The RAVLT (Lezak, 1983) was one of the first word list tests used by neuropsychologists. On Rey's test, a list of 15 unrelated words is presented for five immediate-recall trials. A second interference list of unrelated words is presented for one trial, followed by recall and then recognition testing of the first list. Many examiners test for recall of the first list again after a delay interval (Lezak, 1995). Normative data are available for total number of words recalled on each of the recall trials of the first list (Geffen, Moar, O'Hanlon, Clark, & Geffen, 1990; Ivnik *et al.*, 1992; Lezak, 1995). In addition, the examiner can make qualitative interpretations about various learning parameters, such as primacy-recency effects, learning rate across trials, and increased vulnerability to proactive and retroactive interference. Rey's test has been used to document memory deficits in patients with head injury, Alzheimer's disease, focal lesions, and chronic alcoholism (see Spreen & Strauss, 1991).

The California Verbal Learning Test (CVLT). The CVLT (Delis *et al.*, 1987) is a more recently developed word list test that was modeled after Rey's test. One major difference between the two tests is that the CVLT lists contain four words from each of four semantic categories (e.g., "fruits," "tools"). Words from the same category are never presented consecutively, which affords an assessment of the degree to which an examinee uses an active semantic clustering strategy in recalling the words. Another difference between the tests is that the CVLT's scoring system quantifies and provides normative data for numerous learning and memory variables, in addition to total levels of recall and recognition. These variables include semantic and serial-order clustering strategies, primacy-recency effects, learning rate across trials, consistency of item recall across trials, degree of vulnerability to proactive and retroactive interference, retention of information after short and longer (20-minute) delays, enhancement of recall performance by category cueing and recognition testing, indices of recognition performance derived from signal-detection theory (discriminability and response bias), and frequency of error types (intrusions, perseverations, and false-positives). Computerized scoring automatically computes the multiple learning and memory indices and converts raw scores into standardized scores based on the examinee's age and sex (Fridlund & Delis, 1987). Normative data from 273 normal subjects and 145 carefully diagnosed neurological patients are provided for 26 test variables.

The advantage of the CVLT for clinical practice is that its scoring system quantifies the strategies, processes, and errors an examinee displays in learning verbal material. As such, the test yields a comprehensive profile of a patient's spared and impaired memory components. The clustering indices provide information about the ways in which the examinee processed the to-be-learned information, and primacy-recency scores yield clues about the extent to which the examinee relied on immediate memory versus encoding information into LTM. Inclusion of the recognition condition permits inferences about the integrity of encoding versus retrieval mechanisms. Finally, the long-delay recall conditions and the test's explicit assessment of retroactive and proactive interference effects appraise how well the information was stored and retrieved.

The incorporation of so many relevant learning and memory variables into a clinical procedure has resulted in an increased ability to differentiate various clinical disorders. For example, the presence of high levels of intrusion responses and rapid forgetting rates have been shown to be characteristic of Alzheimer's disease but not other dementias (Delis, Massman, Butters, Kramer, & Cermak, 1991; Kramer, Delis, Blusewicz *et al.*, 1988; Kramer, Levin, Brandt, & Delis, 1989). Distinguishing features of the memory impairment associated with depression (Massman, Delis, Butters, Dupont, & Gillian, 1992; Otto *et al.*, 1994), subcortical dementia (Massman, Delis, Butters, Levin, & Salmon, 1990), and head injury (Crosson, Novack, Trenerry, & Craig, 1989) have been elucidated. Decreased recall, intrusions, and a recency effect have also been found in elderly controls who have a positive history of Alzheimer's disease (Bondi *et al.*, 1994).

An alternate form of the CVLT is available with demonstrated alternative form reliability (Delis, McKee *et al.*, 1991). A children's version is available for children between the ages of 5 and 16 (CVLT-C; Delis, Kramer, Kaplan, & Ober, 1994) with normative data collected from almost 1,000 subjects.

The CVLT has also undergone a recent revision (Delis, Kramer, Kaplan, & Ober, 1997; CVLT-II). The CVLT-II retains the same format as the original CVLT, with 16-item lists and items from several distinct semantic categories. Improvements include: two equivalent alternate forms, utilization of items with higher frequency of use in the English language, a more challenging yes-no recognition condition, a forced-choice recognition designed to be sensitive to malingering, expansion of the process-oriented scoring system, and a normative reference group of over 1,000 adults between the ages of 16 and 90. An abbreviated list learning task called the CVLT-Mental Status version (CVLT-MS) has also been developed for use with patients whose memory impairments may be too severe for the more traditional CVLT. Also with two equivalent alternate forms, the CVLT-MS consists of four learning trials of a 9-item list. The interference list has been omitted, and the delayed recall intervals are 30 seconds and 10 minutes. The two different delayed recall intervals allows assessment of rate of forgetting from LTM, since items recalled after the 30-second delay can be assumed to have been encoded into LTM.

Selective Reminding Procedure. Most immediate free recall tasks draw on some combination of short-term and long-term memory. The selective reminding procedure (Buschke, 1973; Buschke & Fuld, 1974) was designed to differentiate more clearly between these two types of memory. On this task, the to-be-learned list is presented once, followed by an immediate recall trial. On the next trial, only those words the examinee failed to recall on the preceding trial are presented, and the examinee attempts to recall the entire list. This procedure is repeated for several trials until the examinee recalls all of the target words or a predetermined number of trials has been administered. Several memory constructs are operationally defined and quantified by the procedure, including number of items retrieved from STM, number of items retrieved from LTM, number of items stored in LTM, and the consistency with which the examinee retrieves items from LTM.

A number of different versions of the selective reminding procedure have been developed (Buschke & Fuld, 1974; Caine, Ebert, & Weingartner, 1977; Hannay & Levin, 1985; Ober, Koss, Friedland, & Delis, 1985; Randt & Brown, 1983). This procedure has been found to be useful in characterizing the memory deficits of patients with Alzheimer's disease (Buschke & Fuld, 1974; Ober *et al.*, 1985), mood disorder (Caine *et al.*, 1977), and head injury (Levin *et al.*, 1987).

Fuld Object Memory Test. The Fuld Object Memory Test (Fuld, 1981) is a variant of the selective reminding procedure that has some unique properties. As the beginning of the test, the examiner does not present the items orally; rather the examinee must first identify the actual objects by touch alone or, if this is not successful, the examinee is shown the objects and asked to name them. This ensures that the patient's attention is focused and, if the items are successfully named, provides evidence that the information was adequately processed. This procedure offers an important advantage with patients whose attention is questionable. Subsequent learning trials are administered in the same fashion as more traditional selective reminding paradigms with the exception that distractor intervals of 30 to 60 seconds are interposed between presentation and recall. This serves to eliminate the role of immediate memory, thus permitting the inference that any item recalled on any trial must have been encoded and therefore retrieved from LTM. The Fuld was specifically designed for use with older patients undergoing dementia evaluations. There are normative data available for community-dwelling control subjects in their 90's. The test has been fairly successful at discriminating between demented and nondemented patients and even between dementia and depression (la Rue, 1989).

MEMORY FOR STORIES. Other memory tests use a format in which the subject must recall a story.

Logical Memory Subtest of the Wechsler Memory Scale-Revised. The most commonly used story memory test is the Logical Memory subtest of the WMS-R (Wechsler, 1987). In this test, two stories are each read once for immediate recall. After a 30-minute delay, recall of the stories again is assessed.

Other clinical tests of story memory have been developed by Babcock (1930), Kramer, Delis, and Kaplan (1988), Heaton, Grant, and Matthews (1991), Lezak (1995), Randt and Brown (1983), and Talland (1965). The Kramer *et al.* (1988) test includes immediate and 20-minute delayed free recall, cued recall, and recognition testing; the test's scoring system quantifies gist recall, primacy-recency effects, and error types. Quantification of gist recall may have particular clinical relevance, because patients with right hemisphere damage are prone to lose the gist of a story despite remembering an average number of idea units (Wapner, Hamby, & Gardner, 1981). On the test developed by Heaton *et al.* (1991), one paragraph is presented for up to five trials or until a criterion score is reached. Free recall of the story after a 4-hour delay is also assessed.

It is not uncommon for a patient to perform differently on story recall and list recall. Examinees who perform better on stories often have difficulty adopting an active learning strategy (e.g., semantic clustering) during list-learning but are able to benefit from the thematic organization inherent in stories. Examinees who perform better on a word list memory test than on a story memory test may have weaknesses in comprehending complex sentence structure such that their struggle to extract meaning from the story interferes with adequate learning. Another difference between most list-learning and story recall tasks is that stories are typically presented one time only. Patients whose attention and/or processing are compromised will be more vulnerable with story recall, whereas a multiple-trial list-learning paradigm allows them additional opportunities to encode the relevant information.

VISUOSPATIAL MEMORY TESTS

THE REY–OSTERRIETH COMPLEX FIGURE TEST. The Rey–Osterrieth Complex Figure Test (Lezak, 1995) is a favorite visuospatial memory test among many neuropsychologists. The examinee first copies the complex figure (see Figure 2), and then draws it from memory both immediately and after a delay interval (Brooks, 1972; Milberg, Hebben, & Kaplan, 1986; Taylor, 1979; Waber & Holmes, 1986).

One advantage of the Rey–Osterrieth test is that the patient's copy yields information about how well the stimulus was initially processed. A poorly executed copy may suggest that the patient was inattentive (e.g., to one hemifield or to small details) or had underlying visuospatial deficits that dictate caution when interpret-

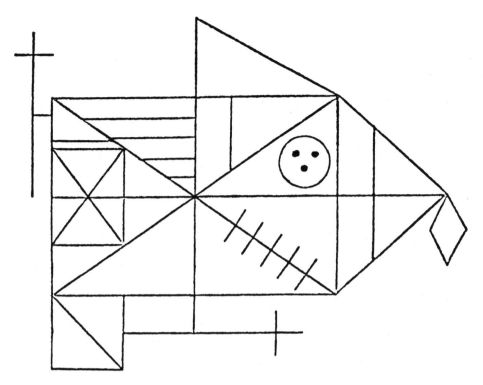

Figure 2. The Rey–Osterrieth Complex Figure.

ing the recall conditions. By reviewing the copy condition, the examiner can also assess the relationship between the examinee's copying strategy and subsequent recall performance. For example, Figure 3a shows the copy of a 54-year-old patient that is fairly accurate. Close examination of the way in which this patient copied the figure, however, suggested a process weakness. First, this patient copied the figure in a very segmented fashion and did not draw any of the larger configural elements as units. Second, in a fairly atypical fashion, he began his copying from the right side of the figure and worked from right to left. This approach to the Rey–Osterrieth test invites the hypothesis that there is relatively less efficient processing in the left hemispace (and damage to the right hemisphere). This hypothesis is borne out by the patient's immediate recall of the design (Figure 3b). The patient had a right frontal tumor.

Just as semantic clustering is the most effective strategy for recalling categorized word lists, *perceptual clustering* can facilitate recall of a complex visual stimulus such as the Rey–Osterrieth figure (Shorr, Delis, & Massman, 1992). Perceptual clustering refers to the perception and copying of a complex figure in terms of perceptual wholes or similar features (e.g., large rectangle, side triangle, diagonal lines, etc.), rather than drawing the figure in a disorganized, piecemeal manner (Paterson & Zangwill, 1944). Perceptual clustering organizes the numerous line segments of a complex figure into a smaller number of perceptual units for more efficient encoding and retrieval. Although perceptual clustering tends to correlate with memory accuracy, impulsive patients may copy the figure in a piecemeal manner, but draw it from memory in a perceptually organized manner. This find-

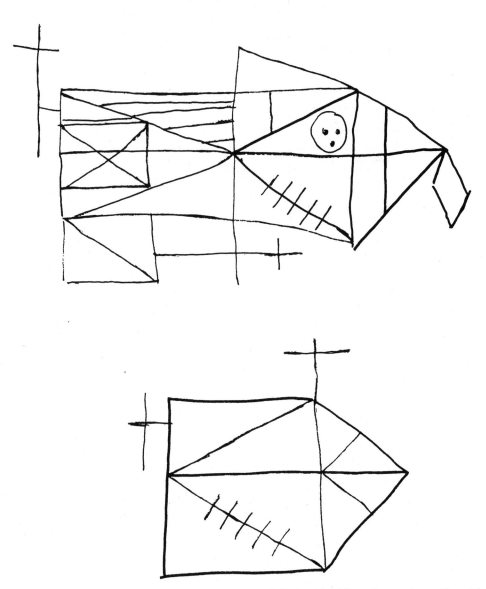

Figure 3. Copy and immediate recall of the Rey–Osterrieth Complex Figure by a patient with a right frontal lesion. The patient's copy, above, appears intact; only the process by which the patient copied the figure suggested a deficit. Below, the absence of information from the left side of the figure in the patient's immediate recall reflects his right hemisphere lesion.

ing indicates that they are slow to adopt a perceptual clustering strategy, but are capable of doing so during consolidation.

Another advantage of the Rey–Osterrieth test is that the figure contains both larger configural features (e.g., the large rectangle) and smaller internal details (e.g., the dots and circle). This configural/detail stimulus parameter helps dissociate the differential processing strategies of unilateral brain-damaged patients. Patients with left hemisphere damage often have difficulty remembering the internal details, whereas patients with right hemisphere damage are impaired in remembering the general configuration (Binder, 1982; Goodglass & Kaplan, 1979; Lezak, 1995; see Figure 4 for examples of patients' drawings).

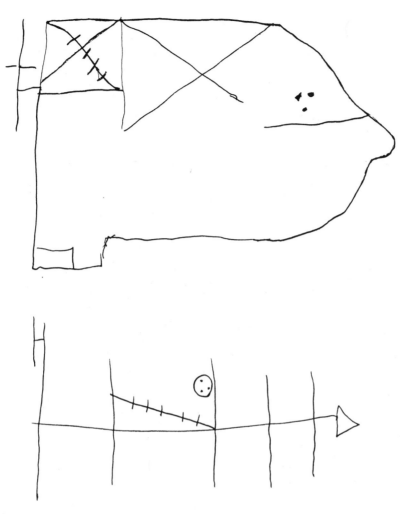

Figure 4. Delayed recall of the Rey–Osterrieth Complex Figure by patients with unilateral lesions. The top figure, drawn by a patient with a left hemisphere lesion, retains the outer configuration but omits most internal and peripheral details. The bottom figure, drawn by a patent with a right hemisphere lesion, omits the outer configuration but retains salient details.

Different scoring systems for the Rey–Osterrieth test are available (Binder, 1982; Meyers & Meyers, 1995; Stern *et al.*, 1994; Taylor, 1959; Waber & Holmes, 1986). The most widely used system was developed by Taylor (Taylor, 1959; Lezak, 1995). Although this system yields a single achievement score that reflects final accuracy only, and does not provide clear scoring criteria for the individual elements of the figure, normative data are available across a broad age range (Boone *et al.*, 1993; Chervinsky, Mitrushina, & Satz, 1992). Both Binder (1982) and Waber and Holmes (1986) have developed scoring systems that quantify an examinee's constructional strategy in addition to overall level of achievement.

Benton Visual Retention Test (BVRT). The BVRT (Benton, 1974) was one of the first visuospatial memory tests developed to assess brain functions. This test involves the presentation of 10 visual stimuli. The first two stimuli consist of one geometric shape each, and the remaining eight stimuli each consist of two large geometric designs and one small one. Standard administration calls for 10-second exposure followed by immediate recall, although other administration conditions are available. There are also three alternate forms to the BVRT. The BVRT is one of the first clinical instruments to use a scoring system that quantifies and provides normative data for multiple variables (e.g., accuracy and error types are analyzed).

The BVRT has been used extensively in neuropsychological investigations (Benton, 1962; Marsh & Hirsch, 1982; Sterne, 1969). It has been found to be sensitive to deficits in hemispatial attention (i.e., perceiving information in the left or right side of space) in unilateral brain-damaged patients (Heilbrun, 1956). However, most of the designs on the test can be easily verbalized (e.g., "large triangle"), and, as with many visuospatial memory tasks, it is difficult to ascertain the degree to which an examinee uses a spatial and/or verbal learning strategies (Lezak, 1995).

If an examinee shows impaired performance on a visuospatial memory test that involves a constructional response (e.g., drawing), the locus of the deficit could be at the perceptual, constructional, or memory level. Larrabee, Kane, Schunck, and Francis (1985), for example, noted in a large factor analytic study that more than half the variance of the BVRT loaded on a visuospatial factor, while less than half loaded on a general memory factor. Asking the examinee to copy the same designs he or she previously drew from memory will often elucidate the nature of the impairment (i.e., whether constructional or memory skills are more impaired; see Kaplan, 1983). If the patient's copies are deficient, then a matching task in which the examinee selects the target design displayed with several similar designs will help determine whether or not the patient has a perceptual problem. Benton and colleagues have developed tests of copying, visual discrimination, and visual recognition memory for greater diagnostic specificity (Benton, 1974; Benton, Hamsher, Varney, & Spreen, 1983).

VISUAL REPRODUCTION SUBTEST OF THE WMS-R. In this test, the examinee studies each of four stimulus cards for 10 seconds and attempts to recall each one immediately after its presentation. The first three cards display one design each, and the last one displays two designs. Following a 30-minute interval, delayed recall for the designs is assessed. Unfortunately, some of the figures are relatively simple and can be easily verbalized. In addition, as with the BVRT, the single, 10-second exposure makes the procedure vulnerable to attentional deficits. Kaplan and colleagues (Milberg *et al.*, 1986) have developed recognition memory, matching, and copying tasks for this subtest to assess the integrity of each of the component functions it taps (i.e., perception, construction, and memory).

CONTINUOUS VISUAL MEMORY TEST (CVMT). As noted above, any visual memory test that requires the examinee to draw a response potentially confounds perceptual and constructional abilities with memory. One means for circumventing this problem is by utilizing a recognition memory test. On the CVMT (Trahan & Larrabee, 1983), for example, examinees are shown a series of 112 abstract figures in sequential order, some of which are repeated several times throughout the series. The examinee's task is to discriminate novel figures from those that had been presented previously. In addition, a forced choice recognition memory condition is administered after a 30-minute delay. Finally, a visual discrimination task (i.e., matching) is administered to assess the examinee's ability to adequately perceive and discriminate the stimuli.

Patients with attentional difficulties also may do poorly on continuous memory tasks, so clinicians should consider attentional factors before asserting that a particular patient's poor performance was due to impaired visuospatial memory. Nonetheless, the CVMT has several strengths. The designs are fairly abstract and difficult to label verbally. The absence of the need to draw also reduces the visuoconstructional demands inherent in most other visuospatial memory tasks. Normative data are provided for two different types of total scores, the total number of items correct (correct hits + correct rejections), and d-Prime, a measure of discriminability based on signal detection theory. Another strength of the CVMT is the delayed recognition condition; after a 30-minute delay, the examinee must select the target figures from arrays with six distractors. Although the change in paradigm from old–new to forced choice makes comparison between the learning and delayed trials problematic, the delayed condition offers information about how well the information remained in storage. Finally, although currently yielding only achievement scores, the test can be adapted to provide information about rates of learning and response bias.

NONDECLARATIVE MEMORY

Assessment of priming and procedural memory has been, for the most part, a research enterprise. Although a sizable literature has accumulated on nondeclarative memory in dementia (Salmon, Shimamura, Butters, & Smith, 1988), amnesia (Shimamura, 1986), and normal development (Light & Singh, 1987; Parkin & Street, 1988), the clinical applications of these paradigms have only begun to be explored. The Colorado Neuropsychology Tests (CNT; Davis *et al.*, 1994) was specifically designed to assess several aspects of implicit and explicit memory using microcomputer technology. Measures of implicit memory (i.e., priming and procedural learning) include the Tower of Hanoi, Tower of London, Tower of Toronto, mirror reading, and a word-completion priming task. Norms are provided on approximately 600 subjects between the ages of 20 and 80 years. The CNT also gives investigators the option of creating their own memory task.

One creative clinical application of nondeclarative memory is in the detection of malingering. The assumption here is that patients who are intentionally trying to present with memory impairment will be naive to the fact that even amnesic patients have intact nondeclarative memory. Consequently, when faced with priming and procedural learning tasks, they are likely to perform at levels far worse than patients with even severe memory disorders. This methodology has been adapted by Davis *et al.* (1994) in the Colorado Malingering Tests, which include measures of priming as well as explicit memory tasks such as forced choice recognition memory.

CONCLUSIONS

Recent years have witnessed major developments in the neuropsychological assessment of learning and memory. Advances in cognitive neuroscience have more clearly delineated the most relevant memory constructs and have identified several neuroanatomical structures that are integral to the process of learning. Great strides have also been made in applying these memory constructs in the differentiation of learning and memory disorders. Thus, while memory impairment is one of the most common complaints or symptoms in neuropsychological patients, differences in the pattern of these deficits are emerging for amnesias, cortical dementias, subcortical dementias, depression, learning disabilities, head injury, and malingering. Finally, more sensitive and sophisticated memory tests have been introduced that enable clinicians to characterize more richly an individual patient's strengths and weaknesses and design interventions and compensatory strategies to maximize the patient's level of functioning. As neuropsychology continues to integrate clinical practice and cognitive neuroscience, we can anticipate exciting new advances in memory assessment with clinical and research applications.

REFERENCES

Albert, M. S., Butters, N., & Brandt, J. (1980). Memory for remote events in alcoholics. *Journal of Studies in Alcoholism, 41*, 1071–1081.

Albert, M. S., Butters, N., & Brandt, J. (1981a). Development of remote memory loss in patients with Huntington's disease. *Journal of Clinical Neuropsychology, 3*, 1–12.

Albert, M. S., Butters, N., & Brandt, J. (1981b). Patterns of remote memory in amnesic and demented patients. *Archives of Neurology, 38*, 495–500.

Albert, M. S., Butters, N., & Levin, J. (1979). Temporal gradients in the retrograde amnesia of patients with alcoholic Korsakoff disease. *Archives of Neurology, 36*, 211–216.

Atkinson, R. C., & Shiffrin, R. M. (1971). The control of short-term memory. *Scientific American, 224*, 82–90.

Babcock, H. (1930). An experiment in the measurement of mental deterioration. *Archives of Psychology, 117*, 105.

Barbizet, J. (1970). *Human memory and its pathology.* San Francisco: Freedman and Company.

Bartlett, F. C. (1932). *Remembering: A study in experimental and social psychology.* Cambridge, England: Cambridge University Press.

Benton, A. L. (1962). The Visual Retention Test as a constructional praxis task. *Confinia Neurologica, 22*, 141–155.

Benton, A. L. (1974). *The Revised Visual Retention Test* (4th ed.). New York: The Psychological Corporation.

Benton, A. L., Hamsher, K. deS., Varney, N. R., & Spreen, O. (1983). *Contributions to neuropsychological assessment.* New York: Oxford University Press.

Binder, L. M. (1982). Constructional strategies on complex figure drawings after unilateral brain damage. *Journal of Clinical Neuropsychology, 4*, 51–58.

Binet, A., & Henri, V. (1895). La memoire des mots. *L'Annee Psychologique, 1*, 1–23.

Blusewicz, M. J., Kramer, J. H., & Delmonico, R. L. (1996). *Interference effects in chronic alcoholism. Journal of the International Neuropsychological Society, 2*, 141–145.

Bondi, M. W., Monsch, A. U., Galasko, D., Butters, N., Salmon, D. P., & Delis, D. C. (1994). Preclinical cognitive markers of dementia of the Alzheimer's type. *Neuropsychology, 8*, 374–384.

Boone, K. B., Lesser, I. M., Hill-Gutierrez, E., Berman, N. G., & D'Elia, L. T. (1993). Rey–Osterrieth Complex Figure performance in healthy, older adults: Relationship to age, education, sex, and IQ. *Clinical Neuropsychologist, 7*, 22–28.

Brandt, J., Corwin, J., & Krafft, L. (1992). Is verbal recognition memory really different in Huntington's and Alzheimer's disease? *Journal of Clinical and Experimental Neuropsychology, 14*, 773–784.

Breen, E. K. (1993). Recall and recognition memory in Parkinson's disease. *Cortex, 29*, 91–102.

Brooks, D. N. (1972). Memory and head injury. *Journal of Nervous and Mental Disease, 155*, 350–355.

Buschke, H. (1973). Selective reminding for analysis of memory and behavior. *Journal of Verbal Learning and Verbal Behavior, 12*, 543–550.

Buschke, H., & Fuld, P. A. (1974). Evaluating storage, retention, and retrieval in disordered memory and learning. *Neurology, 24*, 1019–1025.

Butters, N., & Miliotis, P. (1985). Amnesic disorders. In K. M. Heilman & E. Valenstein (Eds.), *Clinical neuropsychology* (pp. 403–452). New York: Oxford University Press.

Butters, N., Cairns, P., Salmon, D. P., Cermak, L. S., & Moss, M. B. (1988). Differentiation of amnesic and demented patients with the Wechsler Memory Scale-Revised. *The Clinical Neuropsychologist, 2,* 133–148.

Butters, N., Wolfe, J., Granholm, E., & Martone, M. (1986). An assessment of verbal recall, recognition and fluency abilities in patients with Huntington's disease. *Cortex, 22,* 11–32.

Butters, N., Wolfe, J., Martone, E., Granholm, E., & Cermak, L. S. (1985). Memory disorders associated with Huntington's disease: Verbal recall, verbal recognition and procedural memory. *Neuropsychologia, 6,* 729–744.

Caine, E. D. (1981). Pseudodementia: Current concepts and future directions. *Archives of General Psychiatry, 38,* 1359–1364.

Caine, E. D., Ebert, M. H., & Weingartner, H. (1977). An outline for the analysis of dementia: The memory disorder of Huntington's disease. *Neurology, 27,* 1087–1092.

Cannon, B. J. (1997). Like a deer in the headlights: Relative impairment of logical Memory story A versus story B in a clinical sample (abstract). *Journal of the International Neuropsychological Society, 3,* 28.

Chervinsky, A. B., Mitrushina, M., & Satz, P. (1992). Comparison of four methods of scoring the Rey–Osterrieth Complex Figure drawing test on four age groups of normal elderly. *Brain Dysfunction, 5,* 267–287.

Chi, M. T. H. (1981). Knowledge development and memory performance. In M. Friedman, J. P. Das, & N. O'Connor (Eds.), *Intelligence and learning.* New York: Plenum Press.

Corkin, S. (1984). Lasting consequences of bilateral medial temporal lobectomy: Clinical course and experimental findings in H.M. *Seminars in Neurology, 4,* 249–259.

Craik, F. I. M. (1984). Age differences in remembering. In L. R. Squire & N. Butters (Eds.), *Neuropsychology of memory.* New York: Guilford Press.

Craik, F. I. M., & Lockhart, R. S. (1972). Levels of processing: A framework for memory research. *Journal of Verbal Learning and Verbal Behavior, 11,* 671–684.

Crosson, B., Novack, T. A., Trenerry, M. R., & Craig, P. L. (1989). Differentiation of verbal memory deficits in blunt head injury using the recognition trial of the California Verbal Learning Test: An exploratory study. *The Clinical Neuropsychologist, 3,* 29–44.

Cullum, C. M., Thompson, L. L., & Smernoff, E. N. (1993). Three-word recall as a measure of memory. *Journal of Clinical and Experimental Neuropsychology, 15,* 321–329.

Damasio, A. R., Eslinger, P. J., Damasio, H., VanHoesen, G. W., & Cornell, S. (1985). Multimodal amnesic syndrome following bilateral temporal and basal forebrain damage. *Archives of Neurology, 42,* 252–259.

Davis, H. P., Bajszar, G. M., & Squire, L. T. (1994). *Colorado Neuropsychology Test.* Colorado Springs: Colorado Neuropsychology Tests.

Delis, D. C., Kramer, J. H., Kaplan, E., & Ober, B. A. (1987). *The California Verbal Learning Test.* San Antonio, TX: The Psychological Corporation.

Delis, D. C., Kramer, J. H., Kaplan, E., & Ober, B. A. (1994). *The California Verbal Learning Test-Children's Version.* San Antonio, TX, The Psychological Corporation.

Delis, D. C., Massman, P. J., Butters, N., Salmon, D. P., Cermak, L. S., & Kramer, J. H. (1991). Profiles of demented and amnesic patients on the California Verbal Learning Test: Implications for the assessment of memory disorders. *Psychological Assessment, 3,* 19–26.

Delis, D. C., McKee, R., Massman, P. J., Kramer, J. H., Kaplan, E., & Gettman, D. (1991). Alternate form of the California Verbal Learning Test: Development and reliability. *The Clinical Neuropsychologist, 5,* 154–162.

Delis, D. C., Kramer, J. H., Kaplan, E., & Ober, B. A. (1997). *The California Verbal Learning Test-II.* San Antonio, Texas: The Psychological Corporation.

De Renzi, E., Faglioni, P., & Previdi, P. (1977). Spatial memory and hemispheric locus of lesion. *Cortex, 13,* 424–433.

Ebbinghaus, H. (1885). *Uber das Gedachtnis.* Leipzig: Duncker & Humblot.

Fridlund, A. J., & Delis, D. C. (1987). *CVLT research edition administration and scoring program.* San Antonio, TX: The Psychological Corporation.

Fuld, P. A. (1981). *The Fuld Object-Memory Evaluation.* Chicago: Stoelting Instrument Company.

Geffen, G., Moar, K. J., O'Hanlon, A. P., Clark, C. R., & Geffen, L. B. (1990). Performance measures of 16 to 86-year-old males and females on the Auditory Verbal Learning Test. *The Clinical Neuropsychologist, 4,* 45–63.

Goldberg, E., Antin, S. P., Bilder, R. M., Hughes, J. E. O., & Mattis, S. (1981). Retrograde amnesia: Possible role of mesencephalic reticular activation in long-term memory. *Science, 213,* 1392–1394.

Goodglass, H., & Kaplan, E. (1979). Assessment of cognitive deficit in the brain-injured patient. In M. S. Gazzaniga (Ed.), *Handbook of behavioral neurobiology* (Vol. 2). New York: Plenum Press.

Hannay, H. J., & Levin, H. S. (1985). Selective reminding test: An examination of the equivalence of four forms. *Journal of Clinical and Experimental Neuropsychology, 7,* 251–263.

Heaton, R. K., Grant, I., & Matthews, C. G. (1991). *Comprehensive norms for an expanded Halstead–Reitan battery.* Odessa, FL: Psychological Assessment Resources.

Heilbrun, A. B. (1956). Psychological test performance as a function of lateral localization of cerebral lesion. *Journal of Comparative and Physiological Psychology, 49,* 10–14.

Heindel, W. C., Salmon, D. P., & Butters, N. (1989). Neuropsychological differentiation of memory impairments in dementia. In G. C. Gilmore, P. J. Whitehouse, & M. L. Wykle (Eds.), *Memory, aging, and dementia: Theory, assessment, and treatment* (pp. 112–139). New York: Springer Publishing Co.

Ivnik, R. J., Malec, J. F., Smith, G. E., Tangalos, E. G., Petersen, X., Kokmen, X., & Kurland, X. (1992). Mayo's older Americans normative studies: Updated AVLT norms for ages 56 to 97. *The Clinical Neuropsychologist, 6,* 83–104.

James, W. (1890). *Principles of psychology.* New York: Holt.

Kaplan, E. (1983). Process and achievement revisited. In S. Wagner & B. Kaplan (Eds.), *Toward a holistic developmental psychology* (pp. 143–156). Hillsdale, NJ: Earlbaum.

Kaplan, E., Fein, D., Morris, R., & Delis, D. C. (1991). *The WAIS-R as a neuropsychological instrument.* New York: The Psychological Corporation.

Klatzky, R. L. (1980). *Human memory: Structure and processes.* San Francisco: Freeman.

Knee, K., Wittenburg, W., Burns, W. J., DeSantes, M., & Keenan, M. (1990). Memory indices of LD readers using the CVLT-C. *The Clinical Neuropsychologist, 4,* 278.

Kopelman, M. D., Wilson, B., & Baddeley, A. (1989). The autobiographical interview: A new assessment of autobiographical and personal semantic memory in amnesic patients. *Journal of Clinical and Experimental Neuropsychology, 11,* 724–744.

Kramer, J. H., Delis, D. C., Blusewicz, M. J., Brandt, J., Ober, B. A., & Strauss, M. (1988). Verbal memory errors is Alzheimer's and Huntington's dementias. *Developmental Neuropsychology, 4,* 1–15.

Kramer, J. H., Delis, D. C., & Kaplan, E. (1988). *The California Discourse Memory Test.* Manuscript submitted for publication.

Kramer, J. H., Levin, B., Brandt, J., & Delis, D. C. (1989). Differentiation of Alzheimer's, Huntington's, and Parkinson's disease patients on the basis of verbal learning characteristics. *Neuropsychology, 3,* 111–120.

Larrabee, G. J., Kane, R. L., Schunck, J. R., & Francis, D. J. (1985). Construct validity of various memory testing procedures. *Journal of Clinical and Experimental Neuropsychology, 7,* 239–250.

la Rue, A. (1989). Patterns of performance on the Fuld Object Memory Evaluation in elderly inpatients with depression or dementia. *Journal of Clinical and Experimental Neuropsychology, 11,* 409–422.

Levin, H. S., Mattis, S., Ruff, R. M., Eisenberg, H. M., Marshall, L. F., Tabuddor, K., High, W. M., & Frankowski, R. F. (1987). Neurobehavioral outcome following minor head injury: A three-case center study. *Journal of Neurosurgery, 66,* 234–243.

Lezak, M. D. (1995). *Neuropsychological assessment* (3d ed.). New York: Oxford University Press.

Light, L. L., & Singh, A. (1987). Implicit and explicit memory in young and older adults. *Journal of Experimental Psychology: Learning, Memory, and Cognition, 13,* 531–541.

Loftus, E. F. (1979). *Eye witness testimony.* Cambridge, MA: Harvard University Press.

Loge, D. V., Staton, D., & Beatty, W. W. (1990). Performance of children with ADHD on tests sensitive to frontal lobe dysfunction. *Journal of the American Academy of Child and Adolescent Psychiatry, 29,* 540–545.

Marsh, G. G., & Hirsch, S. H. (1982). Effectiveness of two tests of visual retention. *Journal of Clinical Psychology, 38,* 115–118.

Massman, P. J., Delis, D. C., Butters, N., Dupont, R. M., & Gillian, J. C. (1992). The subcortical dysfunction hypothesis of memory deficits in depression: Neuropsychological validation in a subgroup of patients. *Journal of Clinical and Experimental Neuropsychology, 14,* 687–706.

Massman, P. J., Delis, D. C., Butters, N., Levin, B. E., & Salmon, D. P. (1990). Are all subcortical dementias alike?: Verbal learning and memory in Parkinson's and Huntington's disease patients. *Journal of Clinical and Experimental Neuropsychology, 12,* 736–751.

Meyers, J. E., & Meyers, K. R. (1995). *Rey complex figure test and recognition trial.* Odessa, FL: Psychological Assessment Resources.

Milberg, W. P., Hebben, N., & Kaplan, E. (1986). The Boston process approach to neuropsychological assessment. In I. Grant & K. M. Adams (Eds.), *Neuropsychological assessment of neuropsychiatric disorders.* New York: Oxford University Press.

Miller, G. A. (1956). The magical number seven, plus or minus two: Some limits on our capacity for processing information. *Psychological Review, 63*, 81–97.

Milner, B. (1971). Interhemispheric differences in the localization of psychological processes in man. *British Medical Bulletin, 27*, 272–277.

Ober, B. A., Koss, E., Friedland, R. P., & Delis, D. C. (1985). Processes of verbal memory failure in Alzheimer-type dementia. *Brain and Cognition, 4*, 90–103.

Osterrieth, P. A. (1944). Le test de copie d'une figure complexe. *Archives de Psychologie, 30*, 206–356.

Otto, M. W., Bruder, G. E., Fava, M., Delis, D. C., Quitkin, F. M., & Rosenbaum, J. F. (1994). Norms for depressed patients for the California Verbal Learning Test: Associations with depression severity and self-support of cognitive difficulties. *Archives of Clinical Neuropsychology, 9*, 81–88.

Parkin, A. J., & Streete, S. (1988). Implicit and explicit memory in young children and adults. *British Journal of Psychology, 79*, 361–369.

Paterson, A., & Zangwill, O. L. (1944). Disorders of visual space perception associated with lesions of the right cerebral hemisphere. *Brain, 67*, 331–358.

Peterson, L. R., & Peterson, M. J. (1959). Short-term retention of individual verbal items. *Journal of Experimental Psychology, 58*, 193–198.

Postman, L. (1971). Transfer,, interference and forgetting. In J. W. Kling & L. A. Riggs (Eds.), *Experimental psychology*. New York: Holt, Rinehart and Winston.

Randt, C. T., & Brown, E. R. (1983). *Randt Memory Test*. Bayport, NY: Life Science Associates.

Rey, A. (1964). *L'examen clinique en psychologie*. Paris: Presses Universitaires de France.

Salmon, D. P., Shimamura, A. P., Butters, N., & Smith, S. (1988). Lexical and semantic priming deficits in patients with Alzheimer's disease. *Journal of Clinical and Experimental Neuropsychology, 10*, 477–494.

Schacter, D. L. (1992). Understanding implicit memory: A cognitive neuroscience approach. *American Psychologist, 47*, 559–569.

Shimamura, A. P. (1986). Priming effects in amnesia: Evidence for a dissociable memory function. *Quarterly Journal of Experimental Psychology: Human Experimental Psychology, 38*, 619–644.

Shimamura, A. P., Salmon, D. P., Squire, L. R., & Butters, N. (1987). Memory dysfunction and word priming in dementia and amnesia. *Behavioral Neuroscience, 101*, 347–351.

Shorr, J. S., Delis, D. C., & Massman, P. J. (1992). Memory for the Rey–Osterrieth figure: Perceptual clustering, encoding, and storage. *Neuropsychology, 6*, 43–50.

Spreen, O., & Strauss, E. (1997). *A compendium of neuropsychological tests* (2nd ed.). New York: Oxford University Press.

Squire, L. R. (1984). The neuropsychology of memory. In P. Marler & H. S. Terrace (Eds.), *The biology of learning*. Berlin: Springer-Verlag.

Squire, L. R. (1986). The neuropsychology of memory dysfunction and its assessment. In I. Grant & K. M. Adams (Eds.), *Neuropsychological assessment of neuropsychiatric disorders*. New York: Oxford University Press.

Squire, L. R. (1987). *Memory and brain*. New York: Oxford University Press.

Squire, L. R. (1993). The organization of declarative and nondeclarative memory. In O. Taketoshi, L. R. Squire, M. E. Raichle, D. I. Perrett, & M. Fukuda (Eds.), *Brain mechanisms of perception and memory: From neuron to behavior* (pp. 219–227). New York: Oxford University Press.

Squire, L. R., Haist, F., & Shimamura, A. P. (1989). The neurology of memory: Quantitative assessment of retrograde amnesia in two groups of amnesic patients. *Journal of Neuroscience, 9*, 828–839.

Squire, L. R., & Slater, P. C. (1975). Forgetting in very long-term memory as assessed by an improved questionnaire technique. *Journal of Experimental Psychology: Human learning and memory, 1*, 50–54.

Squire, L. R., & Zola-Morgan, S. (1991). The medial temporal lobe memory system. *Science, 253*, 1380–1386.

Stern, R. A., Singer, E. A., Duke, L. M., Singer, N. G., Norey, E., Daughty, E. W., & Kaplan, E. (1994). The Boston qualitative scoring system for the Rey–Osterrieth Complex Figure: Description and interrater reliability. *The Clinical Neuropsychologist, 8*, 309–322.

Sterne, D. M. (1969). The Benton, Porteus and WAIS Digit Span tests with normal and brain-injured subjects. *Journal of Clinical Psychology, 25*, 173–175.

Talland, G. A. (1965). *Deranged memory*. New York: Academic Press.

Taylor, E. M. (1959). *Psychological appraisal of children with cerebral deficits*. Cambridge, MA: Harvard University Press.

Taylor, L. B. (1979). Psychological assessment of neurosurgical patients. In T. Rasmussen & R. Marino (Eds.), *Functional neurosurgery*. New York: Raven Press.

Trahan, D. E., & Larrabee, G. J. (1983). Continuous Visual Memory Test. Odessa, FL: Psychological Assessment Resources.

Tulving, E. (1972). Episodic and semantic memory. In E. Tulving & W. Donaldson (Eds.), *Organization and memory.* New York: Academic Press.

Waber, D. P., & Holmes, J. M. (1986). Assessing children's memory productions of the Rey–Osterrieth complex figure. *Journal of Clinical and Experimental Neuropsychology, 8,* 563–580.

Wapner, W., Hamby, S., & Gardner, H. (1981). The role of the right hemisphere in the apprehension of complex linguistic materials. *Brain and Language, 14,* 15–33.

Waugh, N. C., & Norman, D. A. (1965). Primary memory. *Psychological Review, 72,* 89–104.

Wechsler, D. (1945). A standardized memory scale for clinical use. *Journal of Psychology, 19,* 87–95.

Wechsler, D. (1981). *Manual for the Wechsler Adult Intelligence Scale-Revised.* New York: The Psychological Corporation.

Wechsler, D. (1987). *Wechsler Memory Scale-Revised.* New York: The Psychological Corporation.

Weingartner, H., & Silberman, E. (1984). Cognitive changes in depression. In R. M. Post & J. C. Ballenger (Eds.), *Neurobiology of the mood disorders.* Baltimore, MD: Williams & Wilkins Co.

Wilson, R. S., Kaszniak, A. W., & Fox, J. H. (1981). Remote memory in senile dementia. *Cortex, 17,* 41–48.

Zola-Morgan, S., Squire, L. R., & Amaral, D. G. (1986). Human amnesia and the medial temporal region: Enduring memory impairment following a bilateral lesion limited to field CA 1 of the hippocampus. *Journal of Neuroscience, 6,* 2950–2967.

17

Neuropsychological Assessment of Aphasia

Nils R. Varney

Introduction

Aphasia has been the subject of fascination for some and bewilderment for many since the mid-18th century. Controversy about aphasia originated with the term aphasia itself, which was offered by the French neurologist Trosseau in competition with Broca's personal choice, "aphemia" (Benton, 1981). Although the term aphasia has now gained general acceptance, its definition and implications remain the subject of what is sometimes an acrimonious debate. For example, some view aphasia as a single or unitary disability while others refer to aphasic disorders and view aphasia as a class of disabilities. In addition, controversies about whether specific varieties of aphasia exist remain unresolved. With so much controversy, it is not at all surprising that even experienced neuropsychologists regard aphasia as a very complex area.

A Definition of Aphasia

For the purposes of basic clinical assessment, it is probably best to accept the definition of aphasia as an acquired disturbance of language which occurs as the result of brain disease. There are four critical elements in this definition: (1) Aphasia is a disorder of language, not speech production. Thus, lisps, stuttering, dysarthria, and similar disabilities are not aphasic because central language processing remains intact. Minimally, aphasia requires a disturbance in word finding as well as speech production, and in some aphasias, speech production may appear superficially intact. (2) Aphasia is an acquired disability. That is, in order to become aphasic, one must first have developed normal language. Thus, the language dis-

---

NILS R. VARNEY   Psychology Service, VA Medical Center, Iowa City, Iowa 52246.

Neuropsychology, edited by Goldstein *et al.* Plenum Press, New York, 1998.

turbances of the retarded, autistic, or congenitally deaf are not aphasic. (3) Aphasia is the result of brain damage. The disordered speech of the schizophrenic or Casey Stengle would not count as being aphasic. (4) Aphasia refers to a class of disabilities, a quite heterogeneous class. As the specific locus of lesion changes, so too do the nature, severity, and qualitative aspects of the resulting aphasic disorder.

BASIC ASSESSMENT ISSUES

Prior to making a diagnostic classification of any aphasic disorder, the neuropsychologist must adequately describe the patient's language deficits and areas of preserved function. For those unsure of their reliability in aphasia diagnosis, a thorough and accurate description of the patient's abilities and disabilities is a worthwhile goal in itself. Aphasia assessment can be more or less complex depending on the detail of the examination. Those aspects which are essential are discussed below. A summary of tasks appropriate for a comprehensive work-up of an aphasic patient can be found in Table 1.

FLUENCY

One of the major dichotomous variables by which aphasic disorders are classified is whether spontaneous speech is fluent or nonfluent. This distinction was introduced by Frank Benson in 1967 (Benson, 1967) and has since become universally adopted. In essence, the fluent aphasic is one who speaks a lot but communicates little, while the nonfluent aphasic says relatively little, but can potentially communicate more information than the fluent aphasic. The fluency/nonfluency distinction can be broken down into seven basic elements (Kerchensteiner, Poek, & Brunner, 1972): (1) *Effort:* Obviously, fluent patients show no effort in speech production while nonfluent patients show considerable effort. (2) *Phrase length:* In fluent aphasics, phrase length is normal, whereas in nonfluent aphasics phrase length utterances are rare, or involve only two or three words. (3) *Articulation:* Fluent aphasics articulate their speech normally while nonfluent patients are usually dysarthric. (4) *Rate:* The rate of word production in fluent aphasia is roughly in the 100 words per minute range while in nonfluent aphasia, it is much slower. (5) *Prosody:* This is the melodic intonation of speech. For example, one can tell from prosody whether someone is making a statement or asking a question, regardless of whether one can understand their words. In fluent aphasias, prosody is normal while the speech of nonfluent aphasics is usually dysprosodic. (6) *Word choice:* Nonfluent aphasics tend to speak primarily with substantive words, that is, words that have specific meaning and importance (e.g., hurts, hungry). The speech of fluent aphasics tends to contain considerably more relational words, such as adjective, adverbs, or prepositions, that have little communicative value on their own. (7) *Pauses/blocking:* Fluent aphasics show little in the way of pauses or blocking while these are common in nonfluent aphasics.

The seven elements of fluency and nonfluency are usually found together. Thus, if we were to assign 1 point for each fluent characteristic and no points for the nonfluent characteristics, most aphasics would have scores of 7 and 6 or 1 and 0. That is, the distribution of these symptoms is bimodal rather than continuous, rectangular, or normal (cf. Goodglass, 1993). However, roughly 10% of aphasics fail to be so easily classified and would obtain scores of 3 or 4 (e.g., a patient with

TABLE 1. APHASIA ASSESSMENT: LANGUAGE FUNCTIONS
AND RELATED ABILITIES

I. Spontaneous speech: Fluency[a]

	Fluent	Nonfluent
A. Word choice	Relational	Substantive
B. Rate	Normal	Slow
C. Articulation	Normal	Dysarthric
D. Phrase length	Long	Short
E. Effort	None	Marked
F. Pauses	Few	Many
G. Prosody	Melodic	Disturbed
H. Paraphasia	Frequent	None
I. Perseveration	Rare	Common
J. Press	Marked	Slow

II. Word finding
   A. Observe spontaneous conversational speech
   B. Confrontation naming
      1. Visual
      2. Tactile
   C. Controlled oral word association

III. Repetition
   A. Sentences
   B. Digits

IV. Writing
   A. From copy
   B. From dictation
   C. Spelling (written and oral)
   D. Block spelling
   E. Design copying

V. Aural comprehension
   A. Token Test
   B. Word–object matching
   C. Conversation
   D. Absurd sentences

VI. Reading
   A. Comprehension of single words—word to picture matching
   B. Comprehension of sentences
   C. Oral word reading (substantive vs. relational)(paraphasias)
   D. Letter recognition (vs. letter naming)

VII. Other
   A. Arithmetic
   B. Bucco-facial apraxia
   C. Ideomotor apraxia
   D. Right–left discrimination
   E. Finger localization
   F. Propositional thinking

[a]Rank ordered according to relevance in the fluency vs. nonfluency classification.

dysarthria, dysprosody, and pauses who shows normal phrase length, effort, and rate with frequent relational word choices). It is usually best to simply classify/describe such cases as showing "mixed fluent and nonfluent features."

It has also been found that most fluent aphasics have post-Rolandic lesions and that most nonfluent aphasics have pre-Rolandic lesions (Benson, 1967). However, once again there are occasional exceptions to this, which is why the fluent/nonfluent dichotomy is not referred to as the anterior/posterior dichotomy.

When the fluency distinction was first introduced, it contained three additional elements: paraphasic errors, perseveration, and press of speech. Subsequent study indicated that paraphasias and perseveration can also occur in nonfluent aphasias, and that press of speech was more illusory than real (Kerchensteiner *et al.*, 1972). Nevertheless, the presence of such symptoms should be noted when observed.

WORD FINDING

In order to be aphasic, the patient must demonstrate at least modest problems with word finding, a symptom with the generic title of anomia (i.e., inability to name). In a bedside or simple clinical evaluation, word finding can be evaluated in two basic manners. The first involves having the patient name common objects and their major parts. A wristwatch can be used for this by having the patient name the watch, hands, crystal, dial, stem, and band. Similarly, one could have the patient name a shoe, its laces, sole, and heel, and what it is made of, leather.

The use of the above-mentioned word finding tasks has certain limitations in detecting anomia because they are relatively gross measures. In more formal assessment, one can use black and white line drawings such as that shown in Figure 1 (Benton & Hamsher, 1989). On this item, patients are asked to name the elephant and its ear, tusk, and trunk. In addition to providing a greater variety of stimuli and range of difficulty, line drawing tasks have been shown to be a somewhat more sensitive measure of anomia than the naming of three-dimensional objects. How best to explain this regularly observed clinical phenomenon remains the subject of contemporary investigation and debate.

If the patient's performance in object naming appears to be normal or nearly normal, one can supplement the assessment with a second word finding task in which the patient must list all the words he or she can think of starting with a particular letter of the alphabet such as P, R, W, F, C, or L. The time limit should be

Figure 1. A sample item from visual naming on the Multilingual Aphasia Examination.

60 seconds, and it is usually advisable to use three test trials with different letters, and to tell the patient that he or she should not use any proper nouns. Roughly speaking, a score of 20 or better on three test trials can be regarded as normal. Performances among the more educated, the relatively young, and women tend to be better than those of the less educated, the elderly, or men. However, impaired performance on tasks of this type is also observed among non-aphasics with left frontal lobe disease.

When a patient is impaired in word finding, it is usually appropriate to report descriptive comments on the nature of errors made. There are six basic types of error likely to be observed, and most patients produce more than one type of error. (1) *Blocking:* The patient cannot think of the object's name. (2) *Circumlocution:* A relatively empty sentence or description of the object. For example, in response to being asked to name a watch, the patient might say, "It's, you know, the thing-a-ma-jig you tell time with." (3) *Misarticulation:* Here, the patient's dysarthria results in an incorrect pronunciation of a word. (4) *Semantic paraphasia:* This involves a naming error in which there is a substitution of the correct word with a real word from the same language. For example, a patient might call a cow a horse, or a camera an ice cream cone. (5) *Phonemic paraphasia:* This is a clearly articulated utterance which sounds approximately correct, such as calling an elephant an "eblanant." (6) *Neologism:* A response which makes no sense at all and which is not a real word, such as calling a cup a "plozel" or a "blippot."

In cases with severe anomia and/or nonfluency, one may supplement assessment of word finding by having the patient complete overlearned sentences such as "Mary had a little" or "Jack and Jill went up the." In some otherwise totally anomic patients, responses will be accurate with this type of stimulus. Another cueing trick involves giving the patient the first phoneme of a word, which will many times result in the patient completing the word correctly (e.g., saying "wa" for watch).

REPETITION

Another major factor in the classification of aphasic disorders is whether repetition is intact or impaired. Impaired repetition is a defining feature of some aphasic disorders, and can sometimes be much poorer than spontaneous speech. Similarly, there are other aphasic disorders in which repetition is strikingly well preserved as compared to all other expressive language abilities.

It will usually be appropriate to assess two types of repetition: digit repetition and sentence repetition. While the performances of most aphasics on these two simple tasks are likely to be similar, this is not always the case. When dissociation occurs, it is more likely to be in the direction of digit repetition being poorer than sentence repetition, but there will be occasional, if rare, cases in which the opposite pattern is observed.

The Digit Span section of the Wechsler Adult Intelligence Scale-Revised (WAIS-R) is adequate for assessment of digit repetition, provided that only the digits forward set of stimuli is employed. The lower range of normal with digits forward is six. Because digit reversal places a rather heavy demand on verbal short-term memory, most aphasics will fail this task, whether or not other aspects of repetition are preserved. With repetition of sentences, it is best to employ logical and meaningful sentences which reduce the load on short-term memory as much as possible. One can expect normals, and even amnestics, to repeat nine- or ten-word sentences (e.g., "John used to work here, but retired last year."). In assessment of sentence repetition, one can start with simple three-word sentences, and then increase sentence length by one word per trial.

In evaluation of repetition, it is important to note not only the degree of impairment, but also whether repetition differs from spontaneous speech with regard to fluency and frequency of paraphasic errors. One may also observe a "cliff phenomenon" in which the patient's repetition becomes severely impaired with the addition of a single word to the repetition phrase length (e.g., the patient may be able to repeat "He used to live here" but fail severely with "The apartment was rented last week.").

Many repetition-impaired patients will show no improvement with repeated practice, even if they are able to correctly repeat the first and second half of the stimulus correctly. Finally, one may occasionally observe the other deviant form of repetition which is called "echolalia," in which a patient's *only* preserved language skill may be repetition. Here, repetition occurs slavishly with all types of verbal stimuli, whether repetition is requested or not. Disturbances in repetition are to be expected with lesions in the perisylvian language areas. In transcortical aphasias, lesions lie outside this area, and although other aspects of language can be severely impaired, repetition remains strikingly intact (cf. Damasio, 1981; Goodglass, 1993).

AURAL COMPREHENSION

The third, and probably most important, issue in the classification of aphasic disorders is whether or not the patient can understand what is being said to him or her. The long-standing terminology for this is "receptive" and "expressive" aphasia. Although these terms remain in widespread use, they are somewhat misleading. As was mentioned previously, all true aphasic disorders involve disturbances in language expression, so all could legitimately be termed expressive. In addition, most aphasias involve at least some diminution of language comprehension, making the receptive classification more one of degree than dichotomy. The most frequently employed alternative offers a distinction between "motor" and "sensory" aphasia. This is also potentially cumbersome when applied, for example, to nonfluent aphasias with severe comprehension deficit.

A simple and relatively serviceable procedure for assessing aural comprehension is to place four objects in front of the patient and have him or her point to the objects as they are named by the examiner. Stimuli can be made more difficult simply by offering description of the object's parts or uses. For example, a wristwatch could be identified by the clues watch, tells time, has hands, is worn on the wrist, or has a crystal. As with naming, the use of tests with line drawing as stimuli offers a wider range of difficulty in assessment.

In attempting a bedside examination of aural comprehension, it is important to avoid the following potential pitfalls. (1) Aphasics can appear to be understanding conversational speech by nodding or gesturing appropriately when, in actuality, they are merely attending to the social or prosodic cues of speech. You can check for this rather easily by asking the patient an absurd question in which inflection and gesture imply an affirmative response (e.g., "Has the typewriter under the floor enjoyed lunch?" or "Didn't I see you last elephant?"). (2) Asking aphasics to demonstrate proficiency in aural comprehension by having them engage in gestural responses (e.g., "salute," "pretend to use a hammer," etc.) is complicated by what is called ideomotor apraxia; a disturbance in meaningful gestural expression. This is a frequent correlate of aphasia, including aphasias with normal aural comprehension. Thus, the aphasic with ideomotor apraxia may appear to be impaired in aural comprehension simply because gestural expression is as impaired as speech expression. (3) Having patients point to objects named by the examiner from among many response choices, such as objects in a room, can prove confusing for patients and give the false impression that they are impaired in aural comprehension. It is not uncommon, for example, to find aphasics who cannot point to the light switch in a room with many objects in it, but at the same time, can identify a stethoscope from among other medical instruments when these are placed in front of him or her.

A second and substantially different type of aural comprehension test is the Token Test (De Renzi & Vignolo, 1962). The basic stimuli of this test are 20 Plexiglas or wood squares and circles in two sizes and five colors. Instructions place little demand on aural receptive vocabulary, and consist of commands such as "Pick up the large green circle and the small white square." Although normals, and even many demented patients, find the Token Test to be quite easy, aphasics find the task to be very difficult and confusing, even when aural receptive vocabulary is well preserved. For example, a patient might understand complex, abstract words such as educational or equestrian, but nevertheless fail when asked to put the green circle on the red square. In a series of 100 consecutive patients with left hemisphere lesions seen at our hospital, 97 failed the Token Test while only 52 were impaired in aural word understanding. Of the 52 aphasics with impaired single-word recognition, all were grossly impaired on the Token Test. Thus, while the two test performances are related to some extent, the Token Test is a far more sensitive measure.

How best to interpret the results of the Token Test in the determination of whether an aphasic suffers a receptive language impairment remains the subject of debate in contemporary neuropsychology and neurolinguistics. A number of prominent aphasiologists have interpreted findings from the Token Test and similar measures as demonstrating that all aphasias involve disturbances in language understanding as well as expression. At the same time, however, it is of little clinical value to classify 97% of aphasias as being receptive, particularly when many of these patients understand single words normally. It is probably best, therefore, to consider both aural receptive vocabulary and Token Test performance in making a determination as to whether a patient's aphasia is "receptive." This term is best applied to cases with at least a modest disturbance in aural vocabulary (e.g., below the 5th percentile) and severely impaired performance on the Token Test (i.e., below 50% correct).

READING

The symptom of impaired reading is termed "alexia." It is very important to make a distinction between oral reading and reading comprehension in the assessment of any aphasic patient. Many aphasics who cannot read aloud with even remote proficiency remain quite capable of understanding written language, and a few patients are occasionally observed who can read aloud without comprehending the words they are "reading" (i.e., hyperlexia). It is also critically important not to assume that reading comprehension will be at least as impaired as aural comprehension in the "typical" receptive aphasic. In a few cases, reading comprehension may be preserved while aural comprehension is at least somewhat impaired, and in other cases, reading may be significantly better than aural comprehension even though both are sub-par. This fact is frequently overlooked in works concerned with the classification of aphasic disorders in which it is assumed that reading comprehension is inevitably impaired in the receptive aphasias. It has, of course, been well established that reading comprehension can be grossly impaired in the context of normal aural comprehension (i.e., alexia with and without agraphia). Actually, with the exception of the so-called pure alexias, most aphasia classification systems do not distinguish reading comprehension from aural comprehension.

The most basic procedure for evaluating reading comprehension is to have patients point to objects, or drawings of objects, after silently reading single words or short phrases (i.e., a procedure closely similar to that described previously for aural receptive vocabulary). Oral reading may be assessed as well, and many aphasics will perform substantially differently on oral reading tasks than in visual naming or other expressive tasks.

Reading comprehension defects may be descriptively subdivided into two major types based on whether letter recognition is intact or defective (cf. Varney, 1981). In *verbal alexia,* word reading is impaired while letter reading is normal. In *literal alexia,* both word reading and letter reading are impaired. It is important to note in this regard that some authors have taken to using the term *literal alexia* in reference to what was once called *letter blindness without word blindness* (Hinshelwood, 1900). No such disorder exists. However, some nonfluent aphasics have been observed (cf. Benson, 1977) who are impaired in oral naming of letters while retaining normal oral word reading (i.e., a literal anomia). In what remains a poorly integrated area of aphasiology, three quite distinct procedures have been employed to assess letter recognition. As mentioned above, one of these is oral letter naming. The second is aural comprehension of letter names in which the examiner asks the patient to point to a letter that is named aloud. Failure here is less common than in letter naming, but a majority of aphasics will perform poorly on this task. Finally, there is letter matching, a purely visual task in which patients are required to match similar letters of the alphabet shown in different cases (see Figure 2). This disorder is relatively uncommon, being present in only about 20% of aphasics and approximately half of receptive aphasias in which reading is impaired. However, this is the only type of letter recognition task with direct relevance to reading comprehension, and is the most reliable and useful means of distinguishing literal from verbal alexia (Varney, 1981).

e

E   **A**

Figure 2. Samples from a letter matching task.

Acquired disturbances in writing are referred to as agraphias, and all true aphasias are associated with some disturbance of written expression. If a patient is impaired in language expression, but can communicate effectively in writing, he or she is not aphasic, but is probably dysarthric. The above does not also mean that disturbed writing must parallel disturbances in language expression.

Many fluent and nonfluent aphasics and others show writing disturbances with substantially different qualitative features than their speech. For example, the four writing samples shown in Figure 3 were all obtained from aphasics with fluent, neologistic speech. As can be seen, their writing disturbances are quite different, and only one showed neologistic writing.

In assessment of writing, there are four issues to be considered: written spelling, writing from copy, writing from dictation, and spontaneous writing. It is relatively rare to observe aphasics who are agraphic and still able to write spontaneously in any way. However, an example of this symptom is shown in Figure 4. Writing from dictation and copy are analogous procedures in which the patient is required to copy written material or transcribe sentences offered by the examiner (e.g., with either procedure, sentences such as "The color of the wall is green" can be employed). How one assesses written spelling will depend in large part of the individual aphasic's capabilities. In some, the question will be whether they can spell anything correctly. In others, the question is degree of impairment.

An issue frequently of relevance to the agraphias is whether the writing is or is not dyspraxic, that is, whether the patient is capable of the basic motor elements of written expression. Graphomotor apraxia is common among aphasics, including many with fluent speech. Two examples of apraxic agraphias are shown in Figure 5, along with a sample of non-writing which is nevertheless incomprehensible. (The distinction here is similar to that between dysarthria and phonemic paraphasias. Is the patient producing unintelligible writing because of poor motor control or incorrect selection of fluent writing movements?)

Particularly among the nonfluent aphasics, right hemiparesis will require assessment of writing with the left hand. Because many normals have difficulty writing coherently with the left hand, it is usually desirable to assess "written" spelling with block letters that can be arranged to spell words. For example, one can offer a patient blocks with the letters C, E, I, and R and ask him or her to spell rice. Performance of this type is also sometimes of relevance in cases with apractic agraphia, but disturbances in constructional praxis can also detract from such performance.

SUPPLEMENTARY CONSIDERATIONS

After completing the basic assessment of speech, naming, repetition, aural comprehension, reading comprehension, and writing, the neuropsychologist will have a fairly good idea as to the nature and suitability of supplementary testing. Some specific areas of supplementary language assessment are discussed in the following sections.

ARITHMETIC. Acalculia, poor arithmetic, is a common feature of many aphasias, and may prove nearly impossible to assess in a patient with nonfluent or paraphasic speech and right hemiplegia. However, one can usually assess simple

NILS R.
VARNEY

Figure 3. Samples of agraphia.

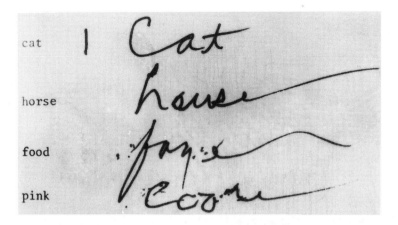

Figure 3. (*Continued*).

arithmetic skills by presenting the patient with written problems of varying complexity. In more severely acalculic patients, one may also find that they cannot count objects reliably and/or that they have lost their sense of the relative magnitude of numbers (e.g., 12 vs. 45; 111 vs. 99). Assessment of arithmetic skills is particularly important in patients with left parietal lesions for whom arithmetic skills may be severely impaired.

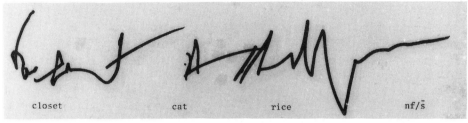

Figure 4. Spontaneous writing from Wernicke's aphasic.

Figure 5. Samples of apraxic agraphia.

VERBAL MEMORY. The use of verbal memory tasks with patients who are severely nonfluent or impaired in repetition or aural comprehension is pointless. However, in cases with mild aphasias or cases with apparent recovery, there may be quite severe problems with verbal short-term memory and other types of verbal learning. While verbal memory disturbances can be a major initial or residual symptom with lesions in a variety of left hemisphere loci, it is highly likely in cases with left temporal damage and/or left temporal seizure focus.

VERBAL IQ. One of the most common errors made in the assessment of aphasics is the use of verbal intelligence tests such as the WAIS-R in the context of significant anomia. If a patient suffers a major word finding deficit, this precludes a correct response to most WAIS Verbal Scale subtests, even when the patient knows the correct answer. The use of Verbal IQ tests is appropriate only as a means of revealing language-related deficits in a patient who is not clinically aphasic. As was true with verbal memory, verbal IQ subtests such as Similarities may indicate impairment even when the patient possesses normal language (cf. Hamsher, 1981).

SINGING. Among nonfluent patients, one will occasionally encounter patients whose fluency and articulation improve when they sing a common or familiar song. For example, the patient might not be able to say or repeat words like happy or birthday, but might be able to say these words clearly as they sing "Happy birthday to you." While this might seem odd at first, it has become a regular part of speech therapy (i.e., melodic intonation therapy) and thus may be of direct relevance to rehabilitation efforts.

BUCCOFACIAL PRAXIS/ORAL APRAXIA. Oral apraxis is a frequent, but not inevitable, correlate of nonfluent aphasias. It can be revealed by having the patient execute simple oral activities such as smiling, whistling, blowing out a match, or making a kissing sound. Aside from being directly relevant to the patient's nonfluency, severe oral apraxia can interfere with the patient's ability to swallow food.

WIDE RANGE ACHIEVEMENT TEST. It is not uncommon to have patients complain of diminished reading or spelling skills, but perform normally or errorlessly on tests of reading and writing from aphasia examinations. The greater range of "difficult" items in the Wide Range Achievement Test will assist in confirming or detecting such deficits.

NONVERBAL TASKS. Anyone assessing an aphasic patient should be aware that there is no such thing as a test that cannot be failed by an aphasic, and that failure on nonverbal tests does not indicate dementia or additional right hemisphere dis-

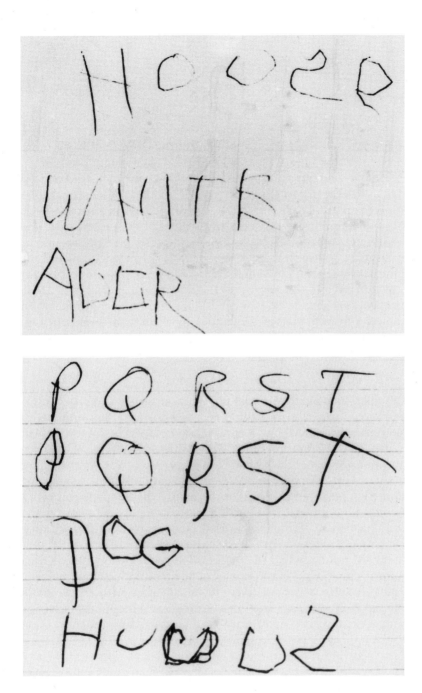

Figure 5. (*Continued*).

ease. Most subtests on the Performance scale of the WAIS-R are particularly un-suited for aphasics (i.e., Digit Symbol, Picture Arrangement).

GRAPHOMOTOR PRAXIS. For aphasics with significant agraphia, and particularly for those whose writing is nonfluent and/or unintelligible, one should determine whether they can do any better in the drawing or copying of geometric shapes. If a significant graphomotor apraxis is found, it will have to be dealt with as part of the remediation of writing.

CLASSIFICATION

Although aphasic disorders are quite rich in nature and present a variety of language symptoms, the six basic aphasic disorders are identified on the basis of three specific symptom areas: fluency, repetition, and comprehension. Consideration of the basic symptoms outlined should enable one to correctly classify about 80% of typical aphasic patients. The remainder fall under the frequently employed heading "mixed aphasia" or one of the relatively uncommon disorders such as alexia without agraphia, pure word deafness, or isolation of the speech area. If one's only goal in assessment is diagnostic classification, the relevant assessments can be performed in less than 30 minutes. However, so brief an assessment will only marginally augment the bedside examination of the clinical neurologist, and will usually provide relatively little in the way of new or useful clinical information. There are numerous other dimensions of aphasia assessment and description that are of considerable if not great importance, and that sometimes have direct relevance for prognosis, placement, and rehabilitation efforts (cf. Kaplan & Goodglass, 1981).

BROCA'S APHASIA

Broca's aphasics typically show all the characteristic features of nonfluent aphasia, are severely impaired in repetition, are severely agraphic, and have relatively well-preserved comprehension of conversational speech. Motor deficits on the right are inevitable. Communication via gesture is likely to be impaired, but a significant minority of Broca's aphasics show no impairments in this area. Aural comprehension at the single-word level should be normal or nearly normal, but Token Test performances are likely to be highly variable between moderate impairment and gross impairment. Writing will also be severely nonfluent and dyspraxic, and spelling will be impaired as well. Not surprisingly, Broca's aphasia is the result of lesions in Broca's area, which is in the **opercular** region of the third frontal gyrus. It is likely, however, that the lesion will extend beyond Broca's area into the frontal, anterior temporal, or anterior parietal areas. In a clinical setting, the diagnosis of Broca's aphasia is most likely to be confused with simple dysarthria or global aphasia. If the patient's writing is normal even though speech is nonfluent, Broca's aphasia is not present. If aural and reading comprehension are severely impaired, the appropriate diagnosis is global aphasia.

It is often of interest in the evaluation of Broca's aphasics to compare oral reading with reading comprehension. Most Broca's aphasics will be quite poor in oral reading, but will comprehend nearly normally. A few, however, may show significant impairments in reading comprehension as well. It has also been observed that many Broca's aphasics cannot name individual letters of the alphabet, a disorder sometimes referred to as the "third alexia." It is not a true alexia, however, because letter recognition will probably be preserved. In evaluating or working with a Broca's aphasic, one may encounter any of the following problems: (1) Broca's aphasics may be highly frustrated because they have ideas or issues they want to communicate, but are unable to do so via any normal expressive ability. As the patient becomes more frustrated, performances in areas of relatively little disability may decline. It is advisable, therefore, to be judicious in the sequential administration of tests that Broca's aphasics are likely to fail, and to assist such patients in their efforts to communicate. (2) In addition to being frustrated or distressed, Broca's aphasics are likely to have periods of emotional lability in which they will cry freely and appear distressed. It is usually best when this occurs to discontinue whatever precipitated the crying spell and converse with the patient as though he or she was not crying. A display of "natural" concern and sympathy may only serve to exacerbate the crying spell and distress the patient to the point that he or she will become untestable or uncooperative. (3) A significant proportion of Broca's aphasics will be occasionally troubled by perseveration. That is, they will continue to offer a particular response over and over in reply to different questions. For example, they might correctly name a cow, and then use this same response in reply to questions about their age or marital status. Once again, this problem can be circumvented by having the patient discontinue all tasks in which oral or motor responses are required in order to "break the set."

WERNICKE'S APHASIA

Wernicke's aphasia is one of the most fascinating and spectacular behavioral disorders in all psychology, either psychiatric or neurologic. The wealth of abnormal language behaviors in such patients can also be intimidating at first, particularly because the typical Wernicke's aphasic is unaware of these deficits. This often creates the false impression that such patients are functionally demented in addition to being aphasic. Although they have a number of specific handicaps and functional limitations, Wernicke's aphasics may be far from being demented.

The speech of Wernicke's aphasics will be fluent, well articulated, prosodic, and quite paraphasic. The frequency of paraphasic errors in speech can be disorienting at first, and in addition, give the false impression that the patient has press of speech. Language understanding of all types will also be impaired, as will be repetition and writing. However, writing may be severely nonfluent or otherwise impaired in a manner quite different from speech. Paraphasic speech is sometimes compared with hebephrenic schizophrenia, but these two disorders are usually quite distinct to the experienced clinician. In this regard, one important consideration is appropriateness of conduct, routine social manners, and facial expression.

In one of the most spectacular and illustrative examples of Wernicke's aphasia an 82-year-old retired attorney had suffered a stroke while driving, and presented with Wernicke's aphasia thereafter. When asked to write a sentence describing his illness, he wrote, "The big and big for my and over again" while saying "We were going down the road, then the cow fell in the water and drowned. No more ice cream." After a short period of contemplation and confusion over his writing, he turned, pointed to an abstract painting, sighed, and said, "It's like me. Some understand. Some don't." His age-corrected WAIS Block Design score was 17.

Wernicke's aphasia is usually the result of lesions involving Wernicke's area in the posterior left temporal lobe (superior temporal gyrus). In most cases, the lesion also compromises much of the white matter below the left posterior temporal area, cortical areas below Wernicke's area, and temporal areas anterior to Wernicke's area. In offering a diagnosis of Wernicke's aphasia, one should be sure that language comprehension is clearly impaired. Slightly more anterior lesions can produce paraphasic speech with normal comprehension. (The diagnosis of Wernicke's aphasia with normal comprehension is not acceptable.)

It is also important to check repetition in order to rule out transcortical sensory aphasia. It is not uncommon for Wernicke's aphasics to become irritable and uncooperative. In part, this reflects the fact that they do not appreciate their deficits or others' responses to their speech. Many also think they understand others' communications to them, with the result that repetition of questions add further frustration. It is usually best when making initial contact with a Wernicke's aphasic to at least occasionally pretend to understand him or her. With prolonged contact, one may be able to grasp what the patient is saying through a combination of occasionally correct words, syntactic structure, and social/prosodic cues. Most Wernicke's aphasics will reject their own utterances as being meaningless, so it is not advisable to ask for clarification by repeating what they say.

Of all the aphasia diagnostic groupings, it is among Wernicke's aphasics that prognosis is most variable. Some, unfortunately, never recover even a limited mastery of language or appreciation of their deficits while others make recoveries which appear, by comparison with their original condition, complete. (No aphasic ever totally recovers.) It is recommended, therefore, that Wernicke's aphasics be routinely reevaluated for at least the first 6 months after onset. This is also important in that with some recovery of language comprehension, they can become considerably more suitable for language rehabilitation even though these efforts proved fruitless in the immediate postacute period.

Global Aphasia

Global aphasia can be viewed either as Broca's aphasia with poor comprehension or Wernicke's aphasia with nonfluent speech. As its name suggests, all language functions should be significantly impaired, including spontaneous speech of all types (i.e., normal or paraphasic). Needless to say, this is the most disabling of the aphasic disorders. However, the exact locus of lesion can be somewhat variable, certainly more so than in Broca's and Wernicke's aphasia.

Although language expression and understanding are severely impaired in global aphasia this does not necessarily also imply that the patient is incapable of any communication and understanding. In assessment of global aphasics, one should place some emphasis on identifying those areas in which the patient is relatively less impaired. For example, global aphasics are sometimes capable of grasping some aspects of conversational speech even though their language comprehension is clearly subnormal. It should also be kept in mind that global aphasics can resolve into Broca's aphasias sometimes as late as 6 to 12 months after onset. Thus, while rehabilitation efforts may prove fruitless in the acute stage of global aphasia, patients should always be reevaluated for rehabilitation potential in the first 6 months after onset. Finally, it should also be mentioned that nursing home placement is not necessarily indicated for global aphasics, even though they appear more impaired than Alzheimer's disease patients. Many global aphasics remain capable of enjoying the company of their family, and can communicate nonverbally in a manner which family members find reinforcing (i.e., affective communication such as pleasure, gratitude, and humor).

Conduction Aphasia

The hallmark of conduction aphasia is that repetition is severely impaired, much more so than spontaneous speech and language comprehension. It would, however, be a mistake to regard conduction aphasia as a "pure" disturbance of repetition. In most cases, the speech will contain regular paraphasic errors, usually of the phonemic type, and word finding deficits will also be present.

There has been, and remains, considerable controversy as to the nature and neuropathologic correlates of conduction aphasia. As recently as 20 years ago, a number of prominent aphasiologists maintained that conduction aphasia was a mythological aphasia "invented" or "predicted" on the basis of human neuroanatomy. That is, because the arcuate fasciculus connects Wernicke's and Broca's areas, it was inferred that lesions in this structure would result in a relatively specific disturbance in repetition. In part, the elusiveness of conduction aphasia is the result of the fact that it is uncommon in the first 2 weeks after onset of aphasia, and requires some time to fully manifest itself. Another problem is that conduction aphasia is relatively more common as a transitory disorder observed for a limited time during the recovery of fluent aphasias in which speech and naming recover more rapidly than repetition. Given the fact that conduction aphasia is at least somewhat elusive during the acute stage of aphasia, there remain important questions about its neuropathologic correlates. It would appear that a substantial number of such cases occur as a result of lesions in the insula, auditory cortex, and adjacent structures. However, it has also been reported that damage to Wernicke's area will produce a chronic disturbance in repetition. Thus, the localizing significance of conduction aphasia may depend, at least in part, on the "age" of the aphasia. One is at relatively little risk to diagnose other fluent aphasias as conduction aphasia because they involve gross language comprehension defects. However, one might miss a conduction aphasia if repetition is not assessed, or if the conduction aphasia emerges from another type of aphasic disorder. Prognosis in conduction aphasia (acute) is good, if for no other reason than that speech is already fluent, word finding is only moderately impaired, and language comprehension is intact.

TRANSCORTICAL APHASIAS

Transcortical aphasia is another term for aphasia with surprisingly well-preserved repetition. It also indicates a lesion that lies outside the perisylvian language zone. Transcortical motor aphasia results from lesions anterior to Broca's area and is, in many ways, simply Broca's aphasia with good repetition. Similarly, transcortical sensory aphasia is Wernicke's aphasia with good repetition, and involves a lesion posterior to Wernicke's area. With lesions of both types, there is isolation of the speech area, where the only functioning aspect of language is repetition. For many such aphasics, repetition cannot be controlled, and is seen as a regular response to any question. This symptom is called echolalia. In transcortical motor aphasia, one can encounter considerable variability in word finding, language comprehension, and ideomotor praxis. That is, some will be relatively less impaired in naming than spontaneous speech while others will be severely anomic. Similarly, comprehension can vary from nearly normal to moderately impaired, and gestural expression can vary from normal to grossly impaired.

The picture in transcortical sensory aphasia is one of frequent, often spectacular semantic paraphasias or neologisms. All aspects of language comprehension will be quite poor. Not only do such patients show good repetition, but many are echolalic and cannot stop themselves from slavish repetition of questions or statements posed to them.

NILS R.
VARNEY

The distinction between anomia and anomic aphasia is like that between dysphoria and major depressive disorder, that is, one is a symptom while the other is a syndrome. Anomia is the symptom of poor word finding, and can be seen in association with lesions in many different areas of the left hemisphere.

In anomic aphasia, word finding is impaired while repetition and language comprehension are normal or nearly so. Written expression may be more or less fluent, but the patient with anomic aphasia should not be able to write the names of objects any better than he or she can say them. It should also be kept in mind that many different aphasias, particularly of the fluent type, resolve into a relatively specific disturbance in word finding, and that lesions of the left frontal lobe which are otherwise "silent" can produce problems with word finding.

ALEXIAS

While most acquired disturbances in reading occur within the context of a receptive aphasia, some do not. One of these is alexia without agraphia in which reading is severely impaired while writing is normal. Such patients present with the remarkable anomaly of not being able to read what they themselves have written. The other is alexia with agraphia, which could also be termed acquired illiteracy because such patients lose both reading and writing abilities. Neither of the above-mentioned disorders is as "pure" as its name would suggest. In alexia without agraphia, visual naming and color naming are likely to be at least somewhat impaired, and usually there is a modest disturbance in writing and written spelling. In alexia with agraphia, anomia is a common finding, as is acalculia (i.e., poor arithmetic) and failing performance on the Token Test. Repetition is also variable in this disorder, and it may be difficult to distinguish alexia with agraphia from other aphasias in which there is significant parietal lobe involvement.

CROSSED APHASIA

Crossed aphasia refers to those language disturbances that result from disease of the right hemisphere. Crossed aphasia is rare among right-handed patients, but is also only a minority characteristic of left-handed patients. That is, only about a third of strongly left-handed patients are strongly right hemisphere-dominant for language. The remainder are either left hemisphere-dominant, or show varying degrees of ambilaterality (i.e., involvement of both hemispheres in the mediation of language). Thus, while most crossed aphasias are found among the left-handed, only a minority of left-handed individuals will show crossed aphasia.

Those crossed aphasias that are observed in individuals with strong right hemisphere dominance for language will, in large part, closely parallel the aphasias mentioned previously. However, because the majority of crossed aphasias occur in individuals with some degree of ambilaterality and/or mixed dominance, one can frequently encounter patients with relatively unusual combinations of symptoms. In addition, the aphasias of such patients may also be relatively subtle and more apparent in testing than in conversation.

The evaluation of a patient in order to determine hemispheric dominance for language will involve a search for signs of crossed aphasia, which may include determining that language symptoms resulting from left hemisphere damage are too minor in relation to the size and locus of lesions (i.e., crossed non-aphasia). In either case, the results will be particularly important for those patients on whom craniotomies will be performed. That is, if the cerebral hemisphere to be operated on plays a significant role in language, the surgical approach may be substantially different.

MIXED APHASIAS

It is important to keep in mind that roughly 25% of aphasias encountered clinically are a weak or poor fit for any of the aphasia diagnostic groupings mentioned above. In large part, this situation is the fault of nature. That is, vascular lesions do not necessarily correspond with those of the classic aphasias, and can partially overlap the areas involved in more than one type of aphasia. In addition, there is the problem of subcortical aphasias, which can appear somewhat like the classic aphasias, but involve completely different lesion loci and sometimes bizarre combinations of symptoms (e.g., dysarthric Wernicke's aphasia with normal repetition). In such cases, the diagnosis can simply be "mixed fluent aphasia," "mixed nonfluent aphasia," or "atypical aphasia."

THE QUESTION OF INTELLIGENCE IN APHASIA

It must be emphasized at the outset of this section that the WAIS Performance Scale does not measure the functional or structural integrity of the right hemisphere. The Digit Symbol subtest is essentially linguistic in nature, frequently indicates impairment among developmental dyslexics, virtually always indicates impairment among left brain-damaged patients (even those with normal motor functioning in their writing hand), and is completely inappropriate for administration to someone with right side motor dysfunction. Similarly, the sequential logic required to perform the Picture Arrangement subtest, while sometimes impaired in association with right-sided lesions, is frequently impaired among aphasic patients. Block Design and Object Assembly are reasonable procedures for demonstrating that the patient retains normal reasoning skills, but impaired performance is meaningless because of constructional apraxia (i.e., they may suffer a primary cognitive disturbance which prevents them from discriminating the stimuli or implementing a solution). Finally, Picture Completion is better suited for verbal than nonverbal response modes, and many aphasics fail to comprehend the task instructions/demands. Thus, it is both possible and not uncommon to have a perfectly sensible, rational aphasic whose Performance IQ is zero.

That aphasics can retain propositional thought while failing standard nonverbal IQ tests can be illustrated by two case reports. One involved a nonfluent aphasic who also suffered from multiple apraxic difficulties, with the result that he was incapable of demonstrating his preserved sensibilities. On the occasion of his 6-month follow-up neuropsychological examination, he was able to describe many of the psychological and neurological tests given earlier, and also remembered individuals who had made patronizing remarks to or about him in staffing. The other case involved a Wernicke's aphasic who, while an inpatient, was judged by many to be a nursing home candidate. His wife, however, took him home, and while still markedly paraphasic, he was able to continue making architectural drawings for commercial use.

If one resorts to nonverbal intelligence measures such as Raven's Progressive

Matrices, which are better suited for administration to someone with language, motor, and sensory handicaps, impaired performance can still occur for reasons other than impaired intelligence. It is best, therefore, to regard psychological tests as a means for demonstrating an aphasic's residual mental assets, but as being inadequate for demonstrating "lame thinking." This is not to say, however, that it is impossible to determine whether an aphasic is legally incompetent. If they are unable to understand spoken and written language or communicate effectively in any way, then they should not be regarded as competent to give consent for surgery and related procedures or to make decisions about financial or legal matters.

Aphasics with no patent language skills are often very adroit in grasping the social cues or normal conversation and in emitting appropriate nonverbal/social responses. It is not uncommon to find aphasics who have "fooled" physicians into accepting their consent for angiograms or other hazardous procedures simply because they realized that an affirmative response was expected.

Nonverbal Deficits in Aphasia

Although the definition of aphasia as an acquired disturbance of language is nearly universally accepted, this was not always the case. In the mid-19th century, the period of aphasiology's birth, a number of early investigators noted that aphasics were frequently impaired in a variety of nonverbal tasks as well as in language. Such observations led some to conclude that aphasia was a disorder of symbolization and/or that impairment in language was inevitably associated with diminished intelligence and reason. However, an equal number of prominent aphasiologists noted that many aphasics performed normally on nonverbal tasks, and that there was no predictable relationship among nonverbal deficits, or between verbal and nonverbal symptomatology. Since nonverbal deficits appeared to occur largely at random, it was eventually concluded that aphasic nonverbal symptoms reflected cognitive deficiencies which had little or nothing to do with the "primary" linguistic symptoms. As a result of this, all aphasic disorders are classified solely according to the nature and severity of language deficits. The one voice of reason in this long-standing mystery was that of Hughlings Jackson, the great 19th-century neurologist. It was his opinion that disturbances limited to language expression were specifically linguistic, but that receptive aphasias rendered their victim "lame in thinking" (Jackson, 1878). He also identified a critical feature of aphasia which he termed propositional thought. That is, even if a patient cannot communicate, he may still have something to say. While Jackson felt that disturbances in propositional thinking would be found primarily among receptive aphasics, he went on to say that some receptive aphasics retain propositional thought. As he put it, the question of whether an aphasic is intellectually competent is like asking the question, "How long is a piece of string?" The fact that Jackson's view is disappointingly complicated does not detract from the fact that it is essentially correct. The modern age of aphasiology has demonstrated that aphasics are frequently impaired on a variety of nonverbal tasks. Some of these tasks are also failed by patients with right hemisphere lesions, but others are failed only by aphasics. For example, the use and understanding of gestural communication is wordless, but defects in this ability are common only among aphasics with alexia (Varney, 1982). Right brain-damaged patients are almost never impaired in gestural "language." The same is true for sound recognition and color association (Varney, 1984; Varney and Risse, 1993).

For most clinicians, it is impractical to assess all of the specific abilities discussed. What follows, therefore, is a brief review of which tasks are most important and in what respects, and which tasks are relevant only for selected patients or subgroups of aphasics. (1) Graphomotor praxis: Because virtually all aphasics show agraphia, and a clear majority are nonfluent in writing, assessment of graphomotor praxis will provide important information about how written expression is impaired, and can give cues as to how speech therapy and occupational therapy should be implemented. (2) Buccofacial apraxia: This disorder in voluntary use of the mouth and larynx is particularly common in nonfluent aphasics. It can be evaluated by having the patient perform various simple oral tasks such as blowing out a match, clicking the tongue, inflating the cheeks with air or making a kissing noise. (3) Ideomotor praxis: Assessment is appropriate for all aphasic patients and can be accomplished by having the patient perform various common gestures such as saluting and pretending to use objects. Impaired performance should be interpreted with caution. Many cases with ideomotor apraxia are mistakenly thought to have receptive language deficits. The impairment can be an important interdisciplinary focus for speech and occupational therapy (OT). (4) Sound recognition: Impaired sound recognition is very rare except in the context of receptive aphasia. Results provide important prognostic information and also offer a more accurate description of the nature of the patient's aural comprehension disturbance. This assessment is unnecessary in patients with normal aural comprehension. (5) Pantomime recognition: Like impaired sound recognition, impaired pantomime recognition is found almost exclusively in the context of receptive aphasia. Findings can potentially indicate a patent channel for communication with some receptive aphrasics. In other cases, they can indicate an area in which speech therapy efforts can be made. Some patients with this disorder enjoy significant improvement in reading and gestural understanding by supplementing reading rehabilitation with 60 minutes of television per day. (6) Phoneme discrimination: Results can potentially have both descriptive and prognostic value. However, about 40% of failures occur because the patient fails to grasp the task's demands or to master the production of same/different responses reliably. (7) Constructional praxis: Assessment of this area is of direct relevance to persons in skilled manual occupations, such as mechanics, artisans, or carpenters. (8) Visuoperceptive tasks: Except in selected cases, performance on these tasks offers relatively little clinical value, particularly given that fact that one can strongly suspect a visuoperceptive disturbance in cases with impaired letter recognition. (9) Color association: This involves having patients identify the characteristic color of objects (e.g., banana: yellow). Impaired performance is common in receptive aphasia and rare in association with other unilateral lesions.

CONCLUSION

This chapter has concentrated on presenting aphasia assessment from a clinical viewpoint, focusing on assessment of basic language skills and clinical issues common in the neuropsychological evaluation of an aphasic patient. Because the chapter has focused on the basics of aphasia assessment from a neuropsychological perspective, it has not been possible to adequately review other assessment perspectives (e.g., speech therapy) or contemporary theoretical issues (e.g., neurolinguistic theory, functional neuroanatomy). For those interested in more extensive coverage of the many facets of aphasia from clinical, historical or theoretical perspectives, a variety of texts are available (Goodglass, 1993; Sarno, 1981).

REFERENCES

Benson, D. F. (1967). Fluency in aphasia: correlation with radioactive scan localization. *Cortex, 3,* 258–271.

Benson, D. F. (1977). The third alexia. *Archives of Neurology, 34,* 327–331.

Benton, A. L. (1981). Aphasia: Historical perspectives. In M. Sarno (Ed.), *Acquired Aphasia* (pp. 1–21). New York: Academic Press.

Benton, A. L., & Hamsher, K. (1989). *The Multilingual aphasia examination.* Iowa City: AJA Associates.

Damasio, H. (1981). Cerebral localization of the aphasias. In M. Sarno (Ed.), *Acquired aphasia* (pp. 27–50). New York: Academic Press.

De Renzi, E., & Vignolo, L. (1962). The Token Test: A sensitive test to detect receptive disturbances in aphasia. *Brain, 85,* 665–678.

Goodglass, H. (1993). *Understanding aphasia.* San Diego: Academic Press.

Hamsher, K. (1981). Intelligence and aphasia. In M. Sarno (Ed.), *Acquired aphasia* (pp. 327–355). New York: Academic Press.

Hinshelwood, J. (1900). *Letter, word and mind blindness.* London: Lewis.

Jackson, H. (1878). On afflictions of speech from disease of the brain. *Brain, 1,* 304–330.

Kaplan, E., & Goodglass, H. (1981). Aphasia related disorders. In M. Sarno (Ed.), *Acquired aphasia* (pp. 303–326). New York: Academic Press.

Kerchensteiner, M., Poek, K., & Brunner, E. (1972). The fluency–nonfluency dimension in the classification of aphasic speech. *Cortex, 8,* 233–247.

Sarno, M. (Ed.). *Acquired aphasia.* New York: Academic Press.

Varney, N. (1981). Letter recognition and visual form discrimination in aphasic alexia. *Neuropsychologia, 19,* 795–800.

Varney, N. R. (1982). Pantomime recognition defect in aphasia: Implications for the concept of asymbolia. *Brain and Language, 15,* 32–39.

Varney, N. R. (1984). Prognostic significance of sound recognition defect in aphasia. *Archives of Neurology, 41,* 181–182.

Varney, N. R., & Risse, G. (1993). Locus of lesion in color amnesia. *Neuropsychology, 7,* 1–6.

18

Assessment of Spatial Abilities

BRUCE M. CAPLAN AND SARAH ROMANS

INTRODUCTION

Analyzing and understanding spatial ability is a complex enterprise. By way of illustration, consider the following daily life activities that contain a spatial aspect: parallel parking an automobile, rearranging living room furniture, reading a blueprint, giving directions to a lost traveler, deciphering the instructions that accompany an item that "requires some assembly," and fitting into the trunk of a car one more suitcase than the trunk was meant to hold. Consider further a professional quarterback about to release a forward pass who (in addition to maintaining spatial awareness of oncoming defensive linemen) must also gauge the speed and path of his intended receiver in order to aim the football so that it will arrive at its intended location at the same time as his teammate. These functions clearly do not lend themselves to precise laboratory measurement. Furthermore, performance is determined not only by whatever spatial skills are relevant, but also by other functions such as language, motor, and planning abilities.

Several authors have attempted to subdivide "spatial abilities" into various subcomponents (e.g., Ardila & Rosselli, 1992; Beaumont & Davidoff, 1992; Benton, 1969, 1985; Benton & Tranel, 1993; Coslett & Saffran, 1992; DeRenzi, 1982, 1985; McCarthy & Warrington, 1990; Newcombe & Ratcliff, 1990; Ratcliff, 1982; Varney & Sivan, 1986). The model proposed by Kritchevsky (1988) gives some idea of the constructs that are commonly included. Kritchevsky defines spatial cognition as "any aspect of an organism's behavior which involves space and is mediated by cerebral activity" (p. 111). He posits five broad categories of spatial function (perception, memory, attention, mental operations, construction) containing nine elementary spatial skills (object localization, line orientation detection, spatial synthesis, short-term spatial memory, long-term spatial memory, attention to left hemispace, atten-

BRUCE M. CAPLAN   Department of Rehabilitation Medicine, Thomas Jefferson University Hospital, Philadelphia, Pennsylvania 19107.   SARAH ROMANS Department of Rehabilitation Medicine, Thomas Jefferson University Hospital, Philadelphia, Pennsylvania 19107; and Drexel University, Philadelphia, Pennsylvania 19104.

Neuropsychology, edited by Goldstein *et al.* Plenum Press, New York, 1998.

tion to right hemispace, mental rotation, spatial construction). Other spatial abilities proposed by various authors include topographical orientation, visual scanning, face perception, visual analysis, perception of movement, identification of incomplete figures, detection of hidden figures, formation of mental images, and spatial reasoning. A further distinction is that between *egocentric* and *allocentric* frames of reference; in the former system, spatial perception occurs with respect to the position of one's self, while the latter pertains to a perspective that is independent of the observer (Beatty & Troster, 1988). Regrettably, the wealth of clinical and experimental literature has not produced a "consensus model" of spatial functions. One cannot argue with the assertion by Bellugi, Wang, and Jernigan (1994) that, in contrast to language, "which includes the well-defined components of phonology, morphology, syntax, semantics, and prosody, spatial cognition has resisted fractionation into components" (p. 35).

The neuropsychological assessment of spatial abilities entails a major emphasis on the visual modality. While measures of auditory–spatial (e.g., sound localization), tactile–spatial (e.g., Tactual Performance Test from the Halstead–Reitan battery), and kinesthetic–spatial (e.g., Rod and Frame Test) abilities exist, contemporary clinical neuropsychological evaluation focuses on the assessment of *visual–spatial* skills; accordingly, this chapter deals primarily with such measures. Newcombe and Ratcliff (1990) have marshaled clinical, neurophysiological, and neuroanatomical evidence to support the separation of spatial problems from other visual–perceptual impairments (see also Newcombe & Russell, 1969); the former appear to involve the parietal-occipital regions while the temporal-occipital areas are implicated in the latter (Damasio, 1985).

Beaumont and Davidoff (1992) have listed several reasons supporting their contention that "the assessment of visuo-perceptual dysfunction is often cursory and inadequate by comparison with other areas of clinical investigation" (p. 115). They cite the absence of a coherent conceptual account of visual–perceptual functions, the blend of assessment strategies that typifies evaluation of this domain, and the multifactorial nature of most tests of visual–spatial skills, making it difficult to isolate the individual's "pure" spatial abilities. Although an interesting speculative article by Ardila (1993) argues that spatial abilities were of primary adaptive significance to our prehistoric ancestors, spatial functions have historically been something of a "poor relation" in neuropsychology, as reflected in the fact that the right cerebral hemisphere (identified 130 years ago as the primary locus of visual–spatial abilities) was long referred to as the "nondominant" side of the brain, that is, less important than the linguistically sophisticated left hemisphere. It is only within perhaps the past two decades that the ecological significance of "right hemisphere functions" has begun to be fully appreciated. While the communication difficulties of a left hemisphere-lesioned patient with aphasia may be more obvious to the casual observer, the limitations imposed by such "right hemispheric" deficits as unilateral neglect, impaired visual memory, and defective perceptual analysis can be extraordinarily disabling. Recent studies by Sea, Henderson, and Cermak (1993), Rapport and colleagues (1993), and Neistadt (1993), among others, reported associations between visual–spatial or constructional impairment and a number of daily life skills including meal preparation, mobility, dressing, and avoidance of falls.

Despite the broad consensus regarding the paramount importance of the right hemisphere in spatial processing (see DeRenzi, 1982, and Young & Ratcliff, 1983, for reviews), recent evidence has suggested a place for the left hemisphere as well. Specifically, certain findings support the notion that the left hemisphere plays an important role in the perception of what are called "local" features of visual–spatial stimuli, that is, the details that comprise "global" patterns (Lamb, Robertson, & Knight, 1989; Robertson, 1986). In a study of patients with lateralized lesions, Delis, Robertson, and Efron (1986) found that the left hemisphere group showed better memory for larger stimuli than for the smaller forms from which they were constructed, while right hemisphere patients displayed the opposite pattern. Furthermore, as DeRenzi (1982), among others, has pointed out, right hemisphere dominance for the more elementary aspects of spatial processing is undisputed, but more complex components may depend on a contribution from the left hemisphere as well. The nature of that contribution has not been specified, but it may involve the need to engage verbal mediation processes. It is interesting to note that research on split-brain patients has suggested that the right hemisphere superiority involves *manipulo* spatial functions, not simply visual–perceptual processing (e.g., Le Doux, Wilson, & Gazzaniga, 1977; Gazzaniga & Le Doux, 1978).

In view of the functional salience of spatial abilities (and deficits), adequate clinical assessment assumes vital importance, especially for patients in the various diagnostic categories (e.g., right hemisphere stroke or tumor, Alzheimer's disease, Parkinson's disease, "nonverbal learning disability") in which impaired capacity for spatial processing is frequently observed. Given the multifactorial nature of virtually all neuropsychological tests, any classification scheme will be at least partially inadequate. We propose to describe a sampling of spatial assessment measures according to the following typology: attention/search, perception/analysis, constructions, and memory. Our discussion of assessment instruments for these domains will be selective, not exhaustive, and will emphasize measures used with adults. We provide a general description of each measure, furnish certain psychometric data such as reliability figures (if available), and illustrate some clinical applications. Interested readers are referred to extensive texts such as those by Lezak (1995) and Sattler (1990) for descriptions of other relevant tests. Discussion of sex differences in spatial abilities (e.g., Caplan, McPherson, & Tobin, 1985; Harris, 1978; McGee, 1979, 1982; Newcombe, 1982; Wittig & Petersen, 1979), developmental trends (Young, 1983), and the impact of handedness (e.g., Carter-Saltzman, 1979; Sanders, Wilson, & Vandenberg, 1982) on spatial functioning can be found elsewhere.

HISTORICAL BACKGROUND

The notion of a link between visual–spatial impairment and lesions of the right cerebral hemisphere appeared in the literature at least 125 years ago (Quaglino & Borelli, 1867), although the patient described in that report was also aphasic, indicating that there may have been left hemisphere damage as well. In both theoretical writings and case descriptions, Jackson (1864, 1874/1915, 1876) argued for the right hemisphere (especially the posterior portion) as the primary cortical residence of visual object recognition and visual memory skills. In support of his position, he offered his findings in a 59-year-old woman (later found to have a right posterior tumor) who demonstrated topographic disorientation, impaired facial recognition, and dressing apraxia. Shortly thereafter, Munk (1878) reported that limited bilateral lesions in dogs produced impaired capacity to grasp the meaning of visual input; the animals failed to recognize their master and showed abnormal response to the presentation of meat or to threatening gestures. Munk labeled this disturbance "mindblindness," a phenomenon that was found in humans by Wilbrand (1887) and Lissauer (1890). Lissauer described two forms of the condition, one deriving from impaired perception of the object or person and the second stemming from difficulty in linking the content of the perception with one's experience; in both instances, impaired visual acuity per se was insufficient to explain the disorder.

In 1888, Badal described a woman with a "disorder of the sense of space" who, in addition to visual–spatial difficulties (e.g., inability to navigate in her neighborhood or even in her own home), also exhibited defective auditory localization as well as poor somesthetic–spatial discrimination (e.g., inability to distinguish right and left body parts). Balint (1909) described the associated phenomena of "psychic paralysis of gaze, optic ataxia, and spatial disturbance of attention," a constellation of symptoms now referred to as Balint's syndrome, in a patient who also exhibited a number of perceptual problems including neglect of the left side of space, difficulty recognizing complex shapes, and impaired visually guided reaching. At autopsy, the patient was found to have bilateral atrophy, especially involving the posterior parietal and temporal areas, but also extending to the occipital lobe.

Benton (1982) points out that by 1910, the scientific literature contained descriptions of at least nine forms of deficit of spatial thought. However, there was no clear consensus about the critical anotomic locus. Despite the frequent reports of bilateral involvement in spatial impairment (e.g., Balint, 1909; Foerster, 1890; Meyer, 1900), some early findings (e.g., Dunn, 1895; Lenz, 1905; Peters, 1896) provided evidence of a special role of the right hemisphere in spatial function.

The "natural experiments" of penetrating brain injuries sustained by soldiers in World War I permitted Holmes (Holmes, 1918; Holmes & Horrax, 1919) to expand the study of the cognitive consequences of such injuries. He described two varieties of visual–spatial disabilities, one based on oculomotor impairment that disrupted the process of visual search, and a second category involving a mixture of higher-order deficits affecting such skills as size estimation, reading, and route learning. Brain (1941) distinguished between two forms of localization disturbance following right hemisphere damage. The first affected nearby objects (i.e., within arm's length) and could be observed in the patient's defective reaching for those objects, while the second form pertained to remote items and would be noticed through misjudgment of distance.

The first to study impairments of constructional skill was Kleist (1912, 1922/1934) who suggested the term "constructional apraxia" to refer to difficulty in "formative activities" in which the underlying problem is neither motoric nor perceptual, but rather an integration of the two processes. Kleist speculated that the responsible lesion was likely located in the left parietal–occipital area, but later studies (many involving soldiers injured in World War II) reported a higher incidence among patients with right hemisphere damage (e.g., Benton, 1962; Hecaen, Ajuriaguerra, & Massonet, 1951; McFie, Piercy, & Zangwill, 1950; Patterson & Zangwill, 1944; Piercy, Hecaen, & Ajuriaguerra, 1960). Nonetheless, constructional apraxia was (and continues to be) found in a substantial proportion of patients with left hemisphere damage as well. Further information on the historical development of neuropsychological understanding of visual–spatial and constructional impairment can be found in Benton (1978, 1982) and Benton and Tranel (1993).

METHODS OF ASSESSMENT

ATTENTION/SEARCH

Mesulam (1985) posits two broad categories of attentional functions: a "matrix" or "state" function responsible for sustained concentration, and a "vector" or "channel" function controlling selective attention to specified targets, which presumably must be sought out. As Mesulam notes, however, this proposed physiologic distinction does not apply at a behavioral level of analysis, "since most attentional behaviors eventually represent an interaction between these two components" (p. 127). Posner (1980) distinguishes between *overt* shifts of attention in which there are discernible eye movements and *covert* shifts that occur while gaze is fixed.

Virtually all neuropsychological tests demand a modicum of attentional capacity. Indeed, many would argue that an inattentive patient (e.g., one in a delerious state or in an agitated condition following brain injury) cannot be tested in a valid and meaningful fashion. The tests discussed below are perhaps most easily distinguished along two parameters: the presence or absence of a motor component, and the extent to which the exploration of space is structured or unstructured; a corollary of the latter is the degree to which the subject's scanning pattern is determined by the configuration of the stimuli or is self-directed.

Weintraub and Mesulam (1985), building on early work by Diller and colleagues (Diller *et al.*, 1974), described four forms of a visual cancellation (target detection) task, two employing letter stimuli and two that used shapes. In one version of each form, the stimuli are arranged in a regular linear fashion, while the other pair of forms has an apparently random unstructured layout. Subjects are required to scan the test sheet and cross out each instance of the designated target. Weintraub and Mesulam reported that right hemisphere-lesioned patients showed better detection of letters than shapes as well as more accurate cancellation with the structured than unstructured forms. As with other cancellation tests (e.g., Diller *et al.*, 1974; Diller & Weinberg, 1977; Caplan, 1985; Vanier *et al.*, 1990), these measures are sensitive to hemispatial neglect, that is, the tendency to ignore or fail to respond to stimuli from the side of space opposite the lesioned hemisphere. Weintraub and Mesulam report that right hemisphere-damaged patients tend to exhibit an erratic scanning strategy (even with the structured tasks), in contrast to left hemisphere-damaged patients whose style tends to be the systematic sort exhibited by normal controls.

Another structured cancellation task, the d2 test, was described by Brickenkamp (1981). In this test, there are 14 lines, each with 47 letters. The letter "D" is shown with a variable number of dots above and/or below; subjects must circle each instance of the letter "D" that has two dots associated with it in any combination. There is a time limit of 20 seconds per line. Spreen and Strauss (1991) report very high test–retest reliabilities (0.89 to 0.92), with predictable practice effects among controls, but no such effect in brain-damaged patients. Correlational and factor analytic studies suggest good construct validity, with significant correlations between the d2 test and other measures of attention, as well as high loadings on an attentional factor. In addition to the skills generally considered to be inherent in cancellation tests (e.g., scanning, motor speed), the d2 test, with its multiple distractors (i.e., "D's" with 1, 3, or 4 dots), also appears to require good capacity for response inhibition; as such, it contains an "executive function" or "mental control" component as well.

Another type of task engaging many of the same abilities requires the individual to trace a path through a printed maze. The two principal examples are the Maze subtest of the Wechsler Intelligence Scale for Children-Revised (WISC-R; Wechsler, 1974) and the Porteus Maze Test (Porteus, 1959, 1965) which contains items suitable for very young children (3 years) through adult levels. There are no time limits on the Porteus Maze Test, in contrast to the WISC-R. Impaired performance on the Porteus Maze Test is often seen following frontal lobe damage (e.g.,

Petrie, 1952; Tow, 1955), although there appears to be some intralobe specificity (Crown, 1952; Lewis, Landis, & King, 1956). The frontal lobe contribution probably accounts for Franzen's (Franzen, Robbins, & Sawicki, 1989) conclusion that the test demands "successful planning, inhibition of impulsive behavior and ability to change set" (p. 225).

Elithorn, Jones, Kerr, and Lee (1964) described a different sort of maze test consisting of a series of 18 lattice-work patterns with black dots located at a number of junctures. Subjects must trace a path through the maze that crosses through as many dots as possible; they are not told that there is a 1-minute time limit, but this is considered in determining performance level. Although certain cognitive demands of the Elithorn resemble those of the Porteus, the former failed to show a selective sensitivity to frontal lobe injury (Benton, Elithorn, Fogel, & Kerr, 1963). Rather, right hemisphere patients more frequently fail the test, but aphasic patients often obtain low scores because of slow execution (Archibald, Wepman, & Jones, 1967; Archibald, 1978; Colonna & Faglioni, 1966). Significant problems with this test are that it is not easily available and that little has been done to establish its psychometric properties.

A related task demanding sustained visual attention and tracking, but without a motor component, is the Visual Pursuit Test of the Employee Aptitude Survey (Ruch & Ruch, 1963). This consists of a number of overlapping lines that zigzag (more or less) horizontally across a page. Subjects must track each line from the right side of the page to the left in order to identify the terminus. There is a 5-minute time limit. Similar tasks have been described by Talland (1965) and Thurstone (1944).

The Trail Making Test (TMT), which is now part of the Halstead–Reitan Neuropsychological Test Battery (Reitan & Wolfson, 1993), was originally one component of the Army Individual Test Battery (1944). Part A requires the individual to draw a line connecting in sequence circled numbers 1 to 25 that are scattered randomly on an $8\frac{1}{2}'' \times 11''$ test form. Part B has a similar structure, but the subject must alternate between numbers and letters. On both parts, performance is enhanced by the use of a systematic scanning strategy and hampered by disorganized search or neglect. The examiner monitors and times the patient's performance, intervening and pointing out errors as they occur and directing the subject to resume from the point at which the error was made. Errors may be tabulated, but time to completion is used as the main score; errors lower performance by lengthening the time required to finish. An intermediate form is available for children 9–14 years, and the adult form may be used with subjects 15 years and older. Lewis and Rennick (1979) developed several alternate versions for use in serial evaluation. Spreen and Strauss (1991) summarize several reports on reliability of the TMT, ranging from a low of 0.3 (a study of epileptic patients by Dodrill and Troupin [1975] using a 6–12-month interval) to 0.94 (Goldstein and Watson's 1989 study of neuropsychiatric patients).

Although earlier authors used a single cutoff point on the TMT to determine "brain damage," these were found to misclassify large numbers of normal individuals, especially the elderly (e.g., Bornstein, Paniak, & O'Brien, 1987; Stuss, Stethem, & Poirier, 1987), and have been replaced by extensive norms with adjustments for sex, age, and educational level (Heaton, Grant, & Matthews, 1991; Yeudall, Reddon, Gill, & Stefanyk, 1987). In the manual by Heaton *et al.*, raw scores are transformed to scaled scores which are further transformed to *T*-scores with demographic adjustments.

The relation between performance on the two parts of the TMT has been studied, but the results have been inconsistent. Reitan and Tarshes (1959) argued that patients with left hemisphere lesions show greater deficit on part B than on part A, an assertion that has received some support (Lewinsohn, 1973), but contradictory data exist as well (Wedding, 1979). Heilbronner, Henry, Buck, Adams, and Fogel (1991) failed to find differences between groups of patients with left or right hemisphere lesions on raw scores on part A or part B, on the difference between the two scores, or on a ratio formed by dividing part B by part A raw score; the latter pair of findings conflicts with earlier reports by Klesges, Fisher, Pheley, Boschee, and Vasey (1984) and Golden, Osmon, Moses, and Berg (1981). Nonetheless, the TMT has proven to be a useful brief screening device to differentiate between brain-injured individuals and controls (Heilbronner *et al.*, 1991; Korman & Blumberg, 1963; Lewinsohn, 1973), and as a predictive factor in successful vocational rehabilitation (Lewinsohn, 1973). However, it has not been consistently effective in distinguishing psychiatric patients from neurologic patients (Heaton, Baade, & Johnson, 1978; Zimet & Fishman, 1970). Performance on the TMT is frequently compromised by closed head injury (e.g., Eson & Bourke, 1982; Levin, Benton, & Grossman, 1982).

In closing this section, we wish to call attention to the ingenious technique employed by Bisiach and Luzzatti (1978) in their study of visual scanning in two patients with unilateral neglect. Bisiach and Luzzatti asked their patients to imagine themselves in the central square of their hometown of Milan, first standing opposite the cathedral and next standing with their back toward the church. They were then asked to identify the buildings located along the two sides of the plaza. In both instances, they failed to mention structures located on the left (neglected) side, given the particular perspective. That they were aware of and could recall the buildings along both sides was demonstrated by their joint performance in the two conditions. Nonetheless, they demonstrated "unilateral neglect of representational space," that is, neglect in scanning their mental image similar to the neglect demonstrated on more conventional testing.

PERCEPTION/ANALYSIS

In this section, we describe a number of measures that make primary demands on the individual's capacity to discriminate, analyze, and reason about visual–spatial stimuli. Motor skills are minimally involved.

The three forms of the Raven Matrices (Raven, 1938, 1965; Raven *et al.*, 1986) permit a wide variety of uses. The major skills required by the Raven are pattern matching, pattern completion, and reasoning by analogy. Each item contains a design or series of designs with a single missing component. The subject is required to identify from a six- or eight-item multiple-choice array the alternative that would complete the pattern or series. Subjects may indicate their selection by pointing, saying the corresponding number, or writing the number on an answer sheet; with severely impaired persons, the examiner may point in turn to each response alternative and the patient can indicate their choice by any available route (e.g., eye blink, head nodding). All three forms are untimed; Spreen and Strauss (1991) estimate that total administration time ranges from about 25 minutes for the easiest version to 40 minutes for the most difficult form. One unique feature of the Raven is that within each of the several subsets of items that comprise each form, the problems become progressively more difficult; thus, after one completes the hardest portions of a set, one then is given some easier problems before the next set, too, becomes more demanding. This contrasts with the more typical test structure in which items of graduated difficulty are presented until the subject commits a certain number of consecutive errors or fails a designated number of items from a consecutive series (e.g., six of eight). The format of the Raven, therefore, helps to minimize mounting frustration and as a result, fosters the subject's motivation.

The 36-item Colored Progressive Matrices test contains three sets of 12 items each. Norms are available for children ages 5–11 and for adults age 65 years and older (Orme, 1966; Raven, Court, & Raven, 1986). Measso *et al.* (1993) recently provided norms for a sample of 894 healthy Italian adults ages 20–79. The Colored Progressive Matrices have been widely used in clinical studies of patients with various forms of brain impairment. Most studies report somewhat lower scores for patients with right lateralized lesions compared to those with left lateralized lesions, with both groups scoring below normal controls. Impairment on the Raven may have somewhat different etiologies in the two lesion groups, with failure in left hemisphere patients deriving from problems with the verbal mediation required to solve some items, and difficulties in the right hemisphere group resulting from fundamental visual–perceptual problems or unilateral neglect (Kertesz & McCabe, 1975; Basso, DeRenzi, Faglioni, Scotti, & Spinnler, 1973). As shown by Costa, Vaughan, Horwitz, and Ritter (1969), patients with neglect may score poorly on the Raven for that reason alone. Furthermore, Drebing and colleagues (1990) found no difference between their left and right hemisphere-damaged groups, *once the effects of neglect were partialed out*. Consequently, some investigators (e.g., Caltagirone, Gainotti, & Miceli, 1977; Caplan, 1988) have suggested that the response alternatives (which are displayed in a 2×3 format in the conventional version) be arrayed in a single vertical column. Caplan reported significant improvement in the performance of neglecting patients who were given the test in this format.

The Standard Progressive Matrices (Raven, 1938) test contains five sets of 12 items each. While norms exist for children as young as 6, the existence of "floor effects" would seem to make the Standard Progressive Matrices a less satisfactory choice than the Colored Progressive Matrices version for elementary school children. Reasonably large-scale norms have been provided by Burke (1958, 1985) and Raven *et al.* (1986). The Standard Progressive Matrices test does not distinguish between left and right hemisphere lesion groups (e.g., Arrigoni & DeRenzi, 1964).

Raven (1965) offered the Advanced Progressive Matrices for use with individuals of above-average intellect, specifically, for those scoring above about 50 on the Standard Progressive Matrices (i.e., 90th percentile for the 36–45-year age range, according to Burke, 1985). There is a 12-item introductory set followed by 36 items of highly challenging materials. Norms for a 40-minute timed version are available in Raven (1965), while untimed norms for college students are found in Paul (1985).

The forms of the Raven possess good psychometric support. Spreen and Strauss (1991) report a median test–retest reliability coefficient of about 0.8 and internal consistency reliability coefficients above 0.7 (Burke, 1985); Carlson and Jensen (1981), however, report unsatisfactory reliability in the youngest of their pediatric groups ($5\frac{1}{2}$–$6\frac{1}{2}$ years). Spreen and Strauss note a correlation of approximately 0.7 between the Raven and conventional IQ measures, suggesting good concurrent validity of the Raven as a measure of general intelligence; correlations with achievement measures are considerably lower. Following an extensive literature review, Court (1983) concluded that sex differences in performance on the Raven cannot be demonstrated. However, Measso *et al.* (1993) reported better performance on the

Colored Progressive Matrices by males; they also found a significant negative effect for age and a positive effect for education. Based on their data, they provide a table of "correction values" for various subgroups to allow the clinician to adjust performance for the effects of these demographic variables.

Other tests that resemble the Raven matrix-type format are the second edition of the Test of Nonverbal Intelligence (TONI-2; Brown, Sherbenou, & Johnsen, 1990) and the Culture Fair Intelligence Test (1973). The TONI-2 contains two equivalent 55-item forms that may be used with subjects from ages 5 to 86 years. The TONI-2 items demand "abstract/figural problem-solving skills," some of which are similar to those engaged by the Raven (e.g., matching, sequence, completion, analogical reasoning), while others require the capacity to classify stimuli or blend two different patterns to form a new figure. The test is inaccurately named, as verbal mediation is clearly necessary to solve many items. Norms are given for 2,764 subjects, most of whom received both forms. Reliability coefficients are quite high; coefficient alpha was determined to be 0.95 and 0.96 for the two forms, and the average alternate form reliability was 0.86. The TONI-2 scores show substantial correlations with a number of measures of intelligence and aptitude and are also significantly correlated with measures of achievement.

The Culture Fair Intelligence Test (CFIT) (1973) consists of three scales, one of which was designed for use with children below age 8 and also with elderly impaired individuals. The other two forms have the advantage of permitting group administration. For each scale 12.5 minutes are allotted, and time limits must be adhered to. Each scale consists of four subtests: sequence analysis, oddity identification, matrix completion, and "conditions," a type of reasoning task. The manual reports reliability and validity data obtained from over 10,000 subjects. Average reliabilities exceed 0.8, and validity coefficients were generally above 0.7. As with the TONI-2, the CFIT correlates more highly with measures of intelligence than with achievement test scores. As the name implies, the CFIT has been successfully employed in studies around the world.

Several measures of spatial perception and analysis have been developed in Arthur Benton's Neuropsychology Laboratory at the University of Iowa. We describe three of these measures in this section; Benton's three-dimensional block construction task is described later in this chapter. Although many of these measures have been widely adopted, it should be acknowledged that the normative samples in some cases are not impressively large. This is especially problematic if one considers that performance may be affected by such demographic factors as age, sex, and educational level, thereby further shrinking the relevant normative group.

In 1978, Benton, Varney, and Hamsher reported on a test of "visual spatial judgment." Five years later (Benton, Hamsher, Varney, & Spreen, 1983), a complete description appeared of two forms of the Judgment of Line Orientation (JLO) test, along with normative data on 137 subjects ages 16–74 years. Subjects are shown a series of cards depicting two line segments of different orientations. Each segment corresponds to the orientation of a complete line appearing on the response card; the latter has an array of lines numbered 1 to 11 that are placed at 18° intervals, all originating from a central point. Subjects must indicate (by naming or pointing) the pair of lines on the response card that lie in the same orientation as the stimulus pair. Split-half and test–retest reliability coefficients were 0.90 and above. In a study of pediatric subjects, Riccio and Hynd (1992) reported good construct and criterion-related validity for the JLO as reflected in its significant correlations with such measures of visual perception as four of the five Performance subtests of the WISC-R and with achievement in mathematics and reading, although not with spelling ability. Raw scores are corrected to take account of the observed consistent sex difference and age-related decline. Later studies by Eslinger and colleagues (Eslinger & Benton, 1983; Eslinger, Damasio, Benton, & van Allen, 1985) reported "steady, moderate decline" with advancing age in normal elderly subjects (65–94 years). Mittenberg, Seidenberg, O'Leary, and DiGiulio (1989) reported a negative correlation between age and JLO score, but Ska, Poissant, and Joanette (1990) found comparable levels of performance among three groups of normal elderly subjects ages 55–64 years, 65–74 years, and 75–84 years. Benton and colleagues provide percentiles and categorical classifications for various levels of performance (Benton, Sivan, Hamsher, Varney, & Spreen, 1994). They also give data for a small sample of normal children ages 7–14 years, replicating the male superiority that was found in adults and documenting a developmental increase in performance, with average adult levels being attained by about 13 years.

Performance on the JLO is felt to be selectively affected by right hemisphere damage (but see Mehta, Newcombe, & Damasio, 1987, for a different view). Benton and colleagues (Benton, Sivan *et al.*, 1994) found that both the incidence and severity of impairment were greater among right hemisphere-damaged patients than among left hemisphere-damaged patients; 46% of the former obtained "defective" scores (36% were "severely defective"), while only 10% of left hemisphere-damaged patients were found to be impaired (only 2% "severely defective"). Hamsher, Capruso, and Benton (1992) confirmed these findings. Carlesimo, Fadda, and Caltagirone (1993) also reported impaired performance among patients with unilateral right hemisphere lesions; the performance of patients with left-lateralized damage was nonsignificantly lower than that of controls. These findings are consistent with the report of Fried, Mateer, Ojemann, Wohns, and Fedio (1982), who conducted electrical stimulation mapping in alert neurosurgery patients. These investigators determined that perception of line orientation was selectively disturbed when the right parietal-occipital region was stimulated, as well as the right hemispheric counterpart of Broca's area. Ska and colleagues (1990) found severely impaired line orientation performance in a small group of patients with dementia of the Alzheimer's type, confirming the earlier report by Eslinger and Benton (1983). Riccio and Hynd (1992) found that the JLO alone did not discriminate among normal controls, children with psychiatric diagnoses, and learning-disabled children.

Another spatial ability that can be tested is face recognition, the impairment of which has long fascinated clinicians (Damasio & Damasio, 1983). The extraordinary phenomenon of prosopagnosia was first observed over 125 years ago (Quaglino & Borelli, 1867). This striking disability, which involves failure to recognize familiar faces, has been more recently described by Sacks (1987) in his study of "the man who mistook his wife for a hat." McNeil and Warrington recently (1993) described the fascinating case of a man who developed prosopagnosia following a stroke that affected both hemispheres, but primarily involved the left frontal, temporal, and occipital regions. This individual became a sheep farmer, and despite his difficulty recognizing human faces, was able to identify the faces of many members of his flock. He also performed better than "profession-matched" controls on a recognition memory test employing faces of unfamiliar sheep, as well as on a sheep face–name paired associate test.

As noted by Benton and colleagues (1983), impaired recognition of familiar faces is quite rare, but difficulty in discrimination and identification of unfamiliar faces occurs rather frequently, particularly in individuals with damage to the posterior region of the right hemisphere. This spurred development of the Facial Recognition Test which in its current form contains three parts requiring matching identical front-view photographs and matching front-views with three-quarter angles and with pictures taken under different lighting conditions. There is a 27-item short form, half the length of the long form. Age and education were reported to have some impact on facial recognition performance; therefore, raw scores are corrected for these factors before percentiles are obtained. Benton and co-workers (1983) provide normative data for 286 normal adults and 266 children ages 6–14. Once again, there was generally an age-related increase in test performance among children, but there is no mention of sex differences. The average performance of 14-year-olds corresponded to that of the adult group. Test–retest reliability was not evaluated, but correlations between the short and long forms were 0.88 and 0.92, respectively, in control and clinical samples. Benton and colleagues reported that deficit in Facial Recognition Test performance was quite common in patients with lateralized posterior lesions, the incidence being slightly higher in right than left hemisphere cases (53% vs. 44%). Egelko and colleagues (1988) reported an average Facial Recognition Test score that fell below the 1st percentile (i.e., "severely defective") in their sample of patients with unilateral right hemisphere stroke. Facial recognition deficit has also been described in both left and right hemispherectomized children (Strauss & Verity, 1983), Parkinson's disease patients (Bentin, Silverberg & Gordon, 1981), and patients who sustained severe closed head injuries rendering them comatose for 24 hours or more (Levin, Grossman, & Kelly, 1977).

Benton and colleagues (1983) also described the Visual Form Discrimination Test, a match-to-sample task employing sets of geometric shapes. Each of the 16 items contains a target set of shapes (two large figures and a small peripheral form) and four response alternatives, one of which precisely matches the target set, one containing a rotation or misplacement of the peripheral figure, one involving rotation of one of the large shapes, and a final distractor with a distortion of the other major design. Subjects may say the number of their selection or point to the chosen set. There are no time limits. Within the normative sample of 85 individuals (ages 19–74 years), no associations were found between test performance and the variables of age, sex, or educational level. In their clinical sample (n = 58), Benton and associates identified a very high proportion of patients in all lesion subgroups with defective performance. This may be attributable, in part, to the high normative level; average control group score was about 30/32 correct, with a score of 26 (achieved by 95% of controls) defining the cutoff point.

Another match-to-sample task is the Matching Familiar Figures Test (MFFT; Kagan, Rosman, Day, Albert, & Phillips, 1964) which was initially applied to the study of problem-solving style (analytic vs. impulsive) in children. The MFFT contains 12 items consisting of a single target figure and six (child version) or eight (adolescent/adult version) response alternatives, one of which precisely matches the target, and distractors that contain minor alterations or distortions. Both accuracy and response latency are recorded. Subjects with high accuracy scores and long latencies (i.e., slow responders) are characterized as "reflective," while those who answer quickly and commit a large number of errors are designated "impulsive." While Salkind (1978) published norms for the MFFT, Walker (1986) found that only 8% of the studies he reviewed from a 5-year period employed that data. Van den Broek, Bradshaw, and Szabadi (1987) published adult norms for the eight-item version. While the evidence is mixed regarding the diagnostic utility of the MFFT in attention deficit disorder (Ault, Mitchell, & Hartmann, 1976; Block, Block, & Harrington, 1974), the test has been found to be helpful in the detection of perceptual deficit in adults, and it appears to be especially sensitive to the impact of unilateral neglect (Caplan & Shechter, 1990).

There are several available tests of figure–ground discrimination (e.g., Thurstone, 1944; Witkin, Oltman, Raskin, & Karp, 1971). Both Thurstone's Hidden Figures Test and the easier Embedded Figures Test developed by Spreen and Benton (1969) require the subject to locate and trace a simple figure that is embedded within (and therefore camouflaged by) a more intricate design. Thurstone and Jeffrey (1956) created a multiple-choice version that can be performed by patients with upper extremity motor impairment. Spreen and Benton impose a 30-second time limit. Spreen and Strauss (1991) give normative data for samples of children and adults, although they note that normal adults make very few errors. Sex differences are largely nonexistent. Spreen and Strauss note that reliability figures for embedded figures tests are generally around 0.9. There is some evidence that patients with right hemisphere lesions are more impaired than those with left hemisphere damage (DeRenzi & Spinnler, 1966), although aphasic patients obtained the lowest scores in one study (Russo & Vignolo, 1967). Size of lesion appears relevant (Corkin, 1979), as individuals with larger areas of damage do worse. Although Teuber, Battersby, and Bender (1951) found that the frontal patients were particularly impaired, Egelko *et al.* (1988) reported that their group with anterior lesions surpassed patients with posterior damage when time limits were eliminated.

Two subtests from the WAIS-R (Wechsler, 1981) should be discussed herein. Picture Completion demands identification of the important component that is missing from each of 20 pictures of objects, people, or scenes. While subjects are instructed to name the missing piece, Lezak (1995) argues that individuals with expressive language difficulties (e.g., Broca's aphasia, dysarthria) should be allowed to point to the location of the missing portion. Picture Completion is the only Performance subtest that does not require a motor response; therefore, it is frequently incorporated in test batteries used with patients with impaired hand and arm function such as those with spinal cord injury (e.g., Morris, Roth, & Davidoff, 1986; Richards, Brown, Hagglund, Bua, & Reeder, 1988; Davidoff, Roth, & Richards, 1992). In addition to the obvious visual–perceptual requirements, Picture Completion appears to demand verbal reasoning and long-term memory skills as indicated by its high correlations with certain Verbal subtests and by the results of factor analyses (e.g., Lansdell & Donnelly, 1977; Russell, 1972; Saunders, 1960). Lezak (1995) argues that Picture Completion may be taken as a good index of premorbid ability, especially among aphasic patients for whom the otherwise highly robust (i.e., lesion-resistant) Information and Vocabulary subtests may not provide valid results. Average split-half reliability across the various age ranges in the standardization sample was 0.81.

The WAIS-R Picture Arrangement subtest consists of 10 sets of cards, each card containing a single picture. Each set contains between three and six cards that, when placed in the proper sequence, tell a story. The examiner lays out the cards in a designated scrambled sequence, and the subject is told to place them in the correct order. Subjects are allowed 1 minute to complete the easier items and 2 minutes for the more difficult ones, but many examiners will allow some additional time if the individual is working diligently and does not appear frustrated. Also, experienced neuropsychologists (e.g., Lezak, 1995; Kaplan, Fein, Morris, & Delis, 1991) recommend that subjects be asked to explain their arrangement of each set of cards, that is, to tell the story that is depicted therein. Patients with unilateral neglect may fail to notice and include cards on the "bad" side; such patients may benefit from being allowed to arrange the cards in a vertical column. Picture Arrangement is something of an "odd duck" among the WAIS-R subtests, as it has relatively low correlations with the other subtests (0.28 to 0.41). Perhaps because of the verbal mediation and reasoning demands, Picture Arrangement performance may be impaired by left hemisphere damage, although right-sided lesions appear to produce a greater degree of impairment (McFie, 1975). Patients with frontal lobe injuries, who may exhibit "organic inertia," may show little inclination to move the cards from the scattered order in which they are presented (Walsh, 1978); if asked to tell the story of the picture sequence, they may fabricate disjointed tales or simply describe each of the cards in turn, ignoring the need to integrate across the sequence. As with most WAIS-R subtests, Picture Arrangement has good split-half reliability (average coefficient across age ranges in the standardization sample = 0.74). As adequate performance requires perception of salient small details, examiners must consider whether it is worthwhile administering the test to subjects who achieve a very low score on Picture Completion.

The Hooper Visual Organization Test (VOT; Hooper, 1958) was originally designed to detect neurological disorders in psychiatric patients. The current manual (Hooper, 1983) stresses that the VOT score must be considered as a single contribution to the diagnostic process. The test consists of pictures of 30 objects (e.g., airplane, basket) that have been cut into pieces and the parts rearranged on the page. Subjects must mentally assemble the components in order to identify the item. Thus, the task essentially requires jigsaw puzzle assembly skills but without the motor component, as there are no pieces to manipulate. If the test is administered individually, a spiral booklet is used and the pictures exposed one at a time with the subject required to name the object. For group administration, each subject receives a test booklet containing the 30 pictures and writes their answers beneath the drawings. Wetzel and Murphy (1991) determined that examiners could discontinue administration following five consecutive failures with little risk to correct classification. Raw scores are adjusted for age and education, and corrected raw scores may be converted to standard scores (*T*-scores). Designated *T*-score ranges are considered to reflect varying levels of "probability of impairment" ranging from "very low" to "very high." Split-half reliabilities have been reported to be 0.78 and above (Gerson, 1974; Hooper, 1948, 1958), and one study of test–retest reliability involving three administrations over a 1-year interval yielded a coefficient of 0.86 (Lezak, 1982b). Although some authors believe the VOT to be reasonably sensitive to brain dysfunction (Wang, 1977; Boyd, 1981, 1982a, 1982b), others dispute the test's value (Rathbun & Smith, 1982; Woodward, 1982). Despite the apparent visual–spatial demands of the VOT, it does not appear to be selectively sensitive to right hemisphere damage, perhaps because of the verbal character of the response (Boyd, 1981; Wang, 1977). A recently developed multiple choice version (Garay & Caplan, 1997) of the VOT may permit more valid findings in persons with anomia. Fitz, Conrad, Hom, Sarff, and Majovski (1992) found a nonsignificant superiority of VOT performance in a small group of patients with unilateral left hemisphere lesions compared to a somewhat larger group with right hemisphere involvement. When raw VOT scores were controlled for age, education, and IQ, the right parietal subjects scored significantly lower than the remaining subjects. Recent studies of pediatric groups (Kirk, 1992; Seidel, 1994) have found developmental improvement in VOT performance between the ages of 5 and 13 years.

Several tasks require identification of incomplete drawings or silhouettes; most are largely of historical interest as they are not easily available for clinical use. The Gestalt Completion Test (Street, 1931) contains drawings of 15 items with portions eliminated to obscure their identity. Gollin's (1960) incomplete drawings test is composed of 20 series of five drawings each. Within each series, the given object is shown in graded degrees of clarity. The increasingly recognizable stimuli are displayed, one at a time, and subjects are to identify the object as soon as they are able. Warrington and Rabin (1970) reported only a nonsignificant trend toward worse performance by right parietal patients on an abbreviated form of the Gollin, with left and right hemisphere groups demonstrating comparable performance. Lezak (1983) describes the Closure Faces Test (Mooney & Ferguson, 1951; Mooney, 1957) which has sometimes, (Lansdell, 1968, 1970; Newcombe, 1969) but not always (Wasserstein, Zappulla, Rosen, Gerstman, & Rock, 1987), shown selective sensitivity to right hemisphere damage, especially in the right temporal parietal region. Subjects are required to sort partially obscured faces into piles depending on whether the face is identified as that of a male or female of one of three different age groups.

CONSTRUCTIONAL ABILITY

Tests evaluating constructional abilities constitute prime examples of the multifactorial nature of most neuropsychological instruments. Among the tasks currently employed are measures requiring copying of simple or complex designs in two or three dimensions, free hand drawing, puzzle assembly, two- and three-dimensional block constructions, and construction of stick designs. The subskills necessary to perform these tasks are varied—attention, motor skills, visual discrimination and analysis, organization, planning, and ability to detect and correct one's errors. As difficulty with any of these individual capacities may produce poor constructional performance (i.e., constructional apraxia), it should not be surprising that such measures are sensitive to dysfunction of either hemisphere of the brain (e.g., Arrigoni & DeRenzi, 1964; Benton, 1973; Benton & Fogel, 1962; Colombo, DeRenzi, & Faglioni, 1976; DeRenzi & Faglioni, 1967; Piercy et al., 1960).

Researchers continue to debate whether constructional deficit occurs more frequently and/or at a greater level of severity following damage to the right or left hemisphere. Although some investigators have argued that constructional deficit results from different mechanisms in the two groups (i.e., a "spatioagnostic" or perceptual deficit factor in right hemisphere patients and an apraxic or dysexecutive form in the left hemisphere group), available data have not always confirmed this hypothesis. Studies by Piercy and Smythe (1962) and Dee (1970) reported an association of constructional deficit with visual–perceptual deficit in both hemisphere groups. However, Carlesimo et al. (1993) failed to replicate this, concluding that the negative impact of visual–perceptual impairment on constructional ability is mediated by difficulty in visually guided manual movement. Carlesimo et al. argued that, among right hemisphere-lesioned patients, constructional apraxia is primarily attributable to a manipulo spatial disorder, but for the left hemisphere patients, "a more severe low-level motor impairment" appears to be at fault. Kirk and Kertesz (1989) inferred from their data that neglect and visual–spatial deficit contribute to constructional apraxia (impaired drawing) among right hemisphere patients, while apraxia among patients with left hemisphere lesions derives from paresis of the dominant hand, as well as a conceptual deficit caused by impaired comprehension.

Although measures of constructional ability can be quite revealing and informative, they are nonetheless often omitted from casual bedside mental status evaluations (Strub & Black, 1988), especially if (as is typically the case) the patient has not complained about difficulties in this sphere. However, those who have developed structured screening devices have recognized the usefulness of even a brief evaluation of constructional ability. For example, two of the most widely used short mental status tests (Mini Mental State Examination; Folstein, Folstein, & McHugh, 1975; Neurobehavioral Cognitive Status Examination: Kiernan, Mueller, Langston, & Van Dyke, 1987) include tasks requiring either figure copying or block constructions. For our discussion, we adopt the distinction (also used by Lezak, 1995) between those tasks that require a drawing response and those that demand assembly of component pieces.

DRAWING TASKS. Tests of drawing may demand copying or "free" (self-initiated) production. Copying tasks may require the individual to reproduce drawings of geometric shapes or of real objects. Examples of the former include the Bender–Gestalt Test (Bender, 1938), the Complex Figure Test (Rey, 1941; Osterrieth, 1944), the Benton Visual Retention Test (Sivan, 1992), and the Beery Developmental Test of Visual-Motor Integration (Beery, 1989).

The Bender–Gestalt, which has also been promoted as a projective measure (e.g., Hutt, 1977), requires the individual to copy a series of nine geometric designs. There are several variations of standard administration procedures, including demanding that patients copy the figures as quickly as possible under distracting conditions; an "interference procedure" (Canter, 1966, 1968) in which subjects must make their drawings on a sheet of paper covered with curved black lines; and several conditions requiring reproduction from memory, either immediate or delayed (Wepman, as cited in Lezak, 1995). Although the Bender–Gestalt has been touted as an excellent measure of "organicity," (i.e., dysfunction of virtually any part of the cerebral cortex), this claim is too broad; while the test shares with other measures of drawing ability a particular sensitivity to right parietal lesions, patients with left frontal involvement may produce normal Bender results (Garron & Cheifetz, 1965). The Koppitz Developmental Bender Scoring System (Koppitz, 1964, 1975) is the most widely used scoring system for children. It includes criteria for "emotional indicators" and for several types of errors including distortion of shape, rotation, integration difficulties, and perseveration. While test–retest reliability ranges from 0.50 to 0.90, reliabilities of the error scores are considerably lower (0.29 to 0.62). Percentile norms are available for ages 5 through 12.

The Complex Figure developed by Rey (1941) is an intricate design consisting of several major core elements and a large number of additional details. The subject's initial task is simply to copy the design as accurately as possible. Many (if not most) examiners follow the approach of giving subjects a different color pencil after they complete each section; this allows the examiner to reconstruct the sequence in which the subject draws the various elements. However, a less disruptive alternative involves the examiner tracking the subject's performance on a photocopy of the figure, numbering the order in which the components are drawn

(Visser, 1973). Several scoring systems (Denman, 1984; Loring, Martin, Meador, & Lee, 1990; Osterrieth, 1944; Taylor, 1969; Tombaugh & Hubley, 1991; Waber & Holmes, 1985) have been proposed for rating the adequacy of reproduction of component parts. Loring *et al.* reported inter-rater reliability of 0.98 for their scoring system. Examiners also evaluate the subject's strategy, noting whether there is awareness of the overall Gestalt (as reflected in initial reproduction of the major elements) or whether the drawing is done in a fragmented "piecemeal" fashion. Binder (1982) reported that patients with left-sided brain damage tended to produce a greater degree of fragmentation in their drawings than did right hemisphere-damaged patients, while the latter omitted more elements. Many examiners subsequently request immediate and/or delayed recall of the Complex Figure (see section on Visual Memory).

The copy administration of the Benton Visual Retention Test (Sivan, 1992) requires the individual to reproduce relatively simple geometric shapes—single figures on the first two cards and three figures (two large shapes and a small peripheral one) on the remaining eight cards. In contrast to the usual administration of the Bender–Gestalt, the subject is given a fresh sheet of paper for each drawing. The manual provides extensive scoring criteria, examples of permissible responses, and an extensive qualitative typology of errors. The Benton offers the advantage of three parallel forms which can be employed in serial assessment. The examiner may also use one set to evaluate copying ability for the purpose of contrasting it with performance on another set administered under "visual memory" conditions. Sivan (1992) summarizes the psychometric properties and research findings. While inter-rater reliability has been found to be quite high (0.94 and above), validity coefficients (i.e., correlations of copying performance with other constructional measures) were relatively low (<0.4); however, this latter finding confirms the need for separate examination of graphomotor and assembly abilities. Benton (1967) reported impaired performance to occur half as often among patients with left hemisphere lesions (14%) as among those with right hemisphere involvement (29%).

The Beery Developmental Test of Visual-Motor Integration (Beery, 1982, 1989) consists of 24 geometric figures of increasing difficulty that the subject must copy. Performance for each design is scored simply as "pass" or "fail." The test was originally normed on over 1,000 children ages 3 to 14 years. Test–retest reliabilities range from 0.63 to 0.92, and inter-rater reliabilities from 0.66 to 0.93 (Sattler, 1990). Concurrent validity is supported by correlations with age (0.89), reading achievement (0.50), and perceptual ability (0.80). Several alternate administration procedures are suggested by the authors to be used for "testing the limits" of children who perform poorly under standard conditions. These include (1) reviewing the first item failed and asking the child whether it looks the same as his or her drawing and then how it differs (visual perception), (2) tracing the figures (motor control), (3) recopying the item and examining any improvement (integration), and (4) having the child imitate the examiner's drawing of the stimulus item (imitation).

Strub and Black (1988) suggest that subjects be asked to copy a series of five object drawings of increasing complexity including a diamond shape, a two-dimensional cross, a three-dimensional block, a three-dimensional pipe, and a pair of triangles, one contained inside the other; their recommended 4-point rating scale is probably adequate for its intended use in a bedside mental status evaluation.

Various items have been proposed for use in "free drawing" or "drawing to command" format including human figures, house, tree, bicycle, and flower

(Lezak, 1995; Strub & Black, 1998). While these may be used to detect difficulties with visual organization or unilateral neglect, some examiners (e.g., Machover, 1948) believe that important information may be revealed about personality characteristics through an individual's spontaneous drawings, especially those of human figures.

The ability to draw a clock face and insert the hands to display a designated time has been studied in a number of neurological conditions. There is some evidence that patients with Alzheimer's disease perform poorly on this task (Sunderland *et al.*, 1989; Wolf-Klein, Silverstone, Levy, Brod, & Breur, 1989), although Libon, Swenson, Barnoski, and Sands (1993) argue that clock drawing does not reliably differentiate Alzheimer's disease from other forms of dementia. Subjects may be asked to draw the clock face either to command or to copy. Goodglass and Kaplan (1983) recommend that the hands be placed at "10 after 11," while Sunderland *et al.* instruct their subjects to make the hands read "2:45." Sunderland describes a 10-point scoring system with very high inter-rater reliability. For the interested reader, a recent book was devoted entirely to the neuropsychology of clock drawings (Freedman *et al.*, 1994).

ASSEMBLY TASKS. The two most commonly used measures requiring assembling of component parts are the Block Design and Object Assembly subtests of the WAIS-R. On the former, the individual must use red-and-white blocks to copy two four-block patterns made by the examiner and then to construct seven designs printed on cards. According to Lezak (1995, p. 592), Block Design is "the best measure of visual–spatial organization in the Wechsler scales," although it also has a very high correlation (0.59) with Verbal IQ, suggesting the importance of verbal mediation. Furthermore, it has the highest reliability coefficient (0.87) and smallest standard error of measurement of any Performance subtest. By contrast, Object Assembly, which requires the individual to put together four jigsaw puzzles (person, profile of a head, hand, elephant), has the lowest reliability figure (0.68) and highest standard error of measurement among the Performance tasks. Both measures are timed, and subjects may earn bonus points for rapid completion. Both tasks show greater sensitivity to posterior lesions, especially those in the right hemisphere (Black & Strub, 1976; Horn & Reitan, 1990; Warrington, James, & Maciejewski, 1986). Although many patients obtain comparable scores on both tests, disparities can be quite informative. For example, Lezak (1995) notes that some individuals who require greater structure may perform relatively well on Block Design, but poorly on Object Assembly, while those who show a selective deficit in abstract conceptualization may do better on the more concrete real objects that comprise the jigsaw puzzles. Interested readers are referred to the work of Kaplan and associates (Kaplan, 1988; Kaplan *et al.*, 1991) for a discussion of their extensive qualitative analyses of Block Design performance.

Less widely known (and far less well supported from a psychometric point of view) are the various two-dimensional stick construction tasks such as those described by Warrington (1969) and Benson and Barton (1970). Warrington proposed requiring constructions of simple shapes and patterns of lines. Butters and Barton (1970), expanding on the work of Benson and Barton (1970), administered two versions of a stick construction task to patients with localized lesions and to a small control group. The examiner initially sits beside the patient and constructs 10 patterns which the subject is required to reproduce. The examiner then moves to the opposite side of the table, facing the patient, and constructs the same series of designs; under this condition, the instruction is to "make your pattern look to you like mine looks to me," thereby requiring mental rotation of the figures. Interestingly, while the copy portion produced the poorest performance among patients with right posterior lesions, those with *left* posterior lesions had the greatest difficulty in the "reversal" condition.

While three-dimensional constructional skills were evaluated by some earlier researchers (Hecaen *et al.*, 1951), Benton and colleagues have conducted the most extensive systematic research in this area, eventuating in a commercially available measure (Benton *et al.*, 1983; Benton, Sivan *et al.*, 1994). There are two forms of the test, each containing three constructions (a six-block pyramid, an eight-block four-level structure, and a 15-block construction also of four levels), each of which must be assembled from an array of 29 blocks; there is a 5-minute limit for each item. Pictorial representations of the model are shown to the subjects. Benton *et al.* provide a typology of errors to be recorded, as well as normative data for both children and adults and their findings in patients with lateralized lesions. They report a much higher failure rate (45%) in their right hemisphere group than in the left hemisphere subjects (20%); the disparity was even more pronounced (30% vs. 5%) when only "severely defective" performances were compared. Furthermore, right hemisphere patients more commonly committed unusual types of errors such as omitting one side of the structure or physically linking the construction to the examiner's model.

An increasingly popular three-dimensional constructional measure is the Tinker Toy Test devised by Lezak (1982a). Subjects are given 50 pieces of a Tinker Toy set and are told to make whatever they want with them. They must spend at least 5 minutes working on the task but may have as much additional time as they desire. Lezak recommends this test as a measure of executive functions (i.e., initiation, planning, and structuring), noting that it differs from other three-dimensional constructional tests, such as that of Benton *et al.*, by eliminating the structure provided by a model to copy. Lezak suggests eight dimensions for evaluation of the subject's product including the number of pieces used, symmetry, presence or absence of three dimensions, mobility of the structure, and complexity. Her preliminary finding that patients with brain lesions who were more functionally independent produced more complex structures containing more pieces was confirmed and expanded by Bayless, Varney, and Roberts (1989). Clinicians who want to employ this test are urged to obtain the same version of the toy used by Lezak. Regrettably, toy manufacturers have little interest in the "test–retest reliability" of their materials; should they modify (or discontinue) the toy, "alternate form reliability" would then need to be demonstrated.

MENTAL IMAGERY AND MENTAL ROTATION

IMAGERY. Although rarely the subject of formal investigation in the neuropsychological clinic, the capacity to form mental images (Tippett, 1992) can have considerable daily life significance. For example, Ardila (1993) described the case of a college chemistry teacher who, following a small right parietal infarction that did not produce "any evident spatial difficulty in her everyday activities" (p. 86), was unable to teach because she could no longer create images of the configurations of molecules. Caplan and Shechter (1991) reported on a woman who developed impairment of the capacity to form sexually explicit images following resection of a right temporo parietal tumor; this skill, which had played a central role in her previously satisfactory sex life, remained impaired more than 1 year following surgery, despite substantial resolution of early visual–perceptual deficits. Certainly, the importance of mental imagery skills to artists, architects, and chess players is apparent.

Much of the assessment of mental imagery ability involves nonstandard measures, many developed on an ad hoc basis such as the previously mentioned task by Bisiach and Luzzatti (1978) that required subjects to visualize the town square of Milan. Standardized measures may be found in Thurstone's (1938) Primary Mental Abilities (PMA) test battery and in certain tasks devised by Guilford and Hoepfner (1971). Schaie (1985), expanding on the PMA, produced the Schaie–Thurstone Adult Mental Abilities Test which includes a Figure Rotation subtest (equivalent to Thurstone's Space Test) and an Object Rotation subtest using common objects; Form OA was designed to be used with older adults (approximately ages 55 and above). One subtest of the Purdue Spatial Visualization Test (Guay, 1977) requires mental rotation. The Stanford–Binet Intelligence Scale Fourth Edition (Thorndike, Hagen, & Sattler, 1986) retains the Paper Folding and Cutting subtest that requires the subject to choose from a multiple-choice array the design that depicts what a folded piece of paper would look like if it were flattened. Goldenberg, Podreka, and colleagues (1989) developed 50-item sets of low and high imagery statements that subjects were to judge as either true or false; the former set primarily involved factual knowledge (e.g., "Bismarck was a German politician."), while the latter demanded generation of visual imagery (e.g., "The letter "W" consists of three lines.").

Perhaps most intriguing about the work on mental imagery is the mounting evidence of a major role of the *left* hemisphere (e.g., Farah, 1984, 1990; Farah, Gazzaniga, Holtzman, & Kosslyn, 1985). Farah (1984) reported on 14 cases with impaired imagery abilities but no significant visual disturbance; these individuals tended to have left posterior damage. Further work by Farah and colleagues (see Farah, 1990, for a review) provided additional support for this position. Goldenberg has reported some degree of selective activation of temporal and occipital regions (as detected by SPECT) on his visual imagery tasks (Goldenberg, Podreka *et al.*, 1989a; Goldenberg, Podreka, Steiner, Franzen, & Deecke, 1991; Goldenberg, Steiner, Podreka, & Deecke, 1992), more often in the left hemisphere, but not invariably so (Goldenberg, Uhl, Willmes, & Deecke, 1989). Following a lengthy critique, Sergent (1990) concluded: "As far as can be inferred from the existing evidence, neither hemisphere has an exclusive competence at generating visual images, and at present, a conservative interpretation of the findings points to a simultaneous involvement of the two hemispheres in this process" (p. 119).

While some portion of the explanation for these conflicting findings probably lies in the oft-cited methodological differences, a more powerful and reasonable explanation is that proposed by Kosslyn (1988), who argued that it might be misleading to consider the creation of mental images as a unitary function; he suggested that there may be different types of image generation abilities and these may have different cortical residences. Kosslyn and Shin (1994) proposed a model containing 24 distinct variations of mental images. Furthermore, tasks that involve the production of images that can be verbally labeled may necessarily engage the left hemisphere to a greater degree than those that are less linguistically based. According to DeRenzi (1978), it is degree of task complexity that determines whether or not the left hemisphere is involved. The issue remains contentious (Tippett, 1992).

ROTATION. Farah (1990) has provided a succinct summary of studies of the ability to manipulate mental images. The most widely studied such skill is that of mental rotation which has been assessed in both normal (e.g., Shepard & Metzler, 1971; Cooper & Shepard, 1973) and impaired (e.g., DeRenzi & Faglioni, 1967; Ratcliff, 1979) individuals. The early studies by Butters and Barton (1970) and Benson and Barton (1970) described previously failed to find consistent impairment among their right hemisphere-damaged subjects. Ratcliff (1979) tested patients with unilateral right or left penetrating missile injuries as well as a bilateral lesion group. Subjects viewed various drawings of a figure in either an upright or inverted position, shown from the front or back; one of the figure's hands was colored black and the other white. As subjects were required to identify whether the blackened hand was on the figure's right or left side, it was necessary in many instances to rotate the figure in order to answer. Ratcliff found no differences between groups for the upright position, but the right posterior group performed significantly worse in the inverted condition.

However, subsequent studies have not been unanimous in their support of right hemispheric specialization for mental rotation, even when highly similar tasks have been employed. Ornstein, Johnstone, Herron, and Swencionis (1980) found greater EEG activity in the left hemispheres of 20 normal subjects who performed a mental rotation task. By contrast, Deutsch, Bourbon, Papanicolaou, and Eisenberg (1988), using a virtually identical task in a normal population, found a significant increase in cerebral blood flow to the right anterior region. Tachistoscopic studies of mental rotation in normal subjects have also produced conflicting results. Corballis and Sergent (1989) reported a right hemifield (i.e., left hemisphere) advantage, while Ditunno and Mann (1990) observed a left visual field advantage; the latter also found that a small group of patients with right parietal lesions made significantly more errors and demonstrated significantly longer response latencies than did patients with left hemisphere lesions or normal controls. It is probably fair to say that, as with studies of mental imagery in general, findings regarding lateralization of mental rotation tasks and abilities are affected by several factors including the subject's cognitive strategy and task complexity (Cook, Fruh, Mehr, Regard, & Landis, 1994). Corballis' (1982) conclusion that "most, but not all, of the evidence" favors right hemisphere dominance for mental rotation likely remains valid.

VISUAL MEMORY

As assessment of the various memory functions is covered in detail in chapter 15, we offer here only a brief consideration of several selected measures of spatial memory. Although the great majority of memory tests are auditory/verbal in nature, several involve visual–spatial presentation. Deficits in visual memory have been reported in a variety of clinical groups such as individuals with right hemisphere damage (Milner, 1971), Parkinson's disease (Netherton, Elias, & Albrecht, 1989; Bradley, Welch, & Dick, 1989), Alzheimer's disease (Trojano, Chiacchio, DeLuca, & Grossi, 1994), those with mild dementia (Robinson-Whelen, 1992), patients with multiple sclerosis (Grant, McDonald, Trimble, Smith, & Reed, 1984), as well as the normal elderly population (Haaland, Linn, Hunt & Goodwin, 1983). Thus, assessment of visual memory should not be ignored.

The Wechsler Memory Scale-Revised (WMS-R; Wechsler, 1987) contains four tasks assessing various aspects of visual learning and retention. On Figural Memory, the subject views one or three abstract designs and then must identify them in a multiple-choice array. As there are only four items in this subtest, it is not surprising that the reliability is quite low (0.4). Visual Reproduction requires the individual to draw from memory four multicomponent designs that are exposed for 10 seconds; there is also a delayed recall trial. Reliabilities for both administrations are considerably better (about 0.7) than that for Figural Memory. Visual Memory Span (an analog of Digit Span) employs a card on which are printed eight squares in quasi-random fashion; each square is associated with a number that is known only to the examiner. Progressively longer series of squares are tapped by the examiner and the subject must then reproduce each sequence. Both forward and backward spans are administered and the two scores are added to obtain a total score. The reliability of this test is the highest of all of the visual subtests (0.75). Finally, the WMS-R includes acquisition and delayed recall of a Visual Paired Associate task consisting of six abstract figures, each of which is linked with a different color. The shape–color pairs are first shown in association; when each figure is then shown in isolation, the subject must supply the associated color. At least three, but not more than six, learning trials are administered, followed later in the session by a single delayed recall trial. Reliabilities for both forms are only 0.58.

Although the WMS-R has been criticized for multiple flaws (e.g., Elwood, 1991; Loring, 1989), it has been widely adopted by many clinicians. Of particular interest for our purposes is the fact that at least two studies (Chelune & Bornstein, 1988; Loring, Martin, & Meador, 1989) were unable to demonstrate selective impairment of visual memory in patients with right hemisphere lesions. Interested readers should note that, as of this writing, a third edition of the WMS is in development.

The Benton Visual Retention Test (BVRT) was first described a half-century ago (Benton, 1945); the current manual (Sivan, 1992) is the fifth edition. As noted above, the BVRT contains three sets of 10 plates. In each set, the first two plates contain a single geometric shape and the remaining eight are composed of two large shapes and a small peripheral figure. In addition to the copy administration, there are three memory administrations involving exposure times of 5 or 10 seconds, with either immediate reproduction from memory or following a 15-second interpolated delay. The current manual provides norms for ages 8 to 69 and also offers normative data from two studies of older adults (Arenberg, 1978; Benton, Eslinger, & Damasio, 1981). Sivan (1992) summarizes a number of reliability studies, virtually all reporting inter-rater reliability figures of 0.9 and above. Benton (1962, 1972) found a higher incidence of impairment on the BVRT among individuals with right hemisphere lesions than among those with left-sided involvement, although the difference was "small and unimpressive" (p. 150). Those with bilateral lesions were more often impaired than either of the unilateral lesion groups.

It is worth pointing out that the parallel forms of the BVRT confer a pronounced advantage over other measures of visual memory; specifically, by contrasting the individual's "copying" with their "memory" performance, one can separate out the contribution of perceptual or graphomotor deficit to observed "visual memory" deficit.

In considering the skills relevant to performance on the BVRT (or any other visual memory task), one should recall the comment made by Moses (1986): "Just because the task presents a nonverbal stimulus to the subject does not prevent one from verbally mediating the task and changing the supposed task demands from a nonverbal task to a verbal one" (p. 155). Moses' two factor analyses including the BVRT (1986, 1989) provide conflicting evidence for the importance of verbal mediation on this task, many stimuli of which have common verbal labels.

Recently, Caplan and Caffery (1992a) described clinical applications of a multiple-choice visual memory test employing stimulus materials from the Visual Form discrimination measure (Benton *et al.*, 1983) which was originally developed as a multiple-choice version of the BVRT (Benton, 1950; Benton, Hamsher, & Stone, 1977). They reported that the multiple-choice version was particularly sensitive to right hemisphere lesions. Normative data are given in Caplan and Caffery (1996).

As mentioned above, many examiners administer a recall trial of the Complex Figure Test, using immediate delay intervals varying from 30 seconds to 3 minutes and delayed recall periods from 20 to 30 minutes. Lezak (1995, pp. 475–480) reviews the various methods and findings. It should be noted that this procedure evaluates "incidental memory." Recently, a recognition format has been developed (Meyers & Meyers, 1995) which consists of 24 designs, half of which are portions of the total figure. The individual indicates by circling or pointing to the items that they believe were contained within the figure.

In an attempt to assess the types of everyday memory problems encountered by neuropsychological patients, Wilson, Cockburn, and Baddeley (1985) developed the Rivermead Behavioural Memory Test (RBMT) which contains several visual memory items including recognition of pictures of both common objects and faces following a brief delay, immediate and delayed recall of a short path around the testing room, and delayed recall of the location of an object hidden by the examiner at the beginning of the session. Wilson, Cockburn, Baddeley, and Hiorns (1989) provide a detailed description of the development of the RBMT, its reliability and validity (both of which appear satisfactory), and clinical interpretation. Recognizing that at least some portions of the RBMT were likely to be failed by individuals with aphasia purely because of their language difficulty, Cockburn and colleagues (1990) identified those items that produced lower scores in persons with mildly to moderately impaired language; an abbreviated form of the RBMT that eliminated those items was reported to be sensitive to memory deficit but not to the impact of aphasia.

"Nonspatial" Tests with a Spatial Component

In this section, we briefly consider several important cognitive functions that are not typically thought of as "spatial" in nature, but nonetheless have a spatial aspect, the selective disruption of which may produce impaired performance. Certainly, any visually presented measure, such as the five subtests that comprise the Peabody Individual Achievement Test-Revised (General Information, Reading Recognition, Reading Comprehension, Mathematics, Spelling) or the two tasks on the Shipley Institute of Living Scale (multiple-choice vocabulary, sequence completion) is subject to sabotage by visuospatial deficit. A recent study by Kempen, Kritchevsky, and Feldman (1994) of neurologically normal subjects ages 55 to 74 years found impaired performance on two of three tests involving higher visual processing among those with diminished visual acuity compared to those with normal vision. Writing ability may be disrupted in a unique fashion following right temporo parietal-occipital lesions, producing "visuospatial agraphia" (Benson & Cummings, 1985) characterized by script that is slanted, located largely on the right side of the page, and unusually spaced. Examiners who fail to make careful behavioral observations or perform adequate qualitative analysis of test protocols run substantial risk of misinterpretation, that is, concluding that a "higher-order" deficit exists when poor performance derives from diminished visual acuity, insufficient scanning, or other "lower-order" functions. We discuss three illustrative examples.

CALCULATIONS

There are several varieties of deficits in mathematical computation called the "acalculias." The type that concerns us here has been referred to as "constructive acalculia" (Krapf, 1937), "parietal alcalculia" (Grewel, 1952, 1969), and "acalculia of the spatial type" (Levin, Goldstein, & Spiers, 1993), and has usually been found far more often following right than left hemisphere damage. For example, Hecaen and Angelergues (1961) identified spatially based arithmetic deficit in 6.7% of their patients with left parietal lesions and 16.3% of those with right parietal damage; 57% of the sample with right temporo parietal-occipital involvement exhibited spatial acalculia (but see Grafman, Passafiume, Faglioni, & Boller, 1982, for contradictory findings).

"Spatial acalculia" may affect either written or mental arithmetic abilities, although the former are generally more impaired (Benton, 1963, 1966, as cited in Levin & Spiers, 1985). The fundamental difficulty involves misalignment or misperception of columns, leading to failure to execute the operation on the proper numbers. Rotation of individual digits or reversal of digit pairs may also occur, as well as incorrect placement of decimal points. The presence of unilateral neglect also contributes to erroneous calculation, causing misperception of the operation symbol. Not surprisingly, spatial acalculia shows a strong association with a variety of spatial deficits including constructional difficulty, visual field defect, and "spatial dyslexia" (Hecaen, Angelergues, & Houillier, 1961). For recent reviews, see Boller and Grafman (1983, 1985), Grafman (1988), Levin *et al.* (1993), and Hartje (1987).

As noted by Levin *et al.* (1993), "acalculia" is not a monolithic disorder; what they refer to as the "calculation system" may be impaired in a variety of ways, and these fractionated deficits may be associated with certain localized lesions. Assessment of mathematical abilities must be correspondingly complex. Suggestions for the range of arithmetic abilities that can be tested have been offered by Collingnon, Leclercq, and Mahy (1977), Boller and Hecaen (1979), Macaruso, Harley, and McCloskey (1992), Benton (1963), and Deloche *et al.* (1994). These authors recommend that examiners evaluate a subject's ability to read numbers aloud, point to numbers named by the examiner, write numbers to dictation and copy, complete written calculations, perform mental arithmetic of both numerical and "word problem" varieties, and describe number concepts (e.g., relations between numbers such as identifying the largest number of a set or the smallest fraction in a series). Standard tests exist for only some of these processes (e.g., arithmetic portion of the Wide Range Achievement Test-III; Arithmetic subtest of the WAIS-R).

READING

Although the majority of reading disorders (alexias) follow damage to the left hemisphere, one form, typically referred to as "neglect alexia" or "spatial alexia" (Ardila & Rosselli, 1994), occurs far more frequently following damage to the right hemisphere (DeRenzi, 1982). Friedman, Ween, and Albert (1993) describe three forms of the disorder. The first occurs in the context of left-side neglect following right parietal damage; patients may omit entire words located on the left side of the page, or produce "near miss" errors on individual words as a result of neglect of the leftmost portion of a word (e.g., reading "prelate" as either "relate" or "late"). In the second form, Friedman *et al.* argue, these types of errors may occur in isolation, that is, with no other manifestations of neglect. The third form involves neglect or misreading of the rightmost portion of words, presumably a manifestation of right-sided neglect resulting from a left hemisphere lesion. The first type appears to be by far the most common (McCarthy & Warrington, 1990). The impaired scanning associated with neglect would also create problems with reading by virtue of the need to refixate leftward from the end of one line to the beginning of the next. Whether the mirror-image problem would have a higher incidence in left than right hemisphere-damaged readers of Hebrew remains to be explored.

Neglect alexia may be detected by a variety of standard measures of oral reading such as the reading portion of the Wide Range Achievement Test-III. However, some individuals with relatively minor degrees of neglect, or those who have learned to compensate, may not encounter any difficulties on such tests, as they are able to form a "spatial set" of the text when the left and right margins are justified. Therefore, Caplan (1987) developed the Indented Paragraph Reading Test (IPRT) as a measure that would be sensitive to mild left-sided neglect. The IPRT consists of a single paragraph of 30 lines with a left-side margin variably indented between 0 and 25 spaces, with the degree of indentation unpredictable from line to line. Subjects with right-sided lesions showed the expected high incidence of neglect (46.5% with "mild" and 25.6% with "moderate-severe" neglect, according to the suggested criteria), findings that were confirmed in a recent study by Bachman, Fein, Davenport, and Price (1993).

DIGIT REPETITION

The Digit Span subtest of the WAIS-R (Wechsler, 1981) is generally considered to be a measure of auditory attention. However, a number of authors have pointed out that the two tasks that contribute to the total score may engage different cognitive mechanisms. While Digits Forward involves auditory attention and "echoic memory," the strategy employed by many subjects on Digits Backward adds a spatial component. When asked to repeat in reverse order a string of digits that has been read by the examiner, many individuals visualize the number series and then read the digits off from right to left, thereby adding a visual scanning aspect. Therefore, it should not be surprising that patients with right hemisphere lesions (in whom scanning deficits are relatively common) may obtain normal forward spans, but impaired backward spans (Costa, 1975; Weinberg, Diller, Gerstmann, & Schulman, 1972; Rapport, Webster, & Dutra, 1994). While this finding has not been invariably replicated (e.g., Black & Strub, 1978) this may be attributable to the mixing within the right hemisphere group of patients with and without perceptual impairment, thereby "washing out" the effect (Heilbronner *et al.*, 1991). Examiners who combine the two subscores and analyze and/or report only the total may lose potentially important information.

SPATIAL ASSESSMENT BATTERIES

Several authors have assembled collections of particular tests for the purpose of evaluating a range of perceptual skills. It is doubtful that any would claim that their "battery" offers a complete assessment of all aspects of visual–spatial skill. Three such batteries are discussed in this section.

The lengthiest of the three is the Rivermead Perceptual Assessment Battery (RPAB; Whiting, Lincoln, Bhavnani, & Cockburn, 1985), a group of 16 subtests that has been found to detect a very high incidence (87%) of perceptual deficit in patients with unilateral stroke (Smith & Lincoln, 1989). The RPAB contains subtests evaluating such skills as visual matching (for colors, objects, and pictures), size recognition, figure–ground discrimination, picture sequencing, copying of words and shapes (both two- and three-dimensional), cancellation, and "body image-self identification." Two studies (Edmans & Lincoln, 1987; Smith & Lincoln, 1989) reported that both left and right hemisphere stroke patients exhibit comparably high frequencies of impairment, although Smith and Lincoln did not use those subtests that had previously (Whiting *et al.,* 1985) been shown to elicit greater impairment among *left* hemisphere-damaged patients. Edmans and Lincoln reported that aphasic patients performed significantly worse than individuals without aphasia. Thus, the impact of language impairment on RPAB performance remains to be clarified.

Warrington and James (1991) published the Visual Object and Space Perception Battery (VOSP) which measures the following skills: identification of degraded (incomplete) letters, silhouette recognition, object decision (selecting the real object from among four silhouettes), silhouette identification in rotated view, dot counting, visual localization, and cube analysis (determining the number of solid blocks that make up a cube). Warrington and James state that the tests were individually developed to investigate theoretical issues, and that inclusion in the battery was determined by "pragmatic strength in terms of their selectivity and sensitivity" (p. 7). Clinicians may use any number and combination of the tests administered in any order. All are untimed. Normative data are provided for two standardization samples, one (n = 200) that received five subtests and another (n = 150) that was given the remaining three. Warrington and James report the results of clinical studies demonstrating the greatest impairment among right hemisphere-damaged subjects on all measures, with left hemisphere-damaged patients showing either a lesser degree of deficit or, in some cases, performance that was indistinguishable from that of the control subjects. In its current form, the VOSP is probably most valuable when individual tests are used on an "as needed" basis by clinicians evaluating patients with selective visual–perceptual impairments.

The Motor Free Visual Perception Test (MVPT; Colarusso & Hammill, 1972; Bouska & Kwatny, 1983) was developed to answer the needs of clinicians who needed to assess visual–perceptual skills in children without introducing the potentially confounding variable of motor skill. Individuals with quadriplegia or cerebral palsy, for example, may or may not have difficulty with certain aspects of perceptual processing; however, if the task (e.g., Visual–Motor Integration) requires a drawing response, the child's ultimate product may be defective for reasons that have nothing to do with impaired perception. Therefore, the MVPT requires oral responses; subjects select their answers from a four-item multiple-choice array. The 36-item test provides brief evaluation of the following skills: spatial relations, visual discrimination of salient features, figure–ground discrimination, visual closure (i.e., completion of fragmented pictures), and visual memory. Colarusso and Hammill report various reliability coefficients for the standardization sample (children ages 4–8), all of which were statistically significant and the majority of which exceeded 0.8. The MVPT demonstrated higher correlations with other measures of visual perception than with intelligence and achievement tests, suggesting a degree of construct validity. Bouska and Kwatny (1983) provided normative data on a sample of 91 adults. Furthermore, they proposed contrasting the number of left- and right-sided responses in order to identify the presence of unilateral neglect. They also recommended that the examiner record response latencies, and they offered a rough cutoff point (more than 6 seconds slower than average) for determining "abnormal" performance. No recommendation was made with respect to "impulsive" (i.e., faster than normal) performance. In addition to the limited normative data, clinical utility of the MVPT suffers from the fact that only a total score is derived rather than separate scores for the individual subskills. Clinicians must then review the actual protocols to attempt to determine the defective domain(s).

NONVISUAL SPATIAL TESTS

Although the bulk of this chapter has focused on spatial impairments and measures in the visual modality, spatial deficits in other modalities are not at all uncommon. Indeed, Benton (1982) discussed 19th-century writings that suggested the possibility of a "supra modal" deficit in spatial cognition. We discuss two measure of tactile–spatial ability and one involving the auditory modality.

The ability to recognize objects by touch (and loss of same) has been studied in neurological patients for more than 100 years (Hoffmann, 1885; Wernicke, 1895). Benton *et al.* (1983) note that most early investigators agreed on the paramount importance of the parietal lobes in the process of tactile recognition. One might imagine that if a verbal response (e.g., naming of the identified object) were required, left hemisphere-damaged patients might encounter difficulty in this domain, despite the obvious spatial requirements. In order to evaluate nonverbal tactile information processing, Benton *et al.* (1983) developed the Tactile Form Perception Test (TFPT). For this task (which has two parallel forms), the subject places one hand through an opening in a box, thereby concealing the hand and the stimuli that the examiner places therein. Subjects are then given a series of 10 geometric shapes cut from sandpaper, instructed to feel the shape for up to 30 seconds and then point to its visual representation among 12 shapes drawn on a multiple-choice response card. Subjects must respond within 45 seconds using the free hand. Each hand is tested in turn, and separate scores are derived for the two hands in addition to the combined score. Benton *et al.* give normative data on 115 subjects ages 15 to 80 years; Spreen and Gaddes (1969) reported norms for 404 children ages 8 to 14. No sex differences were found among the children, and this factor was not studied in the adult population. The pediatric group showed improvement in performance with age, while among the adults, Benton *et al.* reported a slight age-related decline. Overall, there appears to be very little intermanual performance difference.

In their clinical sample of 104 patients, Benton *et al.* (1983) found impairment in over half on the TFPT. The incidence of defective performance was higher in patients with sensory or motor impairment. Patients with aphasia following left hemisphere lesions and non-aphasic patients with right hemisphere involvement exhibited similar high incidence of impairment (58% and 59%, respectively). These two groups also displayed high frequencies of bilateral or ipsilateral impairment in contrast to non-aphasic patients with left hemisphere involvement who showed both a lower incidence of defective performance (38%) and less frequent bilateral or ipsilateral defect. As noted by Benton *et al.* and by Newcombe and Ratcliff (1990), the high frequency of impairment among right hemisphere-damaged patients is consistent with the notion of "supra modal spatial dominance" of the right hemisphere (Corkin, 1965; Dee & Benton, 1970; Semmes, 1965, 1968), but the high frequency of impaired left hemisphere-damaged patients shows that multiple cortical sites are involved in tactile object recognition. Eighty-three percent of bilateral patients were impaired on the TFPT. This may result from the fact that, although the response required is a nonverbal (pointing) one, there may well be verbal components to the processing that is involved (e.g., labeling of the shapes). In this regard, it is worth noting Witelson's (1974) finding of a left hand (i.e., right hemisphere) superiority for tactile recognition of unfamiliar irregular shapes (i.e., nonlinguistic), but equal performance for the two hands in the recognition of linguistic stimuli (letters of the alphabet). Witelson's (normal pediatric) subjects responded by pointing to the matching items from a visual display; no overt verbal response was demanded.

The other tactile–spatial test is the Tactual Performance Test, a component of the Halstead–Reitan Neuropsychological Test Battery (Reitan & Wolfson, 1993). In its original incarnation, the test was a component of the Arthur Point Scale of Performance Tests (Arthur, 1947) called the Seguin Formboard. The test consists of a rectangular piece of wood from which 10 shapes have been cut. Subjects are blindfolded and then required to return the shapes to their proper locations in the board. The two hands are tested separately, and a third trial is then administered using both hands. Time to completion is determined for each trial and a total score is calculated. The blindfold and the formboard are then removed and the subject is required to draw the board from memory including the shapes and their locations relative to each other on the board; this forms the "memory" portion of the TPT. In their recent manual, Heaton *et al.* (1991) recommend a maximum time per trial of 10 minutes, although they allow some latitude for clinical judgment (i.e., permitting subjects to continue if they do not appear frustrated and are progressing well). Heaton *et al.* calculate a "minutes per block" index and use this in their analyses. They provide normative data adjusted for age, sex, and education. To reduce the length and frustration involved in administering the TPT, some investigators (e.g., DeRenzi, 1968, 1978) use a formboard with 6 rather than 10 shapes.

The primary avenue of investigation of auditory–spatial skill has involved studies of sound localization ability in patients with lateralized lesions, a skill that may be preserved in the presence of severe deficit of visual localization (Holmes, 1918). The results are far from consistent. There is some evidence that temporal lobe damage produces difficulty with contralateral sound localization. Sanchez Longo and colleagues (Sanchez Longo, Forster, & Auth, 1957; Sanchez Longo & Forster, 1958) studied patients with unilateral lesions and normal controls. Subjects were blindfolded and asked to identify the location of a sound source by pointing to the direction from which the tone emanated. Although Sanchez Longo concluded that contralateral localization deficit was a marker for temporal lobe dysfunction, Shankweiler (1961), using a comparable technique, found no difference between temporal and nontemporal groups, nor between his brain-injured patients and controls. He did, however, report worse performance in right than left hemisphere-damaged patients. DeRenzi (1982) re-analyzed data reported by Klingon and Bontecou (1966) regarding sound localization (finger snapping) in patients with unilateral lesions and identified a selective deficit among patients with right hemisphere damage compared to those with left-sided involvement. This right hemispheric dominance may be another manifestation of neglect. However, Ruff, Hersch, and Pribram (1981) found bilateral auditory localization impairment in patients with right posterior lesions, suggesting a more global deficit of auditory localization than would be produced by a neglect phenomenon.

Although less frequently investigated, and probably with less drastic clinical implications than visual neglect, auditory neglect has been demonstrated and appears to be more common following right cerebral lesions (e.g., Bisiach, Cornacchia, Sterzi, & Vallar, 1984; Heilman & Valenstein, 1972).

Conclusions

As noted in the introduction, our current understanding of spatial abilities and their assessment is notable for its fragmentation. In large measure, this is attributable to the absence of any widely agreed-upon definitions and, consequently, the lack of a scheme of functions and skills (e.g., naming, repetition, fluency) such as that which has been developed in the sphere of language. As McCarthy and Warrington (1990) observed: "Spatial processing is clearly a highly complex activity which has yet to be integrated into any coherent model of cognitive function" (p. 97). Farah (1988), citing the "dearth of theory," noted that spatial functions have historically "been defined in relatively phenomenon-oriented ways, in terms of specific kinds of tasks failed after brain damage, or in terms of 'what spatial abilities tests test,' rather than in theory oriented ways. . ." (p. 34).

It may be that the search for a coherent and comprehensive typology of spatial abilities is doomed to be a frustrating and ultimately fruitless one, reminiscent of the quest for a "pure" visual memory measure described by Heilbronner (1992). At this stage of development, we see primarily statements of "thesis," occasionally generating an "antithesis" in response, but little movement toward "synthesis." For instance, Bellugi *et al.* (1994), discussing their studies of children with Williams syndrome, divide their presentation between "language abilities" and "spatial cognition." Commenting on this work, Denckla (1994) argues that language ability should be contrasted with "visual–perceptual skills," with spatial functions as a subcategory of the latter. She cites the "disparate tasks" that are "lumped under Spatial," indicting researchers for "failure to fractionate" or develop "a broadly applicable nosology" (p. 291). We doubt whether there is unanimity in support of Denckla's "unimodal" concept of spatial functions. Furthermore, we do not anticipate the imminent appearance of a "consensus model" of spatial skills. Indeed, we hesitated to add to the confusion of the existing partial overlapping by proposing our own four-part subdivision; however, we believe that a scheme with relatively few categories offers considerable flexibility for classification purposes and perhaps goes some distance toward meeting Denckla's desideratum.

Despite these theoretical barriers and pitfalls, major advances continue in the clinic with the inexorable creation of new measures that permit assessment of components of complex spatial skills (e.g., Caplan & Caffery, 1992b). As the literature summarized in this chapter has demonstrated, the investigation of spatial functions and dysfunctions holds tremendous appeal for researchers and clinicians of varied backgrounds. The numerous measures of spatial abilities (however defined), although perhaps seemingly overwhelming and fragmented, provide ample testimony of the vital importance of this functional domain. Fortunately, individuals with congenital or acquired deficits in spatial functioning do not need to wait for delineation and confirmation of an ideal model of spatial abilities, as development continues of effective methods for rehabilitation of spatial deficits (e.g., Klonoff, Sheperd, O'Brien, Chiapello, & Hodak, 1990; Gordon *et al.*, 1985; Weinberg *et al.*, 1977; Hanlon, Dobkin, Hadler, Ramirez, & Cheska, 1991; Weinberg *et al.*, 1981; Young, Collins, & Hren, 1983).

Acknowledgments

The authors wish to thank Patt Williams for her tireless and outstanding secretarial work on this chapter.

REFERENCES

Archibald, Y. M. (1978). Time as a variable in the performance of hemisphere-damaged patients on the Elithorn Perceptual Maze Test. *Cortex, 14,* 22–31.

Archibald, Y. M., Wepman, J. M., & Jones, L. V. (1967). Performance on nonverbal cognitive tests following unilateral cortical injury to the right and left hemispheres. *Journal of Nervous and Mental Disease, 145,* 25–36.

Ardila, A. (1993). Historical evolution of spatial abilities. *Behavioural Neurology, 6,* 83–87.

Ardila, A., & Rosselli, M. (1992). *Neuropsicologia clinica [Clinical neuropsychology].* Medellin, Colombia: Prensa Creativa.

Ardila, A., & Roselli, M. (1994). Spatial alexia. *International Journal of Neuroscience, 76,* 49–59.

Arenberg, D. (1978). Differences and changes with age in the Benton Visual Retention Test. *Journal of Gerontology, 33,* 534–540.

Army Individual Test Battery. (1944). *Manual of directions and scoring.* Washington, DC: War Department, Adjutant General's Office.

Arrigoni, G., & DeRenzi, E. (1964). Constructional apraxia and hemispheric locus of lesion. *Cortex, 1,* 170–197.

Arthur, G. (1947). *Point scale of performance tests. Revised form II.* New York: Psychological Corporation, Chicago: Stoelting.

Ault, R. L., Mitchell, C., & Hartmann, D. P. (1976). Some methodological problems in reflection-impulsivity research. *Child Development, 47,* 227–231.

Bachman, L., Fein, G., Davenport, L., & Price, L. (1993). The indented paragraph reading test in the assessment of left hemi-neglect. *Archives of Clinical Neuropsychology, 8,* 485–496.

Badal, J. (1888). Contribution à l'étude des cécites psychiques: Alexie, agraphie, hemianopsie inferieure, trouble du sens del l'espace. *Archives de l'Ophtalmologie, 8,* 97–117.

Balint, R. (1909). Seelenlahmung des "Schauens," optische Ataxie, raumliche Storung der Auf merksamkeit. *Monatsschrift für Psychiatrie und Neurologie, 25,* 51–81.

Basso, A., DeRenzi, E., Faglioni, P., Scotti, G., & Spinnler, M. (1973). Neuropsychological evidence for the existence of cerebral areas critical to the performance of intelligence tasks. *Brain, 96,* 715–728.

Bayless, J. D., Varney, N. R., & Roberts, R. J. (1989). Tinker toy test performance and vocational outcome in patients with closed-head injuries. *Journal of Clinical and Experimental Neuropsychology, 11,* 913–917.

Beatty, W. W., & Troster, A. I. (1988). Neuropsychology and spatial memory. In H. A. Whitaker (Ed.), *Contemporary reviews in neuropsychology* (pp. 77–108). New York: Springer-Verlag.

Beaumont, J. G., & Davidoff, J. B. (1992). Assessment of visuo-perceptual dysfunction. In J. R. Crawford, D. M. Parker, and W. W. McKinlay (Eds.), *Handbook of neuropsychological assessment* (pp. 115–140). Hillsdale, NJ: Lawrence Erlbaum Associates.

Beery, K. E. (1982). *Revised administration, scoring, and teaching manual for the Developmental Test of Visual-Motor Integration.* Cleveland, OH: Modern Curriculum Press.

Beery, K. E. (1989). *The VMI Test of Visual-Motor Integration, administration, scoring and teaching manual 3rd revision.* Cleveland, OH: Modern Curriculum Press.

Bellugi, U., Wang, P. O., & Jernigan, T. L. (1994). Williams syndrome: An unusual neuropsychological profile. In S. H. Broman & J. Grafman (Eds.), *Atypical cognitive deficits in developmental disorders: Implications for brain function* (pp. 23–56). Hillsdale, NJ: Lawrence Erlbaum Associates.

Bender, L. (1938). A visual motor gestalt test and its clinical use. *American Orthopsychiatric Association Research Monographs, 3.*

Benson, D. F., & Barton, M. I. (1970). Disturbances in constructional ability. *Cortex, 6,* 19–46.

Benson, D. F., & Cummings, J. L. (1985). Agraphia. In J. A. M. Frederiks (Ed.), *Handbook of clinical neurology* (pp. 457–472). Amsterdam: Elsevier Science Publishers.

Bentin, S., Silverberg, R., & Gordon, H. W. (1981). Asymmetrical cognitive deterioration in demented and parkinsonian patients. *Cortex, 17,* 533–544.

Benton, A. L. (1945). A visual retention test for clinical use. *Archives of Neurology and Psychology, 59,* 273–291.

Benton, A. L. (1950). A multiple choice type of the visual retention test. *Archives of Neurology and Psychiatry, 64,* 699–707.

Benton, A. L. (1962). The visual retention test as a constructional praxis task. *Confinia Neurologica, 22,* 141, 155.

Benton, A. L. (1963). *Assessment of number operations.* Iowa City: University of Iowa Hospitals, Department of Neurology.

Benton, A. L. (1966). *Problemi de neuropsicologia* (pp. 147–159). Firenze: Editrice Universitaria.

Benton, A. L. (1967). Constructional apraxia and the minor hemisphere. *Confinia Neurologica, 29,* 1–16.

Benton, A. L. (1969). Disorders of spatial orientation. In P. Vinken & G. Bruyn (Eds.), *Handbook of clinical neurology* (Vol. III, pp. 212–228). Amsterdam: North-Holland.

Benton, A. L. (1972). Abbreviated versions of the visual retention test. *Journal of Psychology, 80,* 189–192.

Benton, A. L. (1973). Visuoconstructional disability in patients with cerebral disease. *Documenta Ophthalmologica, 34,* 67–76.

Benton, A. L. (1978). The interplay of experimental and clinical approaches in brain lesion research. In S. Finger (Ed.), *Recovery from brain damage* (pp. 49–68). New York: Plenum Press.

Benton, A. L. (1982). Spatial thinking in neurological patients: Historical aspects. In M. Potegal (Ed.), *Spatial abilities* (pp. 253–275). New York: Academic Press.

Benton, A. (1985). Visuoperceptual, visuospatial, and visuoconstructive disorders. In K. M. Heilman & E. Valenstein (Eds.), *Clinical neuropsychology* (2nd ed., pp. 151–185). New York: Oxford University Press.

Benton, A., Elithorn, A., Fogel, M., & Kerr, M. (1963). A perceptual maze test sensitive to brain damage. *Journal of Neurology, Neurosurgery, and Psychiatry, 26,* 540–544.

Benton, A. L., Eslinger, P. J., & Damasio, A. R. (1981). Normative observations on neuropsychological test performances in old age. *Journal of Clinical Neuropsychology, 3,* 33–42.

Benton, A. L., & Fogel, M..L. (1962). Three-dimensional constructional praxis. *Archives of Neurology* (Chicago), 7, 347–354.

Benton, A. L., Hamsher, K. deS., & Stone, F. B. (1977). *Visual Retention Test: Multiple choice form I.* Unpublished manuscript, University of Iowa School of Medicine, Division of Behavioral Neurology, Iowa City.

Benton, A. L., Hamsher, K., Varney, N., & Spreen, O. (1983). *Contributions to neuropsychological assessment.* New York: Oxford University Press.

Benton, A. L., Sivan, A. B., Hamsher, K., Varney, N., & Spreen, O. (1994). *Contributions to neuropsychological assessment* (2nd ed.). New York: Oxford University Press.

Benton, A., & Tranel, D. (1993). Visuoperceptual, visuospatial, and visuoconstructive disorders. In K. M. Heilman & E. Valenstein (Eds.), *Clinical neuropsychology* (pp. 165–214). New York: Oxford University Press.

Benton, A. L., Varney, N. R., & Hamsher, K. (1978). Visuospatial judgment: A clinical test. *Archives of Neurology, 35,* 364–367.

Binder, L. M. (1982). Constructional strategies on complex figure drawings after unilateral brain damage. *Journal of Clinical Psychology, 4,* 51–58.

Bisiach, E., Cornacchia, L., Sterzi, R., & Vallar, G. (1984). Disorders of perceived auditory lateralization after lesions of the right hemisphere. *Brain, 107,* 37–52.

Bisiach, E., & Luzzatti, C. (1978). Unilateral neglect of representational space. *Cortex, 14,* 129–133.

Black, F. W., & Strub, R. L. (1976). Constructional apraxia in patients with discrete missile wounds of the brain. *Cortex, 12,* 212–220.

Black, F. W., & Strub, R. L. (1978). Digit repetition performance in patients with focal brain damage. *Cortex, 14,* 12–21.

Block, J., Block, J. H., & Harrington, D. M. (1974). Some misgivings about the matching familiar figures test as a measure of reflection-impulsivity. *Developmental Psychology, 10,* 611–632.

Boller, F., & Grafman, J. (1983). Acalulia: Historical development and current significance. *Brain and Cognition, 2,* 205–223.

Boller, F., & Grafman, J. (1985). Acalculia. In J. A. M. Frederick (Ed.), *Handbook of clinical neurology: Vol. 1 (45). Clinical neuropsychology* (pp. 473–481). Amsterdam: Elsevier Science Publishers.

Boller, F., & Hecaen, H. (1979). L'evaluation des fonctions neuropsychologiques: Examen standard de l'unité de récherches neuropsychologiques et neurolinguistiques (U.111) I.N.S.E.R.M. *Revue de Psychologie Appliquée, 29,* 247–266.

Bornstein, R. A., Paniak, C., & O'Brien, W. (1987). Preliminary data on classification of normal and brain-damaged elderly subjects. *The Clinical Neuropsychologist, 1,* 315–323.

Bouska, M. J., & Kwatny, E. (1983). *Manual for application of the Motor Free Visual Perception Test to the adult population.* Philadelphia: Temple University Rehabilitation Research and Training Center.

Boyd, J. L. (1981). A validity study of the Hooper Visual Organization Test. *Journal of Consulting and Clinical Psychology, 49,* 15–19.

Boyd, J. L. (1982a). Reply to Rathbun and Smith: Who made the Hooper blooper? *Journal of Consulting and Clinical Psychology, 50,* 284–285.

Boyd, J. L. (1982b). Reply to Woodward. *Journal of Consulting and Clinical Psychology, 50,* 289–290.

Bradley, V. A., Welch, J. L., & Dick, D. J. (1989). Visuospatial working memory in Parkinson's disease. *Journal of Neurology, Neurosurgery, and Psychiatry, 52,* 1228–1235.

Brain, W. R. (1941). Visual disorientation with special reference to lesions of the right cerebral hemisphere. *Brain, 64,* 244–272.

Brickenkamp, R. (1981). *Test d2: Aufmerksamkeits-Belastungs-Test* (Handanweisung, 7th ed.) [Test d2: Concentration-Endurance Test: Manual, 5th ed.]. Gottingen: Verlag fur Psychologie Dr. C. J. Hogrefe.

Brown, L., Sherbenou, R. J., & Johnsen, S. K. (1990). *Test of nonverbal intelligence: A language-free measure of cognitive ability. Second edition, manual.* Austin, TX: Pro-Ed.

Burke, H. R. (1958). Raven's Progressive Matrices: Validity, reliability, and norms. *Journal of Psychology, 22,* 252–257.

Burke, H. R. (1985). Raven's Progressive Matrices: More on norms, reliability, and validity. *Journal of Clinical Psychology, 41,* 231–235.

Butters, N., & Barton, M. (1970). Effect of parietal lobe damage on the performance of reversible operations in space. *Neuropsychologia, 8,* 205–214.

Caltagirone, C., Gainotti, G., & Miceli, G. (1977). Una nuova versione delle matrici colorate elaborata specificamente per i pazienti con lesioni emesferiche focali. *Minerva Psichiatrica, 18,* 9–16.

Canter, A. (1966). A background interference procedure to increase sensitivity of the Bender-Gestalt test to organic brain disorder. *Journal of Consulting Psychology, 30,* 91–97.

Canter, A. (1968). BIP Bender test for the detection of organic brain disorder: A modified scoring method and replication. *Journal of Consulting and Clinical Psychology, 32,* 522–526.

Caplan, B. (1985). Stimulus effects in unilateral neglect? *Cortex, 21,* 69–80.

Caplan, B. (1987). Assessment of unilateral neglect: A new reading test. *Journal of Clinical and Experimental Neuropsychology, 9,* 359–364.

Caplan, B. (1988). Nonstandard neuropsychological assessment: An illustration. *Neuropsychology, 2,* 13–17.

Caplan, B., & Caffery, D. (1996). Visual form discrimination as a multiple-choice memory test: Illustrative data. *The Clinical Neuropsychologist, 10,* 152–158.

Caplan, B., & Shechter, J. (1990). Clinical application of the matching familiar figures test: Impulsivity vs. unilateral neglect. *Journal of Clinical Psychology, 46,* 60–67.

Caplan, B., & Caffery, D. (1992a). Development of a normative data base for the Benton Visual Retention Test–multiple choice version. *Journal of Clinical and Experimental Neuropsychology, 14,* 46.

Caplan, B., & Caffery, D. (1992b). Fractionating block design: Development of a test of visual–spatial analysis. *Neuropsychology, 6,* 385–394.

Caplan, B., & Shechter, J. (1991). A neuropsychological basis for sexual fantasy? *The Clinical Neuropsychologist, 5,* 266.

Caplan, P., MacPherson, G. M., & Tobin, P. (1985). Do sex-related differences in spatial abilities exist? A multilevel critique with new data. *American Psychologist, 40,* 786–799.

Carlesimo, G. A., Fadda, L., & Caltagirone, C. (1993). Basic mechanisms of constructional apraxia in unilateral brain-damaged patients: Role of visuo-perceptual and executive disorders. *Journal of Clinical and Experimental Neuropsychology, 15,* 342–358.

Carlson, J. S., & Jensen, C. M. (1981). Reliability of the Raven Colored Progressive Matrices test: Age and ethnic group comparisons. *Journal of Consulting and Clinical Psychology, 49,* 320–322.

Carter-Saltzman, L. (1979). Patterns of cognitive functioning in relation to handedness and sex-related differences. In M. A. Wittig & A. C. Petersen (Eds.), *Sex-related differences in cognitive functioning: Developmental issues* (pp. 97–120). New York: Academic Press.

Chelune, G. J., Bornstein, R. A. (1988). WMS-R patterns among patients with unilateral brain lesions. *The Clinical Neuropsychologist, 2,* 121–132.

Cockburn, J., Wilson, B., Baddeley, A., & Hiorns, R. (1990). Assessing everyday memory in patients with dysphasia. *British Journal of Clinical Psychology, 29,* 353–360.

Colarusso, R. P., & Hammill, D. D. (1972). *Motor Free Visual Perception Test–Manual.* Novato, CA: Academic Therapy Publications.

Collingnon, R., Leclercq, C., & Mahy, J. (1977). Étude de la semiologie des troubles de calcul observés au cours des lesions corticales. *Acta Neurologica Belgica, 77,* 257–275.

Colombo, A., DeRenzi, E., & Faglioni, P. (1976). The occurrence of visual neglect in patients with unilateral cerebral disease. *Cortex, 12,* 221–231.

Colonna, A., & Faglioni, P. (1966). The performance of hemisphere-damaged patients on spatial intelligence tests. *Cortex, 2,* 293–307.

Cook, N. D., Fruh, H., Mehr, A. Regard, M., & Landis, T. (1994). Hemispheric cooperation in visuospatial rotations: Evidence for a manipulation role for the left hemisphere and a reference role for the right hemisphere. *Brain and Cognition, 25,* 240–249.

Cooper, L. A., & Shepard, R. N. (1973). Chronometric studies of the rotation of mental images. In W. G. Chase (Ed.), *Visual information processing.* New York: Academic Press.

Corballis, M. C. (1982). Mental rotation: Anatomy of a paradigm. In M. Potegal (Ed.), *Spatial abilities: Developmental and physiological foundations* (pp. 173–198). Academic Press.

Corballis, M. C., & Sergent, J. (1989). Hemispheric specialization for mental rotation. *Cortex, 25,* 15–25.

Corkin, S. (1965). Tactually guided maze learning in man: Effects of unilateral cortical excisions and bilateral hippocampal lesions. *Neuropsychologia, 3,* 339–351.

Corkin, S. (1979). Hidden-figures-test performance: Lasting effects of unilateral penetrating head injury and transient effects of bilateral cingulotomy. *Neuropsychologia, 17,* 585–605.

Coslett, H. B., & Saffran, E. M. (1992). Disorders of higher visual processing: Theoretical and clinical perspectives. In D. I. Margolin (Ed.), *Cognitive neuropsychology in clinical practice* (pp. 353–404). New York: Oxford University Press.

Costa, L. D. (1975). The relation of visuospatial dysfunction to digit span performance in patients with cerebral lesions. *Cortex, 11,* 31–36.

Costa, L. D. Vaughan, H. G., Horwitz, M., & Ritter, W. (1969). Patterns of behavioral deficit associated with visual spatial neglect. *Cortex, 5,* 242–263.

Crown, S. (1952). An experimental study of psychological changes following prefrontal lobotomy. *Journal of General Psychology, 47,* 3–41.

Court, J. H. (1983). Sex differences in performance on Raven's Progressive Matrices: A review. *The Alberta Journal of Educational Research, 29,* 54–74.

Damasio, A. (1985). Disorders of complex visual processing: Agnosias, achromatopsia, Balint's syndrome, and related difficulties of orientation and construction. In M. M. Mesulam (Ed.), *Principles of behavioral neurology* (pp. 259–288). Philadelphia: F. A. Davis.

Damasio, A. R., & Damasio, H. (1983). Localization of lesions in achromatopsia and prosopagnosia. In A. Kertesz (Ed.), *Localization in neuropsychology* (pp. 417–428). New York: Academic Press.

Davidoff, G. N., Roth, E. J., & Richards, J. S. (1992). Cognitive deficits in spinal cord injury: Epidemiology and outcome. *Archives of Physical Medicine and Rehabilitation, 73,* 275–284.

Dee, H. L. (1970). Visuoconstructional and visuoperceptive deficits in patients with unilateral cerebral lesions. *Neuropsychologia, 8,* 305–314.

Dee, H. L., & Benton, A. L. (1970). A cross-modal investigation of spatial performances in patients with unilateral cerebral disease. *Cortex, 6,* 261–272.

Delis, D. C., Robertson, L. C.,. & Efron, R. (1986). Hemispheric specialization of memory for visual hierarchical stimuli. *Neuropsychologia, 24,* 205–214.

Deloche, G., Seron, X., Larroque, C., Magnien, C., Metz-Lutz, M. N., Noel, M. N., Riva, I., Schils, H. P., Dordain, M., Ferrand, I., Baeta, E., Basso, A., Cipolotti, L., Claros-Salinas, D., Howard, D., Gaillard, F., Goldenberg, G., Mazzucchi, A., Stachowiak, F., Tzavaras, A., Vendrell, J., Bergego, C., & Pradat-Diehl, P. (1994). Calculation and number processing: Assessment battery; role of demographic factors. *Journal of Clinical and Experimental Neuropsychology, 16,* 195–208.

Denckla, M. B. (1994). Interpretations of a behavioral neurologist. In S. H. Broman & J. Grafman (Eds.), *Atypical cognitive deficits in developmental disorders: Implications for brain function* (pp. 283–295). Hillsdale, NJ: Lawrence Erlbaum Associates.

Denman, S. B. (1984). *Denman neuropsychology memory scale.* Charleston, SC: Author.

DeRenzi, E. (1968). Nonverbal memory and hemispheric side of lesion. *Neuropsychologia, 6,* 181–189.

DeRenzi, E. (1978). Hemispheric asymmetry as evidenced by spatial disorders. In M. Kinsbourne (Ed.), *Asymmetrical function of the brain* (pp. 49–85). New York: Cambridge University Press.

DeRenzi, E. (1982). *Disorders of space exploration and cognition.* Chichester, England: Wiley.

DeRenzi, E. (1985). Disorders of spatial orientation. In J. A. M. Frederids (Ed.), *Handbook of clinical neurology: Vol. 1(45). Clinical neuropsychology* (pp. 405–422). Amsterdam: Elsevier Science Publishers.

DeRenzi, E., & Faglioni, P. (1967). The relationship between visuo-spatial impairment and constructional apraxia. *Cortex, 3,* 194–216.

DeRenzi, E., & Spinnler, H. (1966). Visual recognition in patients with unilateral cerebral disease. *Journal of Nervous and Mental Disease, 142,* 515–525.

Deutsch, G., Bourbon, W. T., Papanicolaou, A. C., & Eisenberg, H. M. (1988). Visuospatial tasks compared via activation of regional cerebral blood flow. *Neuropsychologia, 26,* 445–452.

Diller, L., Ben-Yishay, Y., Gerstman, L. J., Goodkin, R., Gordon, W., & Weinberg, J. (1974). *Studies in cognition and rehabilitation in hemiplegia.* (Rehabilitation Monograph No. 50). New York: New York University Medical Center Institute of Rehabilitation Medicine.

Diller, L., & Weinberg, J. (1977). Hemi-inattention in rehabilitation. The evolution of a rational remediation program. In E. A. Weinstein & R. P. Friedman (Eds.), *Advances in neurology.* New York: Raven Press.

Ditunno, P. L., & Mann, V. A. (1990). Right hemisphere specialization for mental rotation in normals and brain damaged subjects. *Cortex, 26,* 177–188.

Dodrill, C. B., & Troupin, A. S. (1975). Effects of repeated administration of a comprehensive neuropsychological battery among chronic epileptics. *Journal of Nervous and Mental Disease, 161,* 185–190.

Drebing, C. E., Takushi, R. Y., Tanzy, K. S., Murdock, G. A., Stewart, J. C., & Majovski, L. V. (1990). Reexamination of CPM performance and neglect in lateralized brain injury. *Cortex, 26,* 661–664.

Dunn, T. D. (1895). Double hemiplegia with double hemianopsia and loss of geographic centre. *Transactions of the College of Physicians of Philadelphia, 17,* 45–56.

Edmans, J. A., & Lincoln, N. B. (1987). The frequency of perceptual deficits after stroke. *Clinical Rehabilitation, 1,* 273–281.

Egelko, S., Gordon, W. A., Hibbard, M. R., Diller, L., Lieberman, A., Holliday, R., Ragnarsson, K., Shaver, M. S., & Orazem, J. (1988). Relationship among CT scans, neurologic exam and neuropsychological test performance in right-brain-damaged stroke patients. *Journal of Clinical and Experimental Neuropsychology, 10,* 539–564.

Elithorn, A., Jones, D., Kerr, M., & Lee, D. (1964). The effects of the variation of two physical parameters on empirical difficulty in a perceptual maze test. *British Journal of Psychology, 55,* 31–37.

Elwood, R. W. (1991). The Wechsler Memory Scale-Revised: Psychometric characteristics and clinical application. *Neuropsychology Review, 2,* 179–201.

Eslinger, P. J., & Benton, A. L. (1983). Visuoperceptual performance in aging and dementia: Clinical and theoretical implications. *Journal of Clinical Neuropsychology, 5,* 213–220.

Eslinger, P. J., Damasio, A. R., Benton, A. L., & van Allen, M. (1985). Neuropsychological detection of abnormal mental decline in older persons. *Journal of American Medical Association, 253,* 670–674.

Eson, M. E., & Bourke, R. S. (1982). Assessment of recovery from serious head injury. *Journal of Neurology, Neurosurgery, and Psychiatry, 41,* 1036–1042.

Farah, M. J. (1984). The neurological basis of mental imagery: A componential analysis. *Cognition, 18,* 245–272.

Farah, M. J. (1988). The neuropsychology of mental imagery: Converging evidence from brain-damaged and normal subjects. In J. Stiles-Davis, M. Kritchevsky, & U. Bellugi (Eds.), *Spatial cognition: Brain bases and development* (pp. 33–56). Hillsdale, NJ: Lawrence Erlbaum Associates.

Farah, M. J. (1990). The neuropsychology of mental imagery. In F. Boller & J. Grafman (Eds.), *Handbook of neuropsychology* (Vol. 2, pp. 395–413). Amsterdam: Elsevier Science Publishers.

Farah, M. J., Gazzaniga, M. S., Holtzman, J. D., & Kosslyn, S. M. (1985). A left hemisphere basis for visual mental imagery? *Neuropsychologia, 23,* 115–118.

Fitz, A. G., Conrad, P. M., Hom, D. L., Sarff, P. L., & Majovski, L. (1992). Hooper Visual Organization Test performance in lateralized brain injury. *Archives of Clinical Neuropsychology, 7,* 243–250.

Foerster, R. (1890). Ueber Rindenblindheit. *Graefes Archiv für Ophthalmologie, 36,* 94–108.

Folstein, M. F., Folstein, S. E., & McHugh, P. R. (1975). Mini-mental state. *Journal of Psychiatric Research, 12,* 189–198.

Franzen, M. D., Robbins, D. E., & Sawicki, R. F. (1989). *Reliability and validity in neuropsychological assessment.* New York: Plenum Press.

Freedman, M., Leach, L., Kaplan, E., Winocur, G., Shulman, K., & Delis, D. C. (1994). *Clock drawing: A neuropsychological analysis.* New York: Oxford University Press.

Fried, I., Mateer, C., Ojemann, G., Wohns, R., & Fedio, P. (1982). Organization of visuospatial function in human cortex. Evidence from electrical stimulation. *Brain, 105,* 349–371.

Friedman, R., Ween, J. E., & Albert, M. L. (1993). Alexia. In K. M. Heilman & E. Valenstein (Eds.), *Clinical neuropsychology* (3rd ed., pp. 37–62). New York: Oxford University Press.

Garay, M., & Caplan, B. (1997). Fractioning the Hooper: Development of a multiple-choice response format. *Archives of Clinical Neuropsychology, 12*, 321–322.

Garron, D. C., & Cheifetz, D. J. (1965). Comment on "Bender Gestalt discernment of organic pathology." *Psychological Bulletin, 63*, 197–200.

Gazzaniga, M. S., & Le Doux, J. E. (1978). *The integrated mind.* New York: Plenum Press.

Gerson, A. (1974). Validity and reliability of the Hooper Visual Organization Test. *Perceptual and Motor Skills, 39*, 95–100.

Golden, C. J., Osmon, D. C., Moses, J. A., & Berg, R. A. (1981). *Interpretation of the Halstead–Reitan Neuropsychological Test Battery.* New York: Grune and Stratton.

Goldenberg, G., Podreka, I., Steiner, M., Franzen, P., & Deecke, L. (1991). Contributions of occipital and temporal brain regions to visual and acoustic imagery: A SPECT Study. *Neuropsychologia, 29,* 695–702.

Goldenberg, G., Podreka, I., Steiner, M., Willmes, K., Suess, E., & Deecke, L. (1989). Regional cerebral blood flow patterns in visual imagery. *Neuropsychologia, 27,* 641–664.

Goldenberg, G., Steiner, M., Podreka, I., & Deecke, L. (1992). Regional cerebral blood flow patterns related to verification of low- and high-imagery sentences. *Neuropsychologia, 30,* 581–586.

Goldenberg, G., Uhl, P. F., Willmes, K., & Deecke, L. (1989). Cerebral correlates of imagining colours, faces and a map—I. SPECT of regional cerebral blood flow. *Neuropsychologia, 27,* 1315–1328.

Goldstein, G., & Watson, J. R. (1989). Test–retest reliability of the Halstead–Reitan battery and the WAIS in a neuropsychiatric population. *The Clinical Neuropsychologist, 3,* 265–273.

Gollin, E. S. (1960). Developmental studies of visual recognition of incomplete objects. *Perceptual and Motor Skills, 11,* 289–298.

Goodglass, H., & Kaplan, E. (1983). *The assessment of aphasia and related disorders* (2nd ed.). Philadelphia: Lea and Febiger.

Gordon, W. A., Ruckdescel-Hibbard, M., Egelko, S., Diller, L., Scotzin-Shaver, M., Lieberman, A., & Ragnarsson, K. (1985). Perceptual remediation in patients with right brain damage. A comprehensive program. *Archives of Physical Medicine and Rehabilitation, 66,* 353–359.

Grafman, J. (1988). Acalculia. In F. Boller & J. Grafman (Eds.), *Handbook of neuropsychology* (Vol. 1, pp. 415–431). Amsterdam: Elsevier Science Publishers.

Grafman, J., Passafiume, D., Faglioni, P., & Boller, F. (1982). Calculation disturbances in adults with focal hemispheric damage. *Cortex, 18,* 37–50.

Grant, I., McDonald, W. I., Trimble, M. R., Smith, E., & Reed, R. (1984). Deficient learning and memory in early and middle phases of multiple sclerosis. *Journal of Neurology, Neurosurgery, and Psychiatry, 47,* 250–255.

Grewel, F. (1952). Acalculia. *Brain, 75,* 397–407.

Grewel, F. (1969). The acalculias. In P. J. Vinken & G. W. Bruyn (Eds.), *Handbook of clinical neurology* (pp. 181–196). Amsterdam: North-Holland.

Guay, R. (1977). *Purdue spatial visualization tests.* West Lafayette, IN: Purdue Research Foundation.

Guilford, J. P., & Hoepfner, R. (1971). *The analysis of intelligence.* New York: McGraw-Hill.

Haaland, K. Y., Linn, R., Hunt, W. C., & Goodwinn, J. S. (1983). A normative study of Russell's variant of the Wechsler Memory Scale in a healthy elderly population. *Journal of Consulting and Clinical Psychology, 51,* 878–881.

Hamsher, K., Capruso, D. X., & Benton, A. (1992). Visuospatial judgment and right hemisphere disease. *Cortex, 28,* 493–495.

Hanlon, R. E., Dobkin, R. H., Hadler, B., Ramirez, S., & Cheska, Y. (1992). Neurorehabilitation following right thalamic infarct: Effects of cognitive retraining on functional performance. *Journal of Clinical and Experimental Neuropsychology, 14,* 433–447.

Harris, L. J. (1978). Sex differences in spatial ability: Possible environmental, genetic, and neurological factors. In M. Kinsbourne (Ed.), *Asymmetrical function of the brain* (pp. 405–522). New York: Cambridge University Press.

Hartje, W. (1987). The effect of spatial disorders on arithmetical skills. In G. Deloche & X. Seron (Eds.), *Mathematical disabilities: A cognitive neuropsychological perspective* (pp. 121–135). Hillside, NJ: Lawrence Erlbaum Associates.

Heaton, R. K., Baade, L. E., & Johnson, K. L. (1978). Neuropsychological test results associated with psychiatric disorders in adults. *Psychological Bulletin, 85,* 141–162.

Heaton, R. K., Grant, I., & Matthes, C. G. (1991). *Comprehensive norms for an expanded Halstead–Reitan battery: Demographic corrections, research findings, and clinical applications.* Odessa, FL: Psychological Assessment Resources, Inc.

Hecaen, H., Ajuriaguerra, J. de, & Massonet, J. (1951). Les troubles visuo-constructifs par lesion pari-eto-occipitale droite. *Encephale, 40*, 122–179.

Hecaen, H., & Angelergues, R. (1961). Etude anatomo-clinique de 280 cas de lesions retrorolandiques unilaterales des hemispheres cerebraux. *Encephale, 6*, 533–562.

Hecaen, H., Angelergues, R., & Houillier, S. (1961). Les varietés cliniques des acalculies au cours de lesions retrorolandiques: Approche statistique du probleme. *Revue Neurologique, 105*, 85–103.

Heilbronner, R. L. (1992). The search for a pure visual memory test: Pursuit of perfection. *The Clinical Neuropsychologist, 6*, 105–112.

Heilbronner, R. L., Henry, G. K., Buck, P., Adams, R. L., & Fogle, T. (1991). Lateralized brain damage and performance on Trail Making A and B, Digit Span forward and backward, and TPT memory and location. *Archives of Clinical Neuropsychology, 6*, 251–258.

Heilman, K. M., & Valenstein, E. (1972). Auditory neglect in man. *Archives of Neurology, 26*, 32–35.

Hoffmann, H. (1885). Stereognostiche Versuche, angestellt zur Ermittelung der Elemente des Gef-uhlssinnes, aus denen die Vorstellungen der Korper im Raume gebildet werden. *Deutsches Archiv Für Klinischen Medizin, 36*, 398–426.

Holmes, G. (1918). Disturbances of visual orientation. *British Journal of Ophthalmology, 2*, 449–486, 506–516.

Holmes, G., & Horrax, G. (1919). Disturbances of spatial orientation and visual attention, with loss of stereoscopic vision. *Archives of Neurology and Psychiatry, 1*, 385–407.

Hom, J., & Reitan, R. M. (1990). Generalized cognitive function after stroke. *Journal of Clinical and Experimental Neuropsychology, 12*, 644–654.

Hooper, H. E. (1948). A study in the construction and preliminary standardization of a visual organiza-tion test for use in the measurement of organic deterioration. Unpublished master's thesis, Univer-sity of Southern California, Los Angeles.

Hooper, H. E. (1958). *The Hooper Visual Organization Test. Manual.* Los Angeles: Western Psychological Services.

Hooper, H. E. (1983). *The Hooper Visual Organization Test. Manual.* Los Angeles: Western Psychological Services.

Hutt, M. L. (1977). *The Hutt adaptation of the Bender-Gestalt test* (3rd ed.). New York: Grune & Stratton.

Jackson, J. H. (1864). Clinical remarks on defects of expression (by words, writing, signs, etc.) in diseases of the nervous system. *Lancet, 1*, 604–605.

Jackson, J. H. (1876). Case of large cerebral tumour without optic neuritis and with left hemiplegia and imperception. *Royal Ophthalmic Hospital Reports, 8*, 434–444.

Jackson, J. H. (1915). On the nature of the duality of the brain. *Brain, 38*, 80–103. (Original work published 1923)

Kagan, J., Rosman, B., Day, D., Albert, J., & Phillips, W. (1964). Information processing in the child: Significance of analytic and reflective attitudes. *Psychological Monographs, 78* (1, Whole no. 578).

Kaplan, E. (1988). A process approach to neuropsychological assessment. In T. Boll & B. K. Bryant (Eds.), *Clinical neuropsychology and brain function: Research, measurement, and practice* (pp. 129–167). Washington, DC: American Psychological Association.

Kaplan, E., Fein, D., Morris, R., & Delis, D. C. (1991). *WAIS-R NI manual.* New York: The Psychological Corporation.

Kempen, J. H., Kritchevsky, M., & Feldman, S. T. (1994). Effect of visual impairment on neuropsycho-logical test performance. *Journal of Clinical and Experimental Neuropsychology, 16*, 223–231.

Kertesz, A., & McCabe, P. (1975). Intelligence and aphasia: Performance of aphasics on Raven's Col-oured Progressive Matrices (RCPM). *Brain and Language, 2*, 387–395.

Kiernan, R. J., Mueller, J., Langston, J. W., & Van Dyke, C. (1987). The neurobehavioral cognitive status examination: A brief but quantitative approach to cognitive assessment. *Annals of Internal Medicine, 107*, 481–485.

Kirk, A., & Kertesz, A. (1989). Hemispheric contributions to drawing. *Neuropsychologia, 27*, 881–886.

Kirk, U. (1992). Evidence for early acquisition of visual organization ability: A developmental study. *The Clinical Neuropsychologist, 6*, 171–177.

Kleist, K. (1912). Der Gang und der gegenwaertige Stand der Apraxieforschung. *Ergebnisse der Neuro-logie und Psychiatrie, 1*, 342–452.

Kleist, K. (1922/1934). Kriegverletzungen des Gehirns in ihrer Bedeutung für die Hirnlokalisation und Hirnpathologie. In O. von Schjerning (Ed.), *Handbuch der Aerztlichen Erfahrung in Weltkriege* (Vol. 4). Leipzig: Barth.

Klesges, R. C., Fisher, L., Pheley, A., Boschee, P., & Vasey, M. C. (1984). A major validation study of the Halstead–Reitan in the prediction of CAT-scan assessed brain-damaged adults. *International Journal of Clinical Neuropsychology, 1*, 29–34.

Klingon, G. H., & Bontecou, D. C. (1966). Localization in auditory space. *Neurology, 16,* 879–886.

Klonoff, P. S., Sheperd, J. C., O'Brien, K. P., Chiapello, D. A., & Hodak, J. A. (1990). Rehabilitation and outcomes of right-hemisphere stroke patients: Challenges to traditional diagnostic and treatment methods. *Neuropsychology, 4,* 147–169.

Koppitz, E. M. (1964). *The Bender-Gestalt test for young children.* New York: Grune & Stratton.

Koppitz, E. M. (1975). *The Bender-Gestalt test for young children: (Vol. 2). Research and application, 1963–1973.* New York: Grune & Stratton.

Korman, M., & Blumberg, S. (1963). Comparative efficiency of some tests of cerebral damage. *Journal of Consulting Psychology, 27,* 303–309.

Kosslyn, S. M. (1988). Aspects of a cognitive neuroscience of mental imagery. *Science, 240,* 1621–1626.

Kosslyn, S. M., & Shin, L. M. (1994). Visual mental images in the brain: Current issues. In M. J. Farah & G. Ratcliff (Eds.), *The neuropsychology of high-level vision: Collected tutorial essays* (pp. 269–298). Hillsdale, NJ: Lawrence Erlbaum Associates.

Krapf, E. (1937). Ueber Akalkulie. *Sweizerische Archiv fur Neurologie und Psychiatrie, 39,* 330–334.

Kritchevsky, M. (1988). The elementary spatial functions of the brain. In J. S. Davis, M. Kritchevsky, & U. Bellugi (Eds.), *Spatial cognition: Brain bases and development* (pp. 111–140). Hillsdale, NJ: Lawrence Erlbaum Associates.

Lamb, M. R., Robertson, L. C., & Knight, R. T. (1989). Attention and interference in the processing of hierarchical patterns: Inferences from patients with right and left temporal-parietal lesions. *Neuropsychologia, 27,* 471–483.

Lansdell, H. (1968). Effect of extent of temporal lobe ablations on two lateralized deficits. *Physiology and Behavior, 3,* 271–273.

Lansdell, H. (1970). Relation of extent of temporal removals to closure and visuomotor factors. *Perceptual and Motor Skills, 31,* 491–498.

Lansdell, H., & Donnelly, E. F. (1977). Factor analysis of the Wechsler Adult Intelligence Scale subtests and the Halstead-Reitan Category and Tapping tests. *Journal of Consulting and Clinical Psychology, 45,* 412–416.

Le Doux, J. E., Wilson, D. H., & Gazzaniga, M. S. (1977). Manipulo-spatial aspects of cerebral lateralization: Clues to the origin of lateralization. *Neuropsychologia, 15,* 743–749.

Lenz, G. (1905). *Beitraege zur Hemianopsie.* Stuttgart.

Levin, H. S., Benton, A. L., & Grossman, R. G. (1982). *Neurobehavioural consequences of closed head injury.* New York: Oxford University Press.

Levin, H. S., Goldstein, F. C., & Spiers, P. A. (1993). Acalculia. In K. M. Heilman & E. Valenstein (Eds.), *Clinical neuropsychology* (3rd ed., pp. 91–122). New York: Oxford University Press.

Levin, H. S., Grossman, R. G., & Kelly, J. (1977). Impairment in facial recognition after closed head injuries of varying severity. *Cortex, 13,* 119–130.

Levin, H. S., & Spiers, P. A. (1985). Acalculia. In K. M. Heilman & E. Valenstein (Eds.), *Clinical neuropsychology* (2nd ed., pp. 97–114). New York: Oxford University Press.

Lewinsohn, P. M. (1973). *Psychological assessment of patients with brain injury.* Unpublished manuscript, University of Oregon.

Lewis, N. D. C., Landis, C., & King, H. E. (1956). *Studies in topectomy.* New York: Grune and Stratton.

Lewis, R. F., & Rennick, P. M. (1979). *Manual for the repeatable cognitive-perceptual-motor battery.* Grosse Pointe Park, MI: Axon Publishing.

Lezak, M. D. (1982a). The problem of assessing executive function. *International Journal of Psychology, 17,* 281–297.

Lezak, M. D. (1982b). The test–retest stability and reliability of some tests commonly used in neuropsychological assessment. Paper presented at the meeting of the International Neuropsychological Society, Deauville, France.

Lezak, M. D. (1995). *Neuropsychological assessment* (3rd ed.). New York: Oxford University Press.

Libon, D. J., Swenson, R. A., Barnoski, E. J., & Sands, L. P. (1993). Clock drawing as an assessment tool for dementia. *Archives of Clinical Neuropsychology, 8,* 405–415.

Lissauer, H. (1890). Ein Fall von Seelenblindheit nebst einem Beitrag zur Theorie derselben. *Archiv für Psychiatrie und Nervenkrankheiten, 21,* 22–70.

Loring, D. A. (1989). The Wechsler Memory Scale-Revised, or the Wechsler Memory Scale-revisited? *The Clinical Neuropsychologist, 3,* 59–69.

Loring, D. W., Martin, R. C., & Meador, K. J. (1989). Verbal and visual memory index discrepancies from the Wechsler Memory Scale-Revised: Cautions in interpretation. *Psychological Assessment, 1,* 198–202.

Loring, D. W., Martin, R. C., Meador, K. J., & Lee, G. P. (1990). Psychometric construction of the Rey–

Osterrieth Complex Figure: Methodological consideration and interrater reliability. *Archives of Clinical Neuropsychology, 5,* 1–14.

Macaruso, P., Harley, W., & McCloskey, M. (1992). Assessment of acquired dyscalculia. In D. I. Margolin (Ed.), *Cognitive neuropsychology in clinical practice* (pp. 405–434). New York: Oxford University Press.

Machover, K. (1948). *Personality projection in the drawing of the human figure.* Springfield, IL: C. C. Thomas.

McCarthy, R. A., & Warrington, E. K. (1990). *Cognitive neuropsychology: A clinical introduction.* San Diego, CA: Academic Press.

McFie, J. (1975). *Assessment of organic intellectual impairment.* London: Academic Press.

McFie, J., Piercy, M. F., & Zangwill, O. L. (1950). Visual-spatial agnosia associated with lesions of the right cerebral hemisphere. *Brain, 73,* 167–190.

McGee, M. G. (1979). Human spatial abilities: Psychometric studies and environmental, genetic, hormonal, and neurological influences. *Psychological Bulletin, 86,* 889–918.

McGee, M. G. (1982). Spatial abilities: The influence of genetic factors. In M. Potegal (Ed.), *Spatial abilities: Development and physiological foundations* (pp. 199–222). New York: Academic Press.

McNeil, J. E., & Warrington, E. K. (1993). Prosopagnosia: A face-specific disorder. *The Quarterly Journal of Experimental Psychology, 46A,* 1–10.

Measso, G., Zappala, G., Cavarzeran, F., Crook, T. H., Romani, L., Pirozzolo, F. J., Grigoletto, F., Amaducci, L. A., Massari, D., & Lebowitz, B. D. (1993). Raven's Colored Progressive Matrices: A normative study of a random sample of healthy adults. *Acta Neurologica Scandinavica, 88,* 70–74.

Measuring intelligence with the culture fair tests. Manual for scales 2 and 3. (1973). Champaign, IL: Institute for Personality and Ability Testing.

Mesulam, M. M. (1985). *Principles of behavioral neurology.* Philadelphia, PA: F. A. Davis Company.

Meyer, O. (1990). Ein- und doppelseitige homonyme Hemianopsie mit Orientierungsstorungen. *Monatsschrift für Psychiatrie und Neurologie, 8,* 440–456.

Meyers, J., & Meyers, K. (1995). *Rey complex figure test and recognition trial.* Odessa, FL: Psychological Assessment Resources.

Milner, B. (1971). Interhemispheric differences in the localization of psychological processes in man. *British Medical Bulletin, 27,* 272–277.

Mittenberg, W., Seidenberg, M., O'Leary, D. S., & DiGiulio, D. V. (1989). Changes in cerebral functioning associated with normal aging. *Journal of Clinical and Experimental Neuropsychology, 11,* 918–932.

Mooney, C. M. (1957). Age in the development of closure ability in children. *Canadian Journal of Psychology, 2,* 219–226.

Mooney, C. M., & Ferguson, G. A. (1951). A new closure test. *Canadian Journal of Psychology, 5,* 129–133.

Morris, J., Roth, E., & Davidoff, G. (1986). Mild closed head injury and cognitive deficits in spinal-cord-injured patients: Incidence and impact. *Journal of Head Trauma Rehabilitation, 1,* 32–42.

Moses, J. A. (1986). Factor structure of Benton's tests of visual retention, visual construction, and visual form discrimination. *Archives of Clinical Neuropsychology, 1,* 147–156.

Moses, J. A. (1989). Replicated factor structure of Benton's tests of visual retention, visual construction, and visual form discrimination. *International Journal of Clinical Neuropsychology, 11,* 30–37.

Munk, H. (1878). Weitere Mittheilungen zur Physiologie de Brosshirnrinde. *Archiv für Anatome und Physiologie, 2,* 161–178.

Neistadt, M. E. (1993). The relationship between constructional and meal preparation skills. *Archives of Physical Medicine and Rehabilitation, 74,* 144–148.

Netherton, S., Elias, J., & Albrecht, N. (1989). Changes in the performance of parkinsonian patients and normal aged on the Benton Visual Retention Test. *Experimental Aging Research, 15,* 13–18.

Newcombe, F. (1969). *Missile wounds of the brain.* London: Oxford University Press.

Newcombe, N. (1982). Sex-related differences in spatial ability: Problems and gaps in current approaches. In M. Potegal (Ed.), *Spatial abilities: Development and physiological foundations* (pp. 223–250). New York: Academic Press.

Newcombe, F., & Ratcliff, G. (1990). Disorders of visuospatial analysis. In F. Boller & J. Grafman (Ed.), *Handbook of neuropsychology* (Vol. 2, pp. 333–356). Amsterdam: Elsevier Science Publishers.

Newcombe, F., & Russell, W. R. (1969). Dissociated visual perceptual and spatial deficits in focal lesions of the right hemisphere. *Journal of Neurology, Neurosurgery and Psychiatry, 32,* 73–81.

Orme, J. E. (1966). Hypothetically true norms for the Progressive Matrices Test. *Human Development, 9,* 222–230.

Ornstein, R., Johnstone, J., Herron, J., & Swencionis, C. (1980). Differential right-hemisphere engagement in visuospatial tasks. *Neuropsychologia, 18,* 49–64.

Osterrieth, P. A. (1944). Le test de copie d'une figure complexe. *Archives de Psychologie, 30,* 206–356.

Patterson, A., & Zangwill, O. L. (1944). Disorders of visual space perception associated with lesions of the right cerebral hemisphere. *Brain, 67,* 331–358.

Paul, S. M. (1985). The advanced progressive matrices: Normative data for an American university population and an examination of the relationship with Spearman's g. *Journal of Experimental Education, 54,* 95–100.

Peters, A. (1896). Ueber die Beziehungen zwischen Orientierungsstoerungen un ein- und doppelseitige Hemianopsie. *Archiv für Augenheilkunde, 32,* 175–187.

Petrie, A. (1952). *Personality and the frontal lobes.* London: Routledge Kegan Paul.

Piercy, M., Hecaen, H., & Ajuriaguerra, J. (1960). Constructional apraxia associated with cerebral lesions: Left and right cases compared. *Brain, 83,* 225–242.

Piercy, M. F., & Smythe, V. (1962). Right hemisphere dominance for certain nonverbal intellectual skills. *Brain, 85,* 775–790.

Porteus, S. D. (1959). *The maze test and clinical psychology.* Palo Alto, CA: Pacific Books.

Porteus, S. D. (1965). *Porteus maze test. Fifty years' application.* Palo Alto, CA: Pacific Books.

Posner, M. I. (1980). Orienting of attention. *Quarterly Journal of Experimental Psychology, 32,* 3–25.

Quaglino, A., & Borelli, G. (1867). Emiplegia sinistra con amaurosi; guaragione; perdita totale della percezione dei colori e della memoria della configurazione degli oggetti. *Giornale Italiano di Oftalmologia, 10,* 106–117.

Rapport, L. J., Webster, J. S., & Dutra, R. L. (1994). Digit span performance and unilateral neglect. *Neuropsychologia, 32,* 517–525.

Rapport, L. J., Webster, J. S., Flemming, K. L., Lindberg, J. W., Godlewski, M. C., Brees, J. E., & Abadee, P. S. (1993). Predictors of falls among right-hemisphere stroke patients in the rehabilitation setting. *Archives of Physical Medicine and Rehabilitation, 74,* 621–626.

Ratcliff, G. (1979). Spatial thought, mental rotation and the right cerebral hemisphere. *Neuropsychologia, 17,* 49–54.

Ratcliff, G. (1982). Disturbances of spatial orientation associated with cerebral lesions. In M. Potegal (Ed.), *Spatial abilities* (pp. 301–311). New York: Academic Press.

Rathbun, J., & Smith, A. (1982). Comment on the validity of Boyd's validation study of the Hooper Visual Organization Test. *Journal of Consulting and Clinical Psychology, 50,* 281–283.

Raven, J. C. (1938). *Progressive matrices: A perceptual test of intelligence: Individual form.* London: H. K. Lewis.

Raven, J. C. (1965). *Advanced progress matrices sets I and II.* London: H. K. Lewis.

Raven, J. C., Court, J. H., & Raven, J. (1986). *Manual for Raven's Progressive Matrices and vocabulary scales: Section 1, General Overview.* London: H. K. Lewis.

Reitan, R. M., & Tarshes, E. L. (1959). Differential effects of the lateralized brain lesions on the Trail Making Test. *Journal of Nervous and Mental Disease, 129,* 257–262.

Reitan, R. M., & Wolfson, D. (1993). *The Halstead–Reitan Neuropsychological Test Battery.* Tucson, AZ: Neuropsychology Press.

Rey, A. (1941). L'examen psychologique dans les cas d'encephalopathie traumatique. *Archives de Psychologies, 28,* 112, 286–340.

Riccio, C. A., & Hynd, G. W. (1992). Validity of Benton's Judgement of Line Orientation test. *Journal of Psychoeducational Assessment, 10,* 210–218.

Richards, J. S., Brown, L., Hagglund, K., Bua, G., & Reeder, K. (1988). Spinal cord injury and concomitant traumatic brain injury: Results of a longitudinal investigation. *American Journal of Physical Medicine and Rehabilitation 67,* 211–216.

Robertson, L. C. (1986). From Gestalt to neo-Gestalt. In T. J. Knapp and L. C. Robertson (Eds.), *Approaches to cognition: Contrasts and controversies* (pp. 159–188). Hillsdale, NJ: Erlbaum.

Robinson-Whelen, S. (1992). Benton Visual Retention Test performance among normal and demented older adults. *Neuropsychology, 6,* 261–269.

Ruch, F. L., & Ruch, M. (1963). *Employee aptitude survey (EAS).* San Diego, CA: Educational and Industrial Testing Service.

Ruff, R. M., Hersch, N. A., & Pribram, K. H. (1981). Auditory spatial deficits in the personal and extrapersonal frames of reference due to cortical lesions. *Neuropsychologia, 19,* 435–443.

Russell, E. W. (1972). WAIS factor analysis with brain-damaged subjects using criterion measures. *Journal of Consulting and Clinical Psychology, 39,* 133–139.

Russo, M., & Vignolo, L. A. (1967). Visual figure-ground discrimination in patients with unilateral cerebral disease. *Cortex, 3,* 118–127.

Sacks, O. (1987). *The man who mistook his wife for a hat and other clinical tales.* New York: HarperCollins Publishers.

Salkind, N. J. (1978). Development of norms for the matching familiar figures test. *JSAS Catalog of Selected Documents in Psychology, 8*(61), No. 1718.

Sanchez Longo, L. P., & Forster, F. M. (1958). Clinical significance of impairment of sound localization. *Neurology, 8,* 119–125.

Sanchez Longo, L. P., Forster, F. M., & Auth, T. L. (1957). A clinical test for sound localization and its applications. *Neurology, 7,* 655–663.

Sanders, B., Wilson, J. R., & Vandenberg, S. G. (1982). Handedness and spatial ability. *Cortex, 18,* 79–89.

Sattler, J. M. (1990). *Assessment of children* (3rd ed.). San Diego: Jerome M. Sattler.

Saunders, D. R. (1960). A factor analysis of the picture completion items of the WAIS. *Journal of Clinical Psychology, 16,* 146–149.

Schaie, K. W. (1985). *Schaie-Thurstone adult mental abilities test manual.* Palo Alto, CA: Consulting Psychologists Press, Inc.

Sea, M. C., Henderson, A., & Cermak, S. A. (1993). Patterns of visual spatial inattention and their functional significance in stroke patients. *Archives of Physical Medicine and Rehabilitation, 74,* 355–360.

Seidel, W. T. (1994). Applicability of the Hooper Visual Organization Test to pediatric populations: Preliminary findings. *The Clinical Neuropsychologist, 8,* 59–68.

Semmes, J. (1965). A non-tactual factor in stereognosis. *Neuropsychologia, 3,* 295–315.

Semmes, J. (1968). Hemispheric specialization: A possible clue to mechanism. *Neuropsychologia, 6,* 11–26.

Sergent, J. (1990). The neuropsychology of visual image generation: Data, method, and theory. *Brain and Cognition, 13,* 98–129.

Shankweiler, D. P. (1961). Performance of brain-damaged patients on two tests of sound localization. *Journal of Comparative and Physiological Psychology, 54,* 375–381.

Shepard, R. N., & Metzler, J. (1971). Mental rotation of three-dimensional objects. *Science, 171,* 701–703.

Sivan, A. B. (1992). *Benton Visual Retention Test manual* (5th ed.). New York: The Psychological Corporation.

Ska, B., Poissant, A. & Joanette, Y. (1990). Line orientation judgment in normal elderly and subjects with dementia of Alzheimer's type. *Journal of Clinical and Experimental Neuropsychology, 12,* 695–702.

Smith, A., & Lincoln, N. B. (1989). The relation between perceptual and language deficits in stroke patients. *British Journal of Occupational Therapy, 52,* 8–10.

Spreen, O., & Benton, A. L. (1969). *Embedded figures test.* Victoria, BC: Neuropsychology Laboratory, University of Victoria.

Spreen, O., & Gaddes, W. H. (1969). Developmental norms for 15 neuropsychological tests age 6 to 15. *Cortex, 5,* 171–191.

Spreen, O., & Strauss, E. (1991). *A compendium of neuropsychological tests: Administration, norms, and commentary.* New York: Oxford University Press.

Strauss, E., & Verity, L. (1983). Effects of hemispherectomy in infantile hemiplegics. *Brain and Language, 20,* 1–11.

Street, R. F. (1931) *A Gestalt completion test.* Contributions to Education, No. 481. New York: Bureau of Publication, Teachers College, Columbia University.

Strub, R. L., & Black, F. W. (1988). *Neurobehavioral disorders: A clinical approach.* Philadelphia: F. A. Davis Company.

Stuss, D. T., Stethem, L. L., & Poirier, C. A. (1987). Comparison of three tests of attention and rapid information processing across six age groups. *The Clinical Neuropsychologist, 1,* 139–152.

Sunderland, T., Hill, J. L., Mellow, A. M., Lawlor, B. A., Gundersheimer, J., Newhouse, P. A., & Grafman, J. H. (1989). Clock drawing in Alzheimer's disease: A novel measure of demential severity. *Journal of the American Geriatric Association 37,* 725–729.

Talland, G. (1965). *Deranged memory.* New York: Academic Press.

Taylor, L. B. (1969). Localization of cerebral lesions by psychological testing. *Clinical Neurosurgery, 16,* 269–287.

Teuber, H. L., Battersby, W. S., & Bender, M. B. (1951). Performance of complex visual tasks after cerebral lesions. *Journal of Nervous and Mental Disease, 114,* 413–429.

Thorndike, R. L., Hagen, E. P., & Sattler, J. M. (1986). *Technical manual, Stanford–Binet Intelligence Scale: Fourth edition.* Chicago: Riverside Publishing.

Thurstone, L. L. (1938). Primary mental abilities. *Psychometric Monographs, 1.*

Thurstone, L. L. (1944). *A factorial study of perception.* Chicago: University of Chicago Press.

Thurstone, L. L., & Jeffrey, T. E. (1956). *Closure flexibility (concealed figures)*. Chicago: Industrial Relations Center, University of Chicago.

Tippet, L. J. (1992). The generation of visual images: A review of neuropsychological research and theory. *Psychological Bulletin, 112,* 415–432.

Tombaugh, T. N., & Hubley, A. M. (1991). Four studies comparing the Rey–Osterrieth and Taylor complex figures. *Journal of Clinical and Experimental Neuropsychology, 13,* 587–599.

Tow, P. M. (1955). *Personality changes following frontal leucotomy.* London: Oxford University Press.

Trojano, L., Chiacchio, L., DeLuca, G., & Grossi, D. (1994). Exploring visuospatial short-term memory defect in Alzheimer's disease. *Journal of Clinical and Experimental Neuropsychology, 16,* 911–915.

van den Broek, M. D., Bradshaw, C. M., & Szabadi, E. (1987). Performance of normal adults on the Matching Familiar Figures Test. *British Journal of Clinical Psychology, 26,* 71–72.

Vanier, M., Gauthier, L., Lambert, J., Pepin, E. P., Robillard, A., Dubouloz, C. J., Gagnon, R., & Joannette, Y. (1990). Evaluation of left visuospatial neglect: Norms and discrimination power of two tests. *Neuropsychology, 4,* 87–96.

Varney, N. R., & Sivan, A. B. (1986). Visual-spatial disability. In T. Incagnoli, G. Goldstein, & C. J. Golden (Eds.), *Clinical application of neuropsychological test batteries* (pp. 383–401). New York: Plenum Press.

Visser, R. S. H. (1973). *Manual of the complex figure test.* Amsterdam: Swets and Zeitlinger B. V.

Waber, D. P., & Holmes, J. M. (1985). Assessing children's copy productions of the Rey–Osterrieth Complex Figure. *Journal of Clinical and Experimental Neuropsychology, 7,* 264–280.

Walker, N. W. (1986). What ever happened to the norms for the Matching Familiar Figures Test? *Perceptual and Motor Skills, 63,* 1235–1242.

Walsh, K. W. (1978). *Neuropsychology: A clinical approach.* New York: Churchill Livingstone.

Wang, P. L. (1977). Visual organization ability in brain-damaged adults. *Perceptual and Motor Skills, 45,* 723–728.

Warrington, E. (1969). Constructional apraxia. In P. J. Vinken & G. W. Bruyn (Eds.), *Handbook of clinical neurology.* Amsterdam: North Holland.

Warrington, E., & James, M. (1991). *The visual object and space perception battery manual.* Bury St. Edmunds, UK: Thames Valley Test Company.

Warrington, E. K., James, M., & Maciejewski, C. (1986). The WAIS as a lateralizing and localizing diagnostic instrument: A study of 656 patients with unilateral cerebral lesions. *Neuropsychologia, 24,* 223–239.

Warrington, E. K., & Rabin, P. (1970). Perceptual matching in patients with cerebral lesions. *Neuropsychologia, 8,* 475–487.

Wasserstein, J., Zappulla, R., Rosen, J., Gerstman, L., & Rock, D. (1987). In search of closure: Subjective contour illusions, gestalt completion tasks, and implications. *Brain and Cognition, 6,* 1–14.

Wechsler, D. (1974). *Wechsler Intelligence Scale for Children-Revised.* New York: Psychological Corporation.

Wechsler, D. (1981). *WAIS-R manual.* New York: Psychological Corporation.

Wechsler, D. (1987). *Wechsler Memory Scale-Revised manual.* New York: Psychological Corporation.

Wedding, D. (1979). A comparison of statistical, actuarial, and clinical models used in predicting presence, lateralization, and type of brain damage in humans. Unpublished doctoral dissertation, University of Hawaii.

Weinberg, J., Diller, L., Gerstmann, L., & Schulman, P. (1972). Digit span in right and left hemiplegics. *Journal of Clinical Psychology, 28,* 361.

Weinberg, J., Diller, L., Gordon, W. A., Gerstman, L. J., Lieberman, A., Lakin, P., Hodges, G., & Ezrachi, O. (1977). Visual scanning training effect on reading related tasks in acquired right brain damage. *Archives of Physical Medicine and Rehabilitation, 58,* 479–482.

Weinberg, J., Diller, L., Gordon, W. A., Gerstman, L. J., Lieberman, A., Lakin, P., Hodges, G., & Ezrachi, O. (1979). Training sensory awareness and spatial organization in people with right brain damage. *Archives of Physical Medicine and Rehabilitation, 60,* 491–494.

Weinberg, J., Piasetsky, E., Diller, L., & Gordon, W. (1981). Treating perceptual organization deficits in nonneglecting RBD stroke patients. *Journal of Clinical Neuropsychology, 4,* 59–76.

Weintraub, S., & Mesulam, M. M. (1985). Mental state assessment of young and elderly adults in behavioral neurology. In M. M. Mesulam (Ed.), *Principles of behavioral neurology* (pp. 71–123). Philadelphia: F. A. Davis Co. Publishers.

Wernicke, C. (1895). Zwei Falle von Rindenlasion. *Arbeiten der Psychiatrischen Klinik in Breslau, 2,* 33–53.

Western Psychological Services. (1983). *Hooper Visual Organization Test (VOT) manual.* Los Angeles: Author.

Wetzel, L., & Murphy, S. G. (1991). Validity of the use of discontinue rule and evaluation of discriminability of the Hooper Visual Organization Test. *Neuropsychology, 5*, 119–122.

Whiting, S. E., Lincoln, N. B., Bhavnani, G., & Cockburn, J. (1985). *Perceptual assessment battery*. Windsor, UK: NFER-Nelson.

Wilbrand, H. (1887). Die Seelenblindheit als Herderscheinung und ihre Beziehungen zur homonymen Hemianopsie. Wiesbaden: Bergmann.

Wilson, B. A., Cockburn, J., & Baddeley, A. D. (1985). *The Rivermead Behavioral Memory Test*. Titchfield, UK: Thames Valley Test Co.

Wilson, B., Cockburn, J., Baddeley, A., & Hiorns, R. (1989). The development and validation of a test battery for detecting and monitoring everyday memory problems. *Journal of Clinical and Experimental Neuropsychology, 11*, 855–870.

Witelson, S. F. (1974). Hemispheric specialization for linguistic and nonlinguistic tactual perception using a dichotomous stimulation technique. *Cortex, 10*, 3–17.

Witkin, H. A., Oltman, P. K., Raskin, E., & Karp, S. A. (1971). *A manual for the Embedded Figures Test*. Palo Alto, CA: Consulting Psychologists Press.

Wittig, M., & Petersen, A. C. (1979). *Sex-related differences in cognitive functioning: Developmental issues*. New York: Academic Press.

Wolf-Klein, G. P., Silverstone, F. A., Levy, A. P., Brod, M. S., & Breur, J. (1989). Screening for Alzheimer's disease by clock drawing. *Journal of the American Geriatric Association, 37*, 730–735.

Woodward, C. A. (1982). The Hooper Visual Organization Test: A case against its use in neuropsychological assessment. *Journal of Consulting and Clinical Psychology, 50*, 286–288.

Yeudall, L. T., Reddon, J. R., Gill, D. M., & Stefanyk, W. O. (1987). Normative data for the Halstead-Reitan neuropsychological tests stratified by age and sex. *Journal of Clinical Psychology, 43*, 346–367.

Young, A. W. (1983). The development of right hemisphere abilities. In A. W. Young (Ed.), *Functions of the right cerebral hemisphere* (pp. 147–169). New York: Academic Press.

Young, G. C., Collins, D., & Hren, M. (1983). Effect of pairing scanning training with block design training in the remediation of perceptual problems in left hemiplegics. *Journal of Clinical Psychology, 5*, 201–212.

Young, A. W., & Ratcliff, G. (1983). Visuospatial abilities of the right hemisphere. In A. W. Young (Ed.), *Functions of the right cerebral hemisphere* (pp. 1–32). New York: Academic Press.

Zimet, C. N., & Fishman, D. B. (1970). Psychological deficit in schizophrenia and brain damage. *Annual Review of Psychology, 21*, 113–154.

19

Neuropsychological Assessment of Motor Skills

Kathleen Y. Haaland and Deborah L. Harrington

Introduction

In this chapter we hope to provide the reader with a greater appreciation for the many different cognitive abilities that support motor performance, even on tasks that superficially appear very simple. This, in turn, has important implications for how the clinician interprets patterns of spared and impaired motor skills. The premise that motor skills are supported by many different cognitive abilities is consistent with the neuroanatomy of the motor system which is composed of a large number of cortical and subcortical areas (Alexander, DeLong, & Strick, 1986; Allen & Tsukahara, 1974; Strick, 1988) that interact to control different movements. Movements such as catching a ball, writing, or dancing clearly differ in their cognitive requirements. It should not be surprising, therefore, that a wide variety of primary and association areas in the cerebral cortex as well as subcortical structures are involved in planning, selecting, initiating, executing, and self-monitoring these movements. The strong interrelationships between subcortical and cortical areas is beautifully illustrated by the various basal ganglia-cortical circuits identified by neuroanatomists in animals (Alexander *et al.*, 1986). These structural interrelationships suggest that different neural systems cooperate in the planning and execution of movements, which is substantiated by human research showing that there are similarities in cognitive deficits in patients with damage to specific basal ganglia-cortical circuits (Cummings, 1993). Equally important is the fact that patterns of spared and impaired cognitive motor functions differ among individuals with basal ganglia disease, cerebellar damage, or damage to the cerebral cortex (Haaland & Harrington, 1990; Keele and Ivry, 1990).

Kathleen Y. Haaland   Research and Psychology Services, Albuquerque VA Medical Center, and Departments of Psychiatry, Neurology, and Psychology, University of New Mexico, Albuquerque, New Mexico 87108.   Deborah L. Harrington   Research Service, Albuquerque VA Medical Center, and Departments of Psychology and Neurology, University of New Mexico, Albuquerque, New Mexico 87108.

Neuropsychology, edited by Goldstein *et al.* Plenum Press, New York, 1998.

KATHLEEN Y.
HAALAND AND
DEBORAH L.
HARRINGTON

In both clinical neuropsychology and clinical neurology, motor skills have been used primarily as indicators of inter- and intrahemispheric location of brain damage. Using motor tasks as indicators of laterality of damage relies upon the well-known finding that damage to the primary motor pathway e.g., motor cortex and its efferent pathways) produces contralateral deficits (Brodal, 1981). While this inference is true for simple motor tasks that have minimal cognitive requirements, it is not true for more complex motor tasks that have greater cognitive demands. This point cannot be overstated because clinicians tend to ignore the significant cognitive requirements of motor tasks and often treat them as unidimensional. This can result in the misleading and often incorrect conclusions that motor deficits reflect direct damage to the primary motor system. Even the motor tasks used as a standard part of the neurologic examination can be reflective of damage to different parts of the motor system. The motor tasks used by neuropsychologists are usually even more difficult to interpret because they are influenced by a variety of cognitive and noncognitive factors (e.g., muscle groups tested, speed and precision requirements, familiarity of the task, practice and fatigue effects, sequencing, simultaneous processing, and strategic requirements) which must be considered when using them for clinical assessment. This is illustrated by comparing the incidence of strictly contralateral and bilateral deficits after unilateral hemispheric damage on motor tasks which vary in their cognitive requirements (Haaland & Delaney, 1981). Strictly contralateral deficits are reflective of damage to the primary motor system and are seen with simple motor tasks, such as finger tapping and grip strength, while bilateral deficits are reflective of damage outside of the primary motor system and are seen with more complex tasks, such as the grooved pegboard or maze coordination tasks.

One focus of this chapter is to emphasize that the multidimensional nature of motor tasks influences intrahemispheric, interhemispheric, and *subcortical* control of movement. Differences in intrahemispheric motor control will be discussed in detail but are illustrated by metabolic data in humans which show that the supplementary motor area (i.e., medial premotor area) is active during the performance of finger sequences but not single finger movements (Roland, Larsen, Lassen & Skinhof, 1980; Rao *et al.*, 1993). As for interhemispheric control, the incidence of motor deficits ipsilateral to left hemisphere damage (LHD) is much greater than after right hemisphere damage (RHD) which is consistent with the view that the left hemisphere regulates the cognitive aspects of a wide variety of motor functions (Haaland & Harrington, 1996a, 1996b). Limb apraxia is the most common example of this dominance (Faglioni & Basso, 1985). Similarly, the basal ganglia and cerebellum are two subcortical structures, both of which have been associated with higher levels of cognitive-motor function than is commonly acknowledged in clinical assessments.

Another focus of this chapter that relates to interhemispheric control of motor function is to discuss the functional and neuroanatomical implications of unilateral and bilateral impairments. Unilateral hemispheric damage can produce deficits strictly in the contralateral limb or in both limbs. Bilateral motor deficits can be seen after damage to either hemisphere implying that both hemispheres are critical to perform the task. Alternatively, damage to the right or left hemisphere alone suggests one hemisphere is dominant for controlling the cognitive aspects of that particular motor task (for reviews see Haaland & Harrington, 1996a, 1996b). The possibility of deficits in the ipsilateral limb must be considered by the neuropsychologist when interpreting results from motor testing. Deficits in both limbs do not necessarily imply bilateral hemispheric damage. Like other tests used in the neuropsychological evaluation, motor tasks must not be treated as unidimensional and dependent only upon the primary motor system.

This chapter begins with a discussion of the motor tasks typically used by neuropsychologists. We describe their administration, procedures, reliability, and validity as well as the impact of demographic factors on normative performance. The objective of this section is to elucidate the many different cognitive abilities that potentially affect performance of each of these tests. Finally, patterns of laterality will be discussed with emphasis on the cognitive factors that influence laterality.

MOTOR TESTS COMMONLY USED BY NEUROPSYCHOLOGISTS

The clinical neurologist relies upon very simple tests of strength as well as walking, rapid alternating movements, or finger-to-nose testing. These tasks are thought to be more reflective of cerebellar functioning. Some neurologists will compare rapid thumb-index finger appositions to sequential appositions of the thumb to each finger to assess involvement of primary motor versus supplementary motor cortex (Roland *et al.*, 1980).

The clinical neuropsychological evaluation of motor skills has also focused on simple motor skills. For example, the Halstead neuropsychological battery initially used the finger oscillation or finger tapping test (Halstead, 1947), and the Halstead–Reitan battery uses grip strength and finger tapping (Boll, 1981). The Klove Matthews Motor Steadiness Battery or Wisconsin Motor Battery (cf. Reitan & Davison, 1974) introduced more complex motor tasks to neuropsychological assessment. Most of these tasks were adaptations and some were similar to tests used to evaluate the military (Fleishman, 1958). This battery included grip strength, finger tapping, maze coordination, resting and static steadiness, horizontal and vertical groove steadiness, and the grooved pegboard (cf. Reitan & Davison, 1974). Luria's methods have been an exception because there is an attempt to separate qualitative differences in complex motor skills due to cognitive deficits from primary motor or sensory deficits. Luria's focus (Luria, 1966) was on the influence of frontal (e.g., primary motor, premotor, and prefrontal) and postcentral areas in producing deficits in rapid alternation of hand movements (e.g., fist/palm alternations with opposing hands), response conflict and inhibition (e.g., "when I knock once you knock twice, and when I knock twice you knock once"), single and repetitive hand sequences (e.g., knock, chop, slap with each hand separately), and limb and buccofacial apraxia. He was very aware that he was using motor performance as a measure of higher-level cognitive processes which influence motor and nonmotor skills alike. The Luria–Nebraska Neuropsychological Battery (Golden, Hammeke, & Purisch, 1980) has quantified these procedures, and all of the items listed above have been put into a motor subscale. This labeling is unfortunate, however, because deficits on this scale are more reflective of cognitive deficits than strictly primary motor deficits. This interpretation is supported because deficits on many of the tests in the motor subscale are associated with deficits on nonmotor testing (Luria, 1966).

KATHLEEN Y.
HAALAND AND
DEBORAH L.
HARRINGTON

This motor test (Lezak, 1995; Spreen & Strauss, 1991) has been used by neuropsychologists most consistently. It was initially part of Halstead's evaluation (Halstead, 1947), was used in the Halstead-Reitan Neuropsychological Test Battery, and is used by many neuropsychologists who eschew the battery approach but incorporate it into their flexible, hypothesis testing approach. A similar, but less quantitative approach, was used by Luria (1966) and is currently used by neurologists by asking patients to rapidly touch thumb to index finger for a 10-second period. The finger tapping test uses a tapping key and requires the patient to move his or her index finger as rapidly as possible over five 10-second trials. The means across five trials for each hand are used. There are a number of variations in administration including the number of trials used, the requirement that additional trials be used if the first five trials are not within five taps of one another, the use of a mandatory rest after three trials, and the testing order of the two hands (i.e., hands alternated across ten trials or one hand tested on five consecutive trials followed by the other hand). These variations may influence performance, hence it is important to ensure that the procedures used by the clinician are the same as those used to derive the normative data. Although on the surface this task appears to have few cognitive requirements, this view can be misleading because the control of speed, rapid alternation of index finger flexion and extension, and consistency of responding over trials can engage several different cognitive abilities.

Table 1 shows that the finger tapping test has variable test–retest reliability. High test–retest reliability has been reported for 10 trials of finger tapping that were administered weekly for 10 weeks (Gill, Reddon, Stefanyk, & Hans, 1986) with more variability using standard procedures. With the exception of one study in which the correlation was low ($r = .24$) (Matarazzo, Wiens, Matarazzo & Goldstein, 1974), reliability is good in stable epileptics ($r = .77$) (Dodrill & Troupin, 1975) and although our study (Haaland, Temkin, Randahl, & Dikmen, 1994) showed statistically significant increases in finger tapping at an 11-month interval, these differences were not clinically significant (i.e., about two taps per 10 seconds).

Finger tapping rates are either not significantly influenced by level of *education* or are minimally influenced (Bornstein, 1985; Finlayson, Johnson, & Reitan, 1977; Heaton, Grant & Adams, 1986; Heaton, Grant, & Matthews, 1991). This is illustrated in Table 1 by comparing the performance of the 20–34-year-old female group with 12 or 18+ years of education, which is similar between the two groups.

TABLE 1. THE INFLUENCE OF SEX, AGE, AND EDUCATION ON MOTOR PERFORMANCE[a]

	Education (years)	Tapping	Grip strength	Grooved pegboard
20–34 Females	12	55	61	41
20–34 Males	12	44	38	47
70–74 Males	12	64	70	61
20–34 Females	18+	51	38	36
Test–retest reliability[b]		0.24–0.94	0.91–0.94	—
Raw score		49.1–51.9	40.0–46.4	63–66

[a]This table is based upon the values provided in the Heaton, Grant, and Matthews (1991) normative data. Values in table are *T*-scores (Mean = 50, SD = 10) for each group based on the specified raw scores for the preferred hand. The pattern of results is comparable for the nonpreferred hand.
[b]Test–retest reliabilities based on neurologically intact individuals.

TABLE 2. PERCENT OF VARIANCE ACCOUNTED FOR BY DEMOGRAPHIC FACTORS[a]

	Combined	Age	Education	Sex
Finger tapping	32–34	9	6	19–20
Grip strength	62–63	2–4	1	55–58
Grooved pegboard	42–47	39–40	13–17	3–4

[a]These figures are adapted from Heaton, Grant, & Matthews (1991) and are based upon stepwise regression analysis with ranges denoting the differences for the preferred and nonpreferred hand.

In fact, across five large studies (Leckliter & Matarazzo, 1989) education accounted for less than 2% of the variance in finger tapping.

Age accounts for about 9% of the variance in finger tapping in one study (Heaton *et al.*, 1991) and about 33% in another (Bornstein, 1985). The reasons for these discrepancies are not clear because the characteristics of the two samples were similar, but most importantly, even in the Heaton *et al.* study age is clinically significant as can be seen in Table 1 when comparing the 20–34-year-old males with the 70–74-year-old males. No significant aging effects have been reported between the ages of 15 to 40 (Yeudall, Reddon, Gill, & Stefanyk, 1987).

Sex differences are most influential, accounting for as much as 20% of the variance in finger tapping (Heaton *et al.*, 1991). Dodrill (1979) has shown that the relationship with sex is associated with hand span such that faster rates are seen with larger hand size. This effect is illustrated in Table 1 by comparing the young males and females with 12 years of education. Table 2 illustrates the results of a stepwise multiple regression analysis assessing the influence of age, education, and gender on three motor tasks in a group of normal individuals ranging in age from 18 to 69. All three variables in combination accounted for about 33% of the variability in tapping rates, with gender accounting for the most unique variance, followed by age and then education (Heaton *et al.*, 1991).

Before the advent of normative tables which control for education, age, and gender, *cutoff scores* were used. Impairment was defined as less than 50 taps per 10 seconds in the preferred hand (Reitan, 1969) and less than 44 taps per 10 seconds in the nonpreferred hand (Russell, Neuringer, & Goldstein, 1970). However, the samples on which these cutoffs were based are not broadly representative of the general population. For instance, Halstead's sample consisted of 29 young individuals (mean age of 28.3) some of whom had minor psychiatric diagnoses (see Boll, 1981). Moreover, cutoff scores do not take critical demographic factors into account and frequently do not account for base rates of impairment in the non-brain-damaged population. For example, in a normal sample ranging in age from 18 to 69, the Reitan cutoff score classified between 70% and 80% as impaired (Bornstein, 1986b).

While absolute performance level is useful, *intermanual* differences in finger tapping rates have been used as indicators of laterality of lesion, especially in individual patients. Caution should be exercised when using difference (i.e., preferred–nonpreferred score) or ratio (i.e., nonpreferred/preferred) scores because of their variability (Bornstein, 1986a, 1986c). This problem underscores the importance of using a variety of measures and identifying consistent patterns of deficits and abilities. Importantly, most neuropsychologists do not use a single measure of laterality because clinical interpretations are improved significantly by examining the pattern of performance, and isolated deficits can occur spuriously (Lezak, 1995).

Initially, clinicians used "rules of thumb" to determine impairment. For example, some would interpret a finding of better performance in the nonpreferred hand as indicative of impairment, which would misclassify many individuals because this pattern of findings occurs in 23% to 34% of normal individuals (Bornstein, 1986a). Another approach assumes that greater than a 10% difference in either hand is reflective of pathology (Boll, 1981). However, in the same normative sample mentioned above, the mean difference between the hands was 6% to 8% so that 25% to 30% of the normative sample would be classified as impaired if this approach was adopted. This appraisal suggests that statements about laterality of lesion in individual patients should not be based on "rules of thumb." A more reasonable approach would be to require consistency of asymmetry across finger tapping and grip strength and an asymmetry of at least one standard deviation based on comprehensive normative data (Heaton *et al.*, 1991). A separate section will focus on validity issues, especially the relationship of finger tapping asymmetries to laterality of lesion.

GRIP STRENGTH

This task was not used by Halstead (1947), but it is part of the Halstead–Reitan battery. It is measured using a Stoelting/Smedley hand dynamometer beginning with the preferred hand and alternating between the two hands for two trials each. Like finger tapping, the exact procedures may vary, including the dependent measures used (e.g., the mean across two trials with each hand, the maximum score across two trials with each hand) and the intertrial variability allowed (e.g., no variability restrictions or the two trials must be within 5 kilograms of one another).

Table 1 shows that the test–retest reliability for grip strength is very good using data collected from 10 weekly trials for 10 weeks (Reddon, Stefanyk, Gill, & Renney, 1985). Although Matarazzo *et al.* (1974) demonstrated poorer test–retest reliability ($r = 0.53$ averaged across the two hands) in normals using two trials per hand across a 20-week interval, which is more comparable to the clinical assessment method, another study (Haaland *et al.*, 1994) found that grip strength did not change significantly·in a normal control group over an 11-month interval using the standard clinical procedure. Test–retest reliability over a 6-month period is also high ($r = 0.96$) for clinical populations in which no deterioration or improvement is expected, such as in stable epileptics (Dodrill & Troupin, 1975), but head-injured patients showed clinically significant improvement from 1 to 11 months postinjury (Haaland *et al.*, 1994).

While correlations of grip strength with *education* and *age* are statistically significant, they account for a small amount of the variance and are of minimal clinical significance. Across three studies, education accounted for an average of 2% of the variance in grip strength (Leckliter & Matarrazzo, 1989). The effect of age was more variable, ranging from 4% (Heaton *et al.*, 1991) to 18% (Bornstein, 1985) in similar samples. Table 1 illustrates the clinical significance of age even in the Heaton *et al.* sample.

The influence of *sex* is clear. Women performed more poorly than men by 18 or 19 kilograms depending on the hand (Bornstein, 1985). These differences vary somewhat with age even within individuals who are 15 to 40 years old (Yeudall *et al.*, 1987). Table 1 illustrates this issue. Grip strength in the 40–46.4 kg range results in a *T*-score of 61 for females and 38 for males. Table 2 shows that sex accounts for the bulk of the variance in grip strength with only a small amount of additional variance associated with age and no unique variance attributed to education (Heaton *et al.*, 1991).

Cutoff scores of impairment have not been used for grip strength, but given the influence of age and sex, appropriate normative data must be used (Heaton *et al.*, 1991). *Intermanual differences* were less affected by age and sex (Bornstein, 1986a, 1986c). Age did not relate to intermanual difference, but education was modestly related ($r = 0.129$, $p < .025$). Males demonstrated a greater difference between hands than females (mean = 3.3, SD = 5.4 for males; mean = 2.3, SD = 2.9 for females), but this is clearly not clinically significant. Therefore, although absolute level of performance is significantly related to age and sex, these variables are not clinically important when using intermanual differences. Furthermore, the percentage of normal individuals demonstrating better performance in the non-preferred hand is about 24%, which suggests caution should be used in interpreting intermanual differences as clinically significant (Bornstein, 1986a, 1986c). In a normal sample the mean intermanual ratio is 0.92 (SD = 0.12) which reflects higher strength in the preferred hand, but the significant variability in ratio scores is problematic from a clinical standpoint.

GROOVED PEGBOARD TEST

This test was initially incorporated into the clinical neuropsychological evaluation as part of the Wisconsin Motor Battery (Matthews & Klove, 1964; Reitan & Davison, 1974). While the grooved pegboard test is widely used there is not as much data available as there is for finger tapping and grip strength. This task is considered an index of fine finger dexterity. The Purdue Pegboard Task is another test of fine finger dexterity, but because it has not been used as extensively by neuropsychologists it will not be reviewed in this chapter (for a review see Spreen & Strauss, 1991). The grooved pegboard task requires the patient to place 25 identical notched pegs into a 5×5 matrix of notched holes. The task is done first with the preferred hand and then with the nonpreferred hand. The subject is asked to go from left to right with the right hand and right to left with the left hand, progressing from top to bottom. The time to place all 25 pegs with each hand plus the number of drops are the dependent measures, but normative data are available only on speed. This test requires reasonable visual acuity and perceptual skills to identify the direction of the notch in the hole, planning to achieve maximum speed by twisting the peg as the arm is moved from the peg bin to the hole on the board, independent finger movement in order to twist the peg with one hand, tactual or visual cues to recognize the direction in which the peg is twisted, and sensory (visual and/or tactual) motor integration in order to accurately hit the hole and modify the response if inaccurate. Unlike finger tapping and grip strength, this measure is highly dependent on sensory motor integration and planning. No test–retest reliability data are available.

Normative data are available on grooved pegboard performance (Heaton *et al.*, 1991). It is modestly associated with education such that faster speeds are associated with higher education (Bornstein, 1985; Heaton *et al.*, 1991); this accounts for 3% to 17% of the variance. Table 1 shows that raw scores in the same range result in *T*-scores of 41 for the females with 12 years of education and 36 for those with more than 18 years of education. Performance deteriorates much more significantly with *age* which accounts for as much as 40% of the variance (Heaton *et al.*, 1991). Table 1 shows that for the same raw score, T-scores of 47 and 61 are received by the young and the elderly groups, respectively. However, this effect is also somewhat inconsistent in that in a similar sample (Bornstein, 1985) age accounts for only 15% of the variance. Performance is also affected by *sex* with poorer performance in males than females, but these differences are small (see Table 1). In contrast to tapping and grip strength, age and education, but not sex, are the most important demographic factors associated with grooved pegboard performance.

Previous work has designated *cutoff scores* of greater than 67 seconds as impaired (Rennick, 1978, cited by Bornstein, 1986b). However, if this cutoff criterion is used, between 40% and 62% of a normal population would be classified as impaired (Bornstein, 1986b). Therefore, similar to finger tapping and grip strength, cutoff scores are highly misleading, and appropriate normative tables should be used to control for the base rate of intermanual differences in the population together with demographic factors (Heaton *et al.*, 1991).

Intermanual differences on the grooved pegboard test are not associated with age or years of education (Bornstein, 1986a, 1986c). Similar to tapping and grip strength, more than 20% of the population would be misclassified as impaired if those with the same or better performance with the nonpreferred hand were considered impaired. In a normative sample (Bornstein, 1986a) the mean intermanual ratio of time to place the 25 pegs is 1.10 (SD = 0.13) reflecting that the nonpreferred hand is typically slower than the preferred hand, but variability is significant and makes clinical decisions regarding impairment more difficult.

WISCONSIN MOTOR BATTERY

This battery is composed of tests of static and resting steadiness, vertical and horizontal groove steadiness, and maze coordination. It was developed empirically by Klove and Matthews when they were asked to examine patients before and after chemopallidectomy, which explains the focus on resting and moving steadiness measures (Matthews, 1994, personal communication). Similar tests were also used to examine Air Force personnel (Fleishman, 1958). In static and resting steadiness the subject places an electric stylus for 15 seconds in each of nine holes ranging in diameter from 3 to 12 mm. Total number of side contacts is recorded for each hand, and testing of the preferred hand is followed by the nonpreferred hand. The difference between the static and resting condition is that the arm is not supported in the static condition. Vertical and horizontal groove steadiness require the subject to run a metal stylus up and down or across a vertical or horizontal groove (0.5 cm wide and 25 cm long) for two trials with each hand. The number of times the stylus touches the side of the groove and the duration of the contacts are automatically recorded for the preferred and nonpreferred hands. The task is done at the subject's pace with no time constraints. Maze coordination uses a maze (Lafayette Instrument Company, #2706A) which has been modified to eliminate all blind alleys. The patient is asked to trace through a maze with a 0.8-cm trough trying not to touch the sides. The subject is given as much time as needed, and the number of contacts and duration of contacts is recorded with the preferred and nonpreferred hand.

Normative data are available for all of these tasks (Beardsley, Matthews, Cleeland, & Harley, 1978; Harley, Matthews, Leuthold, & Bergs, 1980), but test–retest reliability and the influence of age, education, and sex has not been methodically examined. Fleishman (1958) has shown that tests similar to the moving steadiness measures contributed to a single factor in a large factor analytic study of Air Force personnel.

These tasks have been used in a variety of patient groups and have been shown to be the best measures to differentiate patients with multiple sclerosis from patients with other neurologic diseases (Matthews, Cleeland, & Hooper, 1970) and to differentiate parkinsonian patients from non-brain-damaged controls (Matthews & Haaland, 1979). However, despite the similarity in clinical presentation between Parkinson's disease and grain workers who have been exposed to solvent-based fumigant pesticides there was no evidence of motor deficits on the Wisconsin Motor Battery in a sample of these grain workers who demonstrated parkinsonian symptoms (Matthews, Chapman, & Woodard, 1988). Poorer performance was seen on maze coordination and static steadiness but not on finger tapping or grooved pegboard in epileptics who were toxic from their anticonvulsant medication when compared with those who were not toxic (Matthews & Harley, 1975). In addition, XYY males who demonstrate a clinical tremor show a high incidence of deficits on static steadiness and grooved pegboard tests (Daly & Matthews, 1974).

LIMB APRAXIA

Limb apraxia is a deficit in purposeful movement which cannot be attributed to primary motor or sensory deficits, extrapyramidal motor disorders, aphasia, dementia, or lack of cooperation (Geschwind, 1965). Clinically, limb apraxia is assessed by asking the patient to perform various gestures to verbal command, to imitation, or with the object. The most common movements are transitive (e.g., brush teeth); intransitive symbolic movements which do not require an object (e.g., wave good-bye); or meaningless movements. Various forms of limb apraxia have been identified and described (Geschwind, 1965), but the most common forms are ideational apraxia and ideomotor apraxia. Ideational apraxia is identified by conceptual errors (e.g., light candle before striking match; eat with toothbrush) which suggest that the patient does not have a conceptual grasp of the movement's purpose or the appropriate use of a particular object (Ochipa, Rothi, & Heilman, 1989). Ideomotor apraxia is associated with errors in the spatiotemporal aspects of the movement (e.g., misorientation of the hand, body part-as-object errors such as when the patient uses the index finger as a toothbrush).

Limb apraxia is identified more frequently with aphasia, but it can be seen independent of aphasia suggesting the association is due to the overlap in neuroanatomical control. However, this overlap is problematic to the clinician and emphasizes the importance of ruling out aphasia as an explanation of limb apraxia. This is done by assessing aphasia, identifying the limits of a specific patient's comprehension abilities, and using a variety of nonverbal methods to produce the gesture (e.g., imitation, presentation of object). These gestural abilities are examined in the limb ipsilateral to hemispheric lesion to avoid the influence of sensory or motor deficits.

Limb apraxia is more common after left hemisphere damage (for a review see Faglioni & Basso, 1985). While early theories linked it to damage in the parietal or frontal lobes or their interconnections (see Geschwind, 1965, for review), more recent work has emphasized the role of the parietal lobe (DeRenzi, Faglioni, Lodesani, & Vecchi, 1983). One study found no evidence of localization and attributed apraxia to a complex neural network such that damage in many areas could produce limb apraxia for different reasons (Alexander, Baker, Naeser, Kaplan, &

Palumbo, 1992). Luria identified frontal apraxia which is evaluated somewhat differently and is reflective of a problem in identifying the goal of the movement. A recent case study (Schwartz, Mayer, Fitzpatrick, DeSalme, Montgomery, 1993) has demonstrated an elegant functional analysis of frontal apraxia in a head injury patient.

Limb apraxia is not assessed routinely by neuropsychologists for a variety of reasons. First, it has been considered a deficit which is seen in the context of the clinical assessment but has no functional implications. Second, there is no widely accepted limb apraxia battery, and third, most clinicians are not trained or experienced in the administration and scoring of gestural performance. The lack of a functional correlate appears to be a myth as limb apraxia affects the aphasic patient's ability to use gestural compensation, is associated with poorer outcome after stroke, and has been associated with the inappropriate use of objects in the environment (Helm-Estabrook, Fitzpatrick, & Barresi, 1982; Sundet, Arnstein, & Rienvang, 1988; Ochipa, Rothi, & Heilman, 1989). Although limb apraxia batteries are available (Rothi *et al.*, 1988; Haaland & Flaherty, 1984) these batteries have been used primarily for research purposes. Our laboratory is currently developing a limb apraxia battery which will be comprehensive, be easy to use, be validated in unilateral stroke patients, and have high inter-rater reliability.

Cognitive Influences on Motor Tests

In addition to consideration of the different task requirements of finger tapping, grip strength, and grooved pegboard, the variable influence of age, education, and sex on these tasks provides some clues regarding the differential importance of cognitive factors. Table 2 summarizes these results. First, these demographic factors account for a larger percentage of the variance for grip strength than finger tapping or grooved pegboard. While sex was most influential for tapping and grip strength, age and education were clearly more influential for pegboard performance. While a minimal relationship between motor tasks and education is considered an advantage relative to other neuropsychological tests, the greater impact of education upon grooved pegboard performance suggests that higher-level cognitive factors are more influential for this task than for finger tapping or grip strength. While the correlation with age can be attributed partially to decreasing psychomotor speed, grooved pegboard performance also may reflect decreases in strategic processing and sequencing with age. This interpretation is indirectly supported by the fact that age is not as strongly associated with finger tapping as with pegboard performance, and the finger tapping task is particularly dependent on speed. The pegboard's greater dependence on cognitive requirements is supported by the task analysis detailed above and evidence of bilateral deficits after unilateral hemispheric damage on this task, but not on the finger tapping and grip strength tasks (Haaland & Delaney, 1981). However, even though grooved pegboard performance clearly is more strongly related to education, education accounts for a significant percent of the variance in finger tapping but not grip strength. This supports the notion that finger tapping may also be influenced by cognitive factors but to a lesser degree than grooved pegboard performance. This proposal is also reasonable based on task analysis. Finger tapping requires rapid alternation of index finger flexion and extension which is dependent on speed, sequencing, and timing. Many of these requirements are considered cognitive in nature. The greater influence of cognitive factors on finger tapping than grip strength is further illustrate by the finding of residual finger tapping deficits with intact grip strength 12 months after nonfocal closed head injury (Haaland *et al.*, 1994).

The cognitive influences upon the Wisconsin Motor Battery have not been explored, but from a task analysis standpoint, issues such as strategy are clearly important because accuracy, not speed, is the dependent measure. Therefore, subjects will perform optimally if they can perform slowly to minimize side contacts. In addition, attentional factors may also be important.

The bulk of the research in limb apraxia has explored the influence of a variety of cognitive factors including language impairment, the symbolic value of the gesture, the effect of different ways of eliciting the gesture (e.g., imitation, presenting object to cue the gesture, or to use while performing the movement), and the influence of sequencing requirements (Harrington & Haaland, 1992; Haaland & Harrington, 1996a, 1996b). Recent work has examined the patient's knowledge of the object and knowledge of the gesture to differentiate different forms of conceptual apraxia (Ochipa, Rothi, & Heilman, 1992). Our current work is examining how various cognitive abilities (e.g., deficits in auditory comprehension, visual perception, immediate memory, sequencing) may influence apraxia.

LATERALITY OF MOTOR IMPAIRMENT

GENERAL CONSIDERATIONS

From the previous discussion we can see that the cognitive requirements of the motor task critically determine the pattern of intermanual differences and the relationship of those intermanual differences to the laterality of brain damage. Recall that unilateral hemispheric damage can produce strictly contralateral or bilateral deficits. When bilateral deficits occur they can be related to left or right hemisphere damage indicative of hemispheric dominance for the cognitive aspects of the task. Alternatively, they can be related to damage to either hemisphere which implies that both hemispheres are necessary to perform the task with neither hemisphere taking the lead.

It is critical to control several factors when evaluating hemispheric asymmetry of complex motor skills. First, the focus of the evaluation should be the limb ipsilateral to brain damage to minimize the influence of primary motor deficits such as hemiparesis or sensory deficits. Second, the examiner must ensure that peripheral injuries cannot account for motor deficits. This is particularly important in older patients with arthritis or after motor vehicle accidents which are associated with peripheral injuries. Third, each patient must be compared to an appropriate normative group (e.g., right handers using their right hand or right handers using their left hand) to control for hand preference effects and demographic factors. Fourth, studies of movement have focused on right handers because hemispheric dominance of complex motor skills, much like language dominance, are related to hand preference. The relationship is not as strong as that for language dominance, but nonetheless it is more common to see limb apraxia after right hemisphere damage in a left hander than in a right hander (Faglioni & Basso, 1985).

Ideally assessments and studies of motor function should administer a series of tasks which are designed to identify the critical cognitive components which produce various patterns of hemispheric dominance. Unfortunately, most studies have not attempted to isolate the cognitive factors that produce this variation in hemispheric control and thus, there is not a good framework for the clinician to use. Most have administered a single task, and comparison across studies is difficult. Even if the studies were comparable in other ways (e.g., group composition), the tasks used vary on many dimensions, making it impossible to isolate the critical mechanisms. This has led investigators to use very global statements to characterize their effects. For example, Heap and Wyke (1972) contrasted their results with previous studies and stated that "movements of increasing complexity that require high degrees of sensorimotor coordination call for greater degrees of intra- and inter-hemispheric cooperation . . . the more complex a movement the larger areas of the brain that are required for its efficient execution."

Strictly Contralateral Deficits

As can be seen in Table 3, strictly contralateral deficits are seen with tasks that have minimal cognitive requirements, such as grip strength or finger tapping. However, this statement is an oversimplification because finger tapping is not entirely independent of cognitive factors that control the timing and speed of the movement.

Bilateral Deficits after Right or Left Hemisphere Damage

Bilateral deficits after damage to the right or left hemisphere are seen with a variety of tasks. This finding suggests that both hemispheres are necessary to control these tasks and neither hemisphere plays a dominant role relative to the

TABLE 3. MOTOR DEFICITS AFTER UNILATERAL HEMISPHERIC DAMAGE

Strictly contralateral deficits
  Finger tapping (Haaland & Delaney, 1981)
  Grip strength (Haaland & Delaney, 1981)
Bilateral deficits after right or left hemisphere damage
  Single and sequential hand postures (Jason, 1985)
  Grooved pegboard (Haaland & Delaney, 1981)
  Maze coordination (Haaland & Delaney, 1981)
  Vertical groove steadiness (Haaland & Delaney, 1981)
  Static steadiness (Haaland & Delaney, 1981)
  Rotary pursuit (Heap & Wyke, 1972)
Bilateral deficits after right hemisphere damage only
  Slowed paced tapping (Carmon, 1971)
Bilateral deficits after left hemisphere damage only
  Repetitive arm movements to two targets (Wyke, 1967, 1971a; Haaland, Harrington, & Yeo, 1987; Haaland & Harrington, 1994)
  Aiming movements (Wyke, 1968; Haaland & Harrington, 1989)
  Design tracing (Wyke, 1971b)
  Rapid paced tapping (Carmon, 1971)
  Sequential hand postures (Kimura & Archibald, 1974; DeRenzi et al., 1983; Kolb & Milner, 1981; Harrington & Haaland, 1991, 1992)
  Learning hand posture sequence (Kimura, 1977)
  Single hand postures (Kimura, 1982)
  Memory for hand posture sequences (Jason, 1983)
  Gesturing or limb apraxia (Faglioni & Basso, 1985; Haaland & Flaherty, 1984)

other hemisphere. Relative to the tasks which show strictly contralateral deficits, these tasks appear to be more complex, but, as mentioned above, there have been no studies that have methodically attempted to disclose the critical cognitive requirements that produce bilateral deficits.

BILATERAL DEFICITS AFTER LEFT HEMISPHERE DAMAGE ONLY

While there is evidence that damage to the right or left hemispheres produces bilateral motor deficits on different tasks, implying that each hemisphere controls different functions, the left hemisphere in right handers appears to be dominant for a larger number of processes that affect movement (Haaland & Harrington, 1996a, 1996b for review). The basis for this conclusion is that for certain tasks, unilateral left hemisphere damage produces bilateral motor deficits while right hemisphere damage produces only contralateral deficits. The tasks that show left hemisphere dominance are described in Table 3. The mechanisms for left hemisphere dominance are not clear, but work in our laboratory (Haaland & Harrington, 1989) showed greater deficits with left hemisphere damage than right hemisphere damage when performing fast but not slow movements. These results were initially interpreted as reflecting the left hemisphere's greater role in controlling preprogrammed movements which are fast and therefore less dependent upon sensory feedback.

Another interpretation of these results is that the left hemisphere is specialized for controlling rapid, high-frequency responses which require timing and coordination of movements, such as in speech (Kent & Rosenbek, 1983) or nonlinguistic timing tasks (Hammond, 1983). In aiming tasks rapid responses require the precise timing of the different components (e.g., acceleration and deceleration). However, the frequency hypothesis also predicts that slower responses should be controlled by the right hemisphere which has not been shown in the context of arm reaching.

Some have suggested that the greater incidence of complex motor deficits with left hemisphere damage is due to the inclusion of patients with limb apraxia because limb apraxia is more common after left than right hemisphere damage (Fagliano & Basso, 1985). A recent hand posture sequencing study (Harrington & Haaland, 1992) showed that some aspects of sequencing were general to left hemisphere damage whereas other aspects were specific to the disorder of limb apraxia, suggesting that the inclusion of apraxics cannot be the only explanation of left hemisphere dominance.

BILATERAL DEFICITS AFTER RIGHT HEMISPHERE DAMAGE ONLY

Right hemisphere damage occasionally produces bilateral motor deficits. This has been shown with the generation of slower movements (Carmon, 1971), and has been related to impairment in closed-loop movements which are more dependent on sensory feedback by one group (Winstein & Pohl, 1995) but not by another (Haaland & Harrington, 1989). One study (Hom & Reitan, 1982) found greater ipsilateral deficits after right hemisphere damage, but did not control for extraneous factors that could explain the finding, including lack of an analysis of individual motor and sensory tasks (i.e., they only reported the combined results from 11 motor and sensory tests), the possibilities of greater impairment or larger lesions in the right hemisphere group, or different intrahemispheric lesion locations. Nonetheless, the functions of the right hemisphere with regard to the cognitive aspects of movement have not been specified very well and await further evaluation particularly in the context of the frequency hypothesis detailed above.

Conclusions

Motor tasks have been used in clinical neuropsychological evaluation primarily as indicators of laterality. The assumption has been that they reflect the functioning of the primary motor system, and deficits in a particular limb reflect damage to the opposite hemisphere. While this is true for simple motor tasks with minimal cognitive requirements, many motor tasks engage higher-level cognitive abilities which are likely supported by many areas of the brain. Clinicians must be aware of these cognitive factors and how cortical and subcortical structures may interact to subserve these cognitive requirements. These cognitive factors largely determine whether deficits are seen in the limb ipsilateral as well as contralateral to unilateral hemispheric lesion. Ipsilateral deficits are more common after left hemisphere than right hemisphere damage, suggesting that the left hemisphere is dominant for controlling complex motor skills just as it is dominant for language. Exactly what cognitive processes determine left hemisphere dominance for movement is not clear, but this review emphasizes that bilateral motor deficits do not necessarily imply bilateral hemispheric involvement.

Acknowledgments

This work was supported by grants from the Department of Veterans Affairs. The authors would like to thank Dr. Muriel Lezak for helpful discussions.

References

Alexander, G. E., Baker, E., Naeser, M. A., Kaplan, E., & Palumbo, C. (1992). Neuropsychological and neuroanatomical dimensions of ideomotor apraxia. *Brain, 115,* 87–107.

Alexander, G. E., DeLong, M. R., & Strick, P. L. (1986). Parallel organization of functionally segregated circuits linking basal ganglia and cortex. *Annual Review of Neuroscience, 9,* 357–381.

Allen, G. I., & Tsukahara, N. (1974). Cerebrocerebellar communication systems. *Physiological Review, 54,* 957–1006.

Beardsley, J. V., Matthews, C. G., Cleeland, C. S., & Harley, J. P. (1978). Experimental T-score norms for C.A. 34− and 34+ on the Wisconsin Neuropsychology Test Battery. Private publication.

Boll, T. J. (1981). The Halstead Reitan neuropsychological battery. In S. Filskov & T. Boll (Eds.), *Handbook of clinical neuropsychology* (pp. 577–607). New York: John Wiley and Sons.

Bornstein, R. A. (1985). Normative data on selected neuropsychological measures from a nonclinical sample. *Journal of Clinical Psychology, 41,* 651–659.

Bornstein, R. A. (1986a). Consistency of intermanual discrepancies in normal and unilateral brain lesion patients. *Journal of Consulting and Clinical Psychology, 54,* 719–723.

Bornstein, R. A. (1986b). Classification rates obtained with "standard" cut-off measures. *Journal of Clinical and Experimental Neuropsychology, 8,* 413–420.

Bornstein, R. A. (1986c). Normative data on intermanual differences on three tests of motor performance. *Journal of Consulting and Clinical Psychology, 8,* 12–20.

Brodal, A. (1981). *Neurological anatomy.* New York: Oxford University Press.

Carmon, A. (1971). Sequenced motor performance in patients with unilateral cerebral lesions. *Neuropsychologia, 9,* 445–449.

Cummings, J. (1993). Frontal-subcortical circuits and human behavior. *Archives of Neurology, 50,* 873–880.

Daly, R. F., & Matthews, C. G. (1974). Impaired motor function in XYY males. *Neurology, 24,* 655–658.

DeRenzi, E., Faglioni, P., Lodesani, M., & Vecchi, A. (1983). Performance of left brain-damaged patients on initiation of single movements and motor sequences: Frontal and parietal-injured patients compared. *Brain, 19,* 333–343.

Dodrill, C. B. (1979). Sex differences on the Halstead–Reitan neuropsychological test battery and on other neuropsychological measures. *Journal of Clinical Psychology, 35,* 236–41.

Dodrill, C. B., & Troupin, A. S. (1975). Effects of repeated administrations of a comprehensive neuropsychological battery among chronic epileptics. *Journal of Nervous and Mental Diseases, 161,* 185–190.

Faglioni, P., & Basso, A. (1985). Historical perspectives on neuroanatomical correlates of limb apraxia. In E. Roy (Ed.), *Neuro-psychological studies of apraxia and related disorders* (pp. 3–44). New York: Elsevier.

Finlayson, M. A. J., Johnson, K. A., & Reitan, R. M. (1977). Relationship of level of education to neuropsychological measures in brain-damaged and non-brain-damaged adults. *Journal of Consulting and Clinical Psychology, 45,* 536–542.

Fleishman, E. A. (1958). Dimensional analysis of movement reactions. *Journal of Experimental Psychology, 55,* 438–453.

Geschwind, N. (1965). Disconnexion syndromes in animals and man. *Brain, 88,* 237–294; 585–644.

Gill, D. M., Reddon, J. R., Stefanyk, W. O., & Hans, H. S. (1986). Finger tapping: Effects of trials and sessions. *Perceptual and Motor Skills, 62,* 675–678.

Golden, C. J., Hammeke, T. A., & Purisch, A. D. (1980). *The Luria Nebraska Neuropsychological Battery manual.* Los Angeles: Western Psychological Services.

Haaland, K. Y., & Delaney, H. D. (1981). Motor deficits after left or right hemisphere damage due to stroke or tumor. *Neuropsychologia, 19,* 17–27.

Haaland, K. Y., & Flaherty, D. F. (1984). The different types of limb apraxia errors made by patients with left vs. right hemisphere damage. *Brain and Cognition, 3,* 370–384.

Haaland, K. Y., & Harrington, K. Y. (1989). Hemispheric control of the initial and corrective components of aiming movements. *Neuropsychologia, 27,* 961–969.

Haaland, K. Y., & Harrington, D. L. (1990). Complex motor behavior: Toward understanding cortical and subcortical interactions in regulating control processes. In G. Hammond (Ed.), *Advances in psychology: Cerebral control of speech and limb movements.* Amsterdam: Elsevier.

Haaland, K. Y., & Harrington, D. L. (1994). Limb-sequencing deficits after left but not right hemisphere damage. *Brain and Cognition, 24,* 104–122.

Haaland, K. Y., & Harrington, D. L. (1996a). Clinical implications of ipsilateral deficits in complex motor skills. In A. Bruno, F. Chollet, B. J. Vellas, & J. L. Albarede (Eds.), *Stroke in the elderly* (pp. 101–114). New York: Springer Publishing Co.

Haaland, K. Y., & Harrington, D. L. (1996b). Hemispheric asymmetry of movement. *Current Opinion in Neurobiology, 6,* 796–800.

Haaland, K. Y., & Harrington, D. L., & Yeo, R. (1987). The effects of task complexity on motor performance in left and right CVA patients. *Neuropsychologia, 25,* 783–794.

Haaland, K. Y., Temkin, R., Randahl, G., & Dikmen, S. (1994). Recovery of simple motor skills after head injury. *Journal of Clinical and Experimental Neuropsychology, 16,* 1–9.

Halstead, W. C. (1947). *Brain and intelligence.* Chicago: University of Chicago Press.

Hammond, G. (1983). Hemispheric differences in temporal resolution. *Brain and Cognition, 1,* 95–118.

Harley, J. P., Matthews, C. G., Leuthold, C. G., & Bergs, L. E. (1980). *Wisconsin Neuropsychological Test Battery. T-score norms for older Veteran's Administration Medical Center patients.* Madison, WI: University of Wisconsin.

Harrington, D., & Haaland, K. Y. (1991). Hemispheric specialization for motor sequencing: Abnormalities in levels of programming. *Neuropsychologia, 29,* 147–163.

Harrington, D. L., & Haaland, K. Y. (1992). Motor sequencing with left hemisphere damage: Are some cognitive deficits specific to limb apraxia. *Brain, 115,* 857–874.

Heap, M., & Wyke, M. (1972). Learning of a unimanual skill by patients with brain lesions: An experimental study. *Cortex, 8,* 1–18.

Heaton, R. K., Grant, I., & Matthews, C. G. (1986). Differences in neuropsychological test performance associated with age, education and sex. In I. Grant & K. M. Adams (Eds.), *Neuropsychological assessment of neuropsychiatric disorders.* New York: Oxford University Press.

Heaton, R. K., Grant, I., & Matthews, C. G. (1991). *Comprehensive norms for an expanded Halstead Reitan battery.* Odessa, FL: Psychological Assessment Resources.

Helm-Estabrook, N., Fitzpatrick, P. M., & Barresi, B. (1982). Visual action therapy for global aphasia. *Journal of Speech and Hearing Disorders, 47,* 385–389.

Hom, J., & Reitan, R. M. (1982). Effect of lateralized cerebral damage upon contralateral and ipsilateral sensorimotor performances. *Journal of Clinical Neuropsychology, 4,* 249–268.

Jason, G. W. (1983). Hemispheric asymmetries in motor function. I. Left hemisphere specialization for memory but not performance. *Neuropsychologia, 21,* 35–45.

Jason, G. W. (1985). Manual sequence learning after focal cortical lesions. *Neuropsychologia, 23,* 483–496.

Keele, S. W., Ivry, R. (1990). Does the cerebellum provide a common computation for diverse tasks? A timing hypothesis. *Annals of the New York Academy of Sciences, 608,* 179–207.

Kent, R. D., & Rosenbek, J. (1983). Acoustic patterns of apraxia of speech. *Journal of Speech and Hearing Research, 26,* 231–249.

Kimura, D. (1977). Acquisition of a motor skill after left-hemisphere damage. *Brain, 100,* 527–542.

Kimura, D. (1982). Left-hemisphere control of oral and brachial movements and their relation to communication. *Philosophical Transactions of the Royal Society of London, B298,* 135–149.

Kimura, D., & Archibald, Y. (1974). Motor functions of the left hemisphere. *Brain, 97,* 337–350.

Kolb, B., & Milner, B. (1981). Performance of complex arm and facial movements after focal brain lesions. *Neuropsychologia, 19,* 491–503.

Leckliter, I. N., & Matarazzo, J. D. (1989). The influence of age, education, IQ, gender, and alcohol abuse on Halstead-Reitan neuropsychological test battery performance. *Journal of Clinical Psychology, 45,* 484–511.

Lezak, M. (1995). *Neuropsychological assessment* (3rd ed.). New York: Oxford University.

Luria, A. (1966). *Higher cortical function in man.* New York: Basic Books.

Matarazzo, J. D., Wiens, A. N., Matarazzo, R. G., & Goldstein, S. G. (1974). Psychometric and clinical test–retest reliability of the Halstead impairment index in a sample of healthy, young, normal men. *Journal of Nervous and Mental Disease, 158,* 37–49.

Matthews, C. G., Chapman, L. J., & Woodard, A. R. (1988). Differential neuropsychologic profiles in idiopathic versus pesticide-induced parkinsonism. In B. L. Johnson (Ed.), *Advances in neurobehavioral toxicology: Applications in environment and occupational health.* New York: Lewis Publishers.

Matthews, C. G., Cleeland, C. S., & Hopper, C. L. (1970). Neuropsychological patterns in multiple sclerosis. *Journal of Nervous and Mental Disease, 31,* 161–170.

Matthews, C. G., & Haaland, K. Y. (1979). The effect of symptom duration on cognitive and motor performance in Parkinsonism. *Neurology, 29,* 951–956.

Matthews, C. G., & Harley, J. P. (1975). Cognitive and motor-sensory performances in toxic and nontoxic epileptic subjects. *Neurology, 25,* 184–188.

Matthews, C. G., & Klove, H. (1964). *Instruction manual for the Adult Neuropsychology Test Battery.* Madison, WI: University of Wisconsin Medical School.

Ochipa, C., Rothi, L. J. G., & Heilman, K. M. (1989). Ideational apraxia: A deficit in tool selection and use. *Annals of Neurology, 25,* 190–193.

Ochipa, C., Rothi, L. J. G., & Heilman, K. M. (1992). Conceptual apraxia in Alzheimer's disease. *Brain, 115,* 1061–1071.

Rao, S. M., Binder, J. R., Bandettini, P. A., Hammeke, T. A., Yetkin, F. Z., Jesmanowicz, A., Lisk, L. M., Morris, G. L., Mueller, W. M., Estowski, L. D., Wong, E. C., Haughton, V. M., & Hyde, J. S. (1993). Functional magnetic resonance imaging of complex human movements. *Neurology, 43,* 2311–2318.

Reddon, J. R., Stefanyk, W. O., Gill, D. M., & Renney, C. (1985). Hand dynamometer: Effects of trials and sessions. *Perceptual and Motor Skills, 61,* 1195–1198.

Reitan, R. M. (1969). *Manual for administration of neuropsychological test batteries for adults and children.* Private publication.

Reitan, R., & Davison, L. (1974). *Clinical neuropsychology: Current status and applications* (pp. 378–379). Washington, DC: V. H. Winston & Sons.

Roland, P. E., Larsen, B., Lassen, N. A., & Skinhof, E. (1980). Supplementary motor area and other cortical areas in organization of voluntary movements in man. *Journal of Neurophysiology, 43,* 118–136.

Rothi, L. J. H., Heilman, K. M., Mack, L., Verfaellie, M., Brown, P., & Heilman, K. M. (1988). Ideomotor apraxia: Error pattern analysis. *Aphasiology, 2,* 381–387.

Russell, E. W., Neuringer, C., & Goldstein, G. (1970). *Assessment of brain damage: A neuropsychological key approach.* New York: Wiley & Sons.

Schwartz, M. F., Mayer, N. H., Fitzpatrick DeSalme, E. J., & Montgomery, M. W. (1993). Cognitive theory and the study of everyday action disorders after brain damage. *Journal of Head Trauma Rehabilitation, 8,* 59–72.

Spreen, O., & Strauss, E. (1991). *Compendium of neuropsychological tests.* New York: Oxford University Press.

Strick, P. L. (1988). Anatomical organization of multiple motor areas in the frontal lobe: Implications for recovery of function. *Advances in Neurology, 47*, 293–306.

Sundet, K., Arnstein, A., & Rienvang, I. (1988). Neuropsychological predictors in stroke rehabilitation. *Journal of Clinical and Experimental Neuropsychology, 10*, 363–379.

Winstein, C. J., & Pohl, P. S. (1995). Effects of unilateral brain damage on the control of goal-directed hand movements. *Experimental Brain Research, 105*, 163–174.

Wyke, M. (1967). Effects of brain lesions on the rapidity of arm movement. *Neurology, 17*, 1113–1120.

Wyke, M. (1968). The effects of brain lesions in the performance of an arm-hand precision task. *Neuropsychologia, 6*, 125–134.

Wyke, M. (1971a). The effects of brain lesions on the performance of bilateral arm movements. *Neuropsychologia, 9*, 33–42.

Wyke, M. (1971b). The effects of brain lesions on the learning performance of a bimanual coordination task. *Cortex, 7*, 59–72.

Yeudall, L. T., Reddon, J. R., Gill, D. M., & Stefanyk, W. O. (1987). Normative data for the Halstead-Reitan neuropsychological tests stratified by age and sex. *Journal of Clinical Psychology, 43*, 346–367.

20

Assessment Methods in Behavioral Neurology and Neuropsychiatry

Robert M. Stowe

Introduction

The purpose of this chapter is to describe an approach to the clinical assessment of patients presenting to a physician for evaluation of cognitive, behavioral, and/or psychiatric symptoms, in the setting of known or suspected diseases of the central nervous system, as seen from the perspective of a behavioral neurologist working in a psychiatric hospital. Given the increasing convergence of behavioral neurology and neuropsychiatry (guided by recent advances in cognitive neuroscience, and exemplified by the recent joint annual meetings of the American Neuropsychiatric Association and the Behavioral Neurology Society), we believe that the approach described here is representative of contemporary practice in neuropsychiatry as well as behavioral neurology. For convenience and lack of a better term, "neurobehavioral assessment" will be used to refer to this method of clinical evaluation. Objectives of such an assessment are multiple and typically include diagnosis, identification of treatment goals, selection of medications and other treatment modalities, and prognostication. As the examination proceeds, *syndromic, anatomical,* and *etiological* differential diagnoses are narrowed and modified "on the fly," so that by the end of the evaluation the examiner has a relatively narrow set of possibilities to choose from with the aid of laboratory studies, as well as an idea of what treatment strategies are indicated and what prognostic advice can be given to the patient and significant others. We stress the need for clinicians to *explicitly consider* each of the

Robert M. Stowe   Departments of Neurology and Psychiatry, University of Pittsburgh School of Medicine, and VA Pittsburgh Healthcare System, Highland Drive Division, Pittsburgh, Pennsylvania 15206-1297.

Neuropsychology, edited by Goldstein *et al.* Plenum Press, New York, 1998.

above-mentioned "levels" of diagnosis for several reasons. First, they are frequently mutually informative (i.e., a syndromic diagnosis of mania leads one to consider a right frontal lesion, whereas a paranoid psychosis, if associated with a focal lesion, would localize more commonly to the left temporoparietal region). Second, the etiological differential diagnosis is strongly influenced by localization. For example, a progressive dementia syndrome with findings consistent with bilateral temporoparietal involvement, such as episodic and semantic memory impairment, anomia, apraxia, and visuospatial deficits, is likely to be due to Alzheimer's disease (AD), whereas one with relative preservation of these functions but with predominant abulia, bradyphrenia, and a "dysexecutive syndrome" (i.e., deficits in higher-order attention, working memory, abstract reasoning, planning, insight, and judgment) suggests a frontal/subcortical pattern associated with an extrapyramidal degenerative disorder, a dementia syndrome of depression, normal pressure hydrocephalus, HIV infection, etc. Third, the selection of pharmacological treatments is largely based on clinical phenomenology, rather than on lesion location (e.g., apathy or depression following stroke, regardless of lesion location, is likely to respond best to antidepressants and poorly to neuroleptics, whereas the converse is probably true when a pervasive paranoid state is present).

HISTORY-TAKING

HISTORY OF PRESENT ILLNESS (HPI)

As in most areas of medicine, and particularly in neurology, *the history is usually the most informative aspect of neurobehavioral assessment.* Many disorders (e.g., epilepsy, headache, Tourette's syndrome, a right frontal tumor, or an early frontal dementia in a high-functioning individual) may present with completely normal neurological and mental status examinations, and may therefore be diagnosable *only* by careful history-taking. The history is also crucial for framing tentative anatomical, syndromic, and etiological differential diagnoses which help focus the neurological and general physical examinations, as discussed above. For example, in a patient who reports blackouts or "spells" of altered or lost consciousness, brief psychotic or confusional episodes, or a rapidly cycling mood disorder, the examiner should consider the possibility of temporolimbic epilepsy, and therefore inquire in detail about auras (i.e., "funny feelings" or unusual experiences), including specific questioning about déjà vu (an illusory feeling of familiarity in new surroundings or situations) or jamais vu (a peculiar feeling of unfamiliarity in a known environment); epigastric aura (a peculiar "rising" or fearful sensation in the upper abdomen moving toward the throat), olfactory or gustatory hallucinations (illusory smells or tastes), vertigo (a whirling sensation), and "staring spells"; hypergraphia (a need to write excessively); and confusion and drowsiness in the period immediately following a seizure or "spell." These same questions would typically be unproductive in someone presenting for evaluation of dementia (uncomplicated by seizures), in whom more fruitful questioning would include probing for distractability and inattention; deficits in judgment in social and financial situations; deterioration in activities of daily living (ADLs; i.e., meal preparation and eating, shopping, dressing, shaving, bathing, balancing one's checkbook, etc.); loss of bladder and/or bowel control; daytime somnolence and impaired breathing or unusually heavy snoring during sleep (symptoms of sleep apnea syndrome); deterioration in walking and balance (which could be a clue to normal pressure hydrocephalus, a cerebellar degeneration associated with an underlying malignancy, Wernicke–Korsakoff syndrome, or an extrapyramidal disorder); brief stroke-like episodes; and stepwise deterioration in cognition or sensorimotor function (i.e., unilateral weakness and sensory loss, ataxia, double vision, or transient loss or distortion of vision in one eye or one hemifield). In both instances, however, it would be important to ask about headaches (rule out mass lesions, meningeal irritation due to inflammatory or infectious disorders, subdural hematoma); "forgetfulness" and memory loss; traffic or kitchen accidents (which could result from episodic loss of consciousness, or from confusion or memory loss in dementia); temporal and spatial confusion; word-finding difficulty or other changes in language function; and changes in personality (such as apathy, placidity, and social withdrawal, or suspiciousness and hostility; deterioration in social graces and loss of interest in dress and personal hygiene); depressed, elevated, irritable, or labile mood; and "neurovegetative" symptoms (alterations in sleep patterns, appetite and weight, energy level, and interest in sex, social activities, and recreational pursuits). The latter are particularly important from a therapeutic aspect as well, since they may respond to pharmacological intervention with antidepressants, stimulants, or dopamine agonists.

The *temporal profile* of progressive cognitive impairment is particularly critical in differential diagnosis; how (insidious vs. acute onset) and when the first symptoms were noticed and the order in which they appeared, and when ADLs became compromised. Insidious onset (difficult for family members to date, even by year of onset) and slow progression is most typical of a primary degenerative dementia such as AD (especially when memory impairment is the first symptom), or of a frontotemporal dementia. Stepwise deterioration is characteristic of vascular dementias, although occasionally this pattern is also seen in AD. Rapidly progressive cognitive impairment is more typical of delirium, space-occupying lesions (including tumor and subdural hematoma), Lewy body dementia, and some dementias due to infectious agents (including Creutzfeldt–Jakob disease), although HIV encephalitis, neurosyphilis, Lyme encephalopathy, and Whipple's disease may have an insidious or intermediate onset and course. Insidious onset of language disturbance (usually nonfluent), which remains relatively isolated even when the patient has profound deficits in linguistic communication, is seen in the syndrome of primary progressive aphasia without generalized dementia (Mesulam, 1981).

In patients presenting for evaluation and management of aggressive behavior or "spells" suggestive of complex partial or hysterical seizures, a detailed description of episodes is essential for differential diagnosis. Was there a prodromal mood change, alteration in consciousness, or aura, prior to the onset of the behavioral change witnessed by observers? Was there loss or impairment of consciousness (in which case the patient should be amnestic for at least part of the "spell")? In the case of seizures, was there a blank stare followed by automatisms such as lip smacking, swallowing, or other repetitive automatic behavior, followed by confusion, drowsiness, and sometimes headache (as often seen in complex partial seizures), or rhythmic eye blinking or twitching during the period of unresponsiveness, with rapid recovery of consciousness and no postictal confusion or drowsiness (as in primary generalized absence seizures)? When aggressive outbursts are the referral issue, it is important to explore not only the patient's behavior before, during, and after these, but also psychological, physical, and social precipitants, the settings in which violence typically occurs, and whether the patient derives gratification or a sense of relief from the violence (Reid & Balis, 1987). For a more detailed exposition of an approach to diagnosis and management of patients with such outbursts, see Stowe (1994).

PAST MEDICAL AND NEUROPSYCHIATRIC HISTORY (PMHx)

The patient is routinely questioned about any major prior illnesses, hospitalizations, and surgical procedures (including problems in the postoperative period) and psychiatric treatment. Depending on the context, it may be appropriate to include pre- and perinatal and developmental history here as well (see section on social/personal/emotional history). Maternal complications of pregnancy, labor, and delivery of the patient; a history of febrile convulsions in infancy; left-handedness or ambidexterity; delayed motor, language, and/or social milestones; learning disabilities and hyperactivity; autistic behaviors or excessive shyness; or conduct disorder and unusual aggressivity may be clues to the presence of early acquired brain damage or of neurodevelopmental disorders. A childhood history of tics (suggestive of Tourette's syndrome) or of choreiform movements associated with scarlet or rheumatic fever (Sydenham's chorea, "St. Vitus' dance") may be a clue to the etiology of obsessive–compulsive symptomatology (Swedo *et al.*, 1989) or psychotic symptoms (Wilcox & Nasrallah, 1988). Neurologists routinely inquire about a history of meningitis or encephalitis; seizures and blackouts; periods of frequent or severe headache; and, in older patients, stroke or "mini-strokes" (i.e., transient ischemic attacks—episodes of transient neurological dysfunction on a presumed thromboembolic basis). When evaluating patients with "conduct disorder," antisocial behavior, or comportmental deficits, it is particularly important to identify significant head trauma (usually accompanied by loss of consciousness or concussion) in childhood. We have seen several patients whose antisocial and impulsive behavior and poor insight and judgment resulted from early traumatic frontal injuries, the significance of which had been overlooked by multiple clinicians who had assessed and treated them psychiatrically and psychologically (Price, Daffner, Stowe, & Mesulam, 1990).

In patients with rapidly progressive cognitive deterioration, it is particularly important to ask about signs and symptoms of intracranial and systemic infection or disorders associated with immunosuppression; recent head trauma and coagulation disorders (risk factors for intracranial hematomas); malnutrition (rule out Wernicke–Korsakoff syndrome due to thiamine, niacin, or B_{12} deficiency); a history of hepatic, renal, endocrine, and cardiopulmonary dysfunction which could result in a confusional state; seizures; risk factors for Creutzfeldt–Jakob disease (previous intracranial or spinal surgery or myelography, corneal transplantation, administration of human growth hormone, etc.); psychiatric history; and *especially* a careful history of current alcohol, prescription, and nonprescription drug use (including over-the-counter [OTC] sleeping pills and antihistamines, which can induce delirium in the elderly, and illicit drugs).

REVIEW OF SYSTEMS (ROS)

The traditional neurological review of symptoms includes questions about headache; episodes of altered/lost consciousness and "blackouts"; transient ischemic attacks (in older patients), paresthesia (tingling) and numbness; weakness, incoordination, and involuntary movements (see section on the examination of the motor system, below); changes in smell or taste; double vision or other visual changes not corrected by eyeglasses; hearing loss; difficulty with speech or swallowing; lightheadedness on standing; bladder, bowel, and erectile dysfunction (clues to autonomic nervous system and/or spinal cord involvement); vertigo; and balance and walking difficulties. If not already covered in the HPI, it may be relevant to ask about concentration and calculation difficulties, forgetfulness, spatial disorientation, word-finding problems, and difficulty with reading, writing, and ADLs. Psychiatric ROS includes questions about mood and "neurovegetative symptoms" (see previous sections), personality changes, delusions (see below), hallucinations and illusions, phobias, obsessions and compulsions, and (where relevant) suicidal or violent ideation/plan/behavior.

The general medical ROS covers inquiries about the function of each major organ system (cardiovascular, respiratory, gastrointestinal, musculoskeletal, dermatologic, hematopoetic, genitourinary, endocrine), as well as nonspecific symptoms indicative of processes such as infection (fever and chills) or malignancy (fatigue, anorexia, and weight loss, etc). While it is not possible to review these in detail here, it is important to note that many systemic medical illnesses can profoundly affect CNS function and produce confusion, depression, or psychosis, particularly in the elderly or dementing patient. Even in a patient with an established Alzheimer-type dementia, a general medical "tune-up," including medication streamlining aimed at minimizing the use of medications known to impair CNS function in some individuals, may make the difference between living at home with a caregiver, and chronic institutionalization. Therefore, a careful medical history and ROS is imperative in older patients with otherwise unexplained recent cognitive or psychiatric deterioration. Intake of alcoholic beverages (and often illicit drugs) should be probed. When excessive, it is often grossly underreported by patients, and collaterals may provide more accurate information. In younger patients, glue-sniffing (toluene, others) and abuse of paint thinner, gasoline, and other agents containing organic solvents, and use of "recreational" drugs are common potential contributors to cognitive and/or psychiatric impairment. Frequent (and occasionally isolated) use of hallucinogens, PCP, and other "street" drugs may be associated with psychotic symptoms outlasting the period of acute intoxication (Boutros & Bowers, 1996). Amphetamines and cocaine can produce a cerebral vasculitis and multiple strokes or intracerebral hemorrhages, as well as paranoid delusions, aggressive behavior, and severe depression in addicts undergoing withdrawal.

FAMILY HISTORY (FHx)

A history of any neurological and psychiatric problems in biological relatives of the patient is routinely obtained. It is also a good idea to ask if "any diseases run in the family," since a strong family history of autoimmune, endocrine, cardiovascular, or multisystem disorders may be a clue to the presence of a related condition in the patient. It may be necessary to do more detailed probing when conditions that involve genetic risk factors, such as epilepsy, dementia, spinocerebellar degenerations, movement disorders, alcoholism, antisocial personality disorders, obsessive–compulsive disorder, schizophrenia, mental retardation, learning disabilities, and bipolar affective disorder ("manic–depressive illness") are differential diagnostic considerations. When a genetic etiology is suggested by the presence of similar symptoms in relatives or a clinical picture suggestive of a heritable disorder such as

Huntington's disease (HD), Wilson's disease, a leukodystrophy (degenerative disease of white matter), a frontal dementia (dominantly inherited Pick's disease, progressive subcortical gliosis), early-onset AD, or a familial multi-infarct dementia such as CADASIL (Bowler & Hachinski, 1994), the age of onset and clinical course, as well as the ages of death in unaffected relatives, may be helpful both for determining the pattern of inheritance (autosomal dominant, autosomal recessive, X-linked recessive, or mitochondrial—often associated with maternal transmission) and prognosis. Reichman and Cummings (1990) outline a helpful approach to the diagnosis of rare dementia syndromes.

SOCIAL AND OCCUPATIONAL HISTORY

The patient's educational and occupational history should be routinely surveyed. This is important to estimate premorbid intellectual and academic ability, information which is required to guide and interpret the mental status examination. Social functioning and relationships are typically surveyed. In some circumstances, it may be appropriate to inquire about a history of moving violations or criminal charges. Problems in job performance and/or interactions with co-workers and supervisors may be important clues to cognitive and personality changes. Occupational history may also provide clues to a history of exposure to relevant but unsuspected CNS toxins such as lead (associated with anemia, motor neuropathy, and cognitive dysfunction), carbon monoxide, carbon disulfide, manganese (associated with parkinsonian symptoms such as resting tremor, bradykinesia, "masked facies," postural instability, and rigidity or dystonia), organic solvents and paint fumes (neuropathy, encephalopathy, cerebellar syndromes, and rarely parkinsonism), organophosphate insecticides (peripheral neuropathy; acute intoxication may produce irritability and aggression; see Bear, Rosenbaum, & Norman, 1986), organomercurials (visual and psychiatric dysfunction), and organotin compounds (rare but dramatic causes of an acute encephalopathy which could be mistaken for a meningoencephalitis, with seizures, hearing loss, and aggressive behavior, followed by a severe amnestic syndrome). Most other CNS neurotoxins have frank effects on other organ systems and are unlikely to present primarily with cognitive or psychiatric complaints. Evaluating the significance of occupational exposures to potential neurotoxins requires considerable knowledge, expertise, and data, and interpretation of subjective reports of symptoms is often complicated by frank potential for primary and secondary gain in the form of time off work, disability insurance payments, and litigation.

CURRENT MEDICATIONS

Current prescription and OTC drugs recently used by the patient should be surveyed. Compliance with treatment should also be assessed. Medication toxicity (sometimes due to too much medication, but often to an idiosyncratic reaction or side effect at "therapeutic" doses/blood levels) is a very frequent cause of behavioral and/or cognitive decompensation in patients with preexisting CNS dysfunction, but also sometimes in apparently normal elderly individuals. The *Medical Letter* periodically publishes a list of cognitive and psychiatric side effects of prescription drugs (Abramowicz *et al.*, 1993) which can be useful as a reference. As can be seen in Table 1, drugs from almost any class can cause a *delirium* with confusion and/or hallucinations. Probably the most common offenders, however, are drugs

TABLE 1. MEDICATIONS THAT MAY PRODUCE CONFUSION AND/OR HALLUCINATIONS

445

BEHAVIORAL
NEUROLOGY
AND NEURO-
PSYCHIATRY

Anticholinergics/antihistamines/decongestants
  1° (e.g., atropine, benztropine/Cogentin®, trihexiphenidyl/Artane®, oxybutynin/Ditropan®);
  2° (e.g., decongestants/antihistamines (e.g., diphenhydramine/Benadryl®, chlorpheniramine, cimetidine/Tagamet®, famotidine/Pepcid®, scopolamine, etc.; antidepressants and neuroleptics— see below)

Anticonvulsants (phenytoin/Dilantin®, phenobarbital[a], primadone/Mysoline®[a], carbamazepine/Tegretol®, valproic acid/Depakene®/Depakote®, ethosuximide/Zarontin®, gabapentin/Neurontin®, lamotrigine/Lamictal®, felbamate/Felbatol®)

Antidepressants (amitriptyline/Elavil®, imipramine/Tofranil®, desipramine, nortriptyline, clomipramine/Anafranil®, doxepin/Sinequan®, amoxapine/Asendin®, maprotiline/Ludiomil®, MAOIs[2]; SSRIs[2], etc.)

Antihypertensives (alpha-adrenergic drugs, e.g., clonidine/Catapres®; beta-blockers, e.g., propranolol/Inderal®, timolol/Timoptic®; possibly calcium-channel blockers, i.e., nifedipine, diltiazem)

Antimicrobials (cephalosporins; trimethaprim/sulfamethoxazole/Bactrim®/Septra®; chloroquine; floxacins, e.g., Cipro®, Floxin®, gentamycin, tobramycin, cycloserine/Seromycin®, metronidazole/Flagyl®, dapsone)

Antineoplastic drugs (chlorambucil, methotrexate, 5-fluorouracil, L-asparaginase, levamisole)

Antiparkinsonian agents (e.g., amantidine/Symmetrel®, L-dopa+carbidopa/Sinemet®, bromocriptine/Parlodel®, pergolide/Permax®, pramipexole/Mirapex®)

Antivirals (acyclovir/Zovirax®, zidovudine/AZT/Retrovir®, gangcyclovir/Cytovene®)

Barbiturates[a] (e.g., phenobarbital, primidone/Mysoline®)

Benzodiazepines[a] (e.g., clonazepam/Klonopin®, diazepam/Valium®, temazepam, flunazepam/Dalmane®, triazolam/Halcion®, oxazepam/Serax®, alprazolam/Xanax®)

Cardiac drugs (digoxin/Lanoxin®, quinidine, procainamide/Pronestyl®, amiodarone/Cordarone®, disopyramide/Norpace®)

Immunomodulatory drugs (steroids, e.g., prednisone; cyclosporin, OKT3, FK506, interferons [Betaseron®, Avonex®])

Narcotics (morphine, meperidine/Demerol®, buprenorphine/Buprenex®, propoxyphene/Darvon®)

Neuroleptics and antiemetics/prokinetic agents[c] (e.g., chlorpromazine/Largactil®, thioridazine/Mellaril®, trifluoperazine/Stelazine®, halperidol/Haldol®, pimozide/Orap®, molindone/Moban®; clozapine/Clozaril®, risperidone/Risperidol®; metoclopramide/Reglan®, prochlorperazine [Compazine®])

Nonsteroidal anti-inflammatory drugs (e.g., indomethacin/Indocid®, ibuprofen/Motrin®/Advil®)

Miscellaneous (lithium/Lithobid®; buspirone/Buspar®; baclofen [Lioresal®][a]; disulfiram/Antabuse®; cyclobenzaprine/[Flexaril®][a]; metrizamide/Amipaque®—less frequently other iodinated contrast agents)

---

[a]Especially following abrupt withdrawal, or in overdosage.
[b]Selective serotonin reuptake inhibitors (SSRIs: fluoxetine/Prozac®, sertraline/Zoloft®, paroxetine/Paxil®, fluvoxamine/Luvox®) have little anticholinergic activity, but can rarely cause confusion by inducing hyponatremia or a "serotonin syndrome." Monoamine oxidase inhibitors (MAOIs: phenelzine/Nardil, isocarboxazid/Marplan, tranylcypromine/Parnate) can also induce a serotonin syndrome.
[c]Confusion may be due to anticholinergic delirium or neuroleptic malignant syndrome (NMS).

with anticholinergic properties; these include antihistamines contained in OTC sinus and cold medications and sleeping pills, many antidepressants and antipsychotic drugs, and some drugs used to treat ulcers (e.g., cimetidine), diarrhea, and incontinence. Visual hallucinations are a frequent (but not essential) feature of this syndrome. They are also commonly induced by narcotics, and by dopaminergic medications in patients with Parkinson's disease (PD) and diffuse Lewy body dementia. Abrupt *cessation* of sedating drugs such as alcohol, benzodiazepines, barbiturates, and baclofen (a drug used to treat spasticity) is also an important cause of delirium.

THE ASSESSMENT OF MENTAL STATE

The mental status examination (MSE) has many purposes, examples of which include: "screening" for early dementia in elderly patients (with or without cognitive complaints); characterization of the profile of cognitive impairment necessary for localization and differential diagnosis in a patient with clear-cut intellectual deterioration; differentiation of cortical from subcortical, brain stem, or spinal cord lesions in a patient with sensory or motor findings indicative of CNS disease; search for focal lesion(s) in patients with seizures; detection of etiologically salient neurological disease in patients presenting with affective, psychotic, or other psychiatric symptoms; ascertainment of the temporal profile of cognitive decline or improvement for prognostic purposes; and prediction of functional capacity and development of strategies for the remediation and circumvention of cognitive deficits. It should be obvious from such a diverse set of potential goals, as well as from well-known individual differences to cognitive abilities related to constitutional factors, age, and education, and the intimidating breadth of syndromic, etiological, and neuroanatomical differential diagnoses involved in evaluating such patients, that it is clearly impossible to outline a single, rigidly structured and comprehensive examination process suitable for all patients in all contexts. The goal of this section is to present information and guidelines that will help the examiner to conduct an appropriately focused, process-oriented (Kaplan, 1991), and diagnostically informative cognitively oriented "bedside" MSE. Although formalized neuropsychological test instruments are frequently used (often qualitatively, rather than quantitatively) by behavioral neurologists and neuropsychiatrists in office examinations, we have largely omitted reference to these due to space considerations, as they are covered elsewhere in this volume.

To be informative, mental status assessment must (1) sample all cognitive functions using sensitive (but often nonspecific) screening procedures, so that no major unsuspected impairments are overlooked; and (2) employ more specific tests that concentrate on appropriate areas when deficits have been identified. The selection of the procedures to be used is guided by information and unstructured behavioral observations gathered during history-taking, which should allow the examiner to tentatively formulate (1) some *expectations* about the general level of behavioral (see below) and cognitive functioning of the patient, as well as those cognitive domains most involved, which will help determine the most appropriate types, and level of difficulty, of tests to be administered; (2) the *syndromic presentation* of the patient (i.e., frontal or parietotemporal dementia, aphasia, amnestic syndrome, depression, confusional state, etc.); and (3) *etiological and neuroanatomical differential diagnoses.* These formulations may change (and will hopefully narrow) as the examination proceeds; one then "focuses in," using more specific tests to examine in detail those areas which can provide information necessary for the cognitive interpretation, localization, etiology, and functional significance of deficits uncovered by history and screening procedures. The most appropriate order in which to test cognitive functions may vary from patient to patient, but a structured approach helps the inexperienced examiner to avoid missing major areas and to gather further information that will establish expectations for level of performance on subsequent tasks. Regardless of the order in which the examination is performed, logical organization and presentation of the findings is facilitated by following a standard sequence. We recommend beginning with evaluation of "state functions" (Mesulam, 1985b) using behavioral observation and tests of attention, since these will identify deficits which tend to pervade all aspects of cognitive functioning, and will help the examiner apportion his or her time judiciously in the remainder of the exam. We move to language functions and praxis next, since so much of the examination is dependent on intact comprehension, the capacity for verbal thinking and expression, and higher-order motor abilities. Visuospatial functions are then screened, since visual memory and integrative aspects of visuospatial cognition cannot be interpreted in the presence of severe perceptual problems. Memory is tested next, leaving abstraction, insight, and judgment to the end, since these cognitive abilities are at the top of the neurocognitive hierarchy, and their assessment and interpretation require information about the integrity of the more "basic" cognitive functions.

Behavioral Observations

Behavioral observations begin during history-taking and continue throughout the examination process. Documentation should always include specific information as to informal aspects of arousal and attentiveness (and conversely, distractibility to internal and external stimuli); orientation; motivation and cooperation; comportment (appropriateness of conduct and attitude toward the examiner and others, hygiene and grooming); eye contact; and affect, including its range and appropriateness to thought content and the patient's circumstances, as well as to self-reported mood. Aside from the diagnostic importance these observations may have in their own right, such data also "set the stage" for the interpretation of subsequent aspects of the examination (see below) and their documentation is therefore very important. For example, the patient may be apathetic and/or abulic (slow to respond) and akinetic (lacking normal spontaneous movements and gestures) or restless and akathetic (see section on movement disorders, below). Thought processes may be circumstantial and viscous (as in some patients with temporolimbic epilepsy); tangential or disorganized, perhaps with "flight of ideas" and pressured speech; loose, irrelevant, or incoherent; they may reveal delusional content of a paranoid, grandiose, reduplicative, possessive, referential, nihilistic, or bizarre somatic and self-deprecatory or guilty nature (Benson & Gorman, 1996). For more detailed, psychiatrically oriented discussions of behavioral observations and signs important in neuropsychiatric diagnosis, see Trepacz and Baker (1993) and Yager and Gitlin (1995). "Stimulus-bound" or "utilization" behaviors (Lhermitte, Pillon, & Serdaru, 1986), such as automatic reading aloud of printed material and signs in the vicinity of the patient and inappropriate manipulation of objects (i.e., picking up a pen on the examiner's desk and doodling with it) are commonly seen in frontal dementias but may also occur in delirium. Mirroring of the examiner's gestures or phrases (echopraxia and echolalia, respectively) are examples of stimulus-bound behaviors that can occur in other contexts (e.g., Tourette's syndrome, transcortical aphasias, catatonia). Inexperienced examiners frequently overlook or underrate the contribution of states such as drowsiness, hyperarousal, anxiety, depression, or mania, and behavioral tendencies such as poor effort/motivation, impersistence, distractibility, perseveration, and impulsivity, which can result in profoundly defective performance across a wide variety of tasks unrelated to the specific cognitive domain in question. A classic example is the tendency of anxious, impulsive or "stimulus-bound" patients to do poorly on tests of visuoperceptual integration (Lezak, 1995). If performance improves substantially when the examiner constantly reminds the patient to take his/her time and not to jump to conclusions prematurely, a primary visuoperceptual deficit otherwise implied by low scores on such tests becomes less likely.

ATTENTION

Attention can be broadly defined as the ability to maintain and engage salient stimuli in consciousness while extruding irrelevant perceptions, thoughts, and feelings. It is not a unitary concept, and includes both "positive" (orientation, exploration, concentration, and vigilance) as well as "negative" (distractibility, impersistence, confusion, hemispatial neglect) aspects (Mesulam, 1985b). Most bedside measures of attention are intended as screening procedures and are therefore sensitive but nonspecific (i.e., they can be affected by deficits in vigilance, working memory, attentional capacity/"divided attention," or the ability to maintain "cognitive set"). In younger adults, "serial sevens" is an appropriately sensitive screening test for attentional dysfunction. The patient is asked to subtract 7 from 100 and then asked to keep subtracting 7 from the result until the examiner tells him or her to stop (i.e., 100, 93, 86, 79, . . . 16, 9, 2). This may be too sensitive a test for some normal individuals over the age of 70 or those with less than 8 years of education (Trepacz & Baker, 1993). If serial sevens is failed, "serial threes" is an easier task that may be used instead. A screening test for attentional deficits that is useful in older adults or mildly impaired head-injured individuals is the ability to recite the months of the year forward and backward. This test is sensitive to deficits in divided attention, as well as to perseveration, and the inability to inhibit an overlearned response tendency (e.g., "December, November . . . October . . . November, December, January, February . . ." is a typical response of a patient with a frontal dementia). The discrepancy between the time to do months backward and forward can be helpful in determining the mechanism of slowing (i.e., cognitive, motor, or mixed) in patients with apparent "bradyphrenia." It is helpful to find (as a baseline) some measure of divided attention that the patient can perform adequately. Therefore, if "months backward" is failed, days of the week backward can be tried instead.

The forward digit span measures the *auditory span of apprehension* (or auditory "working memory" capacity). The normal forward span is five to seven digits (five being marginal) and is usually recorded as the number of randomly chosen single digits that the patient can repeat correctly. The ability to repeat digits in reverse is a measure of divided attention, and is normally one (and no more than two) digits less than the forward span, with the proviso that it should be at least four for patients with a forward span of five (Lezak, 1995). The examiner should present digits in an evenly spaced manner (not "chunked" like telephone numbers), no faster than two per second. Sequences used should not repeat any digits, and should be free of any obvious logical structure. The "pointing" (from the Wechsler Memory Scale-Revised), or Corsi block span, is the visuospatial analogue of the digit span, measuring spatial attention/immediate memory as the number of locations which the subject can point to in the same (forward) and reverse (backward) order as the examiner. Using a nine-item array of randomly arranged, equal-sized blocks or squares, pointing spans are typically within one item of the corresponding digit spans. Pointing spans are particularly useful for evaluating attention in aphasics and deaf individuals, and might be considered as well in patients with a suspected right hemisphere lesion or learning disability, who might be expected to do disproportionately worse (relative to digit span) on this task.

The spatial distribution of attention can be grossly evaluated by asking the patient to bisect a horizontal line placed squarely in front of him or her. A more sensitive set of measures, which also evaluates visual scanning abilities, includes the Random Letter and Shape Cancellation Tests (Weintraub & Mesulam, 1985). In these tests, targets are symmetrically arranged on the page, while distractors are randomly distributed. When patients with hemispatial neglect miss targets in the left hemifield, they may overlook more and more target stimuli as they progress down the page, because their leftmost fixation point in each line shifts progressively more rightward (see Figure 1). It is important to pay attention to the scanning strategy employed by the patient, since an apparent "right neglect" pat-

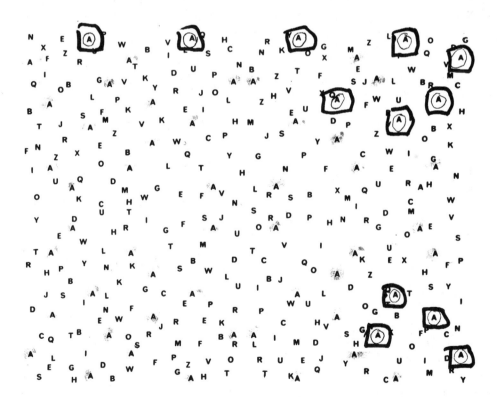

Figure 1. Left-sided neglect in a patient with a right hemisphere infarction. The patient used an unsystematic, right-to-left, top-down search strategy. He identified only 2/30 left-sided, and 10/30 right-sided, target letters (circled "A"s). Omitted targets are circled. Note the progressive rightward detection bias in successive rows.

tern in a subject using a top-to-bottom, left-to-right sequence may merely reflect fatigue.

Response maintenance and perseverance can be assessed using word list generation. In *letter fluency* tasks such as the Controlled Oral Word Association (or "F-A-S") Test, the patient is asked to name as many words as possible (excluding proper nouns) beginning with a particular letter of the alphabet, until told to stop by the examiner. Given 1 minute each for the three letters F, A, and S, normal high school graduates produce a total of 30 to 50 words and maintain their output throughout the entire minute, whereas in depression and frontal/subcortical dementias, there is a tendency for output to be normal in the first 10 seconds or so and then to diminish markedly or stop completely at that point. Semantic *category fluency* can be assessed by asking patients to name types of animals, or items that they could buy in a grocery store or supermarket; normally 15 to 20 can be generated over 1 minute. Impaired word list generation may reflect aphasia and/or left prefrontal pathology in addition to poor perseverance. Rule violations, perseveration, and loss of set on word list generation tasks may be clues to attentional, memory, and/or frontal lobe dysfunction.

Asking the patient to copy and continue graphomotor sequences (see Figure 2) may bring out one form of a response maintenance deficit, namely progressive micrographia (often seen in extrapyramidal disorders), as well as perseverative tendencies, including difficulty shifting response set (e.g., carrying the diagonal line of the triangular portion of a sequence composed of alternating triangle and square wave components into the square wave portion) or continuous perseveration (e.g., adding extra loops with successive copies of a stimulus containing only three linked loops). Patients should be instructed to go as quickly as possible, and not to lift their pencil from the page. Other tests sensitive to response alternation and perseveration are discussed in the section on dexterity and motor sequencing/learning in the general neurological examination.

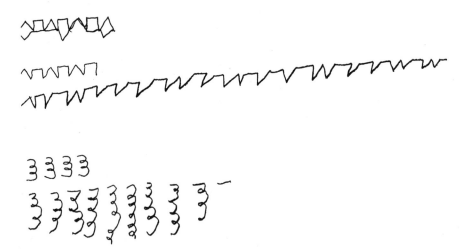

Figure 2. Graphomotor sequences and Luria loops. A 74-year-old man with traumatic bifrontal injuries sustained in a fall at age 70 was asked to "copy and continue" the alternating sawtooth/square wave sequence at the top of the figure. Note "closing in" (drawing on top of the stimulus) on his first, and perseverative features in his second, attempt (carry-over of diagonal and rectangular elements), as well as the duplication of loops when asked to "make a whole row of loops just like these").

The ability to inhibit automatic or prepotent response tendencies (which may be impaired in patients with frontal lobe lesions, especially those involving the orbital and/or mesial components) can be evaluated by a family of tests generically referred to as "go no-go" paradigms (Malloy & Richardson, 1994). These include variants of the Stroop "interference" paradigm and subtests of the Luria–Nebraska Neuropsychological Battery (Golden, Purisch, & Hammeke, 1985) in which the subject is first instructed to mimic the examiner ("When I hold up one finger, you hold up one finger; when I hold up two fingers, you hold up two fingers"). After a series of trials, the subject is instructed that the rules are changing ("Now, when I hold up one finger, you still hold up one finger, but when I hold up two fingers, I don't want you to hold up any"). A popular variant on this theme has been developed by Leimkuhler and Mesulam (1985). The patient is asked first to quickly raise and then lower his or her dominant index finger in response to crisp tapping sounds delivered by the examiner with a ruler or pen (out of view of the patient) about every 2 seconds; once for one tap, twice for two (when two stimuli are presented, the second tap follows immediately after the first). After this simple "vigilance" condition is rehearsed for about a minute, the patient is told now to continue to respond to one tap ("go"), but not to respond to two taps (now the "no-go" signal). The sensitivity of the procedure can be increased by using pseudorandom sequences which contain runs of four or five "go" trials before some of the "no-go" trials. While normals may make one or two errors in the first 10 trials, their errors drop off as testing proceeds. By contrast, impulsive patients remain unable to withhold a response consistently on "no-go" trials (even though they will often be aware that they are making errors, and even comment on them as they continue to respond inappropriately). In our experience, this test is very sensitive to fatigue and subtle attentional deficits.

LANGUAGE

The approach outlined in this section is based on a classical behavioral neurological approach to the diagnosis and classification of acquired language disorders (Benson & Geschwind, 1988). While such an approach remains adequate for clinical purposes, it is somewhat primitive from the standpoint of contemporary cognitive psychology and neurolinguistics. (For an alternative psycholinguistic approach to aphasia and related disorders, see Caplan, 1992). A clinically based language evaluation includes characterization of the following features.

Spontaneous speech should be observed, with attention to parameters such as (1) fluency and grammaticality; (2) information content; (3) spontaneity (not synonymous with fluency: speech output in response to questions may be fluent, yet the patient may produce no speech spontaneously); (4) articulation; and (5) volume and prosody (roughly speaking, the inflection, emphasis, and rhythm or cadence present in normal speech). Prosody is discussed in more detail later. The examiner should listen carefully for cirumlocution, word-finding difficulty, and the presence of paraphasias. The paraphasia may be phonemic (literal) (substitution of a syllable or sound in a word, e.g., "clable" instead of "table") or semantic (verbal) (substitution of a real but incorrect word, e.g., "watch" for "clock" or "shoe" for "necktie"). Neologisms are non-words (i.e., "blixel", "glibbet") which cannot be identified as semantic or phonemic paraphasias by any resemblance to contextually appropriate real words. Highly neologistic output is often referred to as "jargon." Echolalia refers to automatic repetition of verbal input. Palilalia represents a form of verbal perseveration in which the patient repeats syllables, words, or short phrases.

Repetition of sentences rich in closed class words (e.g., "no ifs, ands, or buts"; "He took them to see her"), low-frequency words ("The phantom soared across the foggy marsh"), and passive constructions ("The fighter readied himself for the championship bout") should be recorded. Nonsense words such as "pocotocopetal" are particularly sensitive screens for impairments in repetition.

The clinician should evaluate the patients' ability to *name visually presented objects* (glasses, watch, coat, pen, telephone) and their constituent parts (e.g., frame, lenses; winder or stem, crystal or face of a watch; collar, lapel; nib, clip; receiver, handset); body parts (shin, earlobe, elbow, knuckle, fingers). If abnormal, it is important to document the response to prompting with phonemic (e.g., "fr . . .") and semantic ("stimulus") cues (e.g., "It's the part of the eyeglasses that holds the lenses in place"), and to distinguish between true anomic and agnosic errors. For example, is the patient unable to name a visually presented stimulus item, and to accurately describe its function, but readily able to identify it by touch (suggesting a visual agnosia, rather than an anomia)? Such a patient who can nevertheless provide a visual description of object features may have an "associative" agnosia.

Comprehension can be assessed using "yes/no" questions ("Is a hammer good for cutting wood?"; "Do you put on your socks after your shoes?"; "Does a stone float in water"?); pointing commands ("Point to the ceiling/window/a device used to adjust the room temperature"); syntactically loaded sentences ("Touch the pen to the book"; "With the book, touch the pen"; "Touch the paper with the book and then with the pen"); and commands assessing right–left orientation ("Show me your left/right hand"; "Touch my left/right hand"; "Touch your left ear with your right hand"). In using "yes/no" commands, one should frequently insert questions that require negative responses, and also avoid strings of questions that have the same answer, since some aphasics have positive response biases or perseverative tendencies. If an attentive patient has normal spontaneous speech and shows no evidence of comprehension difficulties in history-taking, the yield of formal tests of comprehension is likely to be low. Syntactically loaded three-step commands which include right–left and finger identification elements ("Touch my left middle finger and then your left ear with your right thumb") can be used as quick and sensitive screens of auditory comprehension.

Writing is usually assessed by asking the patient to write (in script) a grammatical sentence describing the weather, or the reason for hospitalization or consultation. Writing should always be tested if a focal lesion in the language-dominant hemisphere is the differential diagnosis, because it may be impaired even when speech is normal. (Repetition should always be tested for the same reason.) Typically, in aphasias, the content of written productions parallels those of spoken output (i.e., telegraphic and agrammatic with anterior, and fluent but paraphasic and empty in posterior, perisylvian lesions). However, writing may be normal despite nonfluent, dysarthric, paraphasic, and effortful speech (as in the case of aphemia, and in some subcortical aphasias).

Reading a paragraph aloud is sufficient for evaluating reading comprehension. Reading comprehension must be tested in patients with auditory comprehension deficits, since decoding of visual and auditory input may be differentially impaired by posterior left perisylvian lesions (as in the "word blind" and "word deaf" variants of Wernicke's aphasia, respectively), and in pure word deafness. Patients with early AD and those with mild attentional impairments can usually read aloud orally (and repeat) but may have quite impaired reading comprehension. In non-aphasic pa-

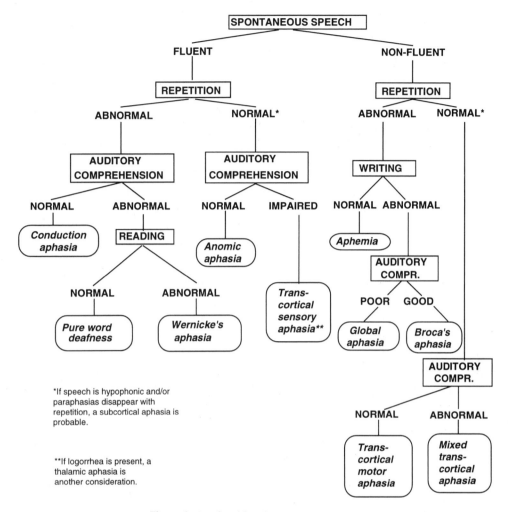

Figure 3. An algorithm for aphasia diagnosis.

tients with reading difficulty, depending on the nature of the disorder, there may be dissociations between the ability to identify single letters and digits and whole words and sentences, between real words and pronounceable non-words, and between phonologically regular and irregular words (Caplan, 1992).

An algorithm for classification of aphasias into classical syndromes is presented in Figure 3. It should be recognized that many patients, even those with cerebral infarctions, will not fit cleanly into the idealized categories of this classical typology.

In patients with nonfluent speech, it may be instructive to evaluate their ability to sing a simple song such as "Happy Birthday" or the "Stars and Stripes Forever." Some severely aphasic patients (to their astonishment) can still sing familiar songs surprisingly well. Patients with aphemia (known also as "apraxia of speech" and "cortical dysarthria") may show improvement in articulatory difficulty if they synchronize word production attempts with limb movements, such as heel-strike or lift-off during walking, or pounding their fist "on-the-tab-le-with-each-syl-la-ble." These observations may have therapeutic implications.

ROBERT M.
STOWE

Assessment of *affective prosody* routinely includes historical and behavioral observations (the patient's ability to communicate affect in facial expression, "body language" and gesticulation, and voice during the evaluation, and his or her responses to emotional aspects of interactions with family members and the examiner). When there is reason to suspect a problem with emotional aspects of communication, one can screen for receptive and expressive aprosodias further at the bedside by asking the patient (1) to guess the emotional state the examiner is trying to impart into an ambiguous phrase like "let's go to the movies" (while the examiner shields his face to avoid giving gestural cues, and intones this statement alternatively in angry, surprised, sad, happy, and sarcastic tones of voice); and (2) to speak a similar sentence provided neutrally by the examiner with these same tones of voice (i.e., "Say this sentence in a happy tone of voice. . ."). If the patient has difficulty with either or both of these tasks, then repetition of affective prosody is tested by asking the patient simply to repeat the sentence with the same emotional intonation as the examiner. Interpretation of facial and limb gestures may also be tested. Formal prosody batteries are available but they are not in widespread clinical use. Ross (1988, 1997) has argued that lesions in the right hemisphere homologous to those in the left which produce classical aphasias result in corresponding disturbances of emotional prosody, although this view is not universally accepted (Tompkins, 1995). Of course, expressive aprosodia is nonlateralizing in psychiatric disorders with blunting of affect, such as schizophrenia, catatonia, and psychotic depression, and in hypokinetic extrapyramidal disorders such as progressive supranuclear palsy (PSP) and Parkinson's disease (PD).

PRAXIS

Apraxia refers to the inability to perform skilled movements or sequences of motor movements on command, in the absence of primary cognitive, motor, and/or sensory abnormalities which would adequately explain this inability. Therefore, the general neurological examination should ideally precede evaluation for apraxia. Praxis should be tested for buccofacial movements (e.g., "Show me how you would lick crumbs off your lips/suck through a straw/blow out a match/blow a kiss"); unilateral and bimanual limb movements involving imagined use of objects, that is, *transitive* commands ("Pretend you're holding a hammer in one hand and a nail in the other; Show me how you'd hammer in the nail; Show me how you would flip a coin; use a saw/egg-beater/comb/toothbrush; kick a ball"), and those that do not involve object use, that is, *intransitive* praxis ("Show me how you would wave goodbye/salute/beckon 'come here'"). Finally, praxis for stance ("Show me how you would . . . stand like a boxer/swing a golf club/a baseball bat") and axial movements ("Stand up, turn around twice, then sit down") should be tested. Since apraxia in the absence of aphasia or dementia is rare, one or two of each of the three types of commands (buccofacial, limb, and stance) suffices as a screen. Other sensitive screens are to ask the patient to pantomine the act of taking a cigarette from a pack, lighting and smoking it, and to fold and insert a letter into an envelope and seal, stamp, and mail it. Heilman and Rothi (1993) point out that some apraxics can pantomime gesture and object use on command but not imitate them; however, the converse dissociation is seen more frequently. It is advisable to test limb praxis in the "nondominant" hand, since callosal lesions which disconnect the dominant supplementary motor area from premotor cortex in the right hemisphere can produce an isolated left-sided apraxia (and sometimes an "alien hand" syndrome in which the "nondominant" hand performs fairly complex but apparently involuntary motor acts, such as crumpling up paper or manipulating objects). Left hemisphere lesions are typically associated with buccofacial and limb apraxia, but may leave praxis for axial commands intact. Common features of apraxic performance include sequencing, timing, and perseverative errors; "body-part-as-object" (using the index finger as a toothbrush, for example), which should not be considered abnormal unless they persist after instruction; and spatial errors which include extensional (failing to imagine the length of an object such as a hammer), orientation, movement, and postural errors (abnormal limb positioning, e.g., waving goodbye with the palm facing the patient). If errors are made, the examiner should pantomime the desired movement and see if the patient can imitate it. Typically in ideomotor and "disassociation" apraxias (Heilman, Watson, & Rothi, 1997), deficits will improve to imitation and especially with object use.

CALCULATIONS

When the history does not suggest calculation difficulties, and language examination is normal, asking the patient to "make change" mentally from $10.00 is a reasonable screen. If acalculia is suspected, written long division, multiplication, addition, and subtraction problems should be given. The most common cause of calculation errors seen in clinical practice is not a true acalculia (anarithmetria), but rather an attentional or working memory deficit. Impairment may also reflect neglect or visuospatial dysfunction (improper positioning of numbers in written calculations), agraphia for numbers, aphasia, alexia, or impaired memory for math table values (frequently seen in adults with a history of dyslexia). For a detailed review of assessment and interpretation of calculation disorders, see Levin, Goldstein, and Spiers (1993).

VISUOSPATIAL FUNCTIONS

Assuming visual acuity is adequate and motor function in the dominant hand is intact, the patient can be asked to draw a cube, or a house in perspective (so that more than one side is visible), and to copy two partially overlapping geometric figures (such as intersecting pentagons) as basic screens for visuospatial impairment. It is important to test both constructions to command and to copy, since these processes may be independently impaired. When asked to draw a cube, many people will use a "trick" and draw and connect two overlapping squares. This, of course, does not require three-dimensional conceptualization, and if the patient does this or is unable to produce an acceptable cube on command, the examiner should provide one for the patient to copy. Many apparently normal elderly individuals have difficulty perceiving and copying the three-dimensional aspect of such drawings, but should be able to copy two-dimensional figures, such as intersecting pentagons or a Greek cross. Inability to do so may reflect perceptual impairment, constructional apraxia, or both. In "apperceptive" visual agnosia, copying of designs and line drawings of simple objects may be impossible or performed slavishly, without recognition of the object being drawn. (Visual object agnosias are discussed briefly above, in the discussion of naming). Differentiating among these possibilities may require more purely "perceptual" tests of visuoperceptual function

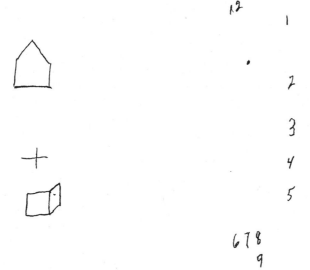

Figure 4. Constructions in a patient with hemispatial neglect. The subject was asked to draw a house, a Greek cross, a cube, and a clock. Note simplification, omission of the external configuration (clock face), and left neglect (clock number placement, cube).

administered by neuropsychologists. Particular note should be made of inattention, perseveration, and impulsivity, which are frequent contributors to failure on these tests. The patient should also be asked to draw a clock and to "place the hands at 10 after 11." (These numbers are not arbitrarily chosen; to place the minute hand correctly, one must abstract from "10" to the "2" which indicates the 10-minute position on an analog clock. The stimulus-bound or concrete patient will place the minute hand at the "10" or even write in "10" adjacent to the "11".)

Examples of constructions in a patient with hemispatial neglect (whose cancellation test results are shown in Figure 1) are provided in Figure 4.

MEMORY

Contemporary cognitive science emphasizes distinctions between different hypothesized systems for memory and learning, leading to dichotomies such as implicit versus explicit, procedural versus declarative, habits versus memories, context versus item, and episodic versus semantic memory, in addition to the postulation of a variety of specialized modules for iconic and modality-specific working memory (Squire, 1992; Nadel, 1992). In clinical practice, explicit testing of memory is aimed primarily at identifying disorders of declarative, *episodic* memory, using tests of what has traditionally been termed "short-term memory," while semantic memory is tapped somewhat more obliquely in other areas of the mental status by assessment of naming and other language abilities, calculations, interpretation of similarities, and so on. Traditional bedside tests for "retrograde amnesia" (such as asking patients to name all the U.S. Presidents in order after Eisenhower, or to give the dates of world wars and the Vietnam conflict, for example) also tap semantic memory, but in our experience they rarely provide useful specific information, since almost all patients with impairments on such tasks are also impaired on anterograde memory tests and/or other measures of semantic memory. Tests of procedural learning are available and have proven differentially sensitive to basal ganglia disease (Phillips & Carr, 1987; Saint-Cyr & Taylor, 1992), but they have not found their way to the bedside, and are unlikely to do so, as the general neurological exam is probably at least as sensitive for identifying basal ganglia involvement, which typically manifests with motor abnormalities. The most clinically useful tests of memory provide information about (1) the subject's ability to register, or *encode* new information; (2) *retrieval,* both immediately and after a delay, without the aid of prompts or cues; (3) cued retrieval and/or *recognition memory;* (4) the effects of rehearsal (which gives an estimate of the rate at which information can be encoded); and (5) any *modality specificity* to observed deficits; typically, verbal and visual/spatial memory are contrasted since the former is usually more affected by left temporal lobe damage, whereas right temporal lesions may produce selective impairment of spatial and figural information, particularly when it cannot be easily verbally encoded. These facets of memory may be differentially affected depending on the locus and etiology of dysfunction; hence tests which fractionate these components of memory performance have particular diagnostic utility. (It should be obvious that, by contrast, the standard clinical practice of asking patients to remember three unrelated words, repeated once, after 5 minutes of distraction, is of little value as a test of memory in many clinical situations.) One such useful "bedside" memory instrument, the Three-Words Three-Shapes Test (Weintraub & Mesulam, 1985; see Figure 5) controls for perceptual/constructional deficits (by requiring patients to copy the stimuli first), and assesses *incidental recall* (immediately after copying, with no forewarning to the patient to memorize the stimuli); the effects of serial study or rehearsal to criterion (the ability to produce five of the six stimuli immediately; subjects get up to five 30-second study periods as needed to reach criterion, and the number of trials necessary to reach it is a measure of the efficiency of encoding); and *delayed recall* (in standard administration, assessed after 5, 15, and finally 30 minutes of intervening distraction). If all six stimuli cannot be accurately produced, the patient is asked to identify the stimuli from a multiple-choice form containing 10 words and 10 shapes. (This test was designed mainly for use in elderly individuals; for younger patients, sensitivity can be increased by adding two words and two shapes from the multiple-choice form to the stimuli to be learned, with criterion being eight of the ten items). Prospective memory can be assessed by asking the patient to "remind" the examiner in 30 minutes that the patient is supposed to redraw the stimuli from memory at that point. An example of performance from a patient with relatively early and prototypical AD is shown in Figure 5. At the time of testing, the patient was fully oriented, socially appropriate, still living independently with some assistance from nearby family members, and was able to recite the months of the year in reverse order in 30 seconds without error. Confrontational naming of line drawings of even uncommon objects was within normal limits for age, but the patient's spontaneous speech was anomic and circumlocutory. Note the spatial errors in copying; impaired encoding of both verbal and visual material; and successive decay of encoded information across delay intervals, both on attempts to reproduce and to recognize the stimuli.

A simple 10-item story can also be used to assess logical memory (e.g., "There was a fire at St. Joseph's School on Fifth Avenue. It happened on a Saturday afternoon at 4:30 p.m., so no school children were injured, but two firemen were treated for smoke inhalation. One was taken to Forbes Regional Hospital, the other to Montefiore Hospital.") The subject is asked to repeat the story immediately, and drilled on those items which cannot be retrieved with cues, until the story has been fully encoded. The patient is asked to recall the story after 5 to 10 or 15 minutes of distraction, and the examiner notes the number of items produced spontaneously and after specific probes (e.g., "What time did the fire occur?") if uncued recall is incomplete, as well as any intrusions (confusion with items from earlier testing) or confabulated material.

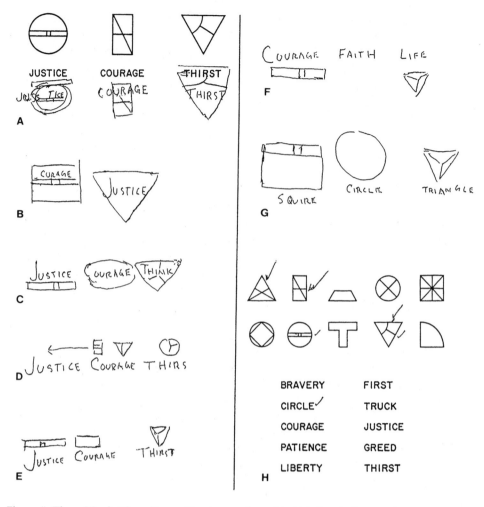

Figure 5. Three Words Three Shapes Test in a patient with Alzheimer's disease. Shown above are (a) copy of stimuli (note spatial errors in placement of the shapes); (b) incidental recall; (c, d, e) recall after successive 30-second study periods; (f, g) recall after 5 and 30 minutes of distraction; and (h) multiple-choice recognition (note false-positive response to the first distractor and second word and inability to recognize any of the previously encoded words).

When patients are too impaired for the Three-Words Three-Shapes and logi-cal memory tests, three objects (preferably, items belonging to the patient) can be hidden in three separate locations around the room, while the patient is asked to watch closely. Repeated drilling is conducted until the patient can immediately recall the names and locations of all three items. Recall is tested again after 5 minutes of intervening distraction, with the aid of cuing and/or multiple-choice questions if spontaneous retrieval is incomplete. In a depressed or poorly moti-vated patient, memory for personally relevant material (such as recent family events) may be better than for pencil-and-paper tests. A discrepancy between the latter and the patient's ability to give a coherent history with accurate details about dates, tests, and names of tests and medical and nursing personnel, and/or to recount recent news items accurately, may also reflect the influence of subjective factors such as motivation and interest.

Prosopagnosia refers to a selective defect in memory for familiar faces. Patients with the pure form of this syndrome can describe facial features and match photographs of faces taken from different orientations, but may be severely impaired in their ability to recognize even immediate family members by sight. Other prosopagnosics have difficulty with facial discrimination and feature identification. Assessment of prosopagnosia requires special materials described elsewhere in this volume. Modality-specific memory deficits have been described for other types of material, including olfactory and tactile stimuli (Ross, 1980). While of theoretical interest, these are not usually evaluated in routine clinical practice.

ABSTRACT REASONING AND PROBLEM-SOLVING

The assessment and interpretation of abstract reasoning abilities (especially verbal ones) is particularly problematic, since these are highly dependent on premorbid intellectual capacity, education, and sociocultural factors, and they are easily perturbed secondarily by attentional, language, and visuoperceptual deficits, as well as normal aging. Traditionally used bedside tests such as interpretation of similarities and proverbs do not assess visuospatial abstraction abilities and are therefore likely to be particularly insensitive to right prefrontal pathology (Lezak, 1995). When subtle frontal impairment is suspected, or questions arise as to the sensitivity and specificity of bedside examination due to high estimated premorbid abilities and/or education, formal neuropsychological testing is strongly advisable. It is also generally acknowledged that while abstract reasoning, problem-solving, and other "executive" functions may be particularly sensitive to frontal dysfunction, high-functioning individuals who develop frontal lobe pathology may perform normally even on many neuropsychological tests of abstraction and problem-solving while manifesting evidence of impaired goal-setting, planning, insight, judgment, and social functioning in "real life" (Damasio & Anderson, 1993). Furthermore, it should be kept in mind that the relative specificity of tests of abstraction abilities and "executive function" for frontal dysfunction only applies when other functions which contribute to task performance such as basic attentional abilities, language, and relevant aspects of visuoperceptual function and declarative memory are intact (Anderson, Damasio, Dallas-Jones, & Tranel, 1991).

The ability to "abstract" *similarities*, a measure of verbal "convergent" reasoning, is assessed by asking "What do (two items belonging to a particular abstract category or semantic class, i.e., an apple and an orange, table and a chair, tree and an ant, poem and a statue) have in common with each other; how are they alike?" *Proverb interpretation*, one measure of "divergent thinking," should proceed first by confirming that the patient understands what a proverb is ("a saying that has a broader meaning/a moral") and trying to find one already familiar to him or her (e.g., "Don't put all your eggs in one basket"; "Don't count your chickens before they hatch"). If even familiar proverbs cannot be interpreted, further examination along these lines is meaningless. On the other hand, the ability to give interpretations of familiar proverbs may reflect rote responses to previously learned material, rather than current abstraction abilities; hence, unfamiliar proverbs should be used next (see Table 2 for examples). The determination of "normal" must take into account the patient's educational and vocational achievements, as well as age (some impairment in abstract reasoning ability being quite common in normal aging). While patients with global intellectual impairment or frontal dysfunction are often very "concrete," schizophrenics may give apparently idiosyncratic or bizarre interpretations of proverbs.

A color/shape/letter sorting task, a crude bedside test of "conceptual flex-

TABLE 2. INTERPRETATION OF PROVERBS

Proverb	Interpretation		
	Very concrete	Mildly concrete	Abstract
Hunger makes the best gravy	"You like gravy when you're hungry"	"Anything tastes good when you're hungry"	"Being deprived of something makes it more enjoyable when you finally get it"
The hot coal burns, the cold one blackens	"Red-hot coal burns you, cold coal is black"	"Both things are bad"	"Extremes of anything are bad"
A gold key opens a steel door	"Keys can open locks"	"Money can buy anything"	"Knowledge is power"

ibility" or "divergent thinking," can be constructed by cutting three circles, squares, and triangles out of three sheets of material, each with a different color. Each of the nine stimuli is then lettered A, B, or C, such that three different groupings of the nine stimuli can be generated by sorting according to color, shape, or letter. The patient is asked to sort the stimuli into three piles "so that all the things in a pile have something in common with each other." After the first sort, they are asked "Is there another way you could have sorted these?", and so on. In our experience, cognitively intact subjects of normal intelligence will achieve at least two of the three possible sorts.

INSIGHT AND JUDGMENT

Insight is usually used to refer to the patient's awareness of his or her deficits. *Anosagnosia* (lack of awareness of a deficit) is particularly striking in hemispatial neglect; fluent aphasias (in contrast to Broca's aphasia, where the patient is frequently very frustrated and depressed by expressive language deficits); cortical blindness associated with bilateral occipitotemporal lesions (Anton's syndrome); bilateral or large right prefrontal lesions; and most amnestic syndromes and dementias. Although bright patients with AD may retain insight in the early stages, true self-awareness and emotional insight is frequently lacking. Lack of concern over, and glib excuses for, failures during testing, and inappropriate jocularity or facetiousness is common. *Judgment* is often best assessed from reports by collateral informants (significant others and supervisors) about the patient's actual behavior. A patient with a very early frontal dementia may behave appropriately in the structured setting of the physician's office, while making errors of judgment and social conduct in other situations. Furthermore, the "ecological validity" of using questions about sensible behavior in paradigmatic situations, commonly used to assess judgment, is dubious. (Examples of such questions, such as those from the Comprehension subtest of the WAIS-R, include "What would you do in a crowded theater if you smelled something burning?", and "What would you do if you found a envelope, bearing an uncanceled stamp, on the street?"). Patients may give appropriate responses (which are frequently overlearned) to such questions, while behaving in an egocentric or impulsive manner in real-life situations. It is widely believed by neuropsychologists and behavioral neurologists that good formal tests of judgment are sorely lacking (Mesulam, 1986).

ASSESSMENT OF RISK TO SELF AND OTHERS

461

BEHAVIORAL
NEUROLOGY
AND NEURO-
PSYCHIATRY

Where relevant, it is very important to document the presence or absence of persecutory delusions, aggressivity, and suicidal or homicidal ideation, and the examiner's estimate of the risk of suicide and aggression toward family members and others. In a cognitively impaired individual, competence to manage finances and to perform activities required for independent living, such as bathing, shopping and cooking, and route-finding, and safety concerns such as whether the patient can safely drive, operate complex or heavy equipment, perform child care unsupervised, or remember to turn off stove burners and unplug electric kettles must be considered. When this information cannot be obtained from a careful history, referral to an occupational therapist or neuropsychologist familiar with ADL assessment may be required.

General Physical and Neurological Examinations

As standard textbooks covering physical examination, general internal medicine, and the finer points of the general neurological examination are readily available, the emphasis here is on providing an overview of standard clinical examination techniques, findings, and in some circumstances, differential diagnostic considerations, with a bias towards those of particular relevance to the examination of patients presenting primarily with cognitive or psychiatric symptoms. Although some attempt is made to explain or simplify medical terminology and basic principles of physical diagnosis, space considerations preclude this in many instances, and nonphysician readers may find the book by Bates (1999) on physical diagnosis helpful. Except where otherwise noted, references for the material that follows may be found in other standard textbooks including Wilson *et al.*, 1991; Ahlskog, J. E., *et al.*, 1991; Brazis, Masdeu & Biller, 1990; DeJong, 1979; Samuels and Feske, 1996; Adams, Victor, & Ropper, 1997; and DeMyer, 1994.

The general physical examination traditionally includes measurement of "vital signs" (blood pressure, heart and respiratory rates, and temperature) as well as systematic examination of the skin and nails, head and neck, lungs, heart and major blood vessels, abdomen, spine and joints, and (where indicated) internal pelvic and rectal examinations. In neurobehavioral consultation, the most critical information is provided by the history and MSE, and usually only highly relevant aspects of the general physical (and in many instances, the neurological) examinations are performed to help confirm tentative diagnoses or adjudicate between a few diagnostic possibilities.

Blood pressure (BP), heart and respiratory rates (HR, RR), and temperature (T) are routinely measured. These parameters are regulated by the autonomic nervous system (ANS) and may be affected by dysfunction of components of the ANS situated in the hypothalamus, brainstem, spinal cord, autonomic ganglia, peripheral nervous system, or peripheral catecholaminergic and cholinergic receptors. Signs and symptoms of ANS dysfunction may include lightheadedness; falling and/or syncope on standing up due to ortohostatic hypotension; severe constipation or diarrhea; urinary retention, impotence, incontinence; sluggish pupillary responses or abnormally large or small pupils; and changes in sweating. ANS dysfunction is common in patients with long-standing diabetes mellitus and degenerative disorders of the extrapyramidal system such as PD and multiple system atrophies. It also occurs acutely in porphyria and neuroleptic malignant syndrome (where lethal cardiac arrhythmias may occur). Orthostatic hypotension (a fall in systolic blood pressure exceeding 10 mg Hg on standing) is also often related to psychotropic and cardioactive medications (especially neuroleptics, tricyclic anti-depressants, anti-Parkinsonian medications, and drugs used to treat high blood pressure) and/or to dehydration. When patients complain of feeling light-headed or "dizzy" on standing, blood pressure (BP) and pulse rate should be checked not only in the standard sitting position, but also after lying supine for five minutes, on standing immediately, and then after standing for two to five minutes. A rise in pulse rate exceeding twenty beats per minute on standing suggests dehydration, or a compensatory tachycardia when normal vasopressor reflexes have failed. The absence of a significant tachycardia when BP falls more than 15 mm Hg suggests autonomic dysfunction. Addison's disease, in which steroid production by the adrenal glands is decreased, is associated with orthostatic hypotension, weight loss, and depression. Conversely, high steroid levels (whether endogenously mediated, as in Cushing's disease, or iatrogenic) may manifest as hypertension, fluid retention, trunkal obesity, acne, and hursuitism. Neuropsychiatrically, these patients may present with headache, insomnia, anxiety, affective lability, irritability, depression, mania, and/or psychosis.

An "irregularly irregular" pulse suggests atrial fibrillation. This is a common cause of cardioembolic stroke, especially in association with rheumatic mitral stenosis. It may also be a clue to hyperthryoidism and atherosclerotic heart disease. Abnormal respiratory rate or breathing patterns, while most commonly associated with cardiopulmonary disease, may reflect abnormalities of central respiratory control, due to metabolic or structural dysfunction of the brainstem reticular formation. Unexplained hypothermia (low temperature) may be a clue to hypothyroidism, acute Wernicke–Korsakoff syndrome, and hypothalamic disease. Fever, while most commonly a sign of infection, may also occur in some malignancies, deep venous thrombosis, and allergic reactions to drugs. In confused psychiatric patients, neuroleptic malignant syndrome and malignant catatonia are important diagnostic considerations (see below).

Changes in the skin may be important clues to acquired systemic diseases or heritable disorders affecting the CNS. In patients with mental retardation, learning disabilities, and/or partial (focal) seizures, it is important to look for the cutaneous and subcutaneous tumors, axillary "freckles," and "cafe-au-lait macules of neuro-fibromatosis (more than five of these light brown, flat spots is considered suggestive), and the facial angiofibromas ("adenoma sebaceum"), lumbosacral "shagreen patch", and the depigmented "ash-leaf macules" of tuberous sclerosis. The latter may only be visible with ultraviolet (Wood's lamp) illumination. A "port wine" stain on the face may be a clue to an intracranial vascular malformation giving rise to seizures (Sturge–Weber syndrome), as may the presence of multiple tiny angiomas and telangectasias of the skin and lips (Osler–Rendu–Weber syndrome, also known as hereditary hemorrhagic telangectasia). "Spider" angiomas of the face and abdomen and palmar erythema raise the possibility of cirrhosis. Associated Dupytren's contractures (thickening and shortening of the palmar flexor tendons) implicate alcoholism as the cause of chronic liver disease. Hyperpigmentation of the skin may be seen in adrenoleukodystrophy ("bronzing") and Addison's disease (in the latter, pigmentation of the buccal mucosa and palmar skin creases differentiates it from a normal suntan). Hypopigmentation and eczema occur in phenylketonuria. Photosensitivity (unusual tendency to burn on mild sun exposure) is commonly a side effect of medications, including some neuroleptics and carbamazepine, and also occurs in rare conditions such as xeroderma pigmentosum, albinism, Hartnup disease (associated with migrainous headaches, ataxia, and sometimes chorea) and some porphyrias (variegate porphyria, and infrequently hereditary coproporphyria) but not others (acute intermittent porphyria and porphobilinogen synthetase deficiency) that have neuropsychiatric manifestations. Reddish or purplish papules, plaques or nodules, often located on the face (especially the earlobes and nose), but also commonly in the oral mucosa and on the soles of the feet, are seen in Kaposi's sarcoma, a common complication of AIDS which may be the presenting feature of the disease. In a patient with confusion or psychosis, the presence of a "malar" rash (over the cheeks) may be a clue to systemic lupus erythematosus (SLE), porphyria (especially in association with unexplained abdominal pain), Hartnup disease, or pellagra (multiple B vitamin deficiencies). Most commonly, however, a rash in this region which prominently involves the scalp is caused by seborrheic dermatitis ("dandruff"). Petechiae (small flat purplish lesions reflecting inadequate numbers of function of platelets, leading to focal hemorrhages under the skin or in the conjunctivae of the eyes) may be a clue to SLE, thrombotic thrombocytopenic purpura (associated with rental dysfunction, microangiopathic hemolytic anemia, and multiple small cortical infarcts), or CNS infection (meningococcal meningitis, Rocky Mountain spotted fever). So-called "splinter hemorrhages" under the nails and small embolic infarcts in the distal extremities are seen in subacute bacterial endocarditis. Xanthomas (subcutaneous accumulations of cholesterol crystals) located over tendons are suggestive of cerebrotendinous xanthomatosis, a rare cause of dementia. Erythema chronicum migricans, an enlarging raised, reddened skin lesion which clears first from the center (in a "bull's eye" fashion), is characteristic of acute infection with the Lyme disease agent, *Ixodes dammani*. Months or years later, this tic-borne spirochetal infection can produce a fluctuating, chronic mild encephalopathy (in addition to headache, arthritis, facial and less commonly other cranial nerve palsies, peripheral neuropathy, and cardiac conduction defects) and rarely, profound dementia.

When the history is unreliable or unobtainable, it may be important to palpate the skull for signs of previous craniotomies and deformities due to trauma. Sinuses should also be palpated and percussed for tenderness in patients with headache. The ears should be inspected if mastoiditis (infection of the petrous temporal bone, which can be associated with intracranial abscess formation) is a consideration. Involuntary resistance to anterior (but not lateral) flexion of the neck suggests irritation of the meningeal coverings of the brain (whether due to acute or chronic infection, inflammation, metastatic tumor, recent subarachnoid hemorrhage, or a chemical meningitis due to medications). When learning disabilities and mental retardation are suspected, head shape and head circumference (the largest occipitofrontal diameter) should be assessed. Asymmetries in facial structure, the size, placement, and shape of the orbits, and abnormal earlobe morphology may be clues to aberrant cortical development. "Coarsened" facial features may be seen in acromegaly (a syndrome seen with pituitary tumors producing excessive growth hormone), chronic phenytoin (Dilantin®) therapy, and a variety of lysosomal storage diseases (see below). The thyroid gland may be diffusely enlarged in hypothyroidism (which may be associated with abulia, apathy, hypersomnolence, and depression, dry skin, hoarseness, obesity, cold intolerance, coarse brittle scalp hair, bradycardia, constipation, and so-called hung up ankle jerk reflexes). It may be nodular in hyperthyroidism (associated with confusion, mania, insomnia, or psychosis; tachycardia, a postural and action tremor, excessive sweating and heat intolerance, weight loss, and diarrhea) or in Hashimoto's thyroiditis, an autoimmune disorder associated in some cases with a fluctuating encephalopathy, myoclonus, and seizures apparently unrelated to thyroid function as measured by standard tests (Shaw, Walls, Newman *et al.*, 1991). Widening of the palpebral fissures, proptosis or exopthalmos (forward displacement of the eyeball) and diplopia (most often on upward gaze) may also be a clue to autoimmune thyroid disease ("thyroid ophthalmopathy" of Grave's disease). Dry eyes and mouth are commonly produced by anticholinergic medications. These signs of the "sicca syndrome," and sometimes parotid gland enlargement, are also seen in Sjogren's syndrome; and oral ulcers, conjunctivitis, and iritis (see below) are features of Behçet's syndrome. Both of these diseases can produce cognitive and neuroradiological manifestations resembling those seen in multiple sclerosis as well as a vascular myelopathy (spinal cord syndrome). Pharyngeal and oral "thrush" (candidiasis), in the absence of other causes such as treatment with antibiotics or immunosuppressive drugs, raises the possibility of HIV infection. Abnormally enlarged lymph nodes in the head and neck (or in the axillae or groin) may be a clue to a lymphoreticular malignancy (e.g., lymphoma, Hodgkin's disease, leukemia, lymphomatoid granulomatosis) or pulmonary malignancy; an inflammatory or autoimmune disorder (e.g., SLE, Sjogren's, rheumatoid arthritis, dermatomyositis); or a chronic systemic infection such as HIV, TB, brucellosis, Whipple's disease, or sarcoidosis. They also occur in chronic fatigue syndrome (CFS) along with low-grade intermittent fever, fatigue, and sometimes sore throat, muscle and joint pains, sleep disturbance, dysphoria and attention, concentration and memory complaints (often not demonstrable on formal tests of cognitive function).

In patients with strokes, TIAs, "blackouts," and focal seizures, the exam commonly includes estimation of pulse volume of the common and external carotid arteries in the neck and their extracranial branches (principally the occipital, superficial temporal, supraorbital and supratrochlear arteries), and arterial auscultation (with a stethoscope) for "bruits" over these vessels, the orbits, and the subclavian and vertebral arteries (in the neck). Highpitched, short bruits may reflect hemodynamically significant narrowing of a blood vessel due to atherosclerosis or inflammation, or increased flow in "collateral" vessels originating from the external carotid system providing alternate routes of blood supply to a compromised arterial territory or to an intracranial arteriovenous malformation. The heart is also auscultated for murmurs and extra sounds (including the opening snap and diastolic rumble of mitral stenosis and the early– to mid-systolic click of mitral valve prolapse, conditions which are risk factors for embolic stroke). Murmurs may also be a clue to rheumatic fever (associated with Sydenham's chorea), and to infective, or nonbacterial ("marantic") endocarditis which can be paraneoplastic (i.e., a geographically remote complication of an underlying malignancy) or associated with antiphospholipid antibody ("lupus anticoagulant") syndrome (with or without SLE). In the latter disorder, a pericardial friction rub may also be present. Auscultation of the lungs may reveal crepitations ("creps"/crackles) in pulmonary edema and disorders causing pulmonary fibrosis, whereas rhochi ("wheezes") are signs of obstruction or constriction of air passages. Dullness to percussion may indicate consolidation (local collapse) of areas of the lung or pleural effusions (fluid collections in the pleural cavities surrounding the lungs). Increased fremitus (transmission of vocal sounds, typically elicited by asking the patient to repeat "ninety-nine") is a sign of consolidation. A pleural "friction rub" may be heard in pleuritis associated with malignancy or SLE. Bronchogenic carcinoma (especially small-cell) may be associated with a paraneoplastic limbic encephalomyelitis.

Palpation and percussion of the abdomen are typically done by neurologists only if there is reason to suspect an "acute abdomen" (due to a perforated viscus, acute pancreatitis, cholecystitis, appendicitis, etc.) as a cause of delerium; pancreatic or colonic malignancies, which can also produce a paraneoplastic limbic encephalitis or CNS metastases; or hepatomegaly or splenomegaly (enlargement of the liver and spleen, respectively). Hepatosplenomegaly occurs in some lysosomal storage diseases, in which substances such as glycolipids produced in the brain and other organs cannot be broken down at a normal rate and accumulated in these organs. Hepatomegaly is seen in acute alcoholic and viral hepatitis (as well as congestive heart failure and sometimes in poorly controlled diabetes). Acute hepatitis and cirrhosis of the liver can lead to "portal-systemic" or hepatic encephalopathy, in which ammonia and other compounds normally detoxified by the liver can bypass it, leading to confusion. A shrunken liver and enlarged spleen are commonly found in chronic cirrhosis with portal hypertension. Splenomegaly may also be a sign of a lymphoreticular malignancy, chronic infection, or an autoimmune disorder.

Inflammatory arthritic changes in the small joints of the hands and feet and the knees can be associated (in the case of SLE and Sjogren's syndrome, for example) with vasculitis or other types of inflammatory and autoimmune pathology in the CNS (and more commonly, the peripheral nervous system). Arthritis of large joints, especially the knees, can be prominent in the later stages of Lyme disease (osteoarthritis, of course, is a much more common cause of chronic noninflammatory knee pain). Arthropathy also occurs in Whipple's disease (see below). Stiff joints and bony abnormalities may be seen in inherited biochemical disorders including mucopolysaccharidoses, sialodoses and mucolipidoses and fucosidosis. Bone pain, fractures, and multiple cystic bony lesions in the hands and feet are seen in membranous lipodystrophy, a rare dementing disorder associated with ataxia, tremor, pyramidal tract signs, seizures, and basal ganglia calcification (Reichman & Cummings, 1990).

Palpation and percussion over the flanks for kidney tenderness is done if renal infection (pyelonephritis) is suspected. Rectal examination is done to rule out prostatic pathology in males with difficulty voiding, and to check the tone of the anal sphincter if there is fecal incontinence. The absence of normal superficial reflexes (bulbocavernosus and cremasteric, anal "wink") may be associated with autonomic neuropathies. Genital (and oral) ulcers in a young patient with dementia or multiple subcortical strokes may be a clue to Behcet's syndrome. Small testicles are seen in Kallman's syndrome (congenital anosmia and hypogonadism, associated sometimes with learning disabilities) and in myotonic dystrophy and Klinefelter's syndrome, both of which are often associated with cognitive deficits and sometimes ill-defined personality disorders. Pelvic, breast, and scrotal examination are indicated if a CNS tumor or a paraneoplastic neurological syndrome is suspected.

The *general neurological examination* is usually performed and recorded in a specific sequence, which helps inexperienced examiners to remember the procedure and findings. We have followed the time-honored division and ordering of "cranial nerves, motor, sensory, cerebellar, and gait" in the discussion that follows. Examining for *soft neurological signs* is traditionally a part of the neurobehavioral examination. This term has most commonly been used to refer to findings on sensory and motor aspects of the neurological examination which disappear in the course of normal development. They have been considered "soft" because (on the whole) they have low predictive and localizing value for the presence of structural pathology of the CNS. Soft signs are of interest in neuropsychiatry as clues to a possible developmental etiology of cognitive impairment (e.g., in learning-disabled children), early brain injury, and subtle neuropathology in psychiatric disorders (e.g., obsessive–compulsive disorder, schizophrenia). In psychiatric disorders they may also have some prognostic significance (Hollander *et al.*, 1991; King, Wilson, Cooper, & Waddington, 1991). (For a detailed review of theoretical and clinical issues associated with soft signs, see Deuel and Robinson, 1987.) From the perspective of systems anatomy and clinical neurology, soft signs constitute a "hodge-podge" of tests which are neither theoretically nor empirically cohesive; therefore, we have incorporated the ones we find most helpful and most commonly used in adult clinical practice into the appropriate components of the examination (e.g., motor, sensory, cerebellar, etc.).

CRANIAL NERVES

There are 12 pairs of cranial nerves: olfactory, optic, oculomotor, trochlear, trigeminal, abducens, facial, stato-acoustic, glossopharyngeal, vagus, accessory, and hypoglossal. Testing of the oculomotor, trochlear, and abducens cranial nerves will be considered jointly, because of their shared role in eye movements.

The *olfactory nerves* mediate smell (which plays a very important role in taste as well). This is typically only tested when an olfactory groove meningioma or orbito-frontal cortex lesion is suspected (in patients with frontal findings) and after head trauma and in suspected cases of Kallman's syndrome. Smell should be tested using non-astringent substances such as powdered coffee, cinnamon, chocolate, or perfumed soaps, since astringents such as ammonia and alcohol also stimulate trigeminal nerve sensory fibers.

The *optic nerves* mediate visual perception and the afferent limb of the pupillary light reflex. *Visual acuity* is frequently degraded by lesions of the optic nerve, macula, and occipital poles and best corrected visual acuity should be assessed in each eye in patients with visuoperceptual dysfunction or complaints. In testing *visual fields* by confrontation, the examiner is seated opposite the patient, using his or her fields as a guide, and small finger movements are made in the extreme periphery of the field and slowly brought in toward the central fields until the patient (who maintains central fixation) detects them. If hemispatial neglect is suspected, detection of bilateral finger movements in homologous areas of each hemifield (double simultaneous stimulation) is performed to check for "extinction," that is, failure to report movement in the neglected hemispace. When visual neglect is severe, patients are often falsely assumed to be hemianopic, because the presence of any visual stimulus (even a static one) in the non-neglected hemifield may result in extinction. It may be helpful in such patients to test the apparently hemianopic field in the dark, presenting illuminated stimuli only in the "bad" field. Optic nerve lesions (see discussion of optic atrophy for some causes) obviously give rise to defects restricted to the affected eye. Homonymous hemianopia (loss of vision in the contralateral visual hemifield of *each* eye) is most likely to be secondary to infarction of the occipital lobe; it may also be seen with lesions involving the optic tract or the lateral geniculate nucleus, the thalamic relay for visual information. Lesions of the optic radiations coursing through the parietal lobe can produce hemianopic or inferior homonymous quadrantic defects, whereas the optic radiations in Meyer's (Flechsig's) loop originating from the lower retina (and therefore subserving the upper quadrants of the visual field, since the retinal image is inverted by the lens) travel next to the temporal horn of the lateral ventricle, and temporal lobe lesions may produce a superior homonymous quadrantic defect.

Either of these defects can be produced by incomplete occipital lesions as well. Hemianopias due to cortical lesions may show interesting dissociations in the affected field, including variable loss of color perception, detection of moving versus static targets, form recognition, reading, and face recognition (the latter condition, "prosopagnosia" is usually associated with bilateral ventral occipitotemporal lesions, right greater than left). "Blindsight" (evidence of "unconscious" visual perception, i.e., the ability to point to a target even when the subject denies having seen it) can also be rarely observed. Conversely, in optic ataxia (another component of Balint's syndrome), a target can be perceived but not accurately reached for, because of damage to the dorsal visual pathway (the "where" visual system) in the dorsal parieto-occipital region, where spatial information about objects is processed and integrated with other sensorimotor modalities.

Ophthalmoscopy of the fundus is routinely done, checking for optic atrophy (the optic disc looks pale), papilledema (swelling of the optic nerve head due to raised intracranial pressure), papillitis (inflammation of the nerve head), and abnormalities related to retinal vascular disease, such as diabetic and hypertensive retinopathies. Fundoscopy may also uncover clues to phakomatosis, lysosomal storage diseases, and AIDS. Ophthalmoscopy of the rest of the eye may uncover relevant abnormalities in the conjunctiva, lens (e.g., cataracts), cornea (greenish-brown Kayser-Fleischer rings of Wilson's disease), sclera, vitreous, and iris (Lisch nodules).

The *oculomotor, trochlear, and abducens nerves* (III, IV, VI) mediate eye movements. The oculomotor nerve also controls elevation of the eyelid and pupillary constriction. Pupillary responses are tested using a bright, focused light, while the patient is asked to look off in the distance (to avoid convergence, which causes reflex pupillary constriction). Asymmetrical pupils are commonly due to cataract extractions, eye drops, or old trauma, but may reflect a Horner's syndrome or compression of the oculomotor nerve (see below). Sluggish or absent pupillary responses to light are seen in autonomic neuropathies and neurosyphilis. In the latter condition, and also in CNS Lyme disease, the pupils may constrict normally on convergence ("light-near dissociation"). Light-near dissociation in either direction may also be seen with pretectal midbrain pathology (intrinsic lesions; dilation of the aqueduct of Sylvius in obstructive hydrocephalus; or compression from a pineal region neoplasm).

A narrowed palpebral fissure and/or *ptosis* (drooping of an eyelid) may reflect a Horner's syndrome (usually associated with a small pupil on the affected side), indicative of a lesion in sympathetic pathways to the eye, which can be involved in the hypothalamus, brain stem, upper thoracic region, or head and neck (commonly seen with occlusion of the carotid artery). In a congenital Horner's syndrome, the iris is not pigmented normally. Ptosis is also seen with a third cranial nerve palsy (large pupil), and as a fluctuating finding in myasthenia gravis. *Blepharospasm* refers to increased frequency and intensity of eye blinking. If not associated with an obvious etiology such as exposure to neuroleptics ("tardive blepharospasm"), ocular disorders, or foreign bodies producing eye discomfort, it is usually termed "essential blepharospasm" (a focal dystonia). This syndrome often begins with an illusory foreign body sensation in both eyes and is exacerbated by bright light. When conjoined with jaw dystonia it is often termed *Meige syndrome*.

The range of *smooth pursuit eye movements* is tested by asking the patient to follow an "H" described by the examiner's finger. Double vision in particular directions of gaze which disappears when one eye is covered implies ocular malalignment, usually due to weakness of an ocular muscle (whether secondary to ocular myopathy or myasthenia, lesions of cranial nerves or cranial nerve nuclei, or the gaze control systems in the brain stem). Failure to fully adduct (bring toward the midline), elevate, or depress the eye is characteristic of an oculomotor palsy, whereas inability to abduct (look laterally) suggests an abducens nerve palsy. A trochlear palsy results in incomplete depression of the eye in adduction (e.g., a left trochlear nerve palsy would produce failure to depress the left eye fully when looking down and to the right). Isolated failure to adduct one eye, or slowing of saccades an adduction in that eye, with or without nystagmus (jerking movements) of the opposite (abducting) eye, is seen with *internuclear ophthalmoplegia* (INO) and reflects a lesion in the medial longitudinal fasciculus (MLF) interconnecting the ipsilateral oculomotor nerve nucleus and the contralateral abducens nerve nucleus. Multiple sclerosis is a common cause of INO. Bilateral internuclear ophthalmoplegia is also associated with vertical nystagmus and sometimes with impaired convergence. Impairment in *conjugate* lateral gaze (i.e., inability of both eyes to look to the left or right) is seen in pontine (rarely midbrain) or frontal eye field lesions. In the latter case, the deviation is to the side opposite any associated hemiparesis and can be overcome with the "doll's eye" maneuver (see below) and other methods of vestibular stimulation, and if mild, may be manifest mainly as a gaze *preference* rather than a total inability to look to one side. An acute oculomotor palsy associated with an ipsilateral hemiparesis is an important clue to the possibility of uncal herniation in a patient with a hemispheric mass lesion and can precede impairment of consciousness. Unilateral or bilateral abducens nerve palsies can also be seen with raised intracranial pressure. Compressive lesions of the oculomotor nerve (neoplastic, traumatic, meningeal inflammatory or infectious processes, or an aneurysm arising at the takeoff of the posterior communicating artery from the circle of Willis) commonly cause mydriasis (pupillary dilation), whereas lesions that spare the pupil and affect only eye movements mediated by the nerve are more likely to be ischemic (i.e., associated with diabetic vasculopathy). Ophthalmoplegia can also result form peripheral neuromuscular disease (i.e., myasthenia gravis, progressive external ophthalmoplegia) or from local pathology in the orbits (i.e., orbital pseudotumor, and thyroid ophthalmopathy, in which the rectus muscles are commonly thickened and trapped). Sudden, forced, intermittent, dystonic conjugate eye deviation may be due to *oculogyric crises*, seen commonly as an acute dystonic reaction to neuroleptics and antiemetics, in Tourette's syndrome, and postencephalitic parkinsonism. Interestingly, they may be accompanied by intrusive compulsive thoughts. Forced eye deviation may also occur in simple and complex partial seizures. "Supranuclear" limitation of conjugate vertical and sometimes horizontal eye movements (i.e., limitation of voluntary but not vestibularly mediated reflex movements) occurs in some conditions associated with dementia (hydrocephalus, PSP, corticobasal degeneration, Niemann-Pick disease, Whipple's disease). Supranuclear limitation of upward gaze is often seen in idiopathic Parkinson's disease and obstructive hydrocephalus, although it has low specificity in very elderly individuals. Downgaze impairment occurs early in PSP. *Ocular apraxia* (the inability to direct one's eyes to a target, even though it is perceived) is also a feature of the spatial–motor processing deficits of Balint s syndrome.

The *trigeminal nerve* supplies sensation to the head and face and innervates the muscles of mastication (chewing). Sensory testing for light touch, pinprick, and temperature is done as outlined for other body parts (below). The trigeminal nerve mediates the afferent arc of the corneal reflex (tested by gently touching the cornea with a wisp of cotton or the rolled tip of a tissue). It also supplies the jaw jerk reflex, which can be useful in distinguishing limb hyperreflexia due to cervical spinal cord compression (commonly due to osteoarthritis of the neck in the elderly), in which case the jaw jerk should be normal, from that due to bilateral suprabulbar corticospinal tract lesions (brisk jaw jerk).

The *facial nerve* supplies the muscles of facial expression, the stapedius muscle of the ear (the efferent component of the acoustic reflex), and the salivary glands, and also carries taste sensation from the ipsilateral anterior two thirds of the tongue. The upper face receives bilateral motor innervation from the facial nerve; unilateral lesions of the corticospinal tract producing facial weakness therefore tend to spare the forehead musculature, whereas in Bell's palsy and other lower motor neuron (LMN) lesions, the forehead is involved and eye closure is weak or absent on the affected side. In subtle facial weakness due to upper motor neuron (UMN) corticospinal tract lesions, there may be only flattening of the nasolabial skin fold and a slight asymmetry of lower facial movements evident during talking, or a delay in smiling or showing the teeth on command. Because limbic structures evidently have direct access to the facial nerve nuclei, *spontaneous* smiling may be symmetrical in such a lesion; conversely, temporolimbic lesions may manifest with decreased spontaneous emotional (mimetic) movements of the contralateral face, but intact strength on formal testing. In *hemifacial spasm* there is twitching of facial musculature on one side of the face (narrowing of the palpebral fissure, simultaneous elevation of the corner of the mouth). This may be due to an irritative lesion or a tortuous blood vessel compressing the facial nerve. Hemifacial spasm needs to be differentiated from facial myoclonus due to an irritative cortical lesion, and from tics (see blow). In *facial myokymia* there are subtle rippling movements of the facial muscles (again usually unilateral), less coarse than fasciculations (see section on movement disorders, below). This can be associated with brain stem lesions, although myokymic twitches of the lower or upper eyelid are commonly due simply to fatigue, stress, caffeine, or nicotine.

The acoustic portion of the *stato-acoustic nerve* carries impulses from the cochlea, whereas the vestibular portion transmits information about head position and movement from the semicircular canals and the otolith organs. Hearing is traditionally screened at the bedside with a whispered voice, ticking watch (not always available in this digital age), and a 512-Hz tuning fork. In the Weber test, the vibrating fork is placed on the vertex and the patient is asked if the sound is louder in one ear than the other; in the Rinne test, the fork is placed on the mastoid process (the bony prominence just behind the ear) to assess bone conduction, and when the vibration is no longer audible, the tip is placed near the patient's ear to see if it can still be perceived by air conduction. If hearing loss is "conductive," due to ear disease, bone conduction will exceed air conduction, and compensatory mechanisms accentuate signal processing from the "bad" ear to increase its sensitivity; thus on the Weber test, the sound will be perceived as originating closer to the involved ear, and the Rinne will be abnormal on that side also. By contrast, in "sensorineural" hearing loss, air conduction retains its normal relative advantage over bone conduction (although both may be reduced), and the Weber lateralizes as one would expect to the normal ear. Since auditory information from both cochlear nuclei reaches each primary auditory cortex, hearing is not clinically damaged by a unilateral lesion of the primary auditory cortex in the temporal lobe, but bilateral lesions of Heschl's gyrus lead to cortical deafness. To test auditory localization, auditory–motor integration, and extinction to double simultaneous auditory stimuli, the patient is seated with eyes closed, and the examiner snaps or rustles fingers on one or both hands in various positions, asking the patient to point to the apparent source(s) of the sounds thus produced. A defect in auditory localization in a patient with intact hearing may suggest a temporal lobe (primary auditory cortex) lesion, whereas left-sided auditory extinction on double simultaneous stimulation could result from temporal (particularly if restricted to the auditory modality) or frontoparietal lesions (if trimodal).

The *glossopharyngeal nerve* receives taste afferents from the posterior third of the tongue, and somatic sensory afferents from the middle ear and the posterior pharynx. It is often considered the afferent end of the "gag" reflex, but the *vagus nerve* also can apparently mediate this reflex alone. Taste can be tested as described for the facial nerve, above. The functions of the vagus nerve commonly tested at the bedside include innervation of the larynx, palate, and pharyngeal muscles, and the gag reflex. It is tested by listening to the patient's voice for nasality, hoarseness, or stridor; and by watching the palate move when the patient says "ah," and in response to touching the left and right posterior pharyngeal walls with a cotton-tip applicator or tongue depressor. Bilaterally hyperactive gag reflexes and dysphagia (difficulty swallowing) are seen in "pseudobulbar palsy." The absence of a gag reflex potentially allows food and liquids to enter the airway, resulting in choking and aspiration pneumonia. Dysphagia can result from pharyngeal pathology, neuromuscular disorders, upper and lower motor neuron lesions, and motor neuron disease, and is a common and difficult problem in many extrapyramidal disorders. If there is a suspicion of vocal cord paralysis or *spastic dysphonia* (a focal dystonic syndrome, formerly thought to be "hysterical," in which the patient's voice is irregularly weak and high-pitched, with a "strangulated" quality, due to abnormal glottic closure during phonation), the patient may be asked to sing "EEEEEEEEE. . . ." During this procedure, the vocal cords may be observed using a laryngeal mirror (local anesthesia of the posterior pharynx may be necessary). *Palatal myoclonus*, a slow (1–4 Hz) rhythmic oscillation of the palate, larynx, pharynx, and occasional cervical muscles, is associated with brain stem and cerebellar disease.

The *spinal accessory nerve* supplies the sternocleidomastoid and trapezius muscles, which are typically screened by asking the patient to turn his head laterally, and to shrug his shoulders, against resistance. When the patient is asked to shrug without resistance, a delay in elevation of the contralateral shoulder may be a subtle sign of a UMN lesion. *Torticollis* refers to sustained deviation and/or jerky rotatory or tilting movements of the head and chin to one side. The etiology is most commonly an idiopathic segmental dystonia in which jerky movements are superimposed on deviation (spasmodic torticollis).

The *hypoglossal nerve* supplies the muscles of the tongue. Strength is tested by having the patient push the tongue into each cheek while the examiner presses against the latter. Weakness or slowed movements may be seen in both UMN and LMN lesions, but the latter are also associated with atrophy (loss of muscle bulk and "scalloping" of the edges of the tongue) and fasciculations (irregular, wormlike or flickering contractions of small portions of the tongue). It is important to examine for these with the tongue resting relaxed inside the mouth (since active contraction will produce movement that may resemble fasciculations), and also to differentiate them from a tongue tremor. Fasciculations of the tongue may be an important sign of amyotrophic lateral sclerosis (ALS) and other motor neuron diseases associated with dementia.

Dysarthria ("slurring," poor articulation of speech) may be due to UMN or LMN lesions of the hypoglossal nerve (as well as to cerebellar or extrapyramidal disease). When dysarthria is present it may be helpful to have the patient wiggle the tongue back and forth as quickly as possible and repeat the syllables "pa" (bilabial), "ta" (lingual), and "ka" (fricative) continuously as evenly as possible, and then to repeat them in combination ("pa-ta-ka"). When associated with the "spastic dysarthria" of bilateral UMN lesions, speech tends to be slow and aprosodic (the exception being apraxia of speech, sometimes called "cortical dysarthria," in which prosody is accentuated), hoarse or "strangulated," and hypernasal. In cerebellar (ataxic) dysarthria there may be inappropriately placed stress and loudness variation ("scanning" speech) and erratic breakdowns in articulatory precision. Combinations of scanning and spastic dysarthria are often seen in multiple sclerosis. Hypokinetic, hypophonic speech with a progressive diminution in volume (degenerating often into mumbling after a second or two) and palilalia is seen in Parkinson's disease and other akinetic-rigid syndromes (see below). Choreiform movements of the tongue and lower face are the hallmark of *tardive dyskinesia* (see section on motor examination). Choreic dysarthria may have an explosive, jerky quality due to associated chorea of the larynx, pharyngeal muscles, and diaphragm.

MOTOR SYSTEM

It is traditional to begin the discussion of the motor examination with observation, inspection, and palpation. The importance of observation cannot be overstressed. Inexperienced examiners frequently miss the *absence* of normal involuntary and "accessory" movements (such as blinking, and "tucking" before attempting to stand up from a chair), and the *presence* of subtle involuntary movements, particularly chorea (often put down to "restlessness") and low-amplitude myoclonic jerks. It is important to observe carefully and specifically at some point during history-taking and/or examination for the presence of such abnormalities.

It may be helpful to begin the exam with some simple screening procedures. After observing the patient getting out of a chair and walking to the examining room, he or she is asked to stand with feet together, eyes closed and arms outstretched in front, with fingers spread wide apart and forearms pronated (palms facing downward). This will often make choreiform movements, myoclonic jerks, and postural and cerebellar outflow tremors more apparent (as well as trunkal sway or tilting due to joint position sense loss or vestibular dysfunction). Asking the patient then to bend back the wrists, as though "stopping traffic," for at least 60 seconds, will help to identify asterixis. With the wrists returned to the neutral position and the forearms supinated (fingers still spread apart), the patient is observed for an additional 30 seconds or so for an inward drift of the little finger and either a pronator, or an upward and outward, drift of the supinated forearm. *Postural stability* can be checked in the same position by standing behind the patient, giving a quick, short tug backward on the shoulders, and again by pushing forward, with the patient trying to remain upright.

INVOLUNTARY MOVEMENTS. Involuntary movements and dystonias restricted to the head and neck (e.g., blepharospasm, oculogyric crises, hemifacial spasm, facial myokymia, palatal myoclonus, spasmodic dysphonia, torticollis, and tardive dyskinesia) are described above in the section on cranial nerves.

Chorea consists of rapid, jerky movements that tend nevertheless to flow into one another, and to involve adjacent regions simultaneously. Choreiform movements are sometimes combined with athetoid (writhing) movements (choreoathetosis).

Chorea is frequently a sign of basal ganglia dysfunction, and often is accompanied by psychiatric and cognitive changes, including affective lability, obsessive–compulsive symptoms, personality changes similar to those seen with frontal lobe lesions, and sometimes psychosis or depression. *Ballistic* movements are wild, flinging movements of a limb, and when due to an infarction in the subthalamic nucleus ("hemiballismus"), they often resolve into chorea. Crossing and uncrossing the legs, "restless" movements of the feet, can be a manifestation of chorea and of restless leg syndrome, but they also occur in *akathisia* (literally "inability to remain seated"), a syndrome in which the movements are most commonly associated with a sensation of inner restlessness and anxiety which is partially relieved by walking. It is very important to identify akathisia since it is extremely distressing to patients, and in our experience it is a frequently unrecognized contributor to agitated behavior in a confused, demented, severely depressed, or psychotic individual. "Piano playing" movements of the fingers while walking, choreiform speech patterns and respiratory abnormalities, motor impersistence (often tested by asking the patient to protrude the tongue continuously for 30 seconds), a "milkmaid's grip" (erratic fluctuations in grip of the examiner's hand by the patient due to superimposed chorea), and moment-to-moment variability in deep tendon reflexes are commonly seen in choreiform movement disorders, and may be helpful in differentiating chorea from "restlessness" and akathisia.

Myoclonic jerks are shock-like, irregular, usually erratic movements of individual muscles or muscle groups in the limbs and/or trunk. These can be physiological (as in hiccups, and the "hypnic jerk" associated with a sensation of falling when drifting off to sleep); pathological but fairly benign (as in periodic limb movements of sleep, with or without the uncomfortable and difficult-to-describe sensations of "restless leg syndrome"), or a sign of serious CNS dysfunction, including toxic/metabolic encephalopathies, seizure disorders, anoxic brain damage, and rapidly progressive forms of prion dementias (in which case stimulus-sensitive or "reflex" myoclonus, including an exaggerated startle response is often prominent). Exaggerated startle responses are also seen in posttraumatic stress disorder and hyperekplexia, a rare hereditary condition linked to a mutation in the inhibitory glycine receptor gene (Rajendar & Schofield, 1995). Myoclonus also can be seen in Alzheimer's disease (especially in the Lewy body and amyloid angiopathy variants) and in Whipple's disease. *Asterixis* ("negative myoclonus") refers to extremely brief, arrhythmic lapses of extensor muscle tone, commonly associated with hepatic, uremic, and other metabolic encephalopathies, including anticonvulsant toxicity and hypercapnea (elevated blood carbon dioxide concentration), although it may also occur in frontal, thalamic, and midbrain lesions. Tics are usually rapid, stereotyped movements, more complex than myoclonus or chorea; examples include shoulder-shrugging, finger-snapping, sniffing, lip-pouting, touching, or punching. Unlike other "involuntary" movements, the patient usually has some ability to suppress tics voluntarily but experiences an "urge" to tic, and/or an unpleasant sensation that builds up until the movement is released, yet the tics are not related to ongoing motor plans and are distressing to the patient. Vocal and phonic tics, including throat-clearing, grunting, and echolalia (imitation of the patient's or other's words or phrases), and in a minority of cases, compulsive swearing (coprolalia) and echopraxia (imitation of the examiner's movements) accompany motor tics in Tourette's syndrome, by far the most common cause of motor and vocal tics.

In psychiatric, developmentally disordered, autistic, and blind or deaf patients, the differential diagnosis of tics includes *stereotypies*. These are coordinated, repetitive, rhythmic and patterned movements, postures, or vocalizations often performed in a ritualistic or compulsive fashion, such as rocking, banging, nodding, touching, tooth-grinding, pelvic thrusting, jumping, walking in circles or back and forth, snorting and blowing, hissing, and various types of self-injurious and self-stimulating behavior (Kurlan & O'Brien, 1992). The distinction between tics and stereotypies may be a question of semantics and determined in part by the overall presentation, since many of these same behaviors would in fact be labeled as tics if observed in a patient with Tourette's syndrome. Stereotypies also are seen with stimulant (especially amphetamine) intoxication. Stereotypic movements that may have particular diagnostic implications include lip-biting in Lesch-Nyhan syndrome; twirling, flapping, rocking, and head-banging in autism, and hand-wringing, "guitar-strumming," and self-grabbing in Rett's syndrome.

Tremors are rhythmic oscillations of a limb, or part of a limb, around a joint. There are four common types of tremor. *Rest(ing) tremor* occurs when the affected limb is placed completely at rest and disappears during attempted movement or active voluntary muscle contraction. The most common form, a Parkinsonian tremor with a frequency of 5–8 Hz, may sometimes be facilitated by flexing the upper limbs and positioning the forearms on the patient's thighs with the palms facing each other and the fingers curled, to produce the so-called pill-rolling tremor. *Postural* (or "sustension") tremor usually has a frequency of around 10–12 Hz, disappears at rest but is brought out by having the patient extend the arms fully with the fingers spread apart, or with the arms extended, elbows flexed to 90 degrees and the index fingers almost touching in front of the nose. Postural tremors often also have an "action" component (see below), especially as the distal limb approaches the target ("terminal tremor"), and most involve primarily the fingers rather than the more proximal upper limb. "Essential tremor" (a familial syndrome typically coming on in later life and often associated with a similar head tremor), orthostatic tremor, and exaggerated physiological tremor are common causes of postural tremors. In *action tremor,* there is a back-and-forth oscillation in the trajectory of movement (rather than perpendicular to it as in an ataxic or "intention" tremor). In both action and ataxic tremors the tremor tends to increase in amplitude at the end of movement. Action and postural tremor are frequently conjoined. *Ataxic* or *intention* tremor is usually associated with disease or drug-induced (alcohol, benzodiazepine intoxication, phenytoin, lithium) cerebellar dysfunction, and is associated with a side-to-side oscillation of the moving limb (brought out by the finger-nose-finger and heel-to-shin maneuvers), and with "past-pointing" or overshoot on reaching the target. When the midline cerebellum is involved there is also trunkal sway ("titubation") and gait ataxia.

Neuroleptics and Parkinson's disease (PD) are the most common cause of resting tremor; both may also be associated with action tremor in some cases and occasionally with a rhythmic lip and jaw tremor (as in "rabbit syndrome"). Exaggerated physiological tremor (primarily postural and terminal) is associated with increased adrenergic tone seen in anxiety; alcohol, benzodiazepine, and barbiturate withdrawal; hyperthyroidism; some antidepressants; and use of stimulants, cocaine, amphetamines, and bronchodilator drugs used to treat asthma. Valproic acid, lithium, and the selective serotonin reuptake inhibitor group of antidepressants (fluoxetine, paroxetine, sertraline, fluvoxamine) commonly produce mild action and postural tremors which may be closer in frequency to a Parkinsonian than to an exaggerated physiological tremor. Wilson's disease is classically associated with a coarse proximal ("wing-beating") tremor or a flapping-hand tremor, but may also produce a resting or action tremor. A combination of resting, action, and postural tremor is sometimes seen with lesions in the midbrain and cerebellar outflow pathways ("rubral" tremor). Orthostatic tremors primarily involve the lower extremities and occur in the standing position. Myoclonic syndromes and dystonia (e.g., primary writing tremor and other action dystonias, spasmodic torticollis) may also be associated with specific tremor subtypes.

Dystonia is used to describe a variety of abnormal postures and movements associated with excessive involuntary muscular contraction (often sustained or writhing, i.e., "athetoid," but occasionally rapid, jerk-like movements, as in myoclonic dystonia). These include primary or idiopathic dystonias such as the head-turning and/or tilting of spasmodic torticollis; action dystonias such as "writer's cramp"; inversion of the feet and hands in hereditary idiopathic generalized torsion dystonias; as well as a variety of "secondary" dystonias due to identifiable pathological or biochemical processes. Dystonic movements are thought to reflect dysfunction of the "extrapyramidal" motor system, and while often developmental, genetic, or idiopathic, they may also be secondary manifestations of a wide variety of neuropsychiatric disorders including cerebral palsy due to birth asphyxia, akinetic-rigid syndromes (see below), lysosomal storage and metabolic disorders, manganese and some other heavy metal intoxications, and basal ganglia calcification or infarction. Potentially treatable causes of secondary dystonia include acute (dystonic reactions, oculogyric crises) and delayed ("tardive") forms associated with neuroleptic drugs and antiemetics; Wilson's disease; dopa-responsive dystonia (and conversely, dystonia related to dopaminergic agonists in Parkinson's disease); vitamin E deficiency, neurosyphilis, tumors, and "pseudodystonias," particularly in torticollis. When dystonic posturing is only present during so-called stress gait examination maneuvers (i.e., asking the patient to walk on the insides and then the outsides of his or her feet may produce corresponding internal or external rotation of the arms and forearms and posturing of the hands) in an adolescent or young adult, it is considered a "soft neurological sign," possibly indicative of an intrauterine or perinatal insult or a developmental syndrome. While dystonic movements are not volitionally induced, they can frequently be suppressed for varying periods of time by a volitional motor act (i.e., a patient with torticollis may discover that if he touches his chin lightly, his head may revert to the normal position) or even by an imagined motor act or mental arithmetic (i.e., silently counting down from 100).

Fasciculations are irregular, worm-like rippling movements of parts of muscles resulting from seemingly random firing of single motor units, reflecting disease of the anterior horn cell in the spinal cord, the ventral (motor) spinal root, or (usually less prominently), peripheral nerve. In a cognitively impaired patient, fasciculations at multiple segmental levels could be a clue to dementia with motor neuron disease (DMND), ALS–parkinsonism–dementia, lead poisoning, HIV infection, a paraneoplastic syndrome, hexosaminidase deficiency, vasculitis or demyelinating disease affecting the brain and spinal cord, or syringomyelia (cavitation of the central spinal cord) in association with an Arnold–Chiari malformation and hydrocephalus.

BULK. Focal atrophy ("wasting") of muscles is a sign of disease of the lower motor unit and is seen in ALS–dementia, along with prominent fasciculations. The tongue and small muscles of the hand are frequently involved early in ALS. Wasting also is prominent in malnutrition and advanced HIV infection. In myotonic dystrophy, which is often associated with intellectual impairment, there are cataracts, ptosis, weakness, and atrophy of facial, jaw, and temporal scalp musculature, in addition to *percussion myotonia* (sustained contraction of muscles on mechanical stimulation, commonly assessed by tapping the abductor muscles of the thumb in the palm with a reflex hammer).

AKINESIA, BRADYKINESIA, AND ASSESSMENT OF TONE. *Akinesia* refers to the failure to initiate movement, including normally "automatic" movements such as blinking, shifting position periodically, swallowing of saliva, and "accessory" movements (e.g., swinging of the arms while walking). Akinesia (or hypokinesia) also is used to describe the inability to maintain the amplitude of a repetitive movement over successive trials. Facial hypomimia ("masking") is another feature of akinesia. Micrographia (an often progressive decrement in the size of handwriting) is an example of hypokinesia and is common in Parkinsonism. *Bradykinesia* refers to slowing of movements. Frequently the term "akinesia" is used to subsume bradykinesia and hypokinesia as well, which are allied features of most cases of the "akinetic rigid" syndromic class of hypokinetic movement disorders. In idiopathic Parkinson's disease and drug-induced Parkinsonism, the most common causes of an akinetic-rigid syndrome, tremor and postural instability are also cardinal manifestations (leading to the mnemonic "TRAP": tremor-rigidity-akinesia-postural instability). The akinetic rigid syndrome has a very large differential diagnosis, however. Akinesia and bradykinesia alone (i.e., *without* lead-pipe rigidity or tremor) may occur in akinetic mutism due to bilateral mesial frontal lobe or hypothalamic lesions, hydrocephalus, severe retarded depressions, and catatonic syndromes.

Tone refers to the resistance to passive movement of limbs by the examiner. Assessment of tone is dependent on complete relaxation of the manipulated limb, which is often difficult for older patients, especially in the lower extremities. In *spasticity* there is a velocity-dependent increase in tone with initial resistance on attempted flexion (greatest with rapid movement) at the elbow, wrist, and knees, and giving way subsequently (so-called clasp-knife phenomenon). Spasticity in a lower extremity is best appreciated by placing one's hands under the knee of a supine patient, asking the patient to relax the leg completely, and picking it up briskly. If a spastic "catch" is present, the leg and foot will "hang" in the air momentarily before falling down. Spasticity classically is taught to signify UMN involvement (of the lateral corticospinal tract and/or motor cortex), but frequently rigidity (see below) and spasticity are conjoined due to involvement of both corticospinal and "extrapyramidal" motor pathways in upper motor lesions. Hyperreflexia, an extensor plantar response, and sometimes flexor spasms are other features of spasticity. *Rigidity* refers to a diffuse (lead-pipe or "plastic") increase in resistance throughout the range of motion. In "cogwheel rigidity" there is a superimposed ratcheting (often best appreciated on rotation of the hand at the wrist) with the frequency of a Parkinsonian tremor. In early parkinsonism, rigidity may only be apparent during passive movement at the wrist when the patient engages in "distracting" movements of the limb contralateral to the one being manipulated by the examiner, such as repetitive clenching and unclenching a fist, or drawing circles in the air. These are sometimes referred to as "contralateral reinforcement" maneuvers. The same maneuvers may produce an increase in tone in the limb on the same side as a developmentally abnormal hemisphere or an old corticospinal tract lesion from which the patient appears to have completely recovered, and "mirror movements" may also be seen in the contralateral limb when the patient moves the previously affected limb. Rigidity is typically due to extrapyramidal disease or dysfunction, although it can occur with neuromuscular diseases, such as neuro-myotonia and "stiff man syndrome," a condition that is frequently misdiagnosed as hysterical. In a patient on dopamine receptor-blocking drugs (especially high potency, long-acting "depot" neuroleptic injections) or a severely parkinsonian patient abruptly withdrawn from anti-parkinsonian drugs, confusion and/or worsening of psychosis, followed or accompanied by marked rigidity, unexplained fever and leukocytosis, elevated liver enzymes and CPK (a muscle enzyme released into the circulation with damage to muscle or leaky muscle membranes), and/or autonomic instability is suggestive of *neuroleptic malignant syndrome* (NMS). NMS is extremely important to recognize since it has a high mortality rate unless recognized and treated early. In *catatonia*, "waxy flexibility" (maintenance of passively induced limb positions after the examiner has withdrawn his or her support for the positioned limbs) often occurs. Other features suggestive of catatonia include bizarre posturing and catalepsy (i.e., standing immobile on one foot for minutes at a time) and unusual gaits, stereotypies, echolalia, marked psychomotor retardation and abulia (sometimes punctuated or heralded by periods of marked hyperactivity and excitement), and automatic obedience. Malignant (lethal) catatonia (Philbrick & Rummans, 1994) is also in the differential diagnosis of NMS. *Paratonic rigidity* and *gegenhalten* (a velocity-independent increase in resistance to movement in every direction, proportional to the force applied by the examiner) may be indicative of mesial frontal lobe involvement. *Hypotonia* (decrease in tone) in adults is characteristically associated with cerebellar hemisphere lesions, although it also occurs in the first few hours or days after a UMN lesion of the brain or spinal cord, before spasticity develops. It is associated with "pendular" knee jerks and *overshoot* (or loss of the cerebellar "check" response).

POWER. *Power* refers to strength and is assessed *relative to expectations* given the patient's age, bulk, effort level, and occupational history (thus, a frail 80-year-old woman easily overcome by the examiner may still be considered to have "normal power"). Power is traditionally graded as follows: 1 = flicker of movement only; 2 = movement only with gravity eliminated; 3 = able to move fully against gravity only; 4 = minimal weakness; 5 = normal strength. Plus and minus signs are sometimes used to indicate gradations within steps (e.g., 5− = barely detectable loss of power).

Corticospinal tract (UMN) weakness (usually the most relevant type in a neurobehavioral context) most often affects distal muscles more than proximal ones, and the extensors of the upper extremities and flexors of the lower (sparing "antigravity" muscles the most). Therefore, in addition to checking for a drift, we test the following: (1) deltoids and iliopsoas (shoulder abduction and hip flexion), which also screens for proximal weakness; (2) wrist and finger extensors/abductors; and (3) hamstrings (knee flexion) and extensor hallicus (dorsiflexion of the great toe) as well as heel-walking (subtle ankle dorsiflexor weakness may only be detectable with this maneuver). UMN weakness is accompanied by corresponding changes in tone, reflexes, and often plantar responses. Distal weakness can also be due to peripheral neuropathies, in which case it is accompanied by hyporeflexia and (when chronic) muscle wasting. Disproportionate weakness of proximal muscles (e.g., deltoids, iliopsoas) may be seen in myopathy (steroids, AZT therapy in HIV infection, alcoholism, thyroid disease, hypokalemia, and inflammatory myopathies such as polymyositis/dermatomyositis being common causes) and in some cases of spinal muscular atrophies (chronic anterior horn cell disorders which may resemble myopathies). Lesions in the high frontal convexity region (commonly infarcts in the border zone between anterior and middle cerebral artery territories, due to a hypotensive insult) tend to affect proximal more than distal muscles. Entrapment and compression of peripheral nerves are common causes of major, focal weakness in a limb. Compression of single cervical and lumbar roots by herniated disks or osteophytes (and less commonly by tumor) usually causes milder degrees of localized weakness, due to the fact that most muscle groups are innervated by more than one spinal root.

The examiner should not document "weakness" (i.e., less than grade 5 power) unless convinced that the patient is giving full effort. A suspicion of incomplete effort should be neutrally recorded along with the apparent level of power at maximal effort, along with any discordant behavioral observations that would lend credence to this hypothesis. Careful observation during history-taking, other activities, and the rest of the exam will often disclose inexplicable preservation of functional capability inconsistent with the degree of "functional" weakness, in such situations. There is one important caveat: Isolated synkinetic movements of an otherwise paralyzed limb during primitive motor stereotypies, such as yawning and stretching, should not be taken to imply a "functional" etiology, since they can occur in hemiplegia due to cortical lesions. Incomplete effort may be due to pain, fear of injury, hemispatial neglect, or poor motivation due to personality or psychiatric disorders, frontal/subcortical dysfunction, hysteria and suggestibility, or frank malingering. In "give-way" weakness, the patient often gives momentary full effort at the onset of movement against the examiner's resistance, then suddenly "lets go" or varies his or her strength, depending on how much force the examiner provides, so that the examiner always appears stronger. This pattern of variable effort is not characteristic of weakness in neurological disease, even myasthenia gravis (in which case weakness may progress fairly quickly on repeated challenge), although some patients with loss of joint position sense appear to have difficulty determining how much force they are applying. In hemispatial neglect, effort on the neglected "side" can be quite variable; sometimes it improves when the head is turned to that side, or the neglected limbs are positioned in the opposite hemispace. Often "give-way" weakness responds to coaxing and encouragement that obliquely and nonconfrontationally indicates the examiner's perception of inadequate effort. When this fails, the examiner can rapidly vary the amount of force applied, which makes it very difficult for a patient with (incomplete) functional weakness to simulate a consistent degree of loss of strength.

DEXTERITY AND MOTOR SEQUENCING/LEARNING/ALTERNATION. The most sensitive screening procedures for motor system dysfunction involve rapid finger-tapping (index finger to thumb), sequential finger–thumb apposition, and finger "drumming." These are sensitive to extrapyramidal (bradykinesia), cerebellar and motor programming (difficulty sequencing), and corticospinal tract dysfunction (impaired dexterity being more sensitive than weakness to subtle UMN involvement). *Motor learning* and *response alternation* can be tested with a "three-step Luria motor sequence" such as fist-palm-side (asking the patient to tap his or her thigh sequentially with the bottom, i.e., ulnar side, of a clenched fist positioned with thumb up, then with the open palm and outstretched fingers, and finally with the ulnar aspect of the vertically positioned outstretched hand). The patient practices continuous repetition of the sequence, trying to obtain a smooth, rhythmic result. In the Oseretsky test of bimanual alternating hand positions, one hand is clenched in a fist, the other opened with outstretching fingers (with palm down in both cases); the hand positions are then repeatedly reversed. In patients without primary motor deficits, difficulty learning and executing the fist-palm-side test may be a clue to parietal and frontal lesions producing apraxia, or to subtle basal ganglia disease. Bimanual alternation may be impaired with callosal and mesial frontal lesions affecting the dominant supplementary motor area (Masdeu, 1990).

REFLEXES. *Deep tendon reflexes* tested routinely are the biceps (C5/6 roots; musculocutaneous nerve), triceps (primary C7; also C6, C8; radial nerve), brachioradialis (C5/6; radial nerve), and quadriceps (L3, 4, some L2; femoral nerve) and gastrocnemius or "ankle jerk" (primarly S1, S2, some L5; tibial nerve) muscle stretch reflexes. These are traditionally graded as 0 (absent), 1 (flicker), 2 (normal), 3 (brisk-normal), 4 (very brisk), or 4+ (sustained clonus). The Hoffman reflex (primarily C8/T1, some C7; median nerve) is tested by briskly flexing the distal phalanx of the middle finger momentarily, while supporting the middle phalanx and is useful if pathological hyperreflexia is suspected (flexion of the distal phalanx of the thumb is indicative of the latter). The jaw jerk reflex (trigeminal nerve) should be tested if both upper and lower limb deep tendon reflexes are brisk, to help ascertain the level of origin of hyperreflexia (if the jaw jerk is normal, the medulla or upper cervical spinal cord is implicated, rather than intracranial segments of the UMN pathway). This is particularly helpful in suspected cervical spondylytic myelopathy, a common cause of lower extremity spasticity and limb hyperreflexia in the elderly.

The *plantar reflex* (L4/5–S1/2) is most commonly tested by bringing a pin or other noxious stimulus lightly across the lateral aspect of the undersurface of the foot from the heel toward the little toe, and then across the ball of the foot toward the big toe. The normal (flexor plantar) response is brisk flexion of the toes. A classic extensor plantar response ("upgoing toe," "Babinski sign") consists of a slow (sometimes more rapid) initial upward deviation of the great toe and fanning of the other toes and is considered virtually pathognomic of a UMN lesion (in the absence of severe metabolic abnormalities such as hepatic coma which can produce reversible CST dysfunction). Rapid voluntary dorsiflexion of the foot, the toe, and the leg is a "withdrawal response" and is not of pathological significance. By contrast, slower, sustained flexion of the great toe, the foot, the knee, and the hip is a "triple flexion" or "spinal defense" response and may also be indicative of an upper motor neuron lesion. (Contraction of tensor facia lata is said by some to be part of a triple flexion response but not of a withdrawal response). When the examiner cannot decide if dorsiflexion of the great toe is due to withdrawal or not, the Chaddock maneuver (applying the stimulus to the lateral edge of the foot rather than the plantar aspect) may be helpful as it is somewhat less likely to induce withdrawal. Numerous other techniques are available for eliciting a Babinski sign but they have not been shown to be more sensitive or reliable when the classic procedure produces equivocal results.

Abdominal reflexes are tested by stroking the abdominal skin gently with a pin (going lateromedially, and upward in the upper abdomen, T8-T9 segments, and downward in the lower, T11-12 segments). The normal response is contraction of the underlying abdominal muscles and deviation of the umbilicus toward the stimulated area. Loss of abdominal reflexes contralateral to the lesion occurs with intracranial upper motor neuron lesions but also ipsilaterally with spinal cord disease at the affected segmental level. Unilateral hyperreflexia and a Babinski sign are classic signs of UMN lesion (when caused by an acute destructive lesion, an initial hyporeflexic phase may last several hours, or longer with spinal cord lesions). If the lesion is in the CST above its decussation in the lower medulla, these signs are contralateral; conversely, if it is in the spinal cord, they are ipsilateral. Common etiologies include infarction, intracranial hemorrhage, demyelination, tumor, and trauma. Bilateral, symmetrical hyperreflexia can be due to bilateral lesions, but it is also seen in a variety of toxic-metabolic encephalopathies and drug intoxications, sedative/hypnotic and alcohol withdrawal, lupus psychosis, leukodystrophies, degenerative diseases and ALS affecting the corticospinal tracts, hyperthyroidism, and anxiety disorders. Distal and symmetrical loss of reflexes usually implies a peripheral neuropathy. The protean differential diagnosis and classification of distal symmetrical peripheral neuropathy is beyond the scope of this chapter, but important considerations in neuropsychiatrically oriented practice include alcoholism; thiamin, B_{12}, folate, niacin, and rarely vitamin E deficiency; diabetes, uremia, and hypothyroidism; medications (particularly phenytoin); connective tissue diseases; lead, organic solvents, organophosphate insecticides, industrial and other toxins; HIV and Lyme disease; manifestations of malignancy including paraneoplastic syndromes, cancer chemotherapy, and paraproteinemic neuropathies (direct involvement by malignant cells and radiation plexopathies tends to be more focal); hereditary and inflammatory demyelinating polyneuropathies. Focal absence of reflexes may be due to nerve root or peripheral nerve entrapment or compression, spinal cord pathology such as syringomyelia, tumor, and hemorrhage involving the anterior horn cell column and root entry/exit zones; lesions of the brachial or lumbosacral plexus; infarctions involving the peripheral nervous system (seen commonly in diabetes, atherosclerotic peripheral vascular disease, vasculitis, connective tissue diseases), and porphyria (which tends to produce proximal motor involvement resembling myopathy).

"Primitive" or "release" reflexes are postulated to represent the re-emergence, or disinhibition, of ontogenetically and phylogenetically older reflexes which disappear in the course of normal development. Landau (1989) has polemically challenged clinical folklore about "primitive reflexes," and certainly we concur that they are grossly insensitive and often misleading signs which have value only in specific contexts. Those traditionally tested by neurologists are discussed in the following paragraphs.

Glabellar Reflex. This is elicited by tapping the bridge of the nose repeatedly (about once a second) from above (so as not to elicit a visual threat-mediated blink). If the patient continues to blink after 10 or more stimuli, this is referred to as "Myerson's sign" and has classically been considered a sign of extrapyramidal or frontal lobe dysfunction. However, in a study of "soft signs" in schizophrenics, 50% of 50 "healthy" control subjects (mean age 32) drawn from the staff of psychiatric facilities reportedly had a Myerson's sign (Buchanan & Heinrichs, 1988).

Palmomental Reflex. The examiner strokes the thenar eminence of the hand with a key or other mildly noxious stimulus and watches for a reflex contraction of the mentus muscle in the chin. Bilateral palmomental reflexes that fatigue after four or five trials have very low specificity, particularly in the elderly (Masdeu, 1990). A persistent, reproducible unilateral palmomental in a young adult, however, might enhance suspicion of a contralateral frontal lesion in the presence of lateralized frontal deficits on cognitive examination.

Grasp Reflex. The grasp reflex is elicited by stroking the palm of the passively outstretched upper limb with the examiner's fingers traversing the lateral side of the palm and curving under the patient's fingers. A reproducible, unilaterally positive reflex (indicated by involuntary, sustained grasping of the examiner's hand) is classically a sign of mesial frontal lobe involvement. When bilateral, however, grasp reflexes are most often the result of a diffuse attentional disturbance. They have localizing value only in an awake, alert, and attentive individual.

Other primitive reflexes include the snout and pout (tapping on the philtrum of the upper lip produces pursing of the lips), sucking, and rooting (head turning toward a rounded object approaching the mouth or stroking the cheek). The snout has low specificity; as in the case of the grasp, the sucking and rooting reflexes are quite insensitive in our experience, and would be very unlikely to uncover a frontal lesion or dementia unsuspected from other results of a good MSE and the standard neurological examination.

SOMATOSENSORY SYSTEM

A detailed sensory examination is only productive in an awake, vigilant, and fully cooperative individual. Before reporting a sensory deficit in a distractible, inattentive, drowsy, or poorly motivated patient, it is important to check for reproducibility. This is especially true when testing primary leminiscal sensory modalities (light touch, vibration, joint position) and graphesthesia, stereognosis, and perception of double simultaneous stimulation.

LIGHT TOUCH. Perception of light touch can be tested using a wisp of cotton or the tip of the examiner's index finger, with the patient's eyes closed. (The skin and hairs should not be brushed since this tickle will stimulate the spinothalamic system, which can also transmit light touch to some extent). A time-saving screening maneuver which simultaneously tests light touch, somatotopic localization, large joint proprioception, and upper limb ataxia is to ask the patient to "touch where I touch," while the examiner touches (in random order) areas on the face, outstretched finger tips, and lower extremities very lightly with a fingertip or a wisp of cotton, while the patient's eyes are closed.

In hemispatial neglect, patients may reliably report a unilateral stimulus delivered to either side of the body, but report only the stimulus on the non-neglected side when homologous areas are simultaneously touched on both sides. This is referred to as "extinction to double simultaneous tactile stimulation." In severe neglect, stimuli delivered to the neglected side may be misattributed to the non-neglected side ("allesthesia").

PAIN (PINPRICK) AND TEMPERATURE. Distinguishing the sharp from the dull end of a disposable safety pin is a good way to test the spinothalamic system. A tuning fork is usually cooler than the examiner's finger and this difference can be used to assess temperature sensation. Loss of these modalities with a hemispheric lesion classically implies involvement at or below the level of the thalamus, but may also be seen with parietal lobe lesions involving thalamocortical radiations (Bassetti, Bogousslavsky, & Regli, 1993). Relatively selective loss of pain and temperature sensation and autonomic dysfunction in a neuropsychiatric patient may also be a clue to a small-fiber sensory neuropathy or neuronopathy, causes of which include diabetes; hereditary systemic amyloidosis (a condition not associated with amyloid angiopathy of the CNS); a paraneoplastic neuropathy (due to lymphoma, or carcinoma of the lung, pancreas, or testicles); megadoses of vitamin B_6 (pyridoxine); autoimmune disorders such as Sjögren's syndrome; porphyria; and hereditary sensory and autonomic neuropathies.

Selective ("dissociated"), "suspended" pain and temperature loss (restricted to affected segmental levels) is also seen with central spinal cord lesions and syringomyelia. Thalamic lesions may also induce hyperpathia (a syndrome in which pain perception thresholds are raised but suprathreshold or repetitive liminal stimuli are unusually painful). Some patients with parietal lesions (typically left-sided) manifest a similar ("pseudothalamic") syndrome. Lesions of the posterior insula do not impair primary sensory abilities but may produce a curious loss of emotional reactivity to painful stimuli ("asymbolia for pain"), a tactile-limbic disconnection (Caselli, 1997). Emotional responses to pain are also blunted by lesions of the anterior cingulate gyrus. Pain hemiagnosia (inability to localize and/or react to painful stimuli on the left side while still showing facial and autonomic manifestations of discomfort) may occur as part of the neglect syndrome associated with right parietal lesions.

VIBRATION, JOINT POSITION SENSATION, AND TWO-POINT DISCRIMINATION. Vibration sense is screened in the digits using a tuning fork applied to the distal phalanges of one finger and toe on each side. The patient is asked to report when the vibration disappears, and the examiner uses her own digits as a control where necessary. If distal sensation is absent, sensation is then tested at more proximal joints to establish a sensory level.

Joint position sense is tested by fixating the proximal phalanx of the same digits and asking the patient to report the direction ("up or down") of subtle movements of the distal phalanx made by the examiner (both phalanges being grasped on either side rather than on the top and bottom so as to avoid giving cues based on differential pressure during movement). Severe loss of joint position sensation may be associated with sensory ataxia, a Romberg's sign (the patient can stand motionless with feet together and eyes open, but sways or falls with eyes closed), and pseudoathetosis (writhing movements of the outstretched fingers, especially with eyes closed). In a patient with cognitive deterioration, impairment of vibration and joint position sensation may be a clue to vitamin B_{12} or folate deficiency, diabetes, multiple sclerosis, a vascular dementia, hypothyroidism, corticobasal degeneration, neurosyphilis, a paraneoplastic syndrome, or AIDS.

Two-point discrimination can be tested on the fingertips with calipers or a paper clip bent into a "U" shape. Normally subjects can distinguish between a touch with one versus two points on the index fingertip if the points are separated by 2 to 3 mm or more. When vibration and joint position sense is impaired at the toes and fingertips but normal at the knees and elbows, this suggests a peripheral neuropathy. Lesions higher up in the lemniscal pathways should not show this dissociation. Similarly, two-point discrimination on the fingertips is quite sensitive to peripheral neuropathies, whereas a loss of two-point discrimination to simultaneous tactile stimuli placed several centimeters apart on the back of the hand, unaccompanied by vibration sense or light touch impairment, is likely to be a cortical phenomenon associated with contralateral parietal lobe involvement (or a parietal-thalamic disconnection due to a lesion in the underlying white matter).

STEREOGNOSIS AND GRAPHESTHESIA. *Stereognosis* is traditionally tested by asking the patient to identify small objects (such as coins, a paper clip, button, screw, etc.) placed in the hand by touch alone. "Astereognosia" can result from peripheral neuropathies and lesions anywhere in the lemniscal system. The more specific syndrome of *tactile agnosia* (the inability to recognize a previously known object by tactile manipulation) occurs in the absence of primary sensory impairment and implies a contralateral inferior parietal lobule lesion.

Graphesthesia refers to the ability to identify numbers or letters traced on the palm with a finger, or on the index fingertip with a stylus. Unilateral agraphesthesia ("graphanesthesia") unaccompanied by more elementary sensory deficits implicates a contralateral parietal lobe lesion. Stereognosis and graphesthesia are part of most "soft sign" batteries, and have been found to be commonly impaired in schizophrenic patients (Buchanan & Heinrichs, 1988). They are also impaired in cortical dementias with prominent parietal lobe involvement (AD, corticobasal degeneration, posterior cortical atrophies). In our experience, they are exquisitely sensitive to motivation and attention and therefore only have localizing significance in a very alert and cooperative patient.

TESTS OF CEREBELLAR FUNCTION AND COORDINATION

Classic tests of cerebellar hemisphere function include the finger-nose-finger and heel-to-shin maneuvers. In the former, the patient alternately touches his nose and the examiner's finger which is placed an arm's length away in various locations. In the heel-to-shin test, the patient lifts his leg in the air, brings the heel down to balance momentarily on the opposite patella (knee cap), then runs his heel down the opposite shin off the foot and back up in the air again (and so on). Ataxic tremor brought out by these maneuvers is described above. Gross reaching errors in a hemifield on finger-nose-finger may also reflect *optic ataxia* due to a contralateral superior parietal lobe lesion. (This occurs bilaterally in Balint's syndrome). To test *rapid alternating movements,* the patient is asked to touch the same spot on the back of one hand alternately with the dorsal and the palmar aspect of the fingertips (i.e., to pronate and supinate the forearm as rapidly as possible), and/or to "pretend to screw and unscrew a light bulb" as rapidly as possible. Cerebellar hemisphere disease classically produces errors in the rhythm and alternation aspects of this task, as well as hypotonia and overshoot (as discussed earlier). Lower extremity cerebellar function and balance can also be tested by asking the patient to hop in the same area and to balance on one foot, to tap the floor as fast as possible with the toes and forefoot, and to make circles on the floor.

The midline cerebellar nuclei, including the vermis, participate mainly in trunkal balance and gait, and tests of these are described below. (Midline cerebellar disease also produces dysarthria and nystagmus, described above in the section on cranial nerves.)

GAIT

First, the patient is observed walking spontaneously. Decreased "arm swing" on one side may be a clue to a contralateral hemispheric lesion. In a "hemiparetic gait" there may also be scuffing of the affected foot and/or abnormal stiffness at the knee or circumduction (lateral swinging of the leg) during the swing phase. Bilateral loss of arm swing, flexed posture, and a short-stepped gait is often associated with parkinsonism, depression, and normal pressure hydrocephalus. In the latter disorder, and in bilateral mesial frontal lobe disease, there may be a "magnetic" quality to the gait (as though the feet are stuck to the floor) with marked difficulty initiating walking. A "steppage" gait (in which the foot is lifted unusually high off the floor) may reflect peroneal nerve distribution weakness, or a sensory ataxia (as in B_{12} deficiency, tabes dorsalis in syphillis, and diabetic "pseudotabes"). Ataxia may also occur in frontal lesions due to involvement of frontocerebellar connections ("Brun's ataxia"). *Festination* or "propulsion" refers to a progressive acceleration of gait associated with a diminution in stride length. *Retropulsion* refers to a tendency to fall or step backward, and is particularly common in PSP.

To assess *tandem gait,* the patient is asked to walk "heel-to-toe, as though walking a tightrope." This is a sensitive test of cerebellar ataxia due to disease or dysfunction of the midline cerebellum (vermis) which is common in alcoholic and phenytoin-induced cerebellar degenerations, as well as with acute intoxication with these substances and benzodiazepines. Causes of cerebellar ataxia and dysarthria include permanent sequelae of lithium toxicity (acute intoxication may also produce downbeating vertical nystagmus); multiple sclerosis; infarctions of the brain stem and cerebellum; Friedreich's ataxia (sometimes associated with mild mental retardation); olivopontocerebellar atrophy and other inherited and degenerative diseases; paraneoplastic syndromes associated with a remote malignancy (especially ovarian, breast, lung, and lymphoma); and some prion dementias.

Acknowledgments

The author wishes to thank Marsel-Mesulam, M.D., David Bear, M.D., and Sandra Weintraub, Ph.D., whose knowledge, guidance and tutelage of the author (during a medical student elective in 1980, and again during a fellowship in 1986–88, in the Behavioral Neurology Unit at Harvard Medical School), together with that of other staff of that unit, profoundly influenced the development of the author's approach to clinical assessment, as outlined in this chapter. We also wish to acknowledge the support of the Highland Drive VA Medical Center and the Department of Veterans Affairs.

REFERENCES

Abramowicz, M. *et al.* (1993). (Eds.) Drugs that cause psychiatric symptoms. *The Medical Letter, 35,* 65–70.

Adams, R. D., Victor, M., & Ropper, A. H. (1997). *Principles of neurology* (6th ed.). New York: McGraw-Hill.

Ahlskog, J. E., *et al.* (1991). *Clinical examinations in neurology.* Rochester, MN: Mayo Foundation for Medical Education and Research.

Anderson, S. W., Damasio, H., Dallas-Jones, R., & Tranel, D. (1991). Wisconsin Card Sorting Test performance as a measure of frontal lobe damage. *Journal of Clinical and Experimental Neuropsychology, 13,* 909–922.

Bassetti, C., Bogousslavsky, J., & Regli, F. (1993). Sensory syndromes in parietal stroke. *Neurology, 43,* 1942–1949.

Bates, B. (1979). *A guide to physical examination* (3rd ed.) Philadelphia: Lippincott.

Bear, D. M., Rosenbaum, J. F., & Norman, R. (1986). Aggression in cat and man precipitated by a cholinesterase inhibitor. *Psychosomatics, 26,* 535–536.

Benson, D. F., & Geschwind, N. (1988). Aphasia and related disorders: a clinical approach. In M.-M. Mesulam (Ed.), *Principles of behavioral neurology* (pp. 193–238). Philadelphia: F. A. Davis.

Benson, D. F., & Gorman, D. G. (1996). Hallucinations and delusional thinking. In B. S. Fogel, R. B. Schiffer & S. M. Rao, (Eds.), *Neuropsychiatry* (pp. 307–323). Baltimore, MD: Williams and Wilkins.

Boutros, N. N., & Bowers, M. B. Jr. (1996). Chronic substance-induced psychotic disorders: state of the literature. *Journal of Neuropsychiatry and Clinical Neurosciences, 8,* 262–269.

Bowler, J. V., & Hachinski, V. C. (1994). Progress in the genetics of cerebrovascular disease: inherited subcortical arteriopathies. *Stroke, 25,* 1696–1698.

Brazis, P. W., Masdeu, J. C., & Biller, J. (1990). *Localization in clinical neurology* (2nd ed.). Boston: Little, Brown & Co.

Buchanan, R. W., & Heinrichs, D. W. (1988). The Neurological Evaluation Scale (NES): A structured instrument for the assessment of neurological signs in schizophrenia. *Psychiatry Research, 27,* 335–350.

Caplan, D. (1992). *Language: structure, processing, and disorders.* Cambridge, MA: MIT Press.

Caselli, R. J. (1997). Tactile agnosia and disorders of tactile perception. In M. J. Farah, & T. E. Feinberg (Eds.), *Behavioral neurology and neuropsychology* (pp. 277–288). New York: McGraw-Hill.

Damasio, A. R., & Anderson, S. W. (1993). The frontal lobes. In K. M. Heilman & E. Valenstein (Eds.), *Clinical neuropsychology* (3rd ed., pp. 409–460). Oxford: Oxford University Press.

DeJong, R. N. (1979). *The neurologic examination* (4th ed.). Philadelphia: Harper and Row.

DeMyer, W. E. (1994). *Technique of the neurological examination: A programmed text* (4th ed.). New York: McGraw-Hill.

Deuel, R. K., & Robinson, D. J. (1987). Developmental motor signs. In D. K. Tupper, (Ed.), *Soft neurological signs.* Orlando, FL: Grune & Stratton, 95–129.

Golden, C. J., Purisch, A. D., & Hammeke, T. A. (1985). *Luria–Nebraska neuropsychological battery: Forms I and II.* Los Angeles: Western Psychological Services.

Heilman, K. M., & Rothi, L. J. G. (1993). Apraxia. In K. M. Heilman & E. Valenstein (Eds.), *Clinical neuropsychology* (3rd ed., pp. 141–163). Oxford: Oxford University Press.

Heilman, K. M., Watson, R. T., & Rothi, L. G. (1997). Disorders of skilled movements: limb apraxia. In M. J. Farah & T. E. Feinberg (Eds.), *Behavioral neurology and neuropsychology* (pp. 227–235). New York: McGraw-Hill.

Hollander, E., DeCaria, C. M., Aronowitz, B., Klein, D. F., Liebowitz, M. R., & Shaffer, D. (1991). A pilot follow-up study of childhood soft signs and the development of adult psychopathology. *The Journal of Neuropsychiatry and Clinical Neurosciences, 3,* 186–189.

King, D. J., Wilson, A., Cooper, S. J., & Waddington, J. L. (1991). The clinical correlates of neurological soft signs in schizophrenia. *British Journal of Psychiatry, 158,* 770–775.

Kurlan, R., & O'Brien, C. (1992). Spontaneous movement disorders in psychiatric patients. In A. E. Lang & W. J. Weiner (Eds.), *Drug-induced movement disorders* (pp. 257–280). Mt. Kisco, NY: Futura Publishing Co., Inc.

Landau, W. M. (1989). Clinical neuromythology VI. Au claire de lacune: Holy, wholly, holey logic. *Neurology, 39,* 725–730.

Leimkuhler, M. E., & Mesulam, M.-M. (1985). Reversible go-no go deficits in a case of frontal lobe tumor. *Annals of Neurology, 18,* 617–619.

Levin, H. S., Goldstein, F. C., & Spiers, P. A. (1993). Acalculia. In K. M. Heilman & E. Valenstein (Eds.), *Clinical neuropsychology* (3rd ed., pp. 91–122). Oxford: Oxford University Press.

Lezak, M. D. (1995). *Neuropsychological assessment* (3rd ed.). Oxford: Oxford University Press.

Lhermitte, F., Pillon, B., & Serdaru, M. (1986). Human autonomy and the frontal lobes. Part I: imitation and utilization behavior: a neuropsychological study of 75 patients. *Annals of Neurology, 19,* 326–334.

Malloy, P. F., & Richardson, E. D. (1994). Assessment of frontal lobe functions. *Journal of Neuropsychiatry and Clinical Neurosciences, 6,* 399–410.

Masdeu, J. C. (1990). The localization of lesions affecting the cerebral hemispheres. In P. W. Brazis, J. C. Masdeu & J. Biller (Eds.) *Localization in clinical neurology* (2nd ed., pp. 361–428). Boston: Little, Brown & Co.

Mesulam, M.-M. (1981). Slowly progressive aphasia without generalized dementia. *Annals of Neurology, 11,* 592–598.

Mesulam, M.-M. (1985b). Attention, confusional states, and neglect. In M.-M. Mesulam (Eds.), *Principles of behavioral neurology* (pp. 126–168). Philadelphia: F. A. Davis.

Mesulam, M.-M. (1986). Frontal cortex and behavior. *Annals of Neurology, 19,* 320–325.

Nadel, L. (1992). Multiple memory systems: what and why. *Journal of Cognitive Neuroscience, 4,* 179–188.

Philbrick, K. L., & Rummans, T. A. (1994). Malignant catatonia. *Journal of Neuropsychiatry and Clinical Neurosciences, 6,* 1–13.

Phillips, A. G., & Carr, G. D. (1987). Cognition and the basal ganglia: a possible substrate for procedural knowledge. *Canadian Journal of the Neurological Sciences, 14,* 381–385.

Price, B., Daffner, K., Stowe, R. M., & Mesulam, M.-M. (1990). The comportmental learning disabilities of early frontal damage. *Brain, 113,* 1383–1393.

Rajendar, S., & Schofield, P. R. (1995). Molecular mechanisms of inherited startle syndromes. *Trends in Neuroscience, 18*, 80–82.

Reichman, W. E., & Cummings, J. L. (1990). Diagnosis of rare dementia syndromes: an algorithmic approach. *Journal of Geriatric Psychiatry and Neurology, 3*, 73–84.

Reid, W. H., & Balis, G. U. (1987). Evaluation of the violent patient. *American Psychiatric Association Annual Review, 6*, 491–509.

Ross, E. D. (1980). Sensory-specific and fractional disorders of recent memory in man. II: Unilateral loss of tactile recent memory. *Archives of Neurology, 37*, 267–272.

Ross, E. D. (1988). Modulation of affect and nonverbal communication by the right hemisphere. In M.-M. Mesulam, (Ed.), *Principles of behavioral neurology* (pp. 239–257). Philadelphia: F. A. Davis.

Ross, E. D. (1997). The aprosodias. In M. J. Farah & T. E. Feinberg (Eds.), *Behavioral neurology and neuropsychology* (pp. 699–709). New York: McGraw-Hill.

Saint-Cyr, J. A., & Taylor, A. E. (1992). The mobilization of procedural learning: the "key signature" of the basal ganglia. In L. R. Squire & N. Butters (Eds.), *Neuropsychology of memory* (2nd ed., pp. 188–202). New York: Guilford Press.

Samuels, M. A., & Feske, S. (Eds.), (1996). *Office practice of neurology*. New York: Churchill-Livingstone.

Shaw, P. J., Walls, T. J., Newman, P. K., Cleland, P. G., & Cartlidge, N. E. F. (1991). Hashimoto's encephalopathy: a steroid-responsive disorder associated with high anti-thyroid antibody titers— report of 5 cases. *Neurology, 41*, 228–233.

Squire, L. R. (1992). Declarative and nondeclarative memory: multiple brain systems supporting learning and memory. *Journal of Cognitive Neuroscience, 4*, 232–243.

Stowe, R. M. (1994). Impulse control disorder. In M. Hersen, R. T. Ammerman & L. A. Sisson (Eds.), *Handbook of aggressive and destructive behavior in psychiatric patients* (pp. 287–304). New York: Plenum Press.

Swedo, S. E., Rapoport, J. L., Cheslow, B. S., Leonard, H. L., Ayoub, E. M., Hosier, D. M., & Wald, E. R. (1989). High prevalence of obsessive–compulsive symptoms in patients with Sydenham's chorea. *American Journal of Psychiatry, 146*, 246–249.

Tompkins, C. A. (1995). *Right hemisphere communication disorders: theory and management*. San Diego, CA: Singular Publishing Group.

Trepacz, P. T., & Baker, R. W. (1993). *The Psychiatric mental status examination*. New York: Oxford University Press.

Weintraub, S., & Mesulam, M.-M. (1985). Mental state assessment of young and elderly adults in behavioral neurology. In Mesulam, M.-M. (Ed.), *Principles of behavioral neurology* (pp. 71–123). Philadelphia: F. A. Davis.

Wilcox, J. A., & Nasrallah, H. (1988). Sydenham's chorea and psychopathology. *Neuropsychobiology, 19*, 6–8.

Wilson, J. D., Baruanwald, E., Isselbacher, K. J., Martin, J. B., Fauci, A. S., & Root, R. K. (Eds.). (1991). *Harrison's principles of internal medicine* (12th ed.). New York: McGraw-Hill.

Yager, Y., & Gitlin, M. J. (1995). Clinical manifestations of psychiatric disorders. In H. I. Kaplan & B. J. Sadock, (Eds.), *Comprehensive textbook of psychiatry* (6th ed., Vol. I, pp. 637–669). Baltimore, MD: Williams and Wilkins.

Index

Color Sorting Test, 323, 324
Coma scales, 141
Commission on Classification and Terminology of the International League Against Epilepsy, 273
Communication, 116–117, 259
Communication disorders, 43
Complex seizures, 273, 274
Comprehension, 452
Computed tomography, 12, 85, 139, 172, 218, 300
Computerized evaluation, 163
Concept formation, 324
Concept identification, 324
Conceptual Level Analogy Test, 319
Concreteness, 317
Conduction aphasia, 372–373
Confusional syndrome, 93–94
Congenital adrenal hyperplasia, 16
Congenital hypothyroidism, 52
Conjugate lateral gaze, 468
Conners' Scale, 25, 48, 128
Consciousness, 140
Consortium to Establish a Registry for Alzheimer's Disease, 192, 202–203
Constructional ability, 45, 379, 392–393
Constructive acalculia, 4, 401
Context, 38
Contingency Naming Test, 45
Continuous Performance Test, 25, 128
Continuous Visual Memory Test, 350–351
Contrecoup injury, 138
Controlled Oral Word Association Test, 97–98, 450
Contusions, 137
Coordination, 482
Coprolalia, 472
Corpus callosum, 13, 85, 252
Corsi block span, 449
Cortical areas, 85
Corticobasal degeneration, 481
Corticosensory capacity, 47
Cranial nerves, 466–471
Creutzfeldt–Jakob disease, 441
Crossed aphasia, 374–375
Crystallized intelligence, 86
Cube Test, 320
Culture Fair Intelligence Test, 387

Declarative vs. nondeclarative memory, 334–335
Decongestants, 445
Deductive reasoning, 238
Deep tendon reflexes, 478
Degenerative disorders, 187
Delayed deficit, 161
Delayed recall, 457
Delirium, 444
Dementia, 4, 181–182, 258, 335, 398
Dementia Rating Scale, 93, 193
Dementia with motor neuron disease, 474
Depression, 179–180, 238, 298–299, 333
Depression Inventory, 157, 243

Depression Scale, 92, 98, 243, 263
Depressive symptomatology, 243
Detroit Tests of Learning Aptitude, 127
Development, 25, 26, 38–39
Developmental Test of Visual–Motor Integration, 45, 394
Diabetes Control and Complication Trial, 240
Diabetes mellitus, 230–243, 296, 481
Diagnostic behavioral clusters, 49
Diaschisis, 174
Diazepam, 275
Differential Abilities Scale, 27, 160
Diffuse axonal injury, 139
Diffuse primary injury, 138
Digit Span Test, 302, 361, 402
Digit Symbol Substitution Test, 241, 257, 263, 302, 375
Digit Vigilance Test, 73, 237, 239
Dodrill's Neuropsychological Battery for Epilepsy, 278
Dorsolateral prefrontal cortex atrophy, 305
Down's syndrome, 53
Drawing tasks, 393–395
Dysarthria, 390, 470, 483
Dyslexia, 13, 14–15, 36, 122
Dysphasia, 15
Dysphonia, 470
Dystonia, 474

Ear infections, 18
Echolalia, 451, 472
Echopraxia, 472
Edema, 137
Electroconvulsive therapy, 299, 340
Electroencephalography, 12, 54, 65, 273, 276
Embedded Figures Test, 239
Encephalitis, 441
Encoding, 339, 457
Environment, 38
Epilepsy, 15, 271–289
Epilepsy syndromes, 279
Episodic memory, 456
Episodic vs. semantic memory, 335
Erythema chronicum migrans, 463
Ethosuximide, 275
Event-related potentials, 12, 218, 219
Executive function, 28, 46–47, 96–97, 118, 305
Exner's Rorschach Comprehensive System, 157
Experimental studies, 11
Explicit memory, 334
Expressive Speech Scale, 76
Eye movements, 467

Facial myokymia, 469
Facial nerve, 469
Facial Recognition Test, 389
Family history, 443–444
Fasciculations, 474
F-A-S Test, 450
Feedback, 158–159, 184

ISBN 0-306-45646-X

90000